History, Philosophy and Theory of the Life Sciences

Volume 24

More information about this series at http://www.springer.com/series/8916

Elena Casetta • Jorge Marques da Silva
Davide Vecchi

Editors

From Assessing to Conserving Biodiversity

Conceptual and Practical Challenges

 Springer Open

Editors
Elena Casetta
Department of Philosophy and Education
University of Turin
Turin, Italy

Davide Vecchi
Centro de Filosofia das Ciências,
Departamento de História e Filosofia das
Ciências, Faculdade de Ciências
Universidade de Lisboa
Lisboa, Portugal

Jorge Marques da Silva
BioISI – Biosystems and Integrative
Sciences Institute and Department of
Plant Biology, Faculdade de Ciências
Universidade de Lisboa
Lisboa, Portugal

ISSN 2211-1948   ISSN 2211-1956   (electronic)
History, Philosophy and Theory of the Life Sciences
ISBN 978-3-030-10990-5   ISBN 978-3-030-10991-2   (eBook)
https://doi.org/10.1007/978-3-030-10991-2

Library of Congress Control Number: 2019936165

This book is an open access publication.

Contents

About the Contributors

Francesco Andrietti is a zoologist interested in insect ethology and biomechanics, as well as in history of biology. He graduated in Biology in 1968 and has carried out his career at the University of Milan (Italy) until retirement. He was a Professor of animal physiology and history of biology. He is actively collaborating in different teaching activities and research projects on bee and wasp behavioural ecology, including from a historical perspective.

Anouk Barberousse is Professor of philosophy of science at Sorbonne Université, Paris (France), where she teaches general philosophy of science, philosophy of biology, and philosophy of scientific expertise. She has recently written on the epistemology of computer simulation, the philosophy of probability, and the role of natural history collections in our knowledge of biodiversity.

Sophie Bary is Doctor in "interdisciplinary biology", working at the National Museum of Natural History of Paris (France). Her thesis is about the analysis of scientific representations of the biodiversity of the seabed. Her approach is taking into account the societal, technical, and economic context. She has recently written on the role of collections in our knowledge of biodiversity, on the place of shell amateurs in the taxonomic knowledge on mollusc, and on the use of a statistical method, i.e. two-mode network, for the exploration of big data set.

Eva Boon Eva Boon's research addresses questions in molecular ecology as well as in philosophy of science. Her biological investigations focused on the use of gene and genome information to delineate and study biological entities. Under these broad questions, she has worked in an array of biological disciplines such as marine biology, speciation and population genetics, fungal genomics, microbial ecology, and human microbiomics. At present, she is applying this expertise to the question of how epistemic and ontological constraints shape inferences in biological and cultural evolution theory.

Luís Borda-de-Água is a Researcher in ecology and conservation biology at CIBIO-InBIO, University of Porto (Portugal). He took his PhD in Ecology at Imperial College London, UK, having studied before electrical engineering and physics. His background had led him to work mainly on theoretical and computational aspects of ecology. Presently, he divides his research activity between studies on global biodiversity patterns and the impacts of linear infrastructures on wildlife.

Andrea Borghini is Associate Professor of philosophy at the University of Milan, Italy, and he previously taught at the College of the Holy Cross, Massachusetts (2007–2017). His research focuses on metaphysical issues in the general areas of metaphysics, philosophy of biology, and philosophy of food. The three main topics he is currently studying are hunger, recipes, and domestication. These topics have ramifications that extend to pressing issues, including food biodiversity, geographical indications, the future of agriculture, food labelling, gender and food, and the ethics and politics of dieting.

Cristina Branquinho is Associate Professor at the Faculty of Sciences of the University of Lisbon (Portugal), in the Centre for Ecology, Evolution and Environmental Changes – cE3c. Her main aim is improving human well-being through the understanding of the effects of anthropogenic factors (e.g. pollution, eutrophication, climate change) on the provision of ecosystem services. She monitors the structure and functioning of natural, semi-natural and urban ecosystems, and the development and implementation of strategies to mitigate, adapt, and restore ecosystems facing global change.

Abigail E. Cahill is an evolutionary ecologist who uses both traditional and molecular tools to understand marine invertebrates, particularly questions relating to dispersal and early life stages. She has also conducted multiple systematic reviews of questions relating to ecology and evolution. She is currently Assistant Professor at Albion College in Albion, Michigan (USA).

Elena Casetta is Assistant Professor at the Department of Philosophy and Education at the University of Turin (Italy), where she teaches philosophy of nature. Trained in theoretical philosophy, she then specialized in philosophy of biology at the IHPST (Institut d'Histoire et de Philosophie des Sciences et des Techniques), CNRS/Paris 1/ENS, Paris, and at the Center for the Philosophy of Sciences of the University of Lisbon. Currently, her main research field is the philosophy of biodiversity.

Anne Chenuil is an evolutionary biologist who studied a variety of marine species with population genetics tools. She found numerous cases of cryptic species without seeking them, which convinced her that the phenomenon needed a closer look and that a theoretical and practical framework was needed before undertaking a comprehensive meta-analysis to better understand morphological evolution, biological diversification, and their relationships. She is a Senior Researcher at the CNRS (Centre National de la Recherche Scientifique) in France.

Sylvain Coq is Assistant Professor at the University of Montpellier (France) and in the Centre for Functional and Evolutionary Ecology. His research interests are at the junction between functional and community ecology. He is particularly interested in soil-plant interactions and more broadly on the role of biotic interactions for ecosystem functioning, in particular at the plant-soil interface.

Vincenzo Crupi is Associate Professor of logic and philosophy of science in the Department of Philosophy and Education and director of the Center for Logic, Language, and Cognition at the University of Turin (Italy). He has been working and teaching in London, Trento, Marseille, Venice, Florence, and Munich. His research interests concern classical issues in general philosophy of science (patterns of scientific inference and scientific rationality) as well a number of interdisciplinary topics in the special sciences (including medicine and the study of human cognition).

Silvia Di Marco is a Postdoctoral Fellow at the Centro de Filosofia das Ciências da Universidade de Lisboa (Portugal) and a member of the FCT research project "BIODECON: Which Biodiversity Definition for Biodiversity Conservation?" She is interested in the analysis of life sciences' concepts, with a focus on conservation biology and biomedicine.

Antoine C. Dussault is a Researcher at the Centre Interuniversitaire de Recherche sur la Science et la Technologie (Montréal, Québec) and philosophy instructor at Collège Lionel-Groulx (Sainte-Thérèse, Québec). His work focuses on philosophical issues related to the science of ecology and the practice of environmental conservation. His current focus is on issues related to the use of functional classifications in ecology, the concept of ecosystem health, and the holism/reductionism debate in ecology. He also pursues research in the fields of environmental ethics and the philosophy of medicine.

Philippe Huneman After having worked and published on the constitution of the concept of organism and Kantian metaphysics, he currently investigates philosophical issues in evolutionary theory and ecology, such as the emergence of individuality, the relations between variation and natural selection in evolutionary theory, the role of the concept of organism, and the varieties of explanations in ecology, as well as the neutral theories in ecology and evolution and the history of the Modern Synthesis. He edited seven books on history and philosophy of biology – among them are *Functions: Selection and Mechanisms* (Synthese Library, Springer, 2012), *From Groups to Individuals: Evolution and Emerging Individuality* (MIT Press, 2013, with F. Bouchard), and *Challenging the Modern Synthesis: Adaptation, Development, and Inheritance* (Oxford UP, 2017, with D. Walsh) – and authored two.

Jorge Marques da Silva is Assistant Professor at the Faculty of Sciences of the University of Lisbon (Portugal). He is a plant physiologist interested on the mechanisms of response to stress, chiefly on the effects of drought stress on photosynthetic metabolism. His main scientific interest is on understanding how interactions of abiotic stresses with photosynthetic metabolism impair primary productivity. Currently, Jorge is working in high-throughput plant phenotyping, aiming to unravel environment-genetic interactions in phenotype development, with the ultimate goal of improving crop responses to climate change. Jorge also developed a parallel interest in teaching and research in global bioethics, mainly environmental ethics.

Paula Matos is a Postdoc at the Faculty of Sciences of the University of Lisbon (Portugal), in the Centre for Ecology, Evolution and Environmental Changes – cE3c. She focusses on understanding the ecological patterns in response to global change to develop, test, and model ecological indicators. Her goal is to develop early warning ecological indicators of the effects of global change that can be applied globally, contributing to a more sustainable future.

Yves Meinard is a Researcher in environmental decision analysis at the CNRS (French National Center for Scientific Research) and PSL Research University, Paris (France). His works aim at developing strategies to make the best of scientific advances in conservation biology, economics, and decision sciences to improve environmental management practices.

Rob Mills is a Postdoctoral scientist and member of the BioISI Institute at the University of Lisbon (Portugal). His research takes synthetic and analytical forms to better understand complex biological systems at various scales in time and space. This work spans the domains of evolutionary computing, theoretical biology, and bio-hybrid systems, with a focus on information-processing capabilities in biological systems. Research interests include collective behaviour, evolutionary simulation modelling, and natural and artificial intelligence.

Alessandro Minelli Full Professor of Zoology at the University of Padova until his retirement in 2011, he has been president of the International Commission on Zoological Nomenclature and vice president of the European Society for Evolutionary Biology and received in 2008 the Sherborn Award for his service to biodiversity informatics, which includes editorship of the Checklist of the Italian Fauna. Since the mid-1990s, his research interests have turned towards evolutionary developmental biology. Among his publications are *Biological Systematics* (1993), *The Development of Animal Form* (2003), *Perspectives in Animal Phylogeny and Evolution* (2009), and *Plant Evolutionary Developmental Biology* (2018).

Alice Nunes is a Postdoc at the Faculty of Sciences of the University of Lisbon (Portugal), in the Centre for Ecology, Evolution and Environmental Changes – cE3c. She is interested in ecosystems' response to global change drivers, particularly of dryland ecosystems. She focused on trait-based ecology to develop ecological indicators of the effects of environmental changes on ecosystem functioning and ecosystem services delivery and as a tool to improve strategies to mitigate, adapt, and restore ecosystems affected by global change.

Markku Oksanen is Senior Lecturer in philosophy at the University of Eastern Finland, Kuopio (Finland). He gained his PhD in philosophy from the University of Turku in 1998. In the year 2000, he was a Postdoctoral Fellow at Lancaster University, UK. Oksanen's current research interests are environmental philosophy, environmental political theory, animal ethics, and human rights. He has coedited several books, such as *Philosophy and Biodiversity* (Cambridge University Press, 2004, edited with Juhani Pietarinen) and *The Ethics of Animal Re-creation and Modification* (Palgrave 2014, edited with Helena Siipi).

Pedro Pinho is an Invited Researcher in the Centre for Ecology, Evolution and Environmental Changes – cE3c – at the Faculty of Sciences of the University of Lisbon (Portugal). He currently works in the study of the effect of environmental changes on ecosystems' functioning and provision of services, considering changes with both natural and human origin: urbanization, desertification and land degradation, climate alterations, eutrophication and land use/cover, using vegetation functional groups, remote sensing and ecological indicators, and spatial characteristics of ecological data.

Carlo Polidori is a zoologist interested in behavioural ecology and evolution of wasps and bees, with special focus on resource use, social evolution, and functional morphology. He completed a PhD degree in natural and environmental sciences at the University of Milan (Italy) in 2007, and then obtained a series of positions as researcher at Italian, Spanish, and Portuguese institutions. He is currently working at the Institute of Environmental Sciences of the University of Castilla-La Mancha (Spain).

Thomas A. C. Reydon is Professor of philosophy of biology in the Institute of Philosophy and in the Centre for Ethics and Law in the Life Sciences (CELLS), Leibniz University Hannover, Germany. His research focuses on biological classification, core concepts in biology ("species", "gene", etc.), the metaphysics of natural kinds, the ontological status of chimeric beings and genetically engineered organisms, and evolutionary explanations within and outside biology ("generalized Darwinism"). He also has research interests in responsibility in science and the integration of philosophy of biology and theoretical biology. He is joint editor in chief of the *Journal for General Philosophy of Science* as well as the Springer book series History, Philosophy and Theory of the Life Sciences.

Sahotra Sarkar is Professor in the Departments of Integrative Biology and Philosophy at the University of Texas at Austin (USA). His work spans conservation science, disease epidemiology, and the philosophy of science. Along with Chris Margules, he is the co-author of *Systematic Conservation Planning* (Cambridge,

2007) and the author of five other books in environmental philosophy and philosophy of biology. He is currently completing a book on the concept of biodiversity.

Bernhard Schmid is Professor of environmental sciences and Dean of the Faculty of Sciences at the University of Zürich (Switzerland). His research interests encompass many aspects of plant sciences and ecosystem ecology, including plant growth and competition, biodiversity-ecosystem functioning relationships, plant-animals interactions, and community genetics and evolution.

Helena Cristina Serrano is a Postdoc Researcher at the Faculty of Sciences of the University of Lisbon (Portugal), in the Centre for Ecology, Evolution and Environmental Changes – cE3c. She is interested in environmental changes, either natural or anthropogenic, associated with ecological conservation and mitigation, towards a more sustainable future.

Georg Toepfer is Head of the Department "Knowledge of Life" at the Centre for Literary and Cultural Research (ZfL) in Berlin (Germany). His principal area of research is the history and philosophy of the life sciences, with a special focus on the history of biological concepts and their transfer between biology and other fields. His major publication is *Historisches Wörterbuch der Biologie. Geschichte und Theorie der biologischen Grundbegriffe* (3 vols., 2011).

Davide Vecchi is a philosopher of biology working in the Department of History and Philosophy of Sciences of the Faculty of Sciences of the University of Lisbon (Portugal). Formed philosophically at Bologna University (Italy) and the LSE (UK) and biologically at the KLI (Austria) and the University of Santiago (Chile), Davide's research interests span historical, philosophical, and theoretical issues in the life sciences, particularly molecular and evolutionary biology. At the moment, his research focuses on making sense of the causal role of DNA, genes, and genomes in development and evolution.

Timo Vuorisalo is Adjunct Professor and Senior Lecturer in environmental science at the University of Turku, Finland, since 1992. In 1989–1990, he was a Postdoctoral Fellow at Indiana University, Bloomington. Vuorisalo's main research interests are evolutionary ecology, urban ecology, and environmental history, and he has coedited/authored several books, including *The Long Shadows: A Global Environmental History of the Second World War* (Oregon State University Press, 2017).

Chapter 1
Biodiversity Healing

Elena Casetta, Jorge Marques da Silva, and Davide Vecchi

With the Convention on Biological Diversity (CBD) that entered into force in 1993, the conservation of biodiversity was recognized for the first time in international law as "a common concern of humankind" and almost the entire world committed to it. Conserving biodiversity, however, is far from being an easy task, as shown by the difficulties to reach the conservation targets articulated in the strategic plans connected to the CBD. The failure of the 2010 Biodiversity Target "to achieve by 2010 a significant reduction of the current rate of biodiversity loss at the global, regional and national level" has been explicitly recognized (Butchart et al. 2010). Moreover, there is a widespread scepticism, at present, concerning the possibility of achieving the Aichi Biodiversity Targets by 2020 (Tittensor et al. 2014), i.e., 20 time-bound targets included into the CBD strategic plan 2011–2020 (such as, for instance, making people aware of the values of biodiversity and the steps they can take to conserve it (Target 1) or identifying and eradicating invasive species (Target 9)).[1] Despite increasing communication, accelerating policy and management responses, and notwithstanding improving ecosystem assessment and endangered species knowledge, conserving biodiversity continues to be more a concern than an

---

[1] For the comprehensive list, see the CBD website: https://www.cbd.int/sp/targets/default.shtml

---

E. Casetta (✉)
Department of Philosophy and Education, University of Turin, Turin, Italy
e-mail: elena.casetta@unito.it

J. Marques da Silva
BioISI – Biosystems and Integrative Sciences Institute and Department of Plant Biology, Faculdade de Ciências, Universidade de Lisboa, Lisboa, Portugal

D. Vecchi
Centro de Filosofia das Ciências, Departamento de História e Filosofia das Ciências, Faculdade de Ciências, Universidade de Lisboa, Lisboa, Portugal

accomplished task. Why is it so? The overexploitation of natural resources by our species is a frequently recognised factor,[2] while the short-term economic interests of governments and stakeholders typically clash with the burdens that implementing conservation actions imply. But this is not the whole story. This book develops a different perspective on the problem by exploring the conceptual and practical challenges posed by conserving biodiversity. By conceptual challenges, we mean the difficulties in defining what biodiversity is and characterising that "thing" to which the word "biodiversity" refers to. By practical challenges, we mean the reasons why assessing biodiversity and putting in place effective conservation actions is arduous. In order to situate the multi-farious conceptual and practical challenges faced when trying to outline the path **From Assessing to Conserving Biodiversity**, we think an interpretive device is useful.

An analogy is generally recognised (see, for instance, Soulé 1985; Sarkar 2002; Casetta and Marques da Silva 2015) between medicine—the discipline whose main mission is human health preservation—and conservation biology—the discipline whose main mission is biodiversity conservation. Unusually for sciences, both have a normative dimension. According to this analogy,

> Conservation biology differs from most other biological sciences in one important way: it is often a crisis discipline. Its relation to biology … is analogous to that of surgery to physiology and war to political science. In crisis disciplines, one must act before knowing all the facts. Crisis disciplines are then a mixture of science and art, and their pursuit requires intuition as well as information. (Soulé 1985: 727)

When biodiversity conservation is at issue, theoretical and practical matters go hand in hand, and practical challenges are intertwined with conceptual ones, requiring the cooperation of the natural sciences and of the humanities in a concerted effort. This book, including contributions from biologists and philosophers from different fields and traditions, reflects this necessary multidisciplinarity.

1.1 Assessing and Diagnosing the Patient. Estimating Biodiversity: Data Collection and Monitoring Challenges

Consider a first aspect of the analogy between medicine and conservation biology. The first thing medical doctors have to do when dealing with patients is to assess their general health state and the severity of the condition affecting them. Diagnosis on the basis of the collection of patients' data and their classification, as well as on the measurement and monitoring of symptoms, comes before treatment prescription and provision. In the case of biodiversity, two main kinds of diagnostic challenges

---

[2] In this sense, Diamond (1989) refers to the "Evil Quartet", the four horsemen of the apocalypse: habitat loss and fragmentation, overharvesting, introduced predators and competitors, and secondary extinction, while E.O. Wilson (2002) expresses similar concerns by characterizing the HIPPO (i.e., Habitat destruction, Invasive species, Pollution, (human) Population growth and Overharvesting).

have to be addressed: on the one hand, the difficulties concerning data collection and systematisation (Chaps. 2, 3 and 4); on the other hand, the choice of the appropriate measurement techniques and ways of monitoring (Chaps. 5, 6 and 7).

Starting with the first challenge, probably the most striking aspect of the living world is its amazing variety, so immense that it eludes even our hardest systematisation attempts. Buffon, in the First discourse of his *Histoire Naturelle* (1749) already highlighted this aspect of the natural world:

> ... it takes a peculiar kind of genius and courage of spirit to be able to envisage nature in the innumerable multitude of its productions without losing one's orientation, and to believe oneself capable of understanding and comparing such productions... The first obstacle encountered in the study of natural history comes from this great multiplicity of objects. But the variety of these same objects, and the difficulty of bringing together the various productions of different climates, is another apparently insurmountable obstacle to the advancement of our understanding, an obstacle which in fact work alone is unable to surmount. It is only by dint of time, care, expenditure of money, and often by lucky accidents, that one is able to obtain well-preserved specimens of each species of animal, plant, or mineral, and thus form a well-ordered collection of all the works of nature. (Buffon, *First discourse*, quoted in Lyon 1976: 145)

Things have not become easier over time. Taxonomic knowledge, as all empirical knowledge, is hypothetical in nature, hence always susceptible to revision as new data become available and new theoretical frameworks replace old ones. In such a context, the challenge posed by taxonomic revisions is a *fil rouge* connecting the first three chapters of this part of the book. The puzzle these contributions pose is effectively a taxonomic version of Kuhn's incommensurability thesis: are the data collected and systematised according to a certain taxonomy translatable, so to speak, into another? Notice that taxonomic revisions, either caused by a change in the theoretical framework or by the availability of new data, have important consequences for biodiversity conservation. As Agapow et al. (2004) argued, for instance, a reclassification of endangered species adopting a phylogenetic species concept (which defines a species as a group of organisms that share at least one uniquely derived character) could increase the cost of recovering all species currently listed in the Endangered Species Act from \$4.6 billion to \$7.6 billion.[3] Counting species, and their members, is not only fundamental for assessing the patient's general health state, but it is also for assessing the severity of the condition affecting the patient and, hence, for determining treatments prioritisation.

Chapter 2, *The hidden biodiversity data retained in pre-Linnaean works: a case study with two important XVII century Italian entomologists* by **Francesco Andrietti**

---

[3] In their article, Agapow and colleagues surveyed the primary literature searching for examples of sets of organisms that had been classified by both the phylogenetic species concept and non-phylogenetic concepts (typically defined, at least for animal species, by means of the biological species concept, according to which species are groups of populations that are reproductively isolated). Reclassifying species under the phylogenetic species concept would lead, this is their conclusion, to an apparent rise in the number of endangered species for two main reasons: the detection of "new" species (for instance by the splitting of "old" ones) and the subsequent reduction in geographic range (a frequently used diagnostic indicator in establishing whether a species is threatened).

and **Carlo Polidori**, and **Chap. 3**, *Marine biodiversity databanks* by **Anouk Barberousse and Sophie Bary** focus respectively on pre-and post-Linnean archival data on biodiversity: what was the pre-Linnean knowledge on the extent of biodiversity? And how to systematise, today, in digital databanks, the incredible—and at the same time insufficient—amount of information on biodiversity in order to help both researchers and conservationists in their respective endeavors? While in Chap. 2 Francesco Andrietti and Carlo Polidori tackle the issue by analysing a case study, i.e., the classification of Hymenoptera in the pre- and post-Linnean taxonomic frameworks (how to make available today data "on species" collected *before* those species were given their contemporary name?), Anouk Barberousse and Sophie Bary bring to attention the taxonomic vicissitudes of earthworms, from Linnaeus' description to the *Barcoding earthworms* programme. Notice that these two case studies are particularly significant for biodiversity conservation. Several members of the Hymenoptera order (bees, wasps, ants, and parasitoids) are major pollinators, and several members of the family Lumbricidae play a fundamental role within the natural soil ecosystem, as Darwin already recognized in his 1881 book, *The formation of vegetable mould*. In **Chap. 4**, **Anne Chenuil and colleagues** discuss the *Problems and questions posed by cryptic species*: to what extent do nominal species (identified through morphological characters and referred to by Linnean binominal names) and biological species (identified instead typically through reproductive isolation) overlap? In this contribution, a rational and practical classification of cryptic species is proposed, based on the crossing of distinct levels of reproductive isolation with distinct levels of morphological differentiation. The focus is on marine biodiversity, and the conceptual challenge of establishing the possible commensurability between morphological and biological species is taken up with the help of genetic tools, such as genome sequencing and the use of genetic markers, whose impressive development (and rapidly decreasing cost) is allowing identification at an increasing rate of cryptic species, i.e., biological species "hidden" within nominal species.

The second challenge addressed in the first part of the book concerns the choice of the appropriate biodiversity measurements and the ways of monitoring the condition of the patient. Measuring biodiversity is a fundamental operation in biodiversity conservation, for instance because, when we need to choose and implement conservation actions, financial resources are usually limited; accordingly, ecologicals systems and/or places—i.e., specific regions on Earth's surface "filled with the particular results of [their] individual story" (Sarkar 2002)—have to be prioritised and, to do so, biodiversity must be measured. There is widespread agreement that biodiversity cannot be measured directly:

> ...conservation biologists almost never measure directly the full range of phenomena that they take to constitute the biodiversity of a system. Rather, they … rely on measurable signs that vary (they believe) with biodiversity itself. Samples and signs are biodiversity surrogates. (Maclaurin and Sterelny 2008, p. 133)[4]

---

[4] On the "surrogacy problem", see for instance Sarkar (2002).

In a similar way as temperature can be measured by means of a substance, mercury, whose characteristics are particularly sensitive to heat fluctuations and easily measurable, a biodiversity surrogate is thought to be a sort of biological thermometer that would allow to measure biodiversity, even though indirectly. Intuitively, the surrogate *par excellence* seems to be species richness:

> Eliminate one species, and another increases in number to take its place. Eliminate a great many species, and the local ecosystem starts to decay visibly. Productivity drops as the channels of the nutrient cycles are clogged. More of the biomass is sequestered in the form of dead vegetation and slowly metabolizing, oxygen starved mud, or is simply washed away ... Fewer seeds fall, fewer seedlings sprout. Herbivores decline, and their predators die away in close concert. (E.O. Wilson 1992, p. 14)

As it can be grasped from the above quote, species richness is considered to be important because it is supposed to be related with the well-functioning or the stability of an ecosystem. But it can be argued that the number of species is not the only surrogate to be taken into account when estimating biodiversity. Another important feature is so-called *evenness*: a biological community, an ecosystem, or a geographical area are said to have evenness when the *abundance* of all species present is similar. However, establishing whether and how surrogates such as species richness and evenness are correlated with one another as well as with patterns of species abundance remains an open theoretical problem that has, moreover, important practical repercussions (for instance, how to infer, from the data collected from an actually sampled area via such surrogates, a possible general estimation of its diagnostic status). **Chapter 5,** by **Luís Borda-de-Água**, *The importance of scaling in biodiversity*, is devoted to these topics, focusing on the species-area relationship (a mathematical expression relating how the number of species changes as a function of the size of the sampled area) and the scaling of species abundance distributions (i.e., the relative abundance of species). **Chapter 6**, *Measures of biological diversity: Overview and unified framework* by **Vincenzo Crupi** is dedicated to diversity indexes, more precisely to the challenge of integrating them in a unified formalism. Here, Crupi presents a unified framework, taken from generalised information theory, to measure biological diversity embedding a variety of statistical measures. While Chaps. 5 and 6 mainly rely on insights coming from information theory, mathematics and statistics to address specific problems primarily related with the choice of the appropriate biodiversity estimation techniques necessary to assess the status of the patient, **Chap. 7**, *Essential biodiversity change indicators for evaluating the effects of Anthropocene in ecosystems at a global scale* by **Cristina Branquinho and colleagues** tackles, from a conservationist point of view, the problem of monitoring the condition of the patient. Once conceded that measuring all forms of biodiversity everywhere and over time is an impossible task, this chapter proposes to broaden the outlook from species diversity to the "essential biodiversity variables" proposed by the Group on Earth Observations—Biodiversity Observation

Network.[5] These include: genetic composition, species populations, species traits, community composition, ecosystem structure and ecosystem function. Putting into practice a global monitoring network to track biodiversity change is far from being an easy endeavour, and the chapter discusses these difficulties as well as suggesting possible solutions.

1.2 Are We Taking Care of the Right Patient? Characterising Biodiversity: Beyond the Species Approach

If measuring biodiversity poses a series of mainly practical challenges, it also opens a Pandora's box of conceptual ones, which are dealt with in the second part of the book. Consider a second aspect of the analogy between medicine and conservation biology. Diagnosis depends on the appropriate characterisation of the biological organism as a unit of medical intervention. Organisms can be decomposed in a variety of entities such as organs, tissues, cells, proteins, genes, microbiotas etc. that interact in the context of metabolic, developmental, immunological, neurological etc. processes. The intervention on the medical patient is thus dependent on the way in which the biological organism is characterised. For instance, we might characterise the organism as made of proteins, and we will be right in treating mad cow disease and Creutzfeldt–Jakob disease. However, not all illnesses are linked to protein abnormalities. Analogously, the biosphere is composed of a variety of entities classifiable in many different ways and interacting in the context of a variety of ecological and evolutionary processes. Species are just one of these entities. To conserve biodiversity by focusing merely on species loss is analogous, let us say, to maintaining the health of an organism intervening on protein deficiencies only.

In this part, two main kinds of foundational issues for diagnosis will be considered: on the one hand, the conceptual challenges in individuating the salient units of biodiversity (Chaps. 8, 9 and 10); on the other, the contrast between entity-based vs. process-based and function-based approaches to biodiversity (Chaps. 11, 12, 13 and 14).

Except for Chap. 7, the implicit underlying assumption of the contributions of the first part of the book is that assessing and measuring biodiversity ultimately amounts to counting species or, at most, taxa. This is probably the most traditional and widely used strategy: counting taxonomic groups and estimating their frequency (Maclaurin and Sterelny 2008, p. 135 ff.). This should come as no surprise. In fact, when it made its appearance in 1986, the term "biodiversity" was, implicitly or explicitly, intended to refer to species diversity. Assessing biodiversity was considered as one and the same thing as inventorying species, and conserving biodiversity consisted in maintaining the inventory. In the words of E.O. Wilson (1992, p. 38):

---

[5] See: https://geobon.org/ebvs/what-are-ebvs/

> … the species concept is crucial to the study of biodiversity. It is the grail of systematic biology. Not to have a natural unit such as the species would be to abandon a large part of biology into free fall, all the way from the ecosystem down to the organism.

Counting species is relatively easy in practice and theoretically well motivated. In fact, we already possess good (even though neither complete nor fully coherent, as particularly emphasised in Chaps. 2 and 3) species inventories, some fairly reliable ways to recognize them in practice as well as methodologically solid ways of counting them. Moreover, there is a widespread agreement that the concept of species refers to a fundamental unit of biological organisation (Mayr 1988) and that species, by speciating, produce new biodiversity. However, two major problems remain: on the one hand, the concept of species is severely flawed (as the persistency of the so-called "Species Problem" shows, cf. Richards 2010; Zachos 2016) and, on the other, it is questionable whether it can be applied across all branches of the tree of life, for instance to bacterial biodiversity.

Given this state of affairs, a question is in order: if biodiversity has to be conserved, are we describing and treating the right patient when we focus on species or, more largely, on other taxonomic groups? If we give a negative answer to this question, then the conceptual challenge consists in proposing a viable characterisation of biodiversity able at the same time to go beyond a mere species-centred approach (whose merits and limits have been mentioned above) and to account for the variety of entities other than species, and of processes other than speciation, that might be considered targets of conservation practice. But then, an entirely new set of basic challenges, both theoretical and practical, opens up: how can we individuate the salient units of biodiversity? How do such units interact among them within the same and other levels, and how can this interaction give rise to novel diversity? Is it possible to keep together, in an ideally comprehensive account, this enormously complex interplay of units belonging to different levels and describable and evaluable at different temporal and spatial scales? How to bridge epistemologies concerning biodiversity conservation? And how to link these epistemologies with the practical concern of conserving biodiversity? These and similar questions are addressed in the second part of the book.

Notice that, to go back to the medicine analogy, in discussing how the patient should be better characterised, and which of its parts, properties and functions should be emphasised, the chapters in this section do not miss to keeping an eye on the relation between diagnosis and treatment, i.e., on the issue of what it would mean, for conservation purposes, to characterise biodiversity in one way rather than another. The three initial chapters in this part couple evolutionary with conservationist considerations in order to go beyond a species-centred approach by individuating different salient units of biodiversity. In **Chap. 8, Thomas Reydon** suggests an answer to the question: *Are species good units for biodiversity studies and conservation efforts?*, embracing a radical approach: species are not good units of biodiversity; yet, a pragmatic notion of species can be used as an epistemic tool in the context of biodiversity studies. Two other contributions try to characterise biodiversity in a more encompassing way by including other entities. In **Chap. 9**, *Why a species-*

based approach to biodiversity is not enough. Lessons from multispecies biofilms, **Jorge Marques da Silva and Elena Casetta** take a look at the microbial world, where the application of the concept of species is particularly controversial. This chapter suggests that entities such as multispecies biofilms, where interaction among parts gives rise to a putative multispecies individual, might play a crucial role in the generation of biodiversity and that, as a consequence, could be adequate targets of conservation. In a similar spirit, in **Chap. 10,** *Considering intra-individual genetic heterogeneity to understand biodiversity,* **Eva Boon** aims at enlightening an unexplored dimension of biodiversity, again focusing on entities other than species, i.e., multicellular life forms characterized by intra-individual genetic heterogeneity (such as, for instance, genetically mosaic and chimeric entities). This chapter argues that studying biodiversity through the lens of intra-individual genetic heterogeneity facilitates thinking in terms of interactions between biological entities rather than in terms of organismal function, allowing a new light on the ecological and evolutionary significance of biological diversity.

The other chapters in this part also couple evolutionary with conservationist concerns. In order to go beyond a species-centred approach, they focus on the role of processes (other than speciation) and functions (such as, for instance, evolvability, evolutionary potential, plasticity). In **Chap. 11,** *Biodiversity, disparity and evolvability,* **Alessandro Minelli** argues that taxic diversity is not necessarily the most important aspect of biodiversity if what most matters is maintenance of ecosystem function. This chapter articulates a rationale for prioritising focus on those species providing the largest contribution to overall phylogenetic diversity, thus proceeding towards an evo-devo approach to conservation focused on evolvability, robustness and phenotypic plasticity. The potential role of the process of phenotypic plasticity in the production of new diversity is also stressed by **Davide Vecchi and Rob Mills** in **Chap. 12,** *Probing the process-based approach to biodiversity: Can plasticity lead to the emergence of novel units of biodiversity?* This contribution aims to provide a model to test the hypothesis that plastic populations of a species might be considered evolutionary significant units amenable to conservation. In addition to Chaps. 11 and 12, Chaps. 13 and 14 also propose a characterisation of biodiversity based on process and function that reveals a common ontological ground. The rationale of a process-based approach to biodiversity is that a mere focus on entities does not address the issue concerning whether evolutionary and ecological processes have the capacity to create novel, salient units of biodiversity. The suggestion is that a process-based approach should integrate an entity-based one. Process-based and function-based approaches are, as a matter of fact, strictly related to the historical roots of the concept of biodiversity. In fact, while it might be argued that the term "biodiversity" only entered the scientific and public discourse the mid-1980s—i.e., on the occasion of the National Forum on Biodiversity that took place in Washington DC in September 1986 (Takacs 1996)—the concept goes back at least to the diversity-stability debate that animated ecology in the middle of the twentieth century (McCann 2000). Yet, at least in its beginning, conservation biology displayed little interest to previous research in ecology, addressing instead the more pragmatic aspects of conservation. It seemed, in other words, that, as it often happens, two scientific disciplines working on the same subject from different perspectives were talking past each other. This stand-off started to unlock with the Harvard Forest Symposium in 1991, where it was explicitly recognised that, in order to effectively conserve biodiversity, a more precise knowledge of the functioning of ecosystems would be needed (Blandin 2014). In this perspective, it becomes clear that a characterisation merely in terms of species diversity does not seem to fully capture the multitude of dynamical interactions at different levels and scales from which biodiversity results. Again, neither do all species play the same role in a community or in an ecosystem, nor do they have the same evolutionary history and potential. However, when species are counted through indexes, they are treated as being equivalent conservation units; in fact, indexes are not easily able to mirror the possibility that a species may be more important than another for the functioning of the ecosystem. Moreover, in biodiversity conservation, it is not sufficient to preserve current biodiversity, but what is also ideally needed is to maintain diversity in the face of possible future losses; but to do so, a metric able to indicate whether diversity in a certain place is mostly constituted by rare species (that are more likely to go extinct) would be needed. Chapters 13, 14 and 15 are mainly dedicated to ecological theoretical perspectives on biodiversity and to the challenge of connecting evolutionary, ecological, and conservation considerations. Through **Chap. 13**, *Between* explanans *and* explanandum: *Biodiversity and the unity of theoretical ecology*, **Philippe Huneman** clarifies the key-role of the concept of biodiversity in ecology as both an *explanans* and an *explanandum*, while **Antoine Dussault,** in **Chap. 14**, *Functional biodiversity and the concept of ecological function*, elucidates some aspects of the concept of functional diversity. Starting from the assumption that measures of biodiversity based on species richness have epistemological limitations, this chapter explores the notion of "ecological function" and characterises it in non-selectionist terms. Finally, in **Chap. 15**, *Integrating ecology and evolutionary theory: A game changer for biodiversity conservation?*, **Silvia Di Marco** spells out the interaction between conservation science, evolutionary biology, and ecology in order to understand whether a stronger integration between evolutionary and ecological studies might help clarifying the interactions between biodiversity, ecosystem functions and ecosystem services.

1.3 Treating the Patient. Conserving Biodiversity: From Science to Policies

In the light of the ongoing complex work of characterisation of the patient articulated in the various contributions of the previous part, it is not surprising that we do not possess a final, universally agreed upon, definition of "biodiversity". The crucial question is therefore whether putting in place effective conservation actions without a satisfactory definition of "biodiversity" makes sense at all. The third part of the book deals with this issue. Consider a third aspect of the analogy between medicine and conservation biology. After diagnosis and proper characterisation of the unit of medical intervention, medical doctors prescribe treatment. The aim of dispensing treatment is always the benefit of the patient as an organism, however successful treatment is. In the case of biodiversity, there exist incompatible ways to characterise the aim of conservation policies. Part of the treatment problem, to which some contributions of this part are dedicated (Chaps. 16 and 17) is that the term "biodiversity", not being well defined, is potentially used differently by the various actors (e.g., scientists, policy-makers and conservationists) devising and implementing conservation policies. A flipside of the treatment problem can be understood if we take into consideration another facet of the definitional conundrum: the term "biodiversity" might have inherited, from the intentions of its original proponents (Takacs 1996), an intrinsic normative element that has to do with biodiversity protection and preservation. However, normativity poses a series of challenges to which the rest of the contributions of this part is dedicated: the first ones (Chaps. 18 and 19) have to do with the characterisation of normativity while the others (Chaps. 20 and 21) concern the global-local tension of conservation aims and constraints.

In his often-quoted review of definitions of "biodiversity", DeLong (1996) listed no less than 85 definitions. It is thus not surprising that the term is recognised by some authors as remarkably vague (Sarkar 2002). Other authors think that the term is clearly defined or, at least, that conservation science possesses perfectly workable operational definitions to prescribe treatments (Bunnell 1998). The role of definitions in science is controversial. On the one hand, it can be said that "definition is one of the most crucial issues in any science; an improper understanding of it can vitiate the success of the whole enterprise" (Caws 1959). On the other hand, it can be claimed that scientific enterprises can proceed quite well even without having clear, univocal and unambiguous definitions of their key terms. After all, focusing on disciplines like biology and ecology, it is widely recognized that no univocal, universally agreed upon, definitions of terms such as 'life' (Benner 2010), 'organism' (J. Wilson 2000), 'species' (Richards 2010), and 'ecosystem' (Sarkar 2002) can be provided. Still, biologists and ecologists successfully go on with their work. Why would the situation be different in the case of the term "biodiversity"? Following Bunnell's (1998) suggestion, whether a definition plays a crucial role or not in scientific endeavours probably depends on the nature of the specific enterprise at issue. For instance, when J. Wilson (2000) wrote that "Biology lacks a central organism concept that unambiguously marks the distinction between organism and non-organism because the most important questions about organisms do not depend on this concept", he was clearly not making reference to conservationist needs. But where the theoretical enterprise is strictly intertwined with pragmatic objectives, as it is the case with biodiversity studies, things are different.

In particular, two main reasons may be given for why a definition of the term "biodiversity" is needed, together with some reasons for why not having it would be a source of impediments in finding agreed-upon methods to evaluate management and conservation strategies and in the implementation of conservation actions. The first reason is that, unlike other scientific terms, "biodiversity" is supposed to play a unifying role for the plethora of discourses (Haila and Kouki 1994) produced by the different disciplines and actors involved in facing the so-called biodiversity crisis. On the one hand, the term performs a unifying function for the scientific disciplines involved in estimating biodiversity and those studying how it is generated (evolutionary biology, genetics, ecology, biogeography, systematics, and so on), for the disciplines involved in its management and conservation (from environmental economics to conservation biology) as well as for all those socio-political disciplines concerned with the interactions between our species and biodiversity exploitation and conservation (from the social sciences to political philosophy and ethics).[6] Furthermore—but not less importantly—this unifying role serves to make scientific discourses uniform for a variety of social and political actors: from the general public to stakeholders, from governments to policy makers. The term "biodiversity" is often used as a flagship, with no explicit definition provided. However, if scientist and the different social and political actors involved in facing the biodiversity crisis define biodiversity in fundamentally different ways, the agreement necessary to perform common actions could be severely impaired and the presumption that common actions are actually oriented towards the same goal could be false. Accordingly, "to create solutions for biodiversity loss, it is essential for natural and social scientists to overcome such language barriers" (Holt 2006).

Georg Toepfer embraces a different view. In **Chap. 16**, *On the impossibility and dispensability of defining "biodiversity"*, Toepfer argues that it is exactly because the term is vague that the concept of biodiversity is able to tie together many different discourses from the fields of biology and bioethics, aesthetics and economics, law and global justice. **Chapter 17**, *The vagueness of "biodiversity" and its implications in conservation practice*, by **Yves Meinard, Sylvain Coq and Bernhard Schmid,** articulates a different argument. A tension emerges here between the theoretical function of the concept and its pragmatic use: providing concrete case studies to support their argument, the authors suggest that the lack of transparency in using the word "biodiversity" can hide profound disagreements on the nature of conservation issues, impairing the coordination of conservation actions, hiding the need to improve management knowledge, and covering up incompatibilities between disciplinary assumptions.

Sarkar (2002) highlighted a second reason for why a definition of "biodiversity" is needed. It concerns the "sociologically synergistic interaction between the use of "biodiversity" and the growth of conservation biology [that] led to the reconfiguration of environmental studies that we see today". In other words, the term would convey the necessity of conserving something we are losing and we care about. In this respect, the vagueness of the term "biodiversity" implies a lack of clarity as to what has to be conserved. At this juncture, a particularly thorny issue is whether a good definition should reflect the normativity that—according to several

---

[6] Many concepts may play a similar unifying role in biology and science at large. For instance, the concept of gene is definable in multifarious ways and has undergone a series of profound conceptual transformations. Nonetheless, it has continued to play an important theoretical and heuristic role in classical and molecular genetics as well as in genomics, constraining and directing both the thoughts and actions of biologists (Kay 2000).

philosophers (Callicott et al. 1999; Norton 2008)—is embedded into the concept of biodiversity. Such normativity was presumed in the early literature on biodiversity, like Soulé's (1985) *manifesto* of conservation biology, which included explicit normative postulates besides scientific ones. In **Chap. 18, Sahotra Sarkar** asks: *What should "biodiversity" be?*, and distinguishes between "scientistic", "normativist", and "eliminativist" approaches to biodiversity. In direct dialogue with **Chap. 19**, *Natural diversity: how taking the bio- out of biodiversity aligns with conservation priorities*, in which **Carlos Santana** embraces a strongly eliminativist approach according to which the concept of biodiversity should just be dismissed and replaced by the more encompassing concept of natural diversity, Sarkar instead advocates a strongly normativist position: biodiversity should be understood as a normative concept, although constrained by a set of adequacy conditions that reflect scientific analyses of biological diversity. The main problem with normativism is, of course, as the chapter underlines, that values are usually culture-dependent: global values ranging across cultures are probably a myth, and local norms (supposedly revealed by the local commitments of people living in their habitats) can be in conflict with each other as well as with alleged global values (Vermeulen and Koziell 2002). The last two chapters of the book are devoted to this tension between local and global values. **Andrea Borghini** in **Chap. 20**, *Ordinary biodiversity. The case of food*, focuses on an often-neglected aspect of biodiversity, which might be called "the edible environment". This chapter poses a series of questions concerning the nature of the criteria for inclusion in conservation effort. The way this contribution tries to answer this question is by asking whether these criteria are global or local, whether they are applicable equally to all living entities, for instance to wild and domesticated species alike. Finally, **Markku Oksanen and Timo Vuorisalo**, in **Chap. 21**, *Conservation sovereignty and biodiversity*, look at the "owners" of wild and domesticated biodiversity: on the one hand, states are self-determining actors and the principal possessors of biological resources in their territories but, on the other hand, the actual fragmentation of conservation labour is not always efficient from the conservation perspective. This contribution tries to address this stand-off.

1.4 The Way Ahead: Interdisciplinary Solutions to Biodiversity Healing

Aiming to cover the entire conceptual and practical pathway that leads from assessing to conserving biodiversity, this book highlights some critical issues that must be addressed to foster effective biodiversity conservation. These include both conceptual and philosophical issues as well as scientific and technological challenges. In Part I, concerned with the assessment of biodiversity, it becomes clear that technical and practical advances are needed. From its origins, the study of living beings was mainly concerned with their phenotypes. The main systematic classification efforts, from Buffon to Linnaeus, were built under a phenotypic paradigm. The concept of gene, formally introduced by the Danish botanist W. L. Johannsen in 1909, started its slow way into biology with Mendel's work in the nineteenth century, received a significant boost in the middle of the twentieth century with the unravelling of DNA structure, ultimately becoming dominant in the 1980s, with the development of the polymerase chain reaction and other molecular techniques. This prevalent role of genes in the conceptual corpus of biology was accompanied by the rapid development of gene sequencing technologies. This had a positive effect on the pursuit of biodiversity inventories—think of the use of DNA barcoding—but at the expense of a decrease on the original focus on the description and characterisation of phenotypes. In fact, the number of "classical" (i.e., non-molecular) taxonomists among professional biologists is sharply decreasing (Coleman 2015). Nonetheless, phenotypic studies are still crucial for the inventory of biodiversity—think about the need to correctly identify cryptic species—and for understanding evolutionary trends, given that selective pressures act on individual phenotypes. Fortunately, in the last few years, the scientific community became aware of the imbalance between genotyping and phenotyping efforts and started an international interdisciplinary venture to fix it (Dayrat 2005; Fiorani and Schurr 2013). Adding machine learning and/or multivariate statistics to digital image and/or spectroscopic analysis led to high throughput phenotyping and the new discipline of phenomics (Houle et al. 2010). High throughput phenotyping has so far been used in biotechnology contexts—both medical (Maier et al. 2017) and agricultural (Crain et al. 2018)—but not yet applied to the inventory of biodiversity, except for the artificial biodiversity of agronomic traditional plant landraces (Costa et al. 2015). For instance, the recent global initiative (Soltis 2017) to digitalize the 350.000.000 specimen stored in 3.500 herbaria all over the world may provide the conditions for automated pipeline image analysis and therefore high throughput phenotyping, fostering the understanding of plant biodiversity. Altogether, these emerging trends suggest that a combination of digital image analysis and artificial intelligence techniques has the potential to boost the phenotypic characterisation of the species inventory.

The use of air-born and satellite images for biodiversity studies is not new, but also here the application of automated artificial intelligence-supported image analysis (Keramitsoglou et al. 2004) may help making operational biodiversity units other than species (communities, habitats), as suggested by some contributions to Part II. Also, the new generation of earth observation satellites, with their increased capacity for remotely estimating ecosystem functions (e.g., photosynthetic production, Joiner et al. 2011) may help making operational the concepts of process-based and function-based approaches to biodiversity conservation defended in other contributions to Part II. Moreover, the emerging extension of bioinformatics to phenotypic analysis, through the development of controlled phenotypic ontologies (Mungall et al. 2010), led to the novel concept of "computable phenotypes" (Lussier and Liu 2007; Deans et al. 2015). Phenotypic ontologies, in conjunction with ecological ontologies (Madin et al. 2007), open new avenues to unravel evolutionary trends. This putative in-depth understanding of phenotypes might contribute to the integration between ecology and evolution, a need emphasized in several contributions to Part II. All these new approaches are strongly interdisciplinary.

Interdisciplinary science still faces, however, a series of constraints (institutional, financial, sociological, epistemological, Vasbinder et al. 2010; Marques da Silva and Casetta 2015) that must be overcome to foster biodiversity knowledge.

Part III discusses the problem of biodiversity definition and its consequences for conservation policies, also addressing the possible relation between mainly epistemological issues (such as measuring, inventorying and, indeed, defining biodiversity) and mainly axiological and moral issues (i.e., concerning, respectively, the possible value of biodiversity and our correct behaviour towards it). This discussion is not new. In the early 1970s Arne Naess developed the concept of "methodological vagueness" (Glasser 1998). The aim was to broaden the support basis of his deep ecology political project. Arguably, the vagueness of the biodiversity concept might play a similar role in biodiversity conservation. The incapacity to clearly identify the object to be conserved might not be an unbearable burden, since targeting a set of closely related objects might have an additive positive effect on biodiversity conservation (albeit this could be a problem when resources are scarce and a prioritisation of conservation targets is needed). Even if we come to a universally shared definition of biodiversity, the ethical question whether we should be committed to biodiversity conservation remains, and the axiological problem persists. Some contributions to Part III address the tension between local and global values. In this respect, moral and political philosophy, for instance theoretical bioethics and environmental ethics, may provide useful resources to enlighten conservation policies. At the same time, part of moral philosophy provides compelling arguments against cultural relativism. This is not to deny, however, that there are cultural differences between societies and that these differences might dictate different biodiversity conservation policies. In fact, institutions such as as the Indigenous Peoples' Biodiversity Network are deeply concerned with this issue. Here, theoretical bioethics might help. For instance, Engelhardt Jr. (1986), working in a medical bioethics context, argued that in multicultural societies there are fundamental incommensurable moral tenets. His bioethical system, therefore, renounces to achieve the ultimate bioethical "truth" and, instead, provides a framework to reach "minimum operational agreements" between multicultural stakeholders, i.e., it becomes a framework for "moral diplomacy". Efforts to clarify the intrinsic value of natural diversity, however, still persist in environmental ethics. Departing from peculiar systems of environmental aesthetics, contemporary authors such as Allen Carlson (2000) and Holmes Rolston III (2002) aim to provide a universal system of recognition of the intrinsic value of natural diversity. Ongoing global scientific, technological, conceptual and normative efforts aim to provide better policies and programs for biodiversity conservation. The success in preventing the so called Big Sixth mass extinction is dependent on this global interdisciplinary collective effort.

Needless to say, the aim of this book is neither to provide a set of contributions that are exhaustive of the virtually unlimited issues raised by biodiversity studies and conservation, nor to solve problems once and for all. The attempt is rather to provide a tool for teaching and for research on a topic that by its own nature is hugely complex and multidisciplinary, without falling into the temptation of simplifying the conceptual and practical challenges involved. On the contrary, we think that bringing such challenges to the fore and thematising them might be a fruitful research approach. We hope the reader will enjoy the book and, above all, find it stimulating and useful.

Acknowledgements We acknowledge the financial support of the **Fundação para a Ciência e Tecnologia** (BIODECON R&D Project. Grant PTDC/IVC-HFC/1817/2014), which also made possible the open-access publication of the book.

References

Agapow, P.-M., et al. (2004). The impact of the species concept on biodiversity studies. *The Quarterly Review of Biology, 79*, 161–179.

Benner, S. A. (2010). Defining life. *Astrobiology, 10*, 1021–1030.

Blandin, P. (2014). La diversité du vivant avant (et après) la biodiversité: repères historiques et épistémologiques. In E. Casetta & J. Delord (Eds.), *La biodiversité en question. Enjeux philosophiques, éthiques et scientifiques* (pp. 31–68). Paris: Les Éditions Materiologiques.

Bunnell, F. L. (1998). Overcoming paralysis by complexity when establishing operational goals for biodiversity. *Journal of Sustainable Forestry, 7*(3/4), 145–164.

Butchart, S., et al. (2010). Global biodiversity: Indicators of recent declines. *Science, 328*(5982), 1164–1168. https://doi.org/10.1126/science.1187512.

Callicott, J. B., Crowder, L. B., & Mumford, K. (1999). Current normative concepts in conservation. *Conservation Biology, 13*, 22–35.

Carlson, A. (2000). *Aesthetics and the environment: The appreciation of nature, art and architecture.* London: Routledge.

Casetta, E., & Marques da Silva, J. (2015). Biodiversity surgery. Some epistemological challenges in facing extinction. *Axiomathes, 25*(3), 239–251. https://doi.org/10.1007/s10516-014-9244-9.

Caws, P. (1959). The functions of definition in science. *Philosophy of Science, 26*(3), 201–228.

Coleman, C. O. (2015). Taxonomy in times of the taxonomic impediment – Examples from the community of experts on amphipod crustaceans. *Journal of Crustacean Biology, 35*(6), 729–740.

Costa, J. M., Garcia Tejero, I. F., Duran Zuazo, V. H., Nunes da Lima, R. S., Chaves, M. M., & Vaz Patto, M. C. (2015). Thermal imaging to phenotype traditional maize landraces for drought tolerance. *Comunicata Scientiae, 6*(3), 334–343.

Crain, J., Mondal, S., Rutkoski, J., Singh, R. P., & Polan, J. (2018). Combining high-throughput phenotyping and genomic information to increase prediction and selection accuracy in wheat breeding. *Plant Genome, 11*(1), 1–14. https://doi.org/10.3835/plantgenome2017.05.0043.

Dayrat, B. (2005). Towards integrative taxonomy. *Biological Journal of the Linnean Society, 85*, 407–415.

Deans, A. R., Lewis, S. E., Huala, E., Anzaldo, S. S., Ashburner, M., Balhoff, J. P., et al. (2015). Finding our way through phenotypes. *PLoS Biology, 13*(1), e1002033.

DeLong, D. C. (1996). Defining biodiversity. *Wildlife Society Bulletin, 24*, 738–749.

Diamond, J. M. (1989). Overview of recent extinctions. In D. Western & M. Pearl (Eds.), *Conservation for the twenty-first century* (pp. 37–41). Oxford: Oxford University Press.

Engelhardt, H. T., Jr. (1986). *The foundations of bioethics: An introduction and critique* (1st ed.). New York: Oxford University Press.

Fiorani, F., & Schurr, U. (2013). Future scenarios for plant phenotyping. *Annual Review of Plant Biology, 64*, 267–291.

Glasser, H. (1998). Demystifying the critiques of deep ecology. In M. Zimmerman et al. (Eds.), *Environmental philosophy: From animal rights to radical ecology* (2nd ed., pp. 212–216). Upper Saddle River: Prentice Hall.

Haila, Y., & Kouki, J. (1994). The phenomenon of biodiversity in conservation biology. *Annales Zoologici Fennici, 31*, 5–18.

Holt, A. (2006). Biodiversity definitions vary within the discipline. *Nature, 444*, 146.

Houle, D., Govindaraju, D. R., & Omholt, S. (2010). Phenomics: The next challenge. *Nature Reviews Genetics, 11*, 855–866.

Joiner, J., Yoshida, Y., Vasilkov, A. P., Yoshida, Y., Corp, L. A., & Middleton, E. M. (2011). First observations of global and seasonal terrestrial chlorophyll fluorescence from space. *Biogeosciences, 8*, 637–651. https://doi.org/10.5194/bg-8-637-2011.

Kay, L. (2000). *Who wrote the book of life? A history of the genetic code.* Stanford: Stanford University Press.

Keramitsoglou, I., Sarimveisb, H., Kiranoudisb, C. T., & Sifakisa, N. (2004). Ecosystem classification using artificial intelligence neural networks and very high spatial resolution satellite imagery. In M. Owe, G. D'Urso, J. F. Moreno, & A. Calera (Eds.), *Remote sensing for agriculture, ecosystems, and hydrology V* (Proceedings of SPIE, Vol. 5232, pp 228–236). Bellingham: SPIE. https://doi.org/10.1117/12.511041.

Lussier, Y. A., & Liu, Y. (2007). Computational approaches to phenotyping: High-throughput phenomics. *Proceedings of the American Thoracic Society, 4*, 18–25. https://doi.org/10.1513/pats.200607-142JG.

Lyon, J. (1976). The "initial discourse" to Buffon's Histoire naturelle: The first complete English translation. *Journal of the History of Biology, 9*(1), 133–181.

Maclaurin, J., & Sterelny, K. (2008). *What is biodiversity?* Chicago: The University of Chicago Press.

Madin, J., Bowers, S., Schildhauer, M., Krivov, S., Pennington, D., & Villa, F. (2007). An ontology for describing and synthesizing ecological observation data. *Ecological Informatics, 2*, 279–296.

Maier, H., Leuchtenberger, S., Fuchs, H., Gailus-Durner, V., & de Angelis, M. H. (2017). Big data in large-scale systemic mouse phenotyping. *Current Opinion in Systems Biology, 4*, 97–104.

Marques da Silva, J., & Casetta, E. (2015). The evolutionary stages of plant physiology and a plea for transdisciplinarity. *Axiomathes, 25*, 205–215. https://doi.org/10.1007/s10516-014-9257-4.

Mayr, E. (1988). The why and how of species. *Biology and Philosophy, 3*, 431–441.

McCann, K. S. (2000). The diversity–stability debate. *Nature, 405*(2000), 228–233.

Mungall, C. J., Gkoutos, G. V., Smith, C. L., Haendel, M. A., Lewis, S. E., & Ashburner, M. (2010). Integrating phenotype ontologies across multiple species. *Genome Biology, 11*(1), R2. https://doi.org/10.1186/gb-2010-11-1-r2.

Norton, B. G. (2008). Toward a policy-relevant definition of biodiversity. In G. D. Dreyer, G. R. Visgilio, & D. Whitelaw (Eds.), *Saving biological diversity* (pp. 11–20). Berlin: Springer.

Richards, R. (2010). *The species problem: A philosophical analysis.* Cambridge: Cambridge University Press.

Rolston III, H. (2002). From beauty to duty: Aesthetics of nature and environmental ethics. In A. Berleant (Ed.), *Environment and the arts: Perspectives on environmental aesthetics* (pp. 127–141). Aldershot/Burlington: Ashgate Publishing.

Sarkar, S. (2002). Defining 'biodiversity'; Assessing biodiversity. *The Monist, 85*(1), 131–155.

Soltis, P. S. (2017). Digitization of herbaria enables novel research. *American Journal of Botany, 104*(9), 1281–1284.

Soulé, M. (1985). What is conservation biology? *Bioscience, 35*(11), 727–734.

Takacs, D. (1996). *The idea of biodiversity. Philosophies of paradise.* Baltimore/London: John Hopkins University Press.

Tittensor, D. P., et al. (2014). Mid-term analysis of progress toward international biodiversity targets. *Science, 346*(6206), 241–244. https://doi.org/10.1126/science.1257484.

Vasbinder, J. W., Nanyang, B. A., & Arthur, W. B. (2010). Transdisciplinary EU science institute needs funds urgently. *Nature, 463*, 876.

Vermeulen, S., & Koziell, I. (2002). *Integrating global and local values: A review of biodiversity assessment*. London: International Institute for Environment and Development.

Wilson, E. O. (1992). *The diversity of life*. Cambridge, MA: Harvard University Press.

Wilson, E. O. (2002). *The future of life*. New York: Knopf.

Wilson, J. (2000). Ontological butchery: Organism concepts and biological generalizations. *Philosophy of Science, 67*(Supplement), S301–S311. Proceedings of the 1998 biennial meetings of the philosophy of science.

Zachos, F. E. (2016). *Species concepts in biology. Historical development, theoretical foundations and practical relevance*. Cham: Springer.

Part I
Estimating Biodiversity: Data Collection and Monitoring Challenges

Chapter 2
The Hidden Biodiversity Data Retained in Pre-Linnaean Works: A Case Study with Two Important XVII Century Italian Entomologists

Francesco Andrietti and Carlo Polidori

Abstract Before Linnaeus published the *Systema Naturae*, in which introduced the modern species concept, a huge amount of information on ecology, behaviour and diversity of many animals had been accumulated. This information, often extremely detailed, suffers from the lack of the assignation of the studied organisms to their modern specific names. Here, we examine in detail the works of Antonio Vallisneri (1661–1730), one of the most important figures of early experimental entomology in Italy. We analyse the ecological and ethological contributions of Vallisneri, as well as those that Diacinto Cestoni (1637–1718), another Italian naturalist, sent to Vallisneri, to the knowledge of parasitoid, predatory and gall-making wasps (Hymenoptera), by studying the *Saggio de' Dialoghi sopra la curiosa origine di molti Insetti* and the *Quaderni di Osservazioni* I-III, trying to assign current taxonomy to the observed insects based on eco-ethological and morphological descriptions. Valuable data have been found in the analysed works on taxonomically diverse ecological webs involving wasps. Information regarded a variety of hymenopteran parasitoids of other Hymenoptera, dipteran parasitoids of Hymenoptera, coleopteran parasitoids of Hymenoptera, and hymenopteran parasitoids associated with non-hymenopteran hosts. Overall, about 20 wasp genera could have been objects of Vallisneri and Cestoni observations, which include the first detailed ecological and ethological data on many of them. Detailed re-examinations of ancient studies may contribute to our knowledge on biodiversity by providing historical distribution data as well as unveiling trophic interactions that may have been modified due to biodiversity loss in the last century.

F. Andrietti (✉)
Dipartimento di Bioscienze, Università degli Studi di Milano, Milan, Italy
e-mail: francesco.andrietti@unimi.it

C. Polidori
Instituto de Ciencias Ambientales (ICAM), Universidad de Castilla-La Mancha, Toledo, Spain

© The Author(s) 2019   21
E. Casetta et al. (eds.), *From Assessing to Conserving Biodiversity*,
History, Philosophy and Theory of the Life Sciences 24,
https://doi.org/10.1007/978-3-030-10991-2_2

Keywords Vallisneri · Cestoni · Hymenoptera · Parasitoid wasp · Gall wasp

2.1 Introduction

Naturalists provided through many centuries a large amount of data on the ecology, behaviour, distribution and morphological diversity of animals. However, before Linnaeus published the *Systema Naturae* (Linnaeus 1735), in which the modern species concept was introduced, this information, often extremely detailed, obviously lacked the assignation of the studied organisms to their modern specific names, making these old studies a great source of hidden biodiversity data. A detailed re-examination of these works could, however, bring these data to surface in a modern context. In particular, studying ancient works is important for several reasons: (1) it may provide historical distribution information on organisms before Linnaeus, a period generally not covered in the analyses of communities' variation through time; (2) it may provide additions on species composition and species interactions in habitats and environments that may have been lost in the last century through human activity; (3) it may unveil information on behavioural ecology of species that could be compared with information on behavioural ecology retrieved from modern populations of the same species. All these points are directly linked with three of the main shortfalls of biodiversity knowledge: the Wallacean shortfall (the knowledge on the geographic distribution of most species is incomplete), the Raunkiæran shortfall (lack of knowledge on species' traits and their ecological functions) and the Eltonian shortfall (lack of enough knowledge on species' interactions and their effects on individual survival and fitness) (Hortal et al. 2015).

The present study makes an attempt to bring to light old "hidden" biodiversity data by primarily analysing the works of Antonio Vallisneri (1661–1730), a naturalist strongly associated with the development of experimental entomology in Italy. Particularly concerning wasp (Hymenoptera) biology, Vallisneri is credited one of the first people who correctly interpreted insect parasitism. Secondarily, because additional important observations on wasp biology were also recovered (and published) by Vallisneri from letters that Diacinto Cestoni (1637–1718), another Italian naturalist, sent to him, we also analysed the Cestoni observations.

The motivation underlying Antonio Vallisneri's analysis of the world of insects proves to be two-fold:

1. The first, that we shall call simply *biological*, relating to its origin, but which, at the time, was considered to be *philosophical*, as it was set within the great debate about the science of life between the supporters of spontaneous generation and those which, instead, maintained that every animal necessarily came from a parent.
2. The second, a more specific motivation, which led him to examine in minute detail the lives and habits of these insects with a curiosity and zeal which made him a prototype of the modern naturalist and entomologist, desirous to directly verify the results of the facts he was describing and to throw off the weight of that erudite stance which still strongly characterized naturalistic studies.

These two points are actually only apparently separate as there are significant elements conjoining them. Although Redi, around the mid XVII century, had shown that the commonly held belief that many insects spontaneously materialized in rotting matter – an opinion given authority by Aristotle himself – was false, the case of insects which were *parasitoids* of others was still problematic, as was that of insects found inside plant galls (Parke 2014). In both cases the various situations seemed to confirm the theory of *equivocal* generation, which many hoped to refute. Only a very careful and thorough examination of these phenomena could clarify the problem satisfactorily.

These authors focused their studies mainly on parasitoid wasps and galler wasps, but also on predatory wasps and bees, thus covering many life-histories shown by Hymenoptera. This insect order includes ants, bees, wasps and sawflies and is one of the richest (115,000 species described worldwide), and represents a diverse and ecologically relevant component of biodiversity, since its members are involved in a wide range of interactions with plants (pollination, herbivory, gall formation), fungi (parasitism), and other animals, both invertebrates and vertebrates (parasitoidism, predation) (La Salle and Gauld 1993). Furthermore, many Hymenoptera interact with other components of environment, such as the soil (e.g. through nesting activities). Thus, any new information on species identities and their biology that could be retrieved from ancient studies may be important for biodiversity studies. Indeed, from one side the species is the most commonly used unit of biological diversity, and from the other side, adding data on behaviour and ecology to the species identification substantially increases the detail of biodiversity studies, since it gives information on the functioning of ecosystem and may also bring to light intraspecific diversity (Boenigk et al. 2015). At last, relying not only on species taxonomic identity but also on their behavioural ecology can be relevant for biodiversity conservation, since these data may help understanding how proximate and ultimate aspects of behaviour can be of value in preventing biodiversity loss (Berger-Tal et al. 2011).

In this chapter, we analyse the contribution of Vallisneri (and Cestoni) to the knowledge of wasps with two different life-style: parasitoid and predatory wasps (whole females lay eggs on or into a host and caused host death by brood feeding on it) and gall wasps (whose females lay eggs into plant tissues and cause tissue deformations (galls) serving as food for the brood). We will confine ourselves to illustrating some of the ecological and ethological aspects in which Vallisneri was particularly interested, and, more specifically, we will try to identify the considered insects. The latter task is not always easy, as the descriptions provided by Vallisneri are very different from those used in the current system, and only in a few cases the insects at issue can be properly identified, at least at the genus-level. Some of the eco-ethological data provided in the work of Vallisneri (and Cestoni) are particularly interesting in the light of biodiversity studies, since they are directly related with food webs (i.e. parasitoid – host relationships), and it is well known how food webs are vulnerable to biodiversity loss (Dunne et al. 2002).

As regards parasitoid and predatory wasps, we have used the *Saggio de' Dialoghi sopra la curiosa origine di molti Insetti* (Vallisneri 1696, 1700) as the main source, whereas, for gall wasps, we have mainly used the *Quaderni di Osservazioni* (Vallisneri 2004; Vallisneri 2007; Vallisneri unpublished manuscript), of which volumes I and II have been recently transcribed.

2.2 Parasitoid and Predatory Wasps

The *Dialoghi sopra la curiosa origine di molti Insetti* take place between two personages, Pliny and Malpighi, the revered scholar who had died only 2 years previously, and who is no other than a stand in for Vallisneri himself. In the first *Dialogo* Pliny, who supports traditional beliefs, says

> May the Modern Gentlemen forgive me, they are depriving themselves of a great, compassionate Mother always ready to succour them in their most urgent needs. They should keep her close as a reference point, because they will eventually find forms of generation which they will not know how to explain honourably, without recourse to this universal benefactor, or some other phantom they have thought up. Amongst other things which may still be hidden, I have seen little flies hatching out of the eggs of caterpillars, and Large Flies, Flies and Wasps from the Chrysalises or Cocoons of Butterflies and moths, inside which I do not know how the mothers could have laid their eggs, or their worms. (Vallisneri 1696, p. 311)

Pliny has in fact hit on some of the most distinctive features of the parasitoids, in other words, the fact that they develop, feed and grow at the expense of eggs or larvae of other insects. In some cases these parasitoids are Diptera, but they are mostly Hymenoptera of the Parasitica (= Terebrantia) group. As to whether they are called flies or wasps or even bees should be disregarded, since Vallisneri often uses these terms interchangeably.

Malpighi replies

> One day, from 40 eggs of a big Butterfly [or moth], of the size and colour of the sort of millet they call "of the "Sun", hatched over a hundred tiny Little Flies [...] in each egg I saw two holes, one big one [...] and ten times bigger than the other [...] out of which the little flies kept on coming out, the other just visible with the Microscope on top of each one [...] The tiny hole was the one made by the worms, which penetrated the egg searching for food, [...] the big hole was made by the Little Flies, after they had become Chrysalises inside the egg, and then flying insects [...] In short, they went in by one door and came out of the other, and nothing was generated by itself inside there, but rather came out of the eggs of the same sort of Little Flies [...] even if they were so small as to look like little flying atoms, I could make out their nubby little antennae, [...] heads, and wasp-like bodies [...] we might call them tiny little wasps. (Vallisneri 1696, pp. 311–312)

In these passages, Malpighi is referring to wasps which live on the eggs of a not specifically identified butterfly or moth ("Parpaglione"), inside which they hatch and develop, presumably belonging to the Trichogrammatidae or Platygastridae as already observed by Pampiglione et al. (2000, p. 16). We are here dealing with a case of endophagous oophagy, because the whole life cycle of the parasite takes place inside the egg of its host, from which the parasite emerges as an adult wasp. The small hole is probably not where the larvae got in, as Vallisneri mistakenly believes, but was made by the ovipositor of the mother wasp when laying the egg.

In another case the butterfly is easier to identify

[...] one day I saw odd little flies hatch out from half of the eggs of a butterfly with eyespots on its wings and, from the other half, black caterpillars. (Vallisneri 1696, p. 312)

The pattern on the wings and the colour of the caterpillar seem to be those of the *Inachis io* (peacock) butterfly. The eggs of this species are parasitized especially by *Telenomus* (Platygastridae) and *Trichogramma* (Trichogrammatidae) (Hondō et al. 1995), which are amongst the most active parasitoids of butterfly and moth eggs (Shaw et al. 2009, p. 147).

Malpighi lengthily discusses other cases of oophagous parasitoids, particularly those emerging from the eggs of Heteroptera (bugs) and Homoptera (Vallisneri 1696, p. 312) observing that, even if there is sometimes a specific relationship between the species of host and parasitoid, some hosts may be parasitized by various parasitoid species. He then goes on to talk of cases in which the parasitoids do not emerge from the eggs of the host, but from its larva or chrysalis or cocoon

Others from large cocoons made of the rough silk of the second Moth, inside which one day I found fourteen empty Chrysalises [...] (Vallisneri 1696, p. 312)

The appearance of the cocoon and the dimension of the moth suggest the possibility that we are dealing with the genus *Saturnia* (*pavonia*?), which is known to have numerous parasites that hatch from its cocoons (Grandi 1984, Vol. II, pp. 249–251; Peigler 1994, p. 3), where they complete their development. Amongst the most common are the dipteran *Masicera* (Grandi 1984, Vol. II, pp. 249, 251, 556; Peigler 1994, p. 81) and *Agrothereutes* (= *Spilocryptus*) (Ichneumonidae) (Grandi 1984, Vol. II, p. 251; Peigler 1994, p. 23) a number of whose pupae can be found in a single moth cocoon.

Parasitoids can also be found in the hard, earthen nests of *Sceliphron* mud-dauber wasps (Sphecidae) or the cocoons of the caddis fly (Trichoptera) (Vallisneri 1696, p. 312): presumably, these are mostly parasitoids from the Ichneumonidae and Braconidae (including from the subfamily Aphidiinae), with perhaps a few dipterans. In the case of the Trichoptera, we are dealing with *Agriotypus* (Ichneumonidae: Agriotypinae).

Vallisneri seems to have noticed the difference between eggs laid on or inside the body of the host i.e. between endo- and ectoparasitoidism

[...] having seen various holes one day under the Microscope in the body of a caterpillar I had just discovered (I am not referring to their "breathing holes" [spiracles]) and, another day, various small eggs amongst their hairs, this led me to understand immediately that those too came from the Mothers [...] (Vallisneri 1696, p. 312)

He also reports

I also saw many Little Flies hatch out of the small, spherical, almost membranaceous, follicles of a little worm, which ought to turn into a certain type of wee aphid [gorgoglioncino], which dwells in the Great Mullein and the Tiny Leaves Figworts, and also out of another worm which, as it was found between the two membranes of the external part of the leaves of an elm, should have hatched out as an aphid similar in form to those found on broad beans, grass peas, and similar legumes. (Vallisneri 1696, p. 312)

This is actually an observation made by Francesco Mattacodi, which was passed on to Vallisneri, in the same words given in the above quote, in a letter sent to him, which was subsequently published by Vallisneri himself (Vallisneri 1726, pp. 61–62). Tremblay and Masutti (2005, p. 37) have apparently wrongly identified this parasitoid ("Little Flies") since, in our opinion, its host, the "wee aphid" ("gorgoglioncino") does not belong to the genus *Aphis* (Homoptera: Aphididae) as "the spherical, almost membranous follicles" remind one more of a scale insect (Homoptera: Coccidae). As concerns its parasitoid, we are spoiled for choice, even if it cannot be a genus specialized in parasitizing aphids (subfamily Aphidiinae) as was proposed (Tremblay and Masutti 2005, p. 37). It might have been a chalcidoid wasp from the subfamilies Aphelininae (Aphelinidae) or Encyrtinae (Encyrtidae). The latter are known to cause multiple infestations: thus a single egg laid by the parasitoid can produce a number of its like, as we can see from the above quote.

Moreover, Vallisneri also subsequently introduces other quotes from the same letter received from Mattacodi into his *Dialoghi,* without any reference as to their source

> From live forest Bugs, and live Beetles ["Cantaridi"] I have observed on several occasions after a single worm has come out of their nether regions, without any detriment to them. It was enclosed in a perfect egg [=cocoon], and then produced a nice little fly or, when enclosed in a little follicle, a fly elongated in shape came out of it. (Vallisneri 1696, pp. 312–313)

And, immediately afterwards,

> Many little worms, which also came out of two caterpillars raised from eggs, wrapped themselves up shortly afterwards in long cocoons, from which emerged little flies, without preventing the said caterpillars from turning into the usual Chrysalises, and afterwards the sort of whitish butterflies which are so harmful to Cabbages [...] (Vallisneri 1696, pp. 312–313)

The two passages quoted above come from Mattacodi's letter (Vallisneri 1726, p. 61) and follow the same wording almost exactly. In the first case this could refer, according to Tremblay and Masutti (Tremblay and Masutti 2005, p. 37), to an infestation of Heteroptera and adult Coleoptera by tachinid Diptera; the second seems to bear witness to parasitoids emerging from the skin of the caterpillar (which, when the infestation is not excessive, may survive and finish its life cycle (e.g. Meijden et al. 2000), which will be further discussed below.

Mattacodi also observed parasitoids emerging from insect eggs, giving emphasis to the case of the "biggest moth" (*Saturnia pyri*) and believing, like Vallisneri after him, that the parasitoid entered in its larval form, not as an egg (Vallisneri 1726, p. 61). He would, moreover, observe parasitoids emerging from butterfly chrysalises, although he admitted that this had already been observed by Vallisneri (Vallisneri 1726, p. 61), who would put the fact into his *Dialoghi.* In particular Mattacodi relates that he had found a cocoon which, in addition to the remains of its maker, contained another cocoon belonging to the parasitoid, which, however, he again believed to have entered as a larva, not as an egg (Vallisneri 1726, p. 62).

The basic problem concerns the birth of the parasitoids. Indeed, Pliny goes on to ask: "And all these, and those, do they come out of an egg, or out of the worm?" ["E tutte queste, e questi, nascono dall'uovo, o dal verme?"] (Vallisneri 1696, p. 312).

There follows Malpighi's answer

> Those who can find the patience to look will always see the two holes described in all the eggs. They will never see Little Flies or Large Flies or Flies emerge from the Chrysalises if they protect them well enough, I mean the cocoons and such like, from which it can be seen that a Mother is always required. It seems far stranger that those which come out of a Caterpillar's body alive should be born of a Mother, and then make their cocoons immediately or harden into Chrysalises [...] (Vallisneri 1696, p. 312)

The last sentence reported above alludes to what happens, for example, with the genus *Apanteles* (Braconidae), a parasite of lepidoptera larvae, in which they lay numerous eggs during the early stages of their host's development. After having grown to full size, the larvae emerge from their caterpillar host, making their own cocoons on or beside it.

Pliny recalls the observations of other two naturalists who had seen the same phenomenon, Aldrovandi and Redi; however, both of them had taken the cocoons of the parasitoids which had emerged from the host for eggs laid by the caterpillar (Vallisneri 1696, p. 313). To which Malpighi replies

> I doubt that both these great men, let it be said with all reverent modesty, could have been mistaken. As you know, it is not caterpillars which lay eggs, but butterflies, and even if they confused eggs with Chrysalises, I still do not believe that the caterpillars made the silk cocoons mentioned by Signor Redi, but that the worms did so as soon as they had hatched [...] That this is what happens is clearly shown by the silk, which they admit to having seen around them, which certainly did not come out of the rear end of the caterpillars, and my eye showed this to me one day, as I had seen them hatch and, then and there, make the little cocoons described. (Vallisneri 1696, p. 313)

In the second *Dialogo* Pliny introduces the matter of the "vespe icneumoni". We can find mention of these in texts by Aristotle (as well as in Pliny and Aldrovandi), even if these are "very unclear, and limited" ["oscurissimi, e scarsi"] (Vallisneri 1700, p. 357), as Malpighi was then made to observe. Indeed, in chapter XX of Book 5 of Aristotle's *Historia animalium*, it is said that these lesser wasps nest in the walls and feed on spiders, even if – it is added – those of a larger size do the same thing (Vallisneri 1700, p. 357). As regards the former, Malpighi reports the following observations

> On the 20th day of June, I observed a swift Little Wasp going frequently in and out a hole made at some point by a nail in a wall in a little frequented Room [...] On the 12 of July, I found it [the hole] closed up on the outside, and with the greatest of diligence daubed with a crusting of fine earth, or mud from the fields. I got the urge to open it up, and I can tell you quite frankly that the plug made of this earthen paste was a good finger in thickness. When I removed it, I saw a little cell with lots of Little Spiders inside and a fat, juicy, whitish-yellow worm, which was greedily devouring the same. When I took this away, there was another little cell further back, with a worm of the same kind, but a bit larger, shut up with other Tiny Little Spiders, and this little cell was in between two other contiguous ones, which also had their respective guests living in them, along with, as one might say, these still steaming corpses. Still further back, there were others without any flaws in them, but I

was so clumsy when I broke [the structure] up that I made a fine mess of the whole thing, and I was unable to make […] any further observations […] I found yet another one amongst the splintered remains of a fallen in house, which I observed to be set up with a total of eleven little cells built in a most orderly manner behind a common passageway, so that almost all of them could get into it to leave [the nest] without going through the cells of the others, by gnawing through just one wall between the aforesaid passageway, and their cell. (Vallisneri 1700, p. 357)

This description could fit with the genus *Trypoxylon* (Crabronidae), which habitually builds its cells in a linear sequence inside empty plant stems or dead branches, but also in tunnels made in the wood by other insects or artificially produced holes. One would be led to think either that the walls of the room were made of wood or that the "nail" had been hammered into a supporting beam. The cells are constructed starting from the bottom of the hole, supplied with a goodly number of little spiders, an egg added and, finally, each unit is separated off from those to be subsequently built with an earth diaphragm. After the construction of the last cell, a plug of earth is applied to isolate the nest from what lies outside. The difference in size of the larvae may be due to the fact that the inner cells are made ready and provisioned first and thus the corresponding larvae would be older (and larger) than the ones nearer the outer part of the nest. Their "common passageway", on the other hand, remains unexplained.

However, the difference relating to size is interconnected with another two: the first is, again, a matter of size but, this time, relating to sex (males are smaller than females); the second is the difference between *wild* "Vespe Icneumoni" and their *domestic* counterparts. The latter, which the reader is reminded were of the type also described by Aldrovandi (Vallisneri 1700, p. 358), were observed

[…] in various places in the house, not only isolated and unfrequented ones, but continuously visited places, even under the smoky old Mantle of a Chimney, where the Kitchen fire was constantly burning. (Vallisneri 1700, p. 358)

This habit matches that of the genus *Sceliphron* (Sphecidae) as this can be confirmed from the description of the wasp which has its abdomen separated from the thorax by a "long pipe" (Vallisneri 1696, p. 358), and has a black head and thorax (which seems to indicate the species *Sceliphron spirifex*) (Pagliano and Negrisolo 2005, p. 82), from the number of cells in a nest (14 but, as would be observed further on, varying between 1 and 22) (Vallisneri 1700, pp. 358, 368), from the spiders contained in a cell (10–12) (Vallisneri 1696, p. 358), from the yellow and black of its legs, the greater length of the back legs and, finally, from its way of collecting mud from a puddle, and also due to its propensity for building its nests in warm places, like a chimney mantel, as well as on beams and under roofs or inside the walls (Vallisneri 1700, p. 359).

The "wild" variety is not so easy to identify. These also have a long waist, but are very different in colour and shape from the former (Vallisneri 1696, p. 359). Here is what Malpighi says with regard to *wild* "vespe icneumoni"

Out walking on the 15th day of March [...] raising my eyes, I saw, on top of a Great Branch of a dead Oak, about eight arm's lengths high, an earthen nest, southern facing, there exposed to every gust of wind. [...] I shut it up jealously in a glass jar [...] it had been made the previous year, and it had been all winter in the snow and ice and wind. On the 12th day of June out hatched a Wasp with a really long pipe in its belly, but with a colour and shape rather different from the aforesaid domestic variety [...] On the 14th and 15th another two [...] And these were all of the same size and very similar in shape. On the 17th, Bigger Ones started hatching out, and these looked stronger and bolder [...] Counting the holes, there were only fourteen, although all the Wasps together numbered seventeen, three of them having pierced the dividing wall of their cells and come out of the little windows already made by their neighbours [...] I supposed, and I swear by your Aristotle that I am making a good faith guess here, that the bigger ones were female and the smaller ones [...] were male [...] (Vallisneri 1700, pp. 359–360)

One has the definite impression that this is a description of a eumenine wasp (Vespidae: Eumeninae), due to the fact that, unlike in the "domestic wasps", "the pipe which divides off the belly bells out like a trumpet" (Vallisneri 1700, p. 360). The observation that the males, which are smaller in size, emerge from the nest before the females is interesting, as this is typical for these Hymenoptera. The earthen nest described would seem that of a *Eumenes*, perhaps *Eumenes unguiculatus* (Eumeninae) (see, for example, Grandi 1961, Fig. 23, p. 39). Although this is correct, the assertion that they probably feed on spiders (Vallisneri 1700, p. 358), like the "domestic" ones (which is more taken for granted than supported by specific observations) is not. Indeed, the wasps in question use Lepidoptera caterpillars. On the other hand, later in the work, Malpighi would say, with regard to other (wild) wasps, which build

[...] a roundish, earthen Nest looking rather like a breast, and like a walnut in size. Once I had detached it and opened it, I found it all empty, in other words, it had just one cell, and with just one solitary worm in it [...] I observed that it ate caterpillars [...] It took the Little Wasp until the 20th of June of the following year to hatch. For the record, this was similar to the females of wild "icneumoni" described, but a good bit smaller [...] with the very same structure, and perhaps, or even actually, there is another one of the same species, which usually builds a rough-looking nest inside walls facing East or south, which also feeds its young on little caterpillars [...] and walls up ten or twelve of them, half-alive, inside each little cell as food relished by the future Little Wasps. I even found two nests in the window of a country House. They were facing East, quite close one to another, and made of hard, white clay, rather pointed towards the top, and roughly rounded off, very much smaller than the afore-described. When I opened one up, I saw just one white worm similar to those mentioned in a round and shiny little cell, which was greedily devouring imprisoned geometer moth larvae a lot smaller than the aforementioned caterpillar [...] (Vallisneri 1700, pp. 364–365)

These passages remove all doubt as to the identity of the wasps in question (and also that of the previous one), *wild* "vespe icneumoni" which, as it is explicitly stated, have the same structure, even if they are smaller. Again we are dealing with wasps of the genus *Eumenes*: the first of this second group of smaller wasps, probably *Eumenes pomiformis*, builds mud nests with a shape reminiscent of a wineskin, is made up of a single cell (unlike *Eumenes unguiculatus*, whose nest was multicellular) and is provisioned with caterpillars of microlepidoptera.

Malpighi also speaks of "Wild (Silvestri) Bees", which he had briefly mentioned before, calling them "Forest (Silvestri) Bees" (Vallisneri 1700, p. 358), about which he says

> [...] even though it works its little honeycomb out of mire and the tiniest of stones, it does still perhaps still retain the nobility of the Bees, as it feeds its little foetuses, inasmuch as I have been able to observe, solely on sweet juices. I would place these amongst the ichneumon owing to the similarity of their earth nests which, from the outside, are hardly distinguishable [...] as we can also find some built by the said Wasps with almost the very same material." (Vallisneri 1700, p. 361)

But, although the structure of the nests is similar to that of the "ichneumon" wasps, the same does not hold as regards the appearance of the insect, seeing that

> The Wasps, or rather the Bees which build these [nests] have really similar features to common Bees, possessing a very different form to that of the aforementioned ichneumon [...] (Vallisneri 1700, p. 362)

Both the descriptions reported above, that of the nest and that of the insect, are consistent with the bee *Megachile*, probably *Megachile parietina* (= *Chalicodoma parietina*) (Megachilidae), which builds earth nests (Grandi 1961, p. 324). Regarding these bees, Malpighi reports the following observation regarding the chrysalis of one of these, which had been separated and placed in a twist of paper so that its development could be followed

> Looking at this chrysalis on the first day of July, I found on top of the same 4 little spherical, white eggs [...] and looking carefully at the twist of paper I found it had been pierced right through by a beetle, which had industriously penetrated through a little split in the first outside wrapping, and this led me to understand even better, how easily the Gentlemen who Defend spontaneous generation can be led astray [...] by not observing that ingenious [...] Mothers have secretly laid their eggs, precisely as happened to the aforesaid unfortunate Chrysalis, or Nymph, from which, most purposefully kept and watched over, hatched four rather hairy little worms with ringed segments and, when it was time, the Little Worms changed into Nymphs, and the Nymphs into Beetles. This did not happen to the other Nymphs of the Bees shut up in their impenetrable nest, seeing that Bees flew out of there, and not Beetles. (Vallisneri 1700, p. 362)

In all probability, we are dealing with *Trichodes* (Coleoptera: Cleridae), a parasite of bees and wasps which attacks the larvae and probably also their food stocks (Grandi 1961, p. 321; Müller et al. 1997, p. 262). Irrespective of the naturalistic interest of this observation, it is also important in a much more general sense, to counterattack the opinions of those who, when faced with the phenomenon of the apparent generation of an insect different from the one it originated from, attributed it to a case of spontaneous generation.

There later appears a consideration made by Pliny regarding a mistake made by Aldrovandi, who had allegedly confused the nests of "domestic ichneumon wasps" (*Sceliphron*) with those of the "wild bees" (*Megachile*) once described by Aristotle, not having taken into account the different colouring of the larvae and having trusted in the mistaken description of a countryman (Vallisneri 1700, pp. 362–363).

Returning to the subject of wild bees (*Megachile parietina*), Malpighi observes that larvae of two sorts come out of the same nest – one which is bigger and black in colour, the other smaller and more colourful – considered to be, respectively, females and males (Vallisneri 1700, p. 363). Indeed, the males of *Megachile parietina* are very different in colour from the females (Grandi 1961 p. 324; Müller et al. 1997, p. 268). As we have seen, Malpighi had already revealed, in connection with wild bees and, in particular, *Megachile*, that the bees, unlike the wasps, which build earthen nests and live "of Little Spiders and Small Caterpillars and Geometers moths larvae" ["di Ragnateli, di Bruchetti, di Geometri"] (Vallisneri 1700, p. 361), feed their own larvae, at least as far as he has been able to ascertain, on a vegetarian diet, consisting of "sweet juices" ["dolci sughi"]. This is an important point, to which Malpighi subsequently returns, with an interesting observation comparing the structure of the tongue in *Megachile* and that of the honey bee

> [...] a long tongue, composed of, as one might say, five little shiny, sharp tongues, almost as if they had teeth, because of some short hairs, which made them look rough and coarse. The one in the middle was twice the length of the others [...] very sharp-looking, and also a bit hairy [...] The other four were made differently from the aforesaid [...] This new discovery as to the tongues led me to the increasing suspicion that that they feed on juices, honey, dew and other like things, because they are very similar to those of ordinary Bees, and they seem much more suited to taking away and manoeuvring all sorts of liquors with their roughnesses, whether these consist of branchy, yielding particles or sweet, sticky ones. And, indeed, I have often seen them on flowers [...] (Vallisneri 1700, p. 364)

As regards the vegetarian diet, Pliny asks for confirmation

> Are you sure that they feed their little ones only on juices, and not sometimes on tender Little Mosquitoes, maggots, tiny flies, small spiders or other such things? (Vallisneri 1700, p. 364)

Malpighi, in view of the lack of other observations to add to those already given, which do not exclude the possibility of a mixed diet, tries to support his hypothesis by putting forward some further information, presumably again regarding *Megachile parietina*, reported to him in one of Cestoni's letters

> [...] in Livorno there are a huge number of earth Nests attached to the stones on the facades of the Houses [...] which look like so many pieces of earth thrown haphazardly by a human hand, inside the little cells of which he [Cestoni] often found a little piece of brown honey intended to nourish the worms, which further supports my suspicion, if we suppose that these were built by Bees of the aforesaid race. (Vallisneri 1700, p. 364)

The persistence on this subject is certainly justified by the importance of the question brought up by Vallisneri. The difference between the diets the larvae are fed on is indeed fundamental when making a distinction, whether ecological or systematic, which separates the large taxon made up of Hymenoptera Aculeata (bees, stinging wasps, ants) into two halves.

Another important note is contained in a passage part of which has already been reported above

> [...] and it walls up ten or twelve of them, half alive, inside each cell as food relished by the future Little Wasps. [...] I found two nests [...] much smaller than the aforesaid. On opening one, I saw just one white worm, curled up, similar to those mentioned, in a shiny rounded little cell, which was greedily devouring imprisoned geometer moth larvae [...] and I saw that there were two still alive, brought there incredibly skilfully without killing

them [...] so that they would continuously provide fresh and tender food for the little one, and so they would not rot or dry up before they [the offspring] reached the required size. And this admirable providence I have also seen practised in some of the Nests of the wild ichneumon Wasps [the *Eumenes* before considered], and perhaps, also, in domestic ones, and in all those which live in the very little holes in the walls [*Trypoxylon*] [...] in which almost all the Small Spiders, which they were going to leave as food for their little ones, were alive. So I saw that it was not always true, as Aristotle said, in the Book 5 Chap. 20 quoted, that the ichneumon wasps, *Phalangia perimunt, occisaque ferunt in parietinas, aut aliquid tale foramine pervium* [they kill spiders, and bring the killed ones in ruins or anything other similar place provided with holes] [...] (Vallisneri 1700, pp. 364–365)

This observation, which is of primary importance, holds with reference to *Eumenes, Sceliphron, Trypoxylon*, in other words to all the observed predatory wasps, and consists in the fact that they do not kill their prey (as Aristotle maintained) "so that they would continuously provide fresh and tender food for the little one, and so they would not rot".

The wasps Aristotle is alluding to are, with all probability, Pompilidae, spider-hunting wasps which Vallisneri does not seem to have observed personally. However, he will report Bellonio's description of how the spider is stung and carried away by the wasp (Vallisneri 1700, p. 361).

Malpighi continues as follows

When they are grown, the Worms make a white cocoon [...] When I looked on the sixth day of June, I found two long Little Wasps had hatched, that is, one from each nest, which had emerged from the back part of the aforesaid, which was already open, by which it was attached and fixed tight to the Wall, having in the meanwhile avoided the bother of gnawing away at the front wall of the same. These are half smaller than the aforementioned ichneumons, but almost exactly the same shape. [...] They have a big back and a broad chest, from which spring six legs, divided up into seven sections, the last of which are really long, and they have a long spine at the end of the third, as have the second legs too. The thorax is connected to the abdomen by a long, hard, black tube, shaped like a trumpet. The abdomen [...] from the end of which is always unsheathed, and ready to strike, a very long but, as I think, innocent tricuspid sting, in the shape of a straight tail. It is almost as long as the whole of the abdomen, and it does not flatten and unsheathe it, as other Wasps do with their harmful, stinging needle. Indeed, it is adorned and hung on one side and the other with two very black and hairy threads, which are usually crooked and contorted like old, intertwined grapevines [...] (Vallisneri 1696, p. 365)

Pliny replies

These may well be the Little Wasps *ex Minuti &c.* mentioned (even if he was then talking more of Wasps), although this was believed in passing, by my Aristotle in the place you mentioned which, as he says, *nomine carent* [have no name] even though *nidos e luto parvos, aut ad sepulcra, aut ad parietinas configunt, atque in iis vermiculos pariunt &c.* [they build small nests from the mud and fix them either to sepulchres or ruins, and inside them they give birth small worms] as industriously as the larger and, if it does not seem overly bold to give a name to an insect which was not given to it by Aristotle, I would call them, at least so as to distinguish them from the other ones, *tailed domestic Ichneumons*. (Vallisneri 1700, p. 365)

The two "Little Wasps" that Vallisneri saw coming out of the back of the nest were ichneumonid parasitoids (in the currently accepted sense of this term) which had fed on the larvae of the host and its prey.

Following a discussion as to whether or not wasps which build earth nests could have wings covered with hard elytra, which seems to have been something maintained by Aristotle, as reported by Pliny, Malpighi is made to reply

Pliny, to be quite frank with you, I really do not think that Aristotle then understood them to be Wasps, even if he was dealing with this issue, and men with a great amount of good sense, and most worthy besides, firmly believed this. And indeed, out of all those Wasps and Forest Bees that I have observed making all or a part of their mud nests, of which, besides those described, I still have many others to describe, I have never found any which have shell wings over their membranous wings [...] I have often found strangers living in the aforesaid earthen nests, and false guests which have either managed to get inside, or have been laid inside by their wise and industrious Mothers, so that they can feed on this juiciest of sweet worms that there is inside[...] I found one of the domestic variety stuck tight under the arch of a public Portico, which [...] contained a beautiful, live Beetle [...] all of it a lovely cinnabar colour, and patterned, with wings dashingly edged with a bright purple [...] (Vallisneri 1700, pp. 365–366)

Urged by Pliny to say whether he has seen other insects "dall'ali superiori di crosta ["with upper wings made out of shell"] in the wasps mud nests, Malpighi replies

Just once, I saw a really odd one, but I imagined that it was (as it indeed turned out to be), an itinerant inhabitant of cells that were not its own. This one had a round Orange, smooth and shiny Head, and was shaped like the bare skull of a dog with a long muzzle [...] This was definitely not a Wasp, as you might hear, but rather some sort of beetle of a kind of its own, quite dashing, and strangely shaped. (Vallisneri 1700, p. 366)

In the last two passages quoted, we have cases of beetles which are using or parasitizing the nests. The latter is probably a beetle belonging to the Ripiphoridae family (Coleoptera), whose genera are all parasites of Hymenoptera Aculeata. To be more precise, it could be *Macrosiagon*, a parasite of some species of predatory wasps, in particular *Sceliphron*, as could be those which build mud nests here considered (Batelka and Hohen 2007).

Malpighi's answer to Pliny's question as to how the wasps' parasites manage to get into their nests and as to whether there are others, apart from the beetles he has already described, is as follows

[...] it is probable that they sneaked in unseen when the Mother Wasp had not yet closed off the top of the cells. I have not only seen the aforesaid Insects, but also a certain sort of Fly, which lays Maggots, or worms that are infamous eaters of live meat [...] and the Mother Wasps being absent, they get into the cells of the latter before they are closed and after having laid their unnoticeable little eggs on the tender little worms, go off. The Maggots, shortly after they have hatched, bore or drill holes into the worm's skin, and sucking out all its white blood [...] and delicate little entrails grow greedily on the ruins, and on the carnage of others. Nor are they content to devour just one, but smelling the nearby prey they pierce the dividing walls with this sort of awl they have, which is really hard and black, attached to their mouths like a spout [...] and they move from one to another, until they are swollen, and sated with those wretched little worms, they reach their destined size [...] from one nest of wild ichneumon Wasps I found hatched [...] four Flies, and just one Wasp. Looking at the nest I saw, in addition to the big hole that the Wasp had emerged from, a little hole also made freshly as could be seen from the detritus above a cell of its home. Opening this, and following the narrow path of the little one, and the unused tiny aperture, I found four cast offs, or empty shells of the Chrysalises of the aforesaid Flies, and two Chrysalises which were still full, along with some excrement and the remains of the devoured Worm

[…] in another wild nest closed in a box […] I found it pierced in three places, that is, with two big holes and one little one. The usual Little Wasps had come out of the big ones and six Flies of the sort described above out of the little one. Again following the trail left by the little hole, I found in the offended cell the shells of the Chrysalises of the six flies which had come out looking the same, with the same features and in the same number as those mentioned. There was also a lateral hole, which led into another cell, and in that one another hole, which led into yet another cell, both of them empty, and bereft of their legitimate owner, just leaving a dry heap of droppings, closed off with the usual webbing in one corner. This led me to suspect that, in this case, the first worm devoured was the one in the first cell, since there were no droppings of any sort there and that, once they had finished, they had got in to the other cells to eat up the others, which they had found fully grown and ready to spin their cocoons, and to change into Nymphs (since they had prepared the cell, and neatly collected the faeces), then they went back into the first one, and there they changed into the usual chrysalises. And if you are itching to know what these bold and ingeniously insolent Flies look like, I can tell you that they are very similar to those which fly and buzz about our houses every day […] although they are a bit more bristly, and a little smaller, rather more ash-grey in colour, more marbled, and they are edged with black and have a silvery head […] from other earth nests, particularly those made by domestic ichneumon, sometimes twenty-five or thirty Little Flies of the same kind as the Small Worm-eating carnivores, which also came out of eggs laid by their Mothers inside the little cell in question before the Wasp which had built the cell had closed it off, so that they might eat up the worm which Owned that cell. (Vallisneri 1700, p. 367)

This last quote is a masterly discourse on behavioural ecology regarding Diptera which parasitize *Eumenes* and *Sceliphron*. In the case of the second wasp ("domestic ichneumon") the parasite is very probably *Pachyophthalmus signatus* (= *Amobia signata*) (Diptera: Sarcophagidae), as pointed out by Tremblay and Masutti (2005, p. 37), which first feeds off the egg and then the prey stored up by the wasp (Grandi 1961, pp. 160–161).

Malpighi also reports some observations he made about bees which reutilize the abandoned nests of *Sceliphron*

[…] a certain nest of *dashing Bee, and tiny* and ingenious inhabitant of holes in the wall, and also of old, empty nests of domestic ichneumon Wasps […] dashing little Bees, the little ones mentioned […] which live not only in holes in the wall, but are also innocent guests in the nests of domestic ichneumon wasps found Empty. (Vallisneri 1700, p. 369)

These are probably bees of the *Osmia* genus (Megachilidae), which may make their nests in the abandoned nests of other Hymenoptera, in particular *Sceliphron* (Cane et al. 2007).

Instead, another of Malpighi's observations can easily be connected with the genus *Trypoxylon* considered above

[…] a sort of Wasp which, finding a Bramble stem hacked off, immediately dug into the yielding, spongy tissue within and, inside that long, dug-out gallery, it laid its eggs separately, neatly spaced out, and together with the eggs, Little Spiders it had caught, and then closed off the space, making a hard earth plug between one egg and another, so that each of the little worms born should have its own little cell and its own store of food. And these old, empty Brambles are […] hidden nests which are very well suited to various Insects which are believed by certain *idle and credulous Putrefactionists* to arise of themselves […] (Vallisneri 1700, p. 369)

In a situation such as the one under consideration, in other words, insects which, in the following season, had apparently sprung from a dead branch, only an analysis as thorough as that provided by Vallisneri could effectively discredit the theory of a spontaneous generation from rotting vegetable matter. This was a further example of the need for a naturalistic analysis which was sufficiently detailed to provide help for the broader "philosophical" questions implicit in the theory of generation.

If we move on from *La Galleria di Minerva* to consider the *Quaderni di osservazioni*, which in part have recently been transcribed, we find reported an important observation, made by Mattacodi, regarding another parasite of the "domestic Ichneumon". It is presumably the wasp *Acroricnus seductor* (Ichneumonidae) found in a nest of *Sceliphron* (Polidori et al. 2011)

> In one smaller little house [cell] there was a little white worm only half the size of the others. However, it was divided up into twelve rings, with a strip of transparent humour not only in its body, but also in its back […] Its little cocoon was all covered with white spittle." (Vallisneri 2004, p. 66: 27 September 1694)

The colour of the "worm", the dorsal line and the appearance of the cocoon indicate that it was an *Acroricnus* larva. In addition to this, Mattacodi reports the presence, in the nest, of the larva of a parasitic (or inquiline) beetle

> This was of a reddish colour and all hairy, with a little flattened, but very hard, head, and with six feet in the three foremost rings, using which it dragged behind it the whole bulk of the rest of its body, which was considerably bigger and longer than the forepart. (Vallisneri 2004, p. 66–67: 27 September 1694)

Judging by the description, this could have been a Ripiphoridae larva.

2.3 Diacinto Cestoni's Letter

Even if the work of this naturalist was well appreciated during his life time (Generali 2004, p. 107), after his death he was relegated to a secondary role or quite ignored, up to recent times. The reason is probably that he never published anything and preferred to confine his observations to letters that he sent mainly to Vallisneri, but also to other important researchers of his time. Only in the last years the importance of his observations has been acknowledged (see, for example, Tremblay and Masutti 2005, p. 37; Generali 2004).

The letter we are concerned here is generally recognised as being of great importance, together with its great value as regards our present interests. This has been particularly emphasized by Tremblay and Masutti (2005, p. 37) who, quite rightly, point out that the descriptions and explanations of behavioural ecology provided by Cestoni are equally as good, if not sometimes even better, than those given by Vallisneri himself. This makes Cestoni's work relevant in the debated importance of building a framework unifying detailed behavioural data and conservation sciences (Berger-Tal et al. 2011). The letter, dated 1692, was sent to Vallisneri in 1698, with the declared intent of confirming "various findings of the aforesaid Gentleman [Vallisneri] as regards the curious Origin of many Insects, described in his First and Second Dialogue" (Vallisneri 1726a, p. 89) ["vari ritrovamenti del suddetto Signore intorno la curiosa Origine di molti Insetti, descritti nel suo Primo, e Secondo Dialogo"], and subsequently published by Vallisneri himself (since the letter was sent before the publication of Vallisneri's Second Dialogue, one must assume that Cestoni had already read it, in some manuscript or preliminary version). We feel it would be appropriate to examine it in detail, as most of it is devoted to parasitoid wasps (and their hosts). Basically, this is a specialized scientific monograph, as it is exclusively dedicated to "little creatures" found on cabbage leaves, which were examined with a thoroughness and precision which could serve as a model for a modern research study.

The first observation of Cestoni relevant to the present paper regards the "fleas" that can be found on cabbages, but also on many other plants and flowers, which Aldrovandi had also mentioned, but only in passing (Vallisneri 1726a, pp. 91–92)

> These horrid little animals are idle, stupid, and very slow to react, and on any plant they live on, they all look the same, or very similar. They have a bold, round little body, very like that of Spiders, six legs, two antennae, or very long horns, two black eyes, a long, thin, sharp rostrum, with which they very often pierce the leaves, to graze on the moist, delicate and tender substances of the plants [...] In short, they take on the colour or of the juices they swallow, having, amongst other things, an extremely thin skin, and being very fragile creatures [...] never having personally seen any one of them up to now about the Act of Mating [...] However this much I have observed that, when they are fully grown, all of them give birth, and produce their little ones alive [...] when they have become as big as they will ever get, these also start to reproduce, and to produce their young alive in the same shape as the others [...] Amongst the little creatures described, there are many which become winged, so I could not help but wonder whether these were of another race, notwithstanding the fact that before they grow wings, not much difference can be seen between them [...] (Vallisneri 1726a, pp. 92–93)

The observation regarding the parthenogenetic viviparous females – which, therefore, do not have to be fertilized to produce their offspring – is of particular importance. This significant biological discovery (subsequently and unfairly attributed to Charles Bonnet), together with the shape of the animals, their universal distribution, the presence of winged generations which alternate with unwinged ones, allows us to identify the "fleas", without a shadow of a doubt, as aphids (Grandi 1984, Vol. I, pp. 827–830). It is thus rather peculiar that, having such precise information at their disposal, Tremblay and Masutti (2005, p. 37) came to the conclusion that the insect was *Aleyrodes proletella*, another homopteran belonging to a different family (Aleyrodidae instead of Aphididae), perhaps deceived by the name "cabbage white fly" as it is commonly called. But the described species is actually only one of the various insects found on cabbage leaves, as Cestoni clearly indicates (Vallisneri 1726a, pp. 89–105). On the other hand, Cestoni does not say he has ever seen the eggs of these *fleas* (Vallisneri 1726a, p. 95), which, are in fact easy to spot in the case of *Aleyrodes*, as they often form a distinctive pattern on the underside of the leaf, alerting the viewer to the presence of the insect in question (Grandi 1984, Vol. I, Fig. 666 p. 824).

Strange things happen to these "fleas"

A few days later, when the aforesaid Insects have given birth, both the winged ones and those without wings, they can mostly be seen to be quite unmoving, and attached with all six feet to the same leaves, and with their rostrum forever stuck in the same, as if, just the same, they were continuing to suck; but once I had had a good look at them, I realised that they were doing anything but suck. They had actually died like that, even if their bodies remained well-preserved, big, fat, round and swollen, as if they were alive, the only difference being that they were starting to turn yellow. I started observing a number of these little creatures; whereupon I found a few of them whose heads and thoraxes had in fact dried up, and the lower part of the abdomen too. However, aside from this, when I squeezed them, I saw and felt that there was still a tiny portion of fresh matter inside. I found yet others that were not only shrivelled and all dried up, but mostly they were completely empty, so that all that remained of them was very simply the outer skin, the husk, or shell, or whatever one prefers to call it, in which a tiny hole could be seen. This observation immediately made me wonder whether there were other animals which went around devouring the insides of these Fleas; whereupon, in order to gain more insight into the matter, I took a large number of these newly-dead animals and, having separated the winged ones from the unwinged sort, I put them separately into two glass jars which I immediately covered very carefully, then, not many days later, looking into these jars again, I saw (to my great amazement), that a lot of Little black Flies had come out of these Fleas. They were very lively and slim, and they walked and flew about inside those jars incredibly quickly. And, at the same time, I observed that the Fleas had been left with just their skins, completely emptied out inside, just as I had seen happening in the other cases I described above. Being unable to imagine the reason, how such a bizarre metamorphosis could come about, and ever more desirous of discovering its cause [...] and after a lot of assiduous research, I was lucky enough to discover how and whereby and why the mentioned transformation should necessarily take place. (Vallisneri 1726a, pp. 94–96)

In the end, the parasite responsible for the "transformation" observed was found and its interesting behaviour described. This was one of the first descriptions made in Europe of a true case of oviposition by a parasitoid (van Lenteren and Godfray 2005, p. 14), something Vallisneri supposed had happened, but never seen for himself.

[...] near these Fleas I saw certain Little Flies buzzing about which, after having walked and flown quite a lot around the Fleas, gradually moved closer to the largest ones, as if identifying those fit for their purpose [...] I armed my eye with a really good Lens, following one of these [...] I observed, that that particular Little Fly had got so close to one of the Fleas that it was almost touching it with its head. Close as it was in this way, I saw that, once it had firmly planted its feet, it raised its wings, as if it wanted to fly and, at the same time, holding its wings raised like that, it plunged its lower abdomen under its chest. As its abdomen was a little longer than the rest of its body, it thus stuck out further than its head; folded over, as its body was in that position, it bent itself so far down, and made so much effort, that it put the very end of its body under the belly of the Flea and, keeping it like that for a very short time, it went away, and I saw that it was going about the very same thing around the other ones. To clarify for myself, and see what this Little Fly might be doing by putting its abdomen under the body of that other Insect, I decided then and there to turn it over, and [...] I found that that Little Fly had very kindly laid an egg under the belly of the other insect, and was doing the same thing with the others. Having this piece of good news, it was not hard for me to find out why Little Flies flew out of those Fleas which looked dead, in view of the fact that lots of little maggots hatch out of these eggs and, as soon as they hatch, the pierce the bellies of the Fleas underneath which they have been laid, and entering their bodies, they use them both as food and as a room: and when they are well enough nourished

> and grown, they make a chrysalis inside the same Flea, then in less than a Month the Little Flies coming flying out, one from each Flea. (Vallisneri 1726a, pp. 96–97)

The parasitoid in question is certainly a genus of the subfamily Aphidiinae, perhaps of the species *Aphidius*, which is characterized by this particular way of laying eggs on the underside of the aphid, bending its abdomen forward, which is made to pass under the thorax and the legs of the wasp (Grandi 1984, Vol. II, Fig. 978 p. 1004) in the way described by Cestoni. When the egg hatches, the larva of the parasite enters the host and eats it up from the inside, turning it into an empty shell, or *mummy*, as it is called in entomological jargon. In this case too, the mistake made by Tremblay e Masutti (2005, p. 37) is hard to understand. These authors hold that the parasitoid was an Aphelininae (Aphelinidae), an identification which might be appropriate, as we have said, in the case of the parasite of the "gorgoglioncino" reported by Mattacodi and discussed above.

But the aphids living on cabbage leaves studied by Cestoni have other two persecutors even if, this time, they are not parasitoid wasps. The first of these is probably a ladybird (Vallisneri 1726a, p. 97), a carnivorous beetle which, both as a larva (as observed by Cestoni) and as an adult insect, is a hearty consumer of aphids. The second (Vallisneri 1726a, p. 97) is the larva of a Syrphidae (Diptera) (hoverfly), as its green colour striped with white indicates (Rotheray 1993) and the fact that the adult of this fly

> [...] which, when it flies, you very often see it hold itself in the air, exactly in the same way that larks do when they hang in the air singing. Nor should you, most Honourable Sir, think that they do this by chance; it is done on purpose, in order to observe, and see where the Insects are, and when it sees some, it lands on that Plant, and lays one or two eggs, then it goes off to fly somewhere else, since it never lays more than two eggs on the same leaf, at least as far as my observations go. (Vallisneri 1726a, p. 98)

Returning now to parasitoid wasps, Cestoni talks of those which parasitize the caterpillars of "lovely big, white cabbage butterflies (= *Pieris brassicae*) (Vallisneri 1726a, p. 98). Cestoni describes some aspects of the biology of these butterflies, such as the fact that they lay their

> [...] eggs under the leaves of the aforesaid Cabbages, and arranging them with a wonderful neatness, they place about fifty, or sometimes seventy in a patch, one next to the other, tidily in one place, about the size of a nail on the hand. I said under the leaf, because very rarely do they lay them on the upper side, but habitually place them on that part of the leaf which faces the ground, so that they remain covered and are not hurt by the rays of the Sun. These eggs look yellow on the outside, and in the space of two or three days a lot of little worms hatch out of them, which immediately start eating the leaf upon which they were laid [...] (Vallisneri 1726a, pp. 98–99)

When they have finished eating, the caterpillars turn into chrysalises which, after ten days, will produce new butterflies. What is more interesting is the fact that, in some cases, instead of a butterfly, "a number of Little Flies" (Vallisneri 1726a, p. 99) comes out. This was a fact that had already been reported, even if more summarily, by Vallisneri in the *Primo Dialogo*, as we have seen above, but here the discussion becomes more precise and detailed

The bizarre birth of the aforesaid Little Flies, noticed also by yourself, Honourable Sir, encouraged me to make a move to determine their origin, as I had then fully recognised the existence [of said phenomenon] [...] To fully understand the metamorphosis mentioned, or the development of so many minute Insects, one needs to know that, during the time that the above-mentioned caterpillars live and feed on the leaves of Cabbages, certain Little Black flies fly around them. They are bigger than those of wine (whose origin You discovered, Sir, in your much praised Dialoghi), very slow when they move forward, but very quick-flying, the females of which land on the most unfortunate caterpillars, and lay a number of tiny eggs on them, almost invisible to the naked eye, out of which in less than two days hatch these really tiny maggots, and then these almost invisible worms, just hatched, like the Small Worms of Mange, poke themselves in under the skin, and they get so far in that, little by little, they eat them up inside. However, these caterpillars keep on eating and growing, just the same. Nonetheless, one can clearly tell which ones are affected by these little maggots, because they start to turn yellow [...] so they no longer think about making a chrysalis, but all of a sudden, when the other healthy and lucky caterpillars set about making their own Chrysalises, the infected ones burst down one side, and all these little maggots, which have fed on the caterpillar's substance, come out of the crevice in each one. These unrestrained little maggots, which have come out of the bodies of the aforesaid caterpillars (which of course then die and dry up) get silk out of their mouths, with which they bundle themselves up, and tangle themselves so much in it that they end up looking like a heap of little cocoons covered with a yellowish silk. Then, at the end of about twelve days, lots of little flies jump out. And this is not a misfortune which happens just to Cabbage caterpillars, but likewise to various other kinds of caterpillars and worms, as I and You yourself, Sir, have observed on a number of occasions, as illustrated in your above-praised Dialogue. (Vallisneri 1726a, pp. 99–100)

Here we are probably dealing with *Cotesia* (= *Apanteles*) *glomerata* (Braconidae: Microgastrinae), which attacks Lepidoptera Pieridae, laying 50 eggs or more (Grandi 1984, Vol. II, pp. 995–996). Cestoni thus clarifies once and for all the aforesaid mistakes made by Redi and Aldrovandi, who had taken the cocoons, made by the larvae of the parasitoid, for "eggs" laid by the caterpillar which could generate animals different from themselves (Vallisneri 1696, p. 313). Vallisneri had arrived at the same conclusion (Vallisneri 1696, p. 312), seemingly without ever having witnessed the act of oviposition, as Cestoni had done (Tremblay and Masutti 2005, p. 36).

Cestoni also observed a female parasitoid wasp smaller than the one previously described laying eggs on the same larvae or those of other Lepidoptera. The size of the parasitoid and the identity of the host indicate that they were, in all probability, *Pteromalus puparum* (Pteromalidae: Pteromalinae), which, as the name tells us, emerges from the chrysalis of *Pieris brassicae*, and also of many other butterflies. Vallisneri had given some cases which, on the surface, appeared to be similar, since they should actually have been connected with "eggs [...] laid upon the Chrysalis" ["uova [...] depositate sopra della Crisalide"] (Vallisneri 1726a, p. 101), as Cestoni observes after carefully reading the already quoted passage of the *Primo Dialogo* in which Malpighi asserts that

Many Flies and Large Flies come out of cocoons, or Pupae made by worms [...] others out of the their Chrysalises [...] They will never see Little Flies or Large Flies or Flies emerge from the Chrysalises, if they protect them well enough, I mean the cocoons and such like [...] (Vallisneri 1696, p. 312)

Instead, the parasitism observed by Cestoni is different and more sophisticated in nature

> Other Little black Flies, smaller than the ones I have described [*Apanteles glomeratus*], less than half the size, land just the same on top of the caterpillars, and offload their eggs on them, and these are so small that they cannot be seen by the eye, unless one has a superb Lens. After some time, longer than it takes for the aforesaid little flies, likewise, little maggots hatch out of these eggs, and they also get into the caterpillar but, as they are quite a bit later in feeding, the caterpillar has time to grow and make its Chrysalis (and this indeed happens) and, during this time, those nasty little maggots continue to feed in the same way on the substance of the caterpillar inside its chrysalis, without emitting faeces. When they have finished feeding, they do not emerge from the caterpillar in order to make a Chrysalis, but they do this inside the caterpillar itself, and they stay there over a Month, and then the Little Flies come out. There are really so, so many that one cannot even imagine that they could have been inside that Chrysalis, wherein the Little Flies make a tiny hole, or perhaps the first Little Fly does so, because it wants to get outside. Indeed, after that, all of them pour out of that same hole. (Vallisneri 1726a, pp. 100–101)

The fact that the egg of the parasite is not laid directly on the pupa or chrysalis, as Vallisneri seems to believe in his *Dialoghi* (something which actually occurs with other parasitoids), is an example of *delayed* parasitism, which makes more difficult to identify the parasite living in the larva of the host until it turns into the chrysalis which will, eventually, be eaten up. Thus the parasite does not emerge from the larva upon which it has been observed laying its eggs, but from the chrysalis into which the larva has turned, a situation which could have been exploited to support spontaneous generation without Cestoni's shrewd analysis.

The last observation regards the "Little white Butterflies" commonly found on the leaves of cabbages (the homopteran *Aleyrodes proletella* (=*brassicae*), already quoted), for which Cestoni gives a painstaking biological description noting, for example, how they mate and the characteristic distribution pattern of the eggs, mentioned above, and observing that they feed "on that juice, which bathes the outer skin of the leaves" (Vallisneri 1726a, pp. 103–104) ["di quel sugo, che viene a irrorare l'esterna buccia delle foglie"] apparently without harming the plant. Let us follow Cestoni's description:

> These Little Butterflies are generated in precisely the same way as happens with most animals, in other words, by means of the male and female [...] which, carrying on with each other, mate, and when the females are pregnant [...] they lay their eggs, which they generally arrange in a semicircle in batches of ten, twelve, fourteen, and sometimes sixteen [...] from each of these emerges a little white animal with six legs and a bit of fuzz on its back [...] I have decided to call them *Little Sheep* from now on [...] and these start walking [...] when they have got to where they have to stay [...] they arrange themselves a little apart from one another, so that, as they grow, they won't touch each other; thus, seen under the Microscope, they look like a lot of little white sheep standing still in a little green field. After that, remaining still and attached, they grow [...] It may seem to You, Honourable Sir, to be an error of judgement, my having given the name of little sheep to the aforementioned creatures, but thinking through what happens to these poor Insects [...] it will not seem so unreasonable [...] just as sheep are subject to being devoured by Wolves, these [...] also have their *Wolves*, which hunt them down. The latter are a breed of Little black Flies, which [...] live on nothing else but the aforesaid dear little sheep ... A copious quantity of the aforementioned Little black Flies constantly buzz around the said little sheep, and some of

them linger around the most tender ones, and slowly but surely, they suck out all their substance, so that in the end they just leave them with their outer skin. Others land on the largest ones, that is, those which have grown to their fullest size or nearly so, and sit on them for a long time; whereupon, having paid particular attention to this, and to what they were doing there, I saw that the Little wolf Flies, after having pierced the back of the little white sheep they were sitting on, proceeded nonchalantly to lay an egg in that hole, from which shortly afterwards I observed that a nasty little maggot had hatched, which started to eat up its poor little sheep [...] It is an easy thing to recognise when the dear little sheep have been unfortunate, and the Little Wolf Flies have laid eggs on them, seeing that they begin to turn from their usual white, to a livid hue, and to go past the time when they would normally emerge as winged insects, which usually takes no longer than twenty days when they have not been ruined by the Wolves [...] when these little maggots have finished feeding, they set about making their Chrysalises, and for their ends, they use the skin of the little white sheep themselves, which they have eaten up, inside which they can be clearly seen to be enveloped and turned into chrysalises. And there they remain for about twenty days or more, before they come out of that, than the aforesaid little butterflies. Then, at the end of that time, tearing that skin, healthy Little winged Flies come out and fly away, to start doing the same thing over again with the other little white sheep, thus continuing to reproduce at the cost of the innards and the flesh of the unfortunates, while the lovely little butterflies feed and multiply under the leaves of the Cabbages without doing any harm to them whatsoever. (Vallisneri 1726a, pp. 102–105)

Cestoni also provides a drawing of this parasite, and its host. We are dealing with *Encarsia* (Aphelinidae: Coccophaginae) which uses predominantly the more fully-grown stages of its host (the "largest ones" [le "più grosse"]) for the purposes of oviposition (Liu and Stansly 1996).

In Table 2.1 is given a summary of the species considered in the previous analysis.

2.4   Gall Wasps and Other Gall Insects

The discussion regarding the origin of galls will not be reported in the present work (see, e.g., Vallisneri 2004, p. 134 and note 653). In Table 2.2, we will confine ourselves to giving a synoptic list of what we were able to find out as regards the identity of the galls and the insects connected with them. The latter can be divided in species which form the galls on the plants, i.e. Hymenoptera (Cynipidae and Symphyta: Tenthredinidae); Diptera (Cecidomyiidae); Hemiptera (Pemphigidae), and species which are parasitoid of the gall-forming species (Pteromalidae, Torymidae, Eurytomidae, Ichneumonidae), and inquilines (Cynipidae and Coleoptera Curculionidae). Due the difficulty to identify species according to the original descriptions (see Discussion), the determination of inquilines and parasitoids was often made very tentatively.

Vallisneri explicitly points out the large number of species that can be found in just one gall

And I did not just see beetles ["cantarelle"] in the said little swellings [galls on willows produced by *Pontania*], but the maggots of various flies, particularly carnivorous ones, all

Table 2.1 Generic or specific identifications

Genus/species	Family	Observed by	Reference and pages*	Host
Hymenopetera: Parasitica				
Agriotypus	Ichneumonidae	Vallisneri	1, p. 312	Trichoptera
Aphidius	Braconidae	Cestoni	2, pp. 96–97	Aphids (Homoptera)
Apanteles glomeratus	Braconidae	Cestoni	2, pp. 99–100	Lepidoptera larvae (Pieridae)
Trichogramma/Teleonomus	Trichogrammatidae/Platygastridae	Vallisneri	1, p. 312	Lepidoptera larvae
Encarsia	Aphelinidae	Cestoni	2, pp. 102–105	*Aleyrodes proletella*
Pteromalus puparum	Pteromalidae	Cestoni	2, pp. 100–101	Lepidoptera larvae (*Pieris?*)
Acroricnus	Ichneumonidae	Mattacodi	3, p. 66	*Sceliphron*
Hymenopetera: Aculeata				
Trypoxylon	Crabronidae	Vallisneri	1, pp. 357, 369	spiders
Sceliphron (spirifex?)	Sphecidae	Vallisneri	1, p. 359	spiders
Eumenes sp.	Vespidae	Vallisneri	1, pp. 359–360	
Eumenes pomiformis	Vespidae	Vallisneri	1, p. 364	Lepidoptera larvae
Megachile (*parietina?*)	Megachilidae	Vallisneri/Cestoni	1, p. 362	plants
Osmia	Megachilidae	Vallisneri	1, p. 369	plants
Diptera				
Pachyophthalmus signatus (=*Amobia signata*)	Sarcophagidae	Vallisneri	1, p. 367	*Sceliphron*
Coleoptera				
Trichodes	Cleridae	Vallisneri	1, p. 362	*Megachile*
Macrosiagon	Ripiphoridae	Vallisneri	1, p. 366	*Sceliphron*

*1: Vallisneri, 1696; 2: Vallisneri, 1726a; 3: Vallisneri, 2004

Table 2.2 Gall–inducing insects, parasitoids and inquilines (the families, when not otherwise indicated, are into Hymenoptera)

Plant and location	Vallisneri's reference	Number of species	Gall (G), Larva (L), Adult (A)	Modern reference	Notes
Quaderno I[a]					
Oak (underleaf)	[99r]: p. 120[b]		(G) Cynipidae: *Neuroterus tricolor* (?)[e]	1 (p. 59 and Fig. X.22 p. 59)	Inducer (reported by Mattacodi)
	[101r]: p. 123		(G) Cynipidae: *Neuroterus quercusbaccarum* (?)	1 (p. 59 and Fig. X.20 p. 59; Pl. 1.7b)	Inducer
	[107v]: pp. 132–133		(G) Cynipidae: *Neuroterus tricolor* (?)	1 (p. 59 and Fig. X.22 p. 59)	Inducer
Oak[d] (leaf buds?) marble gall?	[124r–126v]: pp. 155–158	2	(G+L) Cynipidae: *Andricus kollari?*	1 (p. 62, Fig. 160 p. 35)	Inducer
			(G+L) Cynipidae: *Synergus umbraculus?*		Inquiline
			+ aphids and Lepidoptera larvae		
Oak (inflorescence?)	[173v]: pp. 210–211	1	(G + A) Cynipidae: *Andricus (fecundator? solitarius?)*		Inducer
Rose[d] (bedeguar = cynorrhodon)	[113r–115r]: pp. 141–144	2	(G+L) Curculionidae (Coleoptera): *Curculio (=Balanius) villosus?* or *Anthonomus rubi?*	2, 6	Probable transient inquiline
			(G+L) Cynipidae: *Diplolepis rosae*		Inducer
			(G+L) *Periclistus brandtii?*		Inquiline

(continued)

Table 2.2 (continued)

Plant and location	Vallisneri's reference	Number of species	Gall (G), Larva (L), Adult (A)	Modern reference	Notes
Quaderno II^c					
Oak^d (leaf shoots?) marble gall ?	[72v–77r]: pp. 60–64	>=9	(A) Pteromalidae, Torymidae: *Torymus* sp.?, *Torymus nitens*?, *Megastigmus*?;	1 (p. 87); 3	Parasitoids of *Andricus kollari*
			(A) Eurytomidae: *Eurytoma brunniventris*?		Parasitoid of *Synergus* (inquiline)
Rose^d (bedeguar = cynorrhodon)	[78r–81v]: pp. 64–66	5	(A) Ichneumonidae?: *Orthopelma mediator*?	2 (pp. 220, 223); 1 (p. 85)	Parasitoids of *Diplolepis*
			(A) Torymidae?: *Torymus bedeguaris*?, *Glyphomerus stigma*?		
			(A) Eurytomidae?: *Eurytoma rosae*?		Parasitoid
			(A) Curculionidae (Coleoptera): *Curculio* (=*Balanius*) *villosus*? *Anthonomus rubi*?		Inquiline
Oak			(A) Microlepidoptera	1 (p. 39)	
Quaderno III^i					
Oak (leaf shoots?) oak apple ?	[113r]		(G + L) Cynipidae: *Biorhiza pallida*?	1 (p. 62 and Fig. X.34 p. 62)	Inducer
Oak (underleaf, close to the central vein) oyster gall	[15r]	1	(G + L) Cynipidae: *Andricus anthracina*?	1 (p. 58 and Fig. X.14 p. 58)	Inducer
Idea nuova d'una Division generale degl'Insetti^f					
Beech (leaves)	p. 49		(G) Cecidomyiidae (Diptera): *Mikiola fagi*	4	Inducer
Poplar	p. 51		(G) Pemphigidae (Hemiptera): *Pemphigus spirothecae*?	1 (p. 53)	Inducer

La galleria di Minerva[g]

Willow (leaves)	pp. 316–317	1 + 7	(G + L + A) Tenthredinidae: *Pontania* (*Euura*?) + 7 connected (inquilines, parasitoids)	5; 1 (p. 73)	Vallisneri notes that the fully grown larvae of *Pontania* leave the gall and climb down to earth, where they chrysalize (pupate)
			Curculionidae (Coleoptera)		

(Mattacodi) Nuova giunta di osservazioni, e di esperienze…[h]

Willow (leaves)	pp. 65–66		(G + L + A) Tenthredinidae Microlepidoptera: *Pontania* (*Euura*?)	1 (p. 73)	
			(G + L) (parasites and inducer)		
Elm poplar	p. 66				
Rose	p. 66		(G) ("cappelluta") + (A) Cynipidae: *Diplolepis*	1 (p. 69)	Inducer
Rose	p. 66		(G) ("non cappelluta") + (A) Cynipidae: *Diplolepis*	1 (p. 69)	Inducer

[a]Vallisneri (2004); [b]Numbers between [ ] refer to the original manuscript, the other to the pages of the printed transcript; [c]Vallisneri (2007); [d]Galls which were first opened (to observe the larvae), then re-closed to await the emergence of the adults; [e](?) means that the both genus and species are uncertain; ? after the species name means that only species name is uncertain; [f]Vallisneri (1713); [g]Vallisneri (1696); [h]Vallisneri (1726); [i]Vallisneri (unpublished manuscript); 1 = Redfern and Askew (1992); 2 = László and Tothmeresz (2011); 3 = INTERNET reference at entry "oak marble gall wasp *Andricus kollari*"; 4 = INTERNET reference at entry "*Mikiola fagi*"; 5 = Nyman et al. (2000); 6 = INTERNET reference at entry "*Diplolepis rosae bedeguar*"

of them bastards, foreigners and uninvited guests. I also observed the same thing in the scaly ends of willow shoots throwing out narrow leaves. I found inside, apart from the central worm or fly, which is black with long antennae and a really long sting, all sorts of other little flies, and I counted seven different kinds one day [...] And not only did the said Little Flies which there ought to have been come out, but all these others, of Father unknown, and wild. (Vallisneri 1696, pp. 316–317)

The finding of Vallisneri regarding the large number of insects associated to a single gall is in fact true. For example, Grandi (1984, Vol. II, note 1 p. 1013) reports that Fahringer (1924) had observed 101 species of insects associated with the galls of *Andricus kollari*, which is probably fewer than the number found in more recent studies.

However, it was Mattacodi who identified, more clearly than Vallisneri ever did, the presence of parasitoids in the gall, which live at the expense of the inducer insect

So in the blisters formed on elms, and in any sort of goitres found on poplars, the worms of flies can often be found, which kill the little flies that are born there by sucking at them, and they live in this womb they have appropriated for a long time, and yet those worms did not cause that goitre [...] (Vallisneri 1726, p. 66)

The galls were therefore not caused "by a variety of Insects" but were induced

[...] solely by the aforesaid little caterpillar, and the little worms then got inside it in order to eat up the caterpillar, and gain any other nourishment they could find. (Vallisneri 1726, pp. 65)

Mattacodi therefore holds that it were the larvae of the parasitoid which somehow entered the host gall, as happened when eggs (and chrysalises) were parasitized, an opinion with which Vallisneri himself agreed as we have seen.

Another interesting observation regards the life cycle of *Pontania* (Tenthredinidae) (Table 2.2), whose larvae, once they are fully grown, come out from the gall (confirmed by Mattacodi (Vallisneri 1726, pp. 65)) and go down to earth to pupate (see, for example, Kopelke 1998)

Having carried out some necessary duties, I found that the true caterpillar [the inducer of the gall, not one of the many others connected] of the Willow in question, when it is fully grown, emerges from the Swelling [the gall], and comes down to earth, where it buries itself, and hides itself up, there making its little cocoon [...] to protect itself against the winter cold [...] (Vallisneri 1696, p. 316)

Mattacodi also reports that the inquilines of a gall occupy positions further away from the centre than the larvae of the inducer

[...] up until now, I have always observed the same fly in the middle of the gall, and it nearly always comes out in the Autumn [to pupate, see above], whereas the other inhabitants place themselves between the middle of it and the perimeter, having got into the fully developed gall, where they even later complete their metamorphosis, and they cannot leave the gall before the next Summer. (Vallisneri 1726, p. 66)

In confirmation of the above Vallisneri would write, again as regards oak galls (Quaderno I: [124r-126v])

I have observed many galls, which I collected until September last year, and I found, on the 8th January 1695, many completely empty, many without their original worm, but with the original hole where the worm emerged full of flies and ants and other little insects, with the mouth of the hole closed off with earth. Other galls, although they did not contain the original central worm in it, nonetheless had lots of tiny little worms around the centre, and in one I counted as many as eight. (Vallisneri 2004, p. 155)

One is dealing with galls of *Andricus kollari* (Cynipidae) (Table 2.2), in which the cell of the larva of the inducer insect lies at the heart, whereas the smaller ones, very presumably of the inquiline *Synergus umbraculus* (Cynipidae) (Table 2.2), establish themselves in the peripheral area around it (Redfern and Askew 1992, p. 35, Fig. 160; Pénzes et al. 2012, p. 4).

Another interesting observation, again from Mattacodi, relating to the specificity of the relationship between the gall and the galler insect (Table 2.2), is: two rose galls, both of them spongy, but one of them hairless ("non cappelluta"), the other hairy ("cappelluta"), are actually caused by two different "kinds of little flies" (Vallisneri 1726, p. 67) (presumably two different species of *Diplolepis* (Cynipidae) (Redfern and Askew 1992, p. 69)).

However, there are still a few cases, reported both by Mattacodi and Vallisneri (Vallisneri 1726, p. 66; Vallineri 1696, p. 316), in which a willow gall is used only as a temporary shelter by an insect which lives part of the time outside it and, in Mattacodi's opinion, this can be explained as reuse of a gall which the larva has left in order to migrate into the earth, as we have said.

2.5 Discussion

This work is not the first attempt to produce a systematic determination of the parasitoid wasps identified by Antonio Vallisneri and his cohorts. To our knowledge, it has been preceded by at least two other studies here cited: one by Pampiglione et al. (2000), the other by Tremblay and Masutti (2005). However, these works are not of great help when trying to identify the insects described: the former, because it confines itself to describing the work Vallisneri did on these organisms very generically, with few identifications of the wasps he actually studied; the latter, as it proposes implausible identities, which are sometimes decidedly erroneous, as pointed out above. Moreover, as regards the galler insects described by Vallisneri, we do not know of any other modern analysis. And this should not be surprising, considering that these descriptions are mostly taken from unpublished works, from the *Quaderni di Osservazioni* to be precise (only the first and second of which have recently been transcribed), while just a few are from published works (Table 2.2). Instead, the descriptions of parasitoid wasps are mostly taken from *La Galleria di Minerva* and from a published letter written by Cestoni.

Valuable data have been found in the works we analysed on various aspects of parasitoid-host relationships in taxonomically diverse ecological webs. Vallisneri and Cestoni provided information that could be associated with a variety of hymenopteran parasitoids of other Hymenoptera (e.g. Ichneumonidae attacking Sphecidae, Torymidae attacking Cynipidae), dipteran parasitoids (e.g. Tachinidae,

Sarcophagidae) of Hymenoptera, coleopteran parasitoids (e.g. Ripiphoridae, Cleridae) of Hymenoptera, and hymenopteran parasitoids (e.g. Chalcididae, Pteromalidae) associated with non-hymenopteran hosts such as Homoptera, Lepidoptera and spiders. Overall, about 20 wasp genera could have been objects of Vallisneri and Cestoni observations. Noteworthy, Vallisneri and Cestoni provided about 300 years ago the first detailed ecological and ethological data on many wasp species from a wide range of life-histories.

Our study provides thus an example of the "hidden" biodiversity data that could be brought to light through detailed analysis of pre-Linnean works. By assigning taxonomic names (with the precision that was possible given the descriptions available, see below) we have provided information that could be incorporated in biodiversity databanks, that clearly play a fundamental role in biodiversity science (including conservation science).[1] Contribution to biodiversity data through the inspection of pre-Linnean works was already highlighted for botany, e.g. by analysing ancient herbaria and thus retrieving information on distribution and diversity of plant species >300 years ago (e.g. Pulvirenti et al. 2015), though such studies seem to be scarcer for zoology. In addition, our study revealed many aspects of the behaviour and ecology of a number of species, notably on their interactions with other species, thus providing data on ecological networks at the time of Vallisneri and Cestoni. Geographical data associated with pre-Linnean insect identification have furthermore the potential to better reconstruct a species' distribution more than 300 years ago, thus revealing changes in distribution across the last centuries.

The result of the present work is actually fragmentary, due to the difficult identification of insects based on the morphological descriptions; however, its is interesting in that several ecological and behavioural characteristics of these species could be used to help such identification. However, these aspects are rather lacking in the case of gall insects, apart from the few cases mentioned. Thus the insects have to be identified on the basis of morphological descriptions, which, although painstakingly thorough and rich in detail, often lack the features utilized in modern systematic criteria. Despite such obvious flaws in our species identification, we still believe in the importance of inspecting ancient studies and do the best to assign a taxonomy to the observed individuals. This could in most cases be the genus but not the species. However, for the problems associated with the co-occurrence of cryptic species, a valuable option may be to consider a taxonomic sufficiency approach, i.e. using higher taxa instead of the species in biodiversity analyses.[2]

At the end of the XVII century, as for most of the Eighteenth century, naturalists did not possess systematic methods to precisely identify the majority of insects, which have a variety and a complexity of forms unaccountable by using descriptions of an intuitive nature, even if they are as exhaustive as possible, as are those given by Vallisneri. This was a problem Réamur, too, would come up against in his monumental work devoted to these organisms, around 40 years later. The identifica-

---

[1] See Barberousse and Bary, Chap. 3, in this volume.
[2] See Chenuil et al., Chap. 4, in this volume.

tion of the insects according to their morphological features in the times of Vallisneri, like the one produced by Ulisse Aldrovandi around a century before Vallisneri or the then very recent example given by John Ray, did not go much beyond an elaboration of what Aristotle had said in his *Historia Animalium*. It should be remembered that even the distinction between the two most important groups that Vallisneri is examining in connection with the studies which concern us here – Diptera and Hymenoptera – is often hazy and unclear. Even if, in this case, the difference is very obvious – Diptera are two winged, Hymenoptera four – something which Vallisneri definitely realised; their general appearance, which is often very similar, seems to eclipse this fundamental difference. Francesco Mattacodi proves that this problem exists, in one of the letters already considered

> However, I should think that these little mosquitoes are incorrectly called so, like those from the blisters found on elms and poplars, regarding which I must beg you, when you have leisure to do so, to explain how Lyster makes his distinctions between mosquitoes, wasps and flies etc., as I fail to understand how various sorts of Insects with four wings, even if their features conform to those of flies, should not be called by that name, seeing that flies have, again according to Aristotle himself, only two wings. (Vallisneri 1726, p. 67)

Here we can see direct evidence of the lack of a "methodical" system, which would allow a few, clearly defined distinctive features to be isolated in order to avoid this sort of ambiguity. On the other hand, even Linnaeus, the inventor of the "Method", would not go far beyond subdividing insects into their principal orders. For example, in the tenth edition of *Systema Naturae* (Linnaeus 1758, pp. 553–583), Hymenoptera were divided into eight genera: *Cynipis* (Cynipidae, gall-making Hymenoptera Parasitica), *Tenthredo* (Tenthredinidae and other Hymenoptera Symphyta), *Ichneumon* (Ichneumonidae), *Sphex* (Sphecidae), *Wasp* (social and sphecid wasps), *Apis*, *Formica*, *Mutilla* (a genus of Hymenoptera Aculeata parasitoid of other Hymenoptera Aculeata), giving a total of 229 species for the whole Order.

Vallisneri's and Cestoni's "systematic" classification does not make use of this or other similar terminology, as the only valid grouping they give is actually the individual "species" ["spezie"]. At this level, the number of species described is greater by several orders of magnitude than the corresponding lists provided by Linnaeus or Ray, and their description immeasurably more elaborate and detailed. However, one might ask: what is the taxonomic benefit of such meticulous description, if it does not allow the reader to identify the 'species' in question?" We think it proved particularly useful for Vallsneri himself, as a sort of personal reminder, so he could subsequently recognise a species he had previously observed. This might be one explanation for these minutely detailed descriptions, which are even greater in number in the *Quaderni* than they are in the printed works. However, it seems less likely that they could be used to enable readers to identify the species described, due to the difficulties already indicated, in addition to the use of generic qualitative adjectives, such as "large", "little", "long", "short" etc., lacking any standard reference indication.

In fact, as early as 1713, in *Idea nuova d'una Division generale degl'Insetti*,[3] Vallisneri had set forth a traditional type of systematic classification which he had, admittedly, taken from other contemporary authors (and which, indeed, was based on that proposed by John Ray, 3 years earlier, in his *Historia Insectorum* (Ray 1710)), in which identification was based on the number of wings and "feet". This bore an explicit analogy with the botanical systematics of Tournefort (Vallisneri 1713, p. 67), who had chosen, amongst all other possible bases, to use only the features of the flowers and the seeds to draw up a classification of plants. However, this division, which was "methodical" if a bit rough and ready, seems to have been proposed only as a form of respect for the new methodical systems which were then beginning to arise; Vallisneri thought it should be supported by another method, on which he focussed his energies and interest, which would be ecological in nature. With a strange turnaround, disregarding the classification normally used, perhaps with the aim of diminishing the importance of the "methodical" system, he would use ecological features in order to initially sub-divide the insects into "Classes" (Vallisneri 1713, p. 42). The other method, instead, would confine itself to determining the "specific differences" (Vallisneri 1713, p. 65).

We are now in a position to understand why Vallisneri often seems to treat flies and wasps as altogether similar insects. The mere fact that they possess one or two pairs of wings - a feature considered to be so important by the majority of zoologists who preceded, were contemporaries of or came after Vallisneri -, is instead judged by him to be just a "specific difference"; to Vallisneri, the fact that both flies and wasps shared a habitat is much more significant, justifying their inclusion in the same "Class". Thus, when Vallisneri finally needed to define the species, he would use only a description, set forth in minute detail, of all the morphological features of the insect.

However, it is definitely not the ability of Vallisneri and his circle to systematize which should attract our attention, as much as the wealth of that finely drawn detail in his ecological and behavioural observations, which show the level reached by naturalistic observation, at least in regard to insects, towards the end of the seventeenth century. This would be what ushered in a great era devoted to the study of the behaviour of insects, which would see its crowning achievements with Réamur in France.

It could be argued that Vallisneri considered himself to have been responsible for the discovery of the phenomena relating to parasitoid wasps (Vallisneri 1710, p. 15), apparently not taking into account what other authors had already written about this matter before him.

Even if he was aware of what Swammerdam had already written about insects, "diversae speciei" of which emerged from caterpillars and chrysalises (Vallisneri 1710, pp. 13–14), he believed that this author, who reported the phenomenon "ad

---

[3] See also (Andrietti and Generali 2004, pp. 110–115) for a brief examination of this study.

quartum mutationis ordo" ["fourth order of metamorphosis"], attributed it to "an internal principle" (Vallisneri 1710, p. 14; Swammerdam 1693, p. 131) and not an *ex ovo* origin. Van Lenteren and Godfray (2005, p. 17) opt for a different interpretation: in view of the fact that Swammerdam's "fourth order of metamorphosis" is always based on an egg, they suggest that he viewed parasitoids in the same way. In any case, in his next work, "Bybel der Natur", Swammerdam reports the observation (made by Otto Marsilius van Schriek) of a caterpillar being stung by a "fly", which he connects with the immission of eggs by the parasitoid (van Lenteren and Godfray 2005, p. 18).

Vallisneri certainly could not have known of this second work by Swammerdam (finished in 1679 but only published in 1737–1738) (van Lenteren and Godfray 2005, p. 17). However, he did not even seem to be informed (or at least he did not make any mention) of other researchers which, in the two decades preceding his *Dialoghi*, had already made similar suppositions and, in some cases, actually observed the process of parasitoids laying eggs in their host (for a list of various researchers who reached similar conclusions before Vallisneri, see van Lenteren and Godfray 2005). Before being called by the Università di Padova the sources of Vallisneri's entomological studies were mainly Italian, and his knowledge of European literature was quite limited (Generali, *Introduzione*, in Vallisneri 2004, pp. XXVIII-XXXVI), with the exception of Martin Lister, constantly quoted in Vallisneri's *Quaderni* for his edition and notes to *De Insectis* of Johannes Goedaert (1685). Already in some letters published in the 1670–1671, Lister had suggested the possibility that certain insects could lay their eggs inside living caterpillars (van Lenteren and Godfray 2005, pp. 18–19). In addition, in his note 4 at page 9 of the mentioned work, he writes: "Muscae octoginta duae [observed by Goedaert], hic memoratae, fuerunt cuiusdam muscae Ichneumonis progenies, in Erucae corpus delata; quo pacto autem illuc deferatur, fateor me adhuc non satis ossequi; sed conjecturam facio, illud ab Ichneumone Parente fieri" ["The 82 flies that emerged from the pupa are the progeny of an ichneumon fly, which had gotten into the caterpillar in a manner that is still not entirely clear to me. In all likelihood they were laid right there by the mother fly"] (translation by van Lenteren and Godfray 2005, p. 19).

In this regard, we should recall also Cestoni's observations of the plant flea mentioned above, in which he declared, rather ironically, that the *Aphidius* "had very kindly laid an egg" under its belly. If it is true that his letter is dated 25th July 1698, it appears that it was terminated from the year 1692 (Cestoni 1787, note at page 271; Tremblay and Masutti 2005, p. 37).

Acknowledgements The authors are indebted with Alessandra Cuminale for her help in bibliographical researches and with Dario Generali for his many helpful comments and suggestions. We thank Elena Casetta and Davide Vecchi for the interest shown in our contribution and the invitation to join the book project. Part of the present work has been realized making use of a PRIN 2007 funding (Università degli Studi del Piemonte Orientale, di Milano, Ferrara, Pavia e Insubria) coordinated by Maria Teresa Monti. CP was funded by a Post-Doctoral contract from the Universidad de Castilla-La Mancha.

References

Andrietti, F., & Generali, G. (2004). *Storia e storiografia della scienza: il caso della sistematica.* Milano: Angeli.

Batelka, J., & Hoehn, P. (2007). Report on the host associations of the genus *Macrosiagon* (Coleoptera: Ripiphoridae) in Sulawesi (Indonesia). *Acta Entomoogical Musei Nationalis Pragae, 47*, 143–152.

Berger-Tal, O., Polak, T., Oron, A., Lubin, Y., Kotler, B. P., & Saltz, D. (2011). Integrating animal behavior and conservation biology: A conceptual framework. *Behavioral Ecology, 22*, 236–239.

Boenigk, J., Wodniok, S., & Glücksman, E. (2015). *Biodiversity and Earth History.* Heidelberg: Springer.

Cane, J. H., Griswold, T., & Parker, F. D. (2007). Substrates and Materials Used for Nesting by North American *Osmia* Bees (Hymenoptera: Apiformes: Megachilidae). *Annals of the Entomological Society of America, 100*, 350–358.

Cestoni, D. (1787). *Storia degl'Insetti de' Cavoli in "Opuscoli scelti sulle scienze e sulle arti, Tomo X, Parte IV"* (pp. 260–271). Milano: Marelli.

Dunne, J. A., Williams, R. J., & Martinez, N. D. (2002). Network structure and biodiversity loss in food webs: robustness increases with connectance. *Ecology Letters, 5*, 558–567.

Generali, D. (2004). Uno speciale che "superava la sua condizione". Il caso dell'invisibilità postuma di Diacinto Cestoni. In M. T. Monti & M. J. Ratcliff (Eds.), *Figure dell'invisibilità: Le scienze della vita nell'Italia d'Antico Regime* (pp. 83–118). Firenze: Olschki.

Goadert, J. (1685). Johannes Goedartius de insectis, in methodum redactus; cum notularum additione. Opera M. Lister, e Regia Societate londinensi. Item appendicis ad historiam animalium Angliae, eiusdem M. Lister, altera editio hic quoque exhibetur. Una cum scarabeorum anglicanorum quibusdam tabulis mutis, Londini, Excudebat R.E. sumptibus S. Smith ad Insignia Principis in Coemeterio D. Pauli. London, S. Smith.

Grandi, G. (1961). *Studi di un entomologo sugli Imenotteri superiori.* Bologna: Calderini.

Grandi, G. (1984). *Introduzione allo studio dell'entomologia* (Vol. I and II). Bologna: Edagricole.

Hondō, M., Onodera, T., & Morimoto, N. (1995). Parasitoid attack on a pyramid-shaped egg mass of the Peacock Butterfly, *Inachis io geisha* (Lepidopera: Nymphalidae). *Applied Entomology and Zoology, 30*, 271–276.

Hortal, J., De Bello, F., Diniz-Filho, J. A. F., Lewinsohn, T. M., Lobo, J. M., & Ladle, R. J. (2015). Seven shortfalls that beset large-scale knowledge of biodiversity. *Annual Review of Ecology, Evolution, and Systematics, 46*, 523–549.

Kopelke, J. P. (1998). Oviposition strategies of gall-making species of the sawfly genera *Pontania, Euura* and *Phyllocolpa* (Hymenoptera, Tenthredinidae, Nematinae). *Entomologia Generalis, 22*, 251–275.

La Salle, J., & Gauld, I. D. (1993). Hymenoptera: Their diversity, and their impact on the diversity of other organisms. In J. La Salle & I. D. Gauld (Eds.), *Hymenoptera and Biodiversity* (pp. 1–26). Wallington: CAB International.

László, Z., & Tothmeresz, B. (2011). Parasitoids of the bedeguar gall wasp *Diplolepis rosae*: Effect of host scale on density and prevalence. *Acta Zoologica Academiae Scientiarum Hungaricae, 57*, 219–232.

Linnaeus, C. (1735). *Systema Naturae, sive, Regna Tria Naturae systematice proposita per classes, ordines, genera, & species* (1st ed.). Rotterdam: Theodorum Haak, typography Joannis Wilhelmi de Groot.

Linnaeus, C. (1758). Systema Naturae, Tomus I, Editio Decim. Holmiae, Salvius, 1758 (anastatic copy: London, British Museum, 1956).

Liu, T., & Stansly, P. A. (1996). Oviposition, development, and survivorship of *Encarsia pergandiella* (Hymenoptera: Aphelinidae) in four instars of *Bemisia argentifolii* (Homoptera: Aleyrodidae). *Annals of the Entomological Society of America, 89*, 96–102.

Meijden, E., Veen-Van Wijk, C. A. M., & Ginneken, V. J. T. (2000). Cotesia (Apanteles) popularis L. parasitoids do not always kill their host. Entomologist's. *Monthly Magazine, 136*, 117–120.

Müller, A., Krebs, A., & Amiet, F. (1997). *Bienen*. München: Naturbuch Verlag.

Nyman, T., Widmer, A., & Roininen, H. (2000). Evolution of gall morphology and host-plant relationships in willow-feeding sawflies (Hymenoptera: Tenthredinidae). *Evolution, 54*, 526–533.

Pagliano, G., & Negrisolo, E. (2005). *Hymenoptera Sphecidae: Fauna d'Italia* (Vol. XL). Calderini: Ozzano dell'Emilia.

Pampiglione, G., Quicke, D. L. J., & Brandt, A. (2000). Observations of insect parasitoids by Antonio Vallisneri in the XVII century. *Antenna, 24*, 11–17.

Parke, E. C. (2014). Flies from meat and wasps from trees: Reevaluating Francesco Redi's spontaneous generation experiments. *Studies in History and Philosophy of Science Part C: Studies in History and Philosophy of Biological and Biomedical Sciences, 45*, 34–42.

Peigler, R. S. (1994). Catalogue of Parasitoids of Saturniidae of the World. *Journal of Research on the Lepidoptera, 33*, 1–121.

Pénzes, Z., Tang, C.-T., Bihari, P., Bozsó, M., Schwéger, S., & Melika, G. (2012). *Oak associated inquilines (Hymenoptera, Cynipidae, Synergini)* (TISCIA monograph series, Vol. 11).

Polidori, C., Federici, M., Mendiola, P., Selfa, J., & Andrietti, F. (2011). Host detection in *Acroricnus seductor* (Scopoli, 1786) (Hymenoptera: Ichneumonidae), a natural enemy of *Sceliphron caementarium* (Drury, 1773) (Hymenoptera: Sphecidae). *Animal Biology, 61*, 57–73.

Pulvirent, S., Indriolo, M. M., Pavone, P., & Costa, R. M. S. (2015). Study of a pre-Linnaean Herbarium attributed to Francesco Cupani (1657–1710). *Candollea, 70*, 67–99.

Ray, J. (1710). *Historia Insectorum*. London: Churchill.

Redfern, M., & Askew, R. R. (1992). *Plant Galls. Naturalists' Handbook 17*. Slough: Richmond Publishing.

Rotheray, G. E. (1993). Colour guide to hoverfly larvae (Diptera, Syrphidae) in Britain and Europe. *Dipterists Digest, 9*, 1–156.

Shaw, M. R., Stefanescu, C., & Van Nouhuyse, S. (2009). Parasitoids of European butterflies. In J. Settele, T. Shreeve, M. Konviĉka, & H. Van Dyck (Eds.), *Ecology of butterflies in Europe* (pp. 130–156). Cambridge: Cambridge University Press.

Swammerdam, J. (1693). Historia Insectorum Generalis, Ed. Secunda. Ulltrajecti, ex officina Otthonis de Vries. In *ex officina Otthonis de Vries*. Utrecht.

Tremblay, E., & Masutti, L. (2005). History of insect parasitism in Italy. *Biological Control, 32*, 34–49.

Vallisneri, A. (1696). Saggio de' Dialoghi sopra la curiosa origine di molti Insetti del Medico Filosofo Antonio Valisnieri ecc. In *"La Galleria di Minerva", 1696, I* (pp. 297–322). Venezia: Albrizzi.

Vallisneri, A. (1700). Secondo Dialogo...In *"La Galleria di Minerva", 1700, III*, (pp. 297–318 and 353–372). Venezia: Albrizzi.

Vallisneri, A. (1710). *Considerazioni ed Esperienze intorno alla Generazione de' Vermi ordinari del corpo umano... Stamperia del Seminario, at Gio*. Padova: Manfrè.

Vallisneri, A. (1713). *Idea nuova d'una Division generale degl'Insetti in "Esperienze ed Osservazioni intorno all'Origine, e costumi di vari Insetti, con altre spettanti alla Naturale, e Medica Storia...", Stamperia del Seminario, at Gio* (pp. 40–76). Manfrè: Padova.

Vallisneri, A. (1726). *Osservazioni spettanti alla Storia naturale degl'Insetti mandate al nostro Autore dal Signor Francesco Mattacodi, ecc. in "Nuova Giunta di Osservazioni, e di Esperienze intorno all'Istoria Medica, e Naturale, non solamente del Signor Vallisnieri, ma di altri celebri Autori, a lui scritte, con Annotazioni, e Riflessioni del medesimo", Stamperia del Seminario, at Gio*, 60–63 (first letter) and 64–67 (second letter). Padova: Manfrè.

Vallisneri, A. (1726a). *Nuove, e meravigliose scoperte dell'Origine di molti Animalucci su le foglie de' Cavoli, e di molti Insetti dentro gl'Insetti candidamente partecipate, e dedicate all'Illustriss. Sig. Antonio Vallisneri...da Diacinto Cestoni Livornese, colle quali si confermano vari ritrovamenti del suddetto Signore intorno la curiosa Origine di molti Insetti, descritti nel suo Primo, e Secondo Dialogo in "Nuova Giunta di Osservazioni, e di Esperienze intorno all'Istoria Medica, e Naturale, non solamente del Signor Vallisnieri, ma di altri celebri Autori, a lui scritte, con Annotazioni, e Riflessioni del medesimo", Stamperia del Seminario, at Gio*, 89–105. Padova: Manfrè.

Vallisneri, A. (2004). Quaderni di Osservazioni, Vol. I (C. Pennuto Ed.). Firenze: Olschki.

Vallisneri, A. (2007). *Quaderni di Osservazioni*, Vol. II (M. Bresadola Ed.). Firenze: Olschki.

Vallisneri, A. Quaderni di Osservazioni, Vol. III (unpublished manuscript present at the Biblioteca Estense Universitaria di Modena, segnatura "Ms Campori 703, γ.D6.38").

van Lenteren, J. C., & Godfray, H. C. J. (2005). European science in the Enlightenment and the discovery of the insect parasitoid life cycle in The Netherlands and Great Britain. *Biological Control, 32*, 12–24.

Chapter 3
Marine Biodiversity Databanks

Anouk Barberousse and Sophie Bary

Abstract This chapter presents the contribution of databanks to the development of biodiversity knowledge through the example of marine biodiversity databanks. Focusing on the marine field allows us to insist on the imbalance of the unknown vs. the better known part. The chapter emphasizes the role of taxonomic and genetic databanks as well as the ongoing transformations that databanks are submitted to in order to answer pressing demands due to the biodiversity crisis. It aims to analyse the requirements biodiversity databanks have to satisfy in order to help both researchers and conservationists in their respective endeavors. It begins by pointing out the main characteristics and limits of biodiversity knowledge and defend the view that databanks are well-suited to overcome these limits as soon as they are widely accessible and interoperable. These constraints are analysed as both technical and scientific. Their dynamic dimension is emphasized as databanks must comply with the rapid evolution of scientific knowledge. We also propose a view on the relationships between biodiversity knowledge, assessment, and conservation.

Keywords Databanks · Genetic data · Interoperability · Taxonomy

3.1 Introduction

Assessing biodiversity, according to the comparison developed in the Introduction of this volume, is like assessing the state of a patient, that is, trying to infer how bad she is from observations. The double aim of this operation is to guess how her state will evolve in the future and to design actions to improve it. Assessment is thus

A. Barberousse (✉)
Sciences, Normes, Démocratie, UMR 8011, Sorbonne Université, Paris, France
e-mail: anouk.barberousse@sorbonne-universite.fr

S. Bary
UMR 7205 ISYEB – Institut de Systématique Evolution et Biodiversité,
Muséum National d'Histoire Naturelle de Paris, Paris, France
e-mail: sophie.bary@mnhn.fr

© The Author(s) 2019
E. Casetta et al. (eds.), *From Assessing to Conserving Biodiversity*,
History, Philosophy and Theory of the Life Sciences 24,
https://doi.org/10.1007/978-3-030-10991-2_3

future- and action-oriented. It benefits from available scientific knowledge and may in turn contribute to its development, but assessment and knowledge do not evolve at the same pace as assessment is submitted to the pressure of time. The relationship between assessment of biodiversity in a given geographical area and scientific knowledge of biodiversity, as built up in evolutionary studies, taxonomy, phyloge-netics, population genetics, and ecology, is unbalanced since the corpus that has been scientifically established is usually far from sufficient for the assessment task. The diversity of living beings, of their behaviours and interactions is so huge that what is known about it may be compared to ancient maps in which the size of the known world is much smaller than the breath of the unknown world. Thus, in most cases, currently available knowledge of biodiversity cannot provide but a small con-tribution to the assessment of biodiversity. However, some parts of this knowledge are more readily useful for assessment tasks than others: the knowledge of biodiver-sity that is contained in databanks can be easily harnessed. But to what extent can the data in databanks be treated as knowledge? We discuss this question in the fol-lowing, and emphasize again that retrieving this knowledge cannot be anything else than a small part of the assessment task.

Learning about biodiversity is a complex endeavor. First, because biodiversity itself is a complex object of knowledge. Second, because it extends on virtually every region of our planet, however small. Third, biodiversity knowledge comes from heterogeneous sources: taxonomic, evolutionary (including phylogenetic), genetic, and ecological research. This results in a confused picture in need of clari-fication. However, in the last few decades, biodiversity knowledge has immensely benefited from the establishment of international databanks. They play a fundamen-tal role in the improvement of the current, confused picture of biodiversity because they provide scientists with elements from which they can develop knowledge of biodiversity at a the global scale.

Our aim in this paper is to examine how biodiversity databanks contribute to developing current knowledge of biodiversity. We do so by putting forward an epis-temological analysis of the structure and functioning of databanks, both at the indi-vidual and collective (network) level. The epistemological analysis of scientific databanks has greatly benefited from Leonelli's work (2010, 2013a, b, 2016). Whereas she focused on various fields within biology, she never addressed the topic of biodiversity databanks. With this chapter, we wish to fill this gap and participate in her effort to put databanks in their right place in contemporary biological science and assessment and conservation policies.

The chapter is focused on marine biodiversity databanks and their role in the development of knowledge and assessment of biodiversity. Marine biodiversity is even less known than terrestrial biodiversity, thus illustrating the "ancient map fla-vour" of this domain of scientific knowledge. By describing how marine biodiver-sity databanks are developed, we show what kind of knowledge they promote and how it may be used in assessment and conservation tasks.

We begin by analyzing what it means to know about biodiversity. This question is raised because biodiversity is an unusual object of knowledge, crossing spatial and temporal scales. It is thus important to explore how a reasonably unified picture of biodiversity can be achieved by combining various components. Databanks contribute to this picture in an important way. We try to uncover which complex processes result in the "data" that are included in databanks and available for assessment and conservation tasks. We show that these data, far from being "brute data", are pieces of scientific knowledge subject to constant revision. The second part of the paper is devoted to the current uses of biodiversity databanks and associated requirements for databank designers. Finally, we put forward insights about how to build up biodiversity databanks that could improve our current knowledge of biodiversity.

3.2 What Does It Mean and What Does It Take to Know Biodiversity?

Because biodiversity knowledge involves evolutionary, taxonomic, and ecological research, as well as attempts at unifying insights from these three domains, it has to face a major conceptual and theoretical challenge. In recent times, this epistemic task has been shaped by a major external factor: urgency. Biologists can no more consider themselves as free of taking their time: they have to hurry up because of the severe crisis biodiversity is currently suffering (Western 1992; Grehan 1993; Takacs 1996; Olson et al. 2002; Singh 2002). The urgency of designing conservation policies induces an acceleration of assessment endeavours, which themselves increase the pressure on knowledge-building.

In this part, we examine in what sense the biodiversity crisis shapes the way we conceive of biodiversity knowledge and of biodiversity itself as an object of scientific knowledge. We first present the main features of the knowledge of biodiversity in its current available form and the difficulties it faces (3.2.1). We then review how it may be improved by the development of appropriate cyber-infrastructures, by examining the question: What are data in biodiversity databanks? (3.2.2) and then by giving examples of cyber-infrastructures (3.2.3).

3.2.1 Our Current Knowledge of Biodiversity and the Difficulties It Faces

The aim of this section is to describe the main features of the current knowledge of biodiversity. We do so by focusing on the position of taxonomic knowledge within biodiversity knowledge because the slow pace of its development is a major hindrance of assessment and conservation endeavours. We focus on the following tension: On the one hand, many attempts at biodiversity assessment try to circumvent the delays of taxonomic identification, but on the other, taxonomic knowledge appears as an indispensable component of biodiversity knowledge. We complete our description of the current state of biodiversity knowledge and the difficulties it faces by emphasizing how heterogeneous and patchy it is.

From the point of view of taxonomists, taxonomic knowledge, namely, the association of organisms with species names (or at least with genus names, and maybe with variety names), is an indispensable component of biodiversity knowledge and assessment as the descriptions of most components and processes of biodiversity rely on species identification. For sure, some studies are not taxonomy-dependent, like measurements of mass or energy transfer during biological cycles, but taxonomy appears as a main gateway to understanding what is going on at the various spatial and temporal scales where biological processes take place. As such, taxonomic knowledge may be seen to serve as the ground on which other epistemic endeavours within biodiversity studies, including conservation biology, can flourish. To what extent should the taxonomists' point of view be taken into consideration? In order to answer this question, we present and discuss an example illustrating the position of taxonomic knowledge within biodiversity knowledge: the earth worm example.

Earth worms are well-known, and for long, because they are ecologically important; however, their phylogeny and taxonomic status has long been unclear. Earth worms have been briefly described by Linnaeus in the eighteenth century. He gave them the species name *Lumbricus terrestris*. At the beginning of the nineteenth century, Savigny put forward a taxonomic revision based on the study of morphological characters (this story is told in James et al. 2010). He hypothesized that the organisms that Linnaeus had called *Lumbricus terrestris* actually form two species and introduced the name *Enterion herculeus* to designate the newly recognized species. In 1900, Savigny's morphological data have been re-interpreted and his revision rejected (James et al. 2010). At that time, the difference between the two sets of characters that Savigny relied upon in favor of taxonomic revision was interpreted as intra-specific polymorphism. However, this was not the end of the story: in 2009, genetic analysis by Richard et al. (2009) detected two homogeneous genetic groups within the set of organisms called *Lumbricus terrestris*. This led James and co-authors (2010) to begin a new, systematic study that included 230 fresh specimens from Europe and North America (belonging to *L. terrestris* and other species in the *Lombricus* genus) and specimens that had been preserved by Savigny. This new study was both genetic and morphological; it took part in the *Barcoding earthworms* programme and its results have been integrated in BoLD and GenBank, which are two major international, genetic databanks (see below for more details about these databanks). James et al. showed that Savigny was right: there are two diverging groups within *L. terrestris*. A new revision, similar to the one put forward by Savigny, thus occurred. We are now left with two species of earth worms: *Lumbricus terrestris* and *Enterion herculeus*.

What has been the upshot of this history of successive taxonomic revisions? It is strikingly different within and outside taxonomy. Within taxonomy, the earthworm episode is just another example showing that taxonomic knowledge, as all empirical knowledge, is of hypothetical nature and is thus susceptible of being criticized and revised as new data are available. For the specialists of the *Lumbricus* genus, the state of knowledge has been upgraded in such a way that there are now two well-established species where there used to be only one. Outside taxonomy, the situation is utterly different. Few biologists have realized that the state of taxonomic knowledge has changed within the *Lumbricus* genus.[1] The main reason why this is so is worth emphasizing: it is that non-taxonomists do not consider taxonomic knowledge as hypothetical, but rather as established once and for all, which is obviously erroneous. This ignorance has negative consequences: they may believe they have reached firm, well-established results for *Lumbricus terrestris* whereas they were actually studying *Enterion herculeus*. Unless they preserved (parts of) the specimens from which they have extracted genetic or physiological materiel, it is impossible to know what species they are talking about in their publications, *L. terrestris* or *Enterion herculeus*. As a result, their studies are simply pointless as they cannot possible lead to any useful conclusion. Imagine the same error about a species of mosquitos—the economical and medical consequences would be huge.

The lesson we may draw from the earth worm example is that taxonomic knowledge cannot be ignored by non-taxonomists. When they happen to use outdated taxonomic knowledge, their results are threatened. In some cases, the impact of taxonomic revisions may only bear on taxonomy itself, but in others, the chain of consequences may affect other parts of biology, including conservation biology and biodiversity assessment. Thus, with respect to taxonomy, biodiversity knowledge is not what it should be, as non-taxonomists do not use it appropriately. This appears as a major difficulty facing the development of biodiversity knowledge. It is not the only one. In the remaining of this section, we briefly discuss two other features of biodiversity knowledge that forbid it to provide us with a clear and unified picture of biodiversity, namely its heterogeneity and patchiness.

As it is built up from elements coming from various origins (taxonomy, phylogenetic studies, ecology, macro-ecology, biogeography, and evolution studies), biodiversity knowledge is dis-unified in such a way that it is unable to provide people in charge of assessment or conservation policies with any kind of firm ground. The main reason for this lack of unity is that each involved discipline has its own units, which are difficult to compare with one another. For instance, in population genetics, a gene may be considered a unit of biodiversity (but as is well-known, a gene in population genetics is not exactly the same entity as a gene in molecular biology and genetic databanks, which renders things even more complicated (Baetu 2012; Carlson 1991; Falk 1986; Fogle 2000; Gerstein et al. 2007; Kitcher 1982; Sterelny and Kitcher 1988; Waters 1994). In conservation biology, it is not uncommon to count organisms as units of biodiversity. Populations, species, communities, ecosystems, or even landscapes are *also* relevant units of biodiversity in ecology and conservation biology. But how do these different units compare? There is no general theory yet that would be able to precisely connect genes with organisms, or *a forti-*

---

[1] In order to provide the reader with evidence for this claim, we searched for "earth worm" on current search engines and found out that, outside taxonomy, scientific publications about earth worms exceptionally mention James et al. 2010 paper and the revision it contains. This suggests that non-taxonomists simply do not pay attention to the way scientific knowledge develops and changes within taxonomy and that taxonomic revisions are not generally considered mandatory elements within their own knowledge of biodiversity, whereas they play a central role within taxonomic knowledge of biodiversity.

ori with communities or ecosystems. With respect to spatial units, things are not better, as fragments of millimeters are as important as regional scales or even the whole surface of the Earth. The same is true with temporal units. Even though, from the theoretical point of view, all the involved disciplines are somehow unified by the theory of evolution, in practice, unification faces many obstacles due to the diversity of relevant units.

Besides being heterogeneous, the knowledge of biodiversity is also patchy in many ways. First, the conceptual links among taxonomy, phylogenetic studies, ecology, macro-ecology, biogeography, and evolution studies are not strong enough to provide biodiversity knowledge with firm theoretical structure (see for instance Leonelli 2009; Sarkar 2016). For instance, the relationships between ecology and evolution studies are notoriously difficult to assess,[2] besides other difficulties, like differences in temporal scales. Second, a bunch of other difficulties affect the development of biodiversity knowledge, like the large diversity of involved spatial and temporal scales, the difficulty to access certain zones, like deep sea, and the crude fact that certain taxonomic groups, like tunicates (marine invertebrates, sub-group of the Chordates), are much less known[3] that fish or crustacean decapods (Bouchet 2006).

The upshot of all this is that our knowledge of biodiversity, in its current state, is unable to provide biodiversity assessments and conservation policies with the firm epistemic ground that they might hope for. Besides being hindered by the awkward position of taxonomy among the other disciplines of biology, biodiversity knowledge is heterogeneous and patchy whereas it should be as unified as possible because biodiversity is an object of knowledge that extends over the whole planet instead of being the object of a series of local, disconnected pieces of knowledge. The latter is testified by the very foundation of GBIF, the Global Biodiversity Information Facility, as described below. The existence of these difficulties forces biologists and conservationists to work at strategies of improvement. In the next section, we present how such improvement may be achieved.

3.2.2 Improving Our Knowledge of Biodiversity via Cyber-Infrastructures

In order to improve our knowledge of biodiversity, it is first necessary to establish what its vehicles are. By "vehicles", we mean the devices that are used by researchers to acquire and develop biodiversity knowledge. Among these vehicles, scientific papers play a major role, but they are far from being the only way to build up and

---

[2] The relationship between the most developed attempt to provide ecology with a theory, namely the Neutral Theory of Ecology (Hubbell 2001), illustrates this point: the Neutral Theory is not yet unified with evolutionary theory, according to its proponent himself.

[3] The fact that knowledge of taxonomic groups is not uniformly spread is a major problem for biodiversity assessment and conservation (see e.g. http://www.iucnredlist.org/about/summary-statistics).

complement biodiversity knowledge. Expert reports and outcomes of inventories and assessment tasks are also among the means that researchers can rely on, as well as gene sequences and specimens in natural history collections.

Inventories, gene sequencing, collective scientific expertise, results of assessment endeavours, and collection management are key vehicles of biodiversity knowledge. As most of their outcomes are made available as "data" within databanks, databanks are indispensable vehicles of biodiversity knowledge as well. The "data" they contain are immensely diverse, from gene sequences to species descriptions, taxonomic revisions, geographical localisations, etc. "Data" is thus an ambiguous term that has to be further analyzed. We do so at the end of this section. In the mean time, we put forward a brief history of biodiversity databanks, which is part of the history of cyber-infrastructures Bastow and Leonelli (2010) have analyzed. These authors rightly emphasize that "databases and other online resources have become a central tool for biological research". Hereby, we present some historical elements relative to biodiversity databanks and emphasize their specific international features.

3.2.2.1   A Brief History of Biodiversity Databanks

Less than 10 years after the ratification of the Convention on Biological Diversity (CBD),[4] several databanks were created, whose main objective was to collect geographic and taxonomic data. Among these, we will focus our discussion on the ones relative to the marine field. They are the result of intense collective work during the 1990s aiming at the standardization and organization of data types. We present the development of this collective work in Fig. 3.1. Figure 3.2 shows its results.

The very first databanks have been devoted to taxonomical classification, like the Integrated Taxonomic Information System (ITIS), itself derived from the National Oceanographic Data Center, a former databank from US NOAA (National Oceanographic and Atmospheric Administration). The Taxonomic Database Working Group (TDWG), which first worked on plants, then had an important role in the elaboration of data standards, as well as the bioinformatics working group called "Global Initiative Taxonomy" (OECD 1999). The Global Initiative Taxonomy was the first step toward the creation of the GBIF databank, a few years later (Wieczorek et al. 2012).

Let us now review some outputs of the collective work aimed at the constitution of biodiversity databanks. First, it must be emphasized that the taxonomic impediment (here presented in all its urgent details: https://www.cbd.int/gti/problem.shtml) is a crucial problem facing the development of biodiversity databanks: taxonomists are too few and too slow to cope with the urgency of the biodiversity crisis and do not manage to catch up with extinction rates, thus leaving many extinct species unnamed and un-described. In the 1990s, in the same period in which standard-

---

[4] https://www.cbd.int/convention/text/ (On the CBD, see also Oksanen and Vuorisalo, Chap. 21, in this volume).

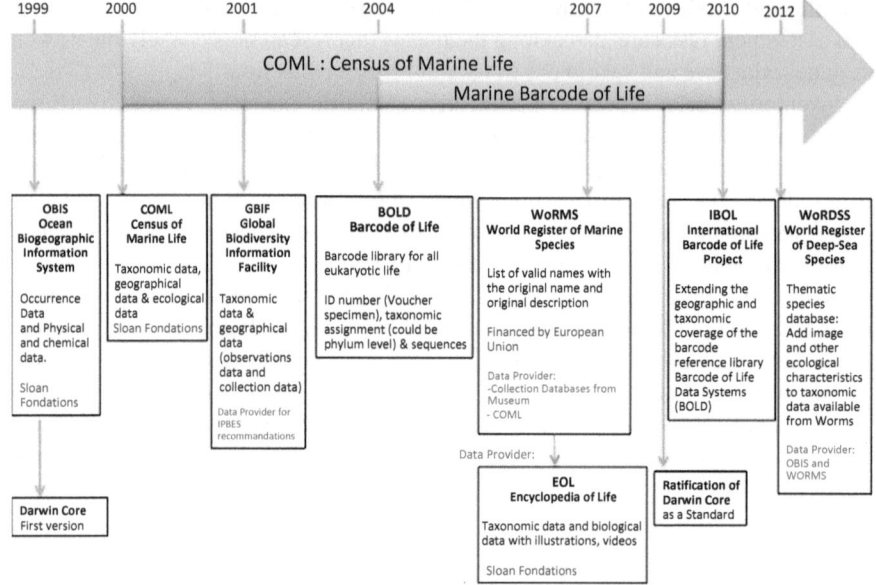

Fig. 3.1 Before databanks: collective, scientific work

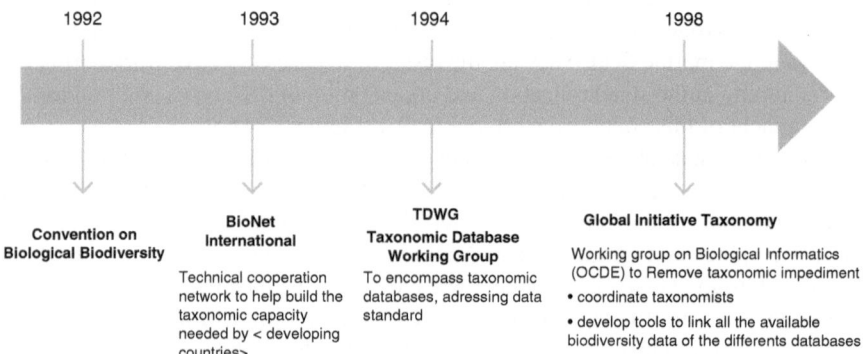

Fig. 3.2 Main databanks used for the assessment of the marine biodiversity. In red: funding sources; in purple: location of data

ization of biodiversity data was occurring, Bionet International (https://www.uia.org/s/or/en/1100052951), a technical cooperative network for taxonomy, was launched in order to foster exchanges of taxonomic knowledge among different countries and help facing the taxonomic impediment. This was the pre-condition for developing unified knowledge by trying to overcome the heterogeneity of taxonomic expertise among countries.

Against this historical background, it is important to emphasize that biodiversity knowledge relies on the development of genetic as well as taxonomy-based databanks. The most important genetic databank is GenBank (see below). It is commonly used as a resource for biodiversity knowledge and assessment even though its primary goal is not biodiversity-oriented.

By contrast, BoLD (Barcoding of Life Databank) focuses on the link between genetic sequences and taxonomic practices. It aims at being part of a genetic tool for quick species identification. BoLD, and its marine component MarBol, complement traditional taxonomy-oriented databanks like Ocean Biogeographic Information System (OBIS, http://www.iobis.org/), World Register of Marine Species (WoRMS, http://www.marinespecies.org/), and Encyclopedia of Life (EOL, http://eol.org/), which is not restricted to marine organisms. These databanks have been fueled by data gathered during the Census of Marine Life project (CoML 2000–2010) aimed at collecting data and providing researchers with bioinformatics tools. This was an international project gathering 2700 scientists who collected taxonomic, geographic, and ecological data from 540 oceanographic expeditions.

At the beginning, databanks focused on taxonomic classification and upgrading thereof, but they soon had to face a new challenge: connecting taxonomic data with data coming from other kinds of classifications, like genetic, biological and ecological classifications. For instance, biological classifications deliver data like attributes of life stages, reproduction, body size, behavior, feeding method, and diet (Costello et al. 2015). Establishing easy-to-retrieve connections among taxonomic data and biological or ecological traits is an important means to providing conservation biologists with indicators of the not-well-being of ecosystems (e.g., impact of pollution, of fishery, of climate change). To do so, some traits in ecosystems (like reproduction rates and features of habitats) must be described and named on a standardized basis. This cannot be achieved unless a robust consensus has been reached within the scientific community of both ecologists and taxonomists. Marine Species Traits (http://www.marinespecies.org/traits/) aims at the generation of these traits from taxonomic (WoRMS) and geographic databanks (OBIS), which requires and important work of coordination and terminological standardization, as emphasized by Costello et al. (2015). The latter recall that "a rich terminology surrounds descriptions of a species biology and ecology, with sometimes different definitions for the same terms, synonymous terms, and context dependent (e.g., habitat) terms. This terminology has developed over several hundred years of natural history, in different languages, and often terms have multiple meanings in common use." This requires databanks' designers to perform scientifically-informed terminological regimentation. The example of Marine Species Traits illustrates how biodiversity databanks, first developed by taxonomists, then connected with genetic databanks, now tend to diversify in order to account for other aspects of biodiversity. This tendency is however difficult to implement because elements of knowledge are much less standardized in ecology than they are in taxonomy and in molecular genetics.

3.2.2.2 Biodiversity Cyber-Infrastructures

How do the biodiversity databanks we have mentioned so far play contribute to the development of biodiversity knowledge? First, they make data easily accessible. Second, they allow for data being interoperable in a sense to be discussed below.

More generally, they organize the vast amounts of data that are relevant to biodiversity study. We discuss and illustrate this three aspects in the following.

Let us begin with accessibility of data. As already emphasized by Leonelli (2010), accessibility of data through internet-based databanks is a major precondition for knowledge-building. However, common accessibility, namely, the capacity of a piece of data to be easily retrieved at any scientific institution in the world, without overcoming outrageous paywalls, is not enough to define a *useful* biodiversity databank. Common accessibility is just the baseline condition of any useful scientific databank. We define scientific "usefulness" in this context as the capacity of databanks to facilitate and accelerate the work of researchers and increase the validity of their results. This cannot only be done by gathering data; *organizing* data of various origins and nature is instrumental. This involves links *among*, not only within databanks. Let us emphasize what organizing data and data flows means. Biodiversity databanks, considered collectively, are not as efficient as they could be when information is scattered and when data come from heterogeneous origins. A good biodiversity databank thus should provide researchers with a unique entry to a variety of types of data, e.g., genetic and taxonomic data from various geographical zones. We may illustrate this point with sampling-event data (https://www.gbif.org/sampling-event-data). They report the presence of an organism of such or such species (usually rare or endangered, but not always) together with its spatiotemporal location. These taxonomic, geographical, and temporal data may be provided by amateurs or professional taxonomists. Multiple programmes contribute to produce sampling-event data but they do so for different purposes, some of which might be related with conservation efforts, others with the development of ecological models or the study of migrations and the effects of climate change. This heterogeneity of purpose may generate confusion as to how these data should be stored and used. A scientifically useful sampling-event databank (in our sense) should organize access to information and information flow in such a way as to diminish heterogeneity of origin and facilitate integration of data into more structured pieces of knowledge.

We now turn to interoperability, which can be defined as "the ability of two or more systems or components to exchange information and to use the information that has been exchanged" (Covitz 2004, quoted by Leonelli 2013a). The increased number of databanks involved the ratification of data standards in order to facilitate interoperability among databanks. This allows any individual databank to function as data provider for other databanks, thus facilitating accessibility. The major requirement for interoperability is that databank designers organize data along common rules: in 2009, the Darwin Core version, which defines minimum standard data (with glossary and synonymy) related with biodiversity, has been internationally adopted for that purpose.

To sum up, accessibility, organization of information, and interoperability contribute to improving the usefulness of biodiversity databanks.

3.2.2.3   What Are Data in Biodiversity Databanks?

Databanks are not just (organized) collections of data, but also participate in the very definition of what may count as data in the knowledge of biodiversity, a point already made by Leonelli (2013b) about the field of plant science. As emphasized above, different elements are called "data" in biodiversity databanks, like gene sequences, species descriptions, taxonomic revisions, and geographical information. What do all these elements have in common? For what reasons can they be defined as "data" and qualify for being included in databanks? We shall present and discuss the following working hypothesis about data in biodiversity databanks: what they have in common is not that they are all basic or fundamental in the same sense (this would correspond to a definition of "data" as intrinsically basic pieces of information), but that they may play the same epistemic role: they can be relied upon in the further steps of a scientific inquiry. This corresponds to a *functional* analysis of data. In the remaining of this section, we argue in favor of this hypothesis by comparing species descriptions and gene sequences when they are both categorized as "data".

Let us first indicate easy-to-notice differences between species descriptions and gene sequences. On the one hand, the respective situations of species descriptions and gene sequences among life sciences are utterly different. Whereas many other pieces of biological knowledge depend on species descriptions, the use of gene sequences within the process of knowledge production depends on other elements of knowledge that may be found in genomics, proteomics, the study of gene regulation, phylogenetic history, etc. However, despite this difference, both species descriptions and gene sequences, once they have been validated as *bona fide* data, can be considered a sound floor on which one can step in order to go on and explore further research topics. This functional way of understanding data may be contrasted, in both cases, with the view according to which data are defined by their simplicity or easiness of acquisition. Neither gene sequences nor species descriptions are simple or easy to acquire. They are both issued from complex processes. First, a gene sequence is the result of an interpretative judgment with respect to the result of a biochemical experience; the judgement is about which nucleotides appear in the sequence and their order. Second, when a gene sequence is integrated in a databank, for instance in BoLD, where it is associated with a species name (which can be temporary), researchers may also provide the file containing information about the relevant genetic material, the Polymerase Chain Reaction (PCR) primers used to generate the sequenced amplicon, the identifier of the specimen, as well as the collection record, i.e., the location of the original specimen in a collection of natural history. These items may help other scientists check whether the sequence has been associated with the right species name. They illustrate how complex the transformation of a gene sequence into useful data is. On the side of species description, it should be emphasized that associating a specimen with a species name is also the result of a complex process of hypothesis assessment. This association itself possesses a hypothetical status, as it can be changed (via taxonomic revision) when a new set of characters is taken into account or when new specimens are collected. The information-processing facilities that are currently operated within databanks allow databank designers to create links between a specimen and the various taxonomic hypotheses (species names) that have been associated with it over time. In the marine field, this revision process can be followed in the literature and more easily in the WoRMS databank (see below).

The above shows that biodiversity data, the components of biodiversity databanks, are not called "data" because of their simplicity or because they are easily obtained. They are *bona fide* data thanks to the robust scientific processes on which their production relies. This means that even though some of them require complex material devices for their generation or years of scientific education, these processes are judged reliable enough to be bracketed as the inquiry develops. To put it in another way, the components of biodiversity databanks are so firmly established that, even though they do have a hypothetical character, as any item of knowledge within empirical science, this hypothetical character can be ignored as far as we know. For sure, they do not have the same status as proven mathematical theorems, but they are sufficiently well established to count as firm grounds for knowledge production.

For all the above-presented reasons: international effort of standardization, accessibility, interoperability, robust processes of data production, databanks appear as an efficient way to overcome the discrepancy between the reality of biodiversity knowledge, which is heterogeneous and patchy, and the hope that it may become more and more homogeneous and united. In the next part, we present how existing biodiversity databanks are used by scientists in order to make clear in which ways the organizational logic of databanks relates to the dynamics of knowledge development.

3.3   Uses of Biodiversity Databanks

This section is devoted to studying the various ways scientists, as distinguished from conservation practitioners and policy makers, use biodiversity databanks in order to develop biodiversity knowledge. We try to disclose the requirements useful databanks have to fulfill and how the data they include are operated in the production of knowledge. In Sect. 3.3.1, we describe, based on examples, what scientists do with the data they retrieve from databanks as well as the quick evolution of this scientific practice. In Sect. 3.3.2, we systematically compare catalogs and databanks in order to explore the specificities of the relationships between a databank and its expected users. At last, we try to show the underlying organizational principles of biodiversity databanks and how they may foster the evolution of scientific knowledge.

3.3.1   What Do Scientists Do with the Data They Retrieve from Biodiversity Databanks?

As explained above, "data" in biodiversity databanks are already complex units of knowledge that are used to work out other types of scientific results, usually more general and systematic. These data are used to express general hypotheses that cannot be formulated unless two important features of data in biodiversity databanks are realized: they have (i) to cover large geographical regions or large taxonomic groups and (ii) to be valid. Scientists interested in biodiversity study, assessment or conservation may be familiar with a taxonomic group or geographical area and have personal estimates of species abundance in this group or of biodiversity in this area; however, they cannot only rely on their personal experience to put forward general hypotheses and submit them to empirical test. Usually, personal experience, however valuable, is not robust enough to allow for hypothesis testing. By contrast, data in databanks possess the quality that personal connection with biodiversity will always lack: they have been validated by the scientific community, and as such, as explained in Sect. 3.2, they can be relied upon to explore new, more general hypotheses and build up quantitative models.

In order to illustrate, first, how databanks provide researchers with valid pieces of knowledge that they can rely upon and second, the quick transformations of this practice, we shall now present a recent databank that has been created at the Paris Muséum National d'Histoire Naturelle. It is called "BasExp": Databasis for scientific Expeditions (https://expeditions.mnhn.fr/) and has been designed to gather data related to a 40-year-long programme of marine expeditions initiated by Paris Muséum National d'Histoire Naturelle and the Institut de Recherche pour le Développement, first called "Musorstom", then "Tropical Deep Sea Benthos". BasExp collects scientific papers, monographs, and reports issued from this programme. It combines information from these papers, books and reports with data relative to the collected specimens that are preserved at the Paris Muséum. It also collects information about the marine expeditions themselves (not only about the scientific information they have contributed to establish), like who was on board, the main objectives of the expedition, its location, origin of funding, sampling sites, quantity of associated publications, etc. Its main purpose is to allow scientists to overcome two major biases affecting the study of biodiversity, the taxon sampling bias, and the geographic sampling bias, by allowing researchers to know more about the context in which specimens have been sampled. The taxon sampling bias is the tendency to focus on a particular group of organisms and ignore organisms from other groups. The geographical sampling bias is the tendency to go again and again in the same geographical areas to sample specimens instead of exploring other areas. For instance, information about the various researchers on board (specialists of fish, of crustaceans, etc.) may reveal why some taxa were more extensively collected (and on the contrary, the absence of any specialist of a given taxon on board, in a given expedition, may explain why no specimen of this taxon was collected). In a similar way, its being funded by the fish industry might explain the over-representation of fish specimens in another expedition, and so on. There is no doubt that it is important to take these factors into account when looking for biodiversity patterns. As BasExp provides researchers with key elements of the context of sampling, it may be an adequate tool to avoid these common biases.

We now turn to the context in which BasExp has been created. This will shed more light on its purpose and potential benefits for researchers. Most first-generation biodiversity databanks have emerged as answers to inventory requirements. The practice of biodiversity inventory is characterized by its being static in the sense that it is blind, by nature, to changes in biodiversity. Inventories may now be considered of limited interest for biodiversity knowledge because changes in biodiversity are currently a major epistemic challenge, either with respect to conservation or in order to assess the effects of climate change. By contrast with inventory-based databanks, more up-to-date databanks aim at tracking biodiversity change. The best way to do so is by connecting several databanks together in order to be able to follow the evolution of spatiotemporal data in as much details as possible. BasExp allows for such connections.

Among other changes in biodiversity databanks, another one is worth mentioning: they are currently evolving toward less taxa-centered architectures. Many databanks, especially those related with natural history museums, are organised along taxonomic group divisions: one databank for flowering plants, one for crustaceans, etc. However, as emphasized above, there is an increasing need to access a synthetic representation of biodiversity that overcomes the intrinsic limits of taxa-oriented databanks. As biodiversity has to be captured according to many different aspects (geographical, dynamical, taxonomic, genetic, etc.) at once, some databanks offer scientists the means to question their data according to several criteria. BasExp nicely illustrates this possibility. In particular, as there has been a huge effort within BasExp to homogenize geographical data about sampling locations, which were difficult to find out and exploit in the past, it provides researchers with a new and long-awaited type of ready-to-use information. Outside BaxExp, geographical information varies in format and degree of precision from one collection to the next within the Paris Muséum. Gathering all information on a given location and standardizing its format is thus an important advance in itself. Moreover, before the establishment of BasExp, each collection databank had its own data system: they did not use the exact same names for expeditions and did not have the same degree of precision for geographic location. By now, BasExp is the geographical data provider for all the Paris Muséum's collection databanks. Because it is not taxa-centered, BasExp provides researchers with a synthetic representation of what has been studied and what remains to be investigated about deep-sea fauna.

3.3.2   Databanks vs. Catalogs

In this section, we put forward a systematic comparison between library or collection catalogs and databanks in order to make clear what the specific features of databanks are from the users' point of view. We shall show that far from being improved catalogs, allowing for gain in time, databanks are flexible tools that play new roles in the development of biodiversity knowledge.

An important difference between databanks and catalogs is that databanks allow for more than one guiding principle with respect to organisation of information. Let us first explain what we mean by a "guiding principle" with respect to organization of information. In a museum of natural history, catalogs are usually designed according to the way specimens' identifications are produced. By contrast, an internet-connected databank may be organised according to several guiding principles: specimens' names, geographical localization, gene sequence, date of discovery, name of discoverer, etc. The organisation of information along multiple dimensions, all of them of scientific interest, is an efficient way to disclose connections that remain inaccessible to catalog users. A nice illustration of this important feature of databanks is that whereas the *absence* of a species at some place cannot be inferred from consulting a catalog, it may be discovered by cleverly questioning relevant databanks. The main reason for that difference is that catalogs based on specimen identifications cannot but register the presence of specimens without leaving any opportunity to discover information about absence, whereas a databank organised by geographical location may disclose information about absence.

Let us mention another difference between databanks and catalogs, which revolves about their users. Usually, the users of a catalog are determined *before* its implementation. For instance, the users of the catalog of a museum's collection are often meant to be people working at the museum, most of them taxonomists: catalogs are most devised for local use. By contrast, internet-based databanks are usually used by different users, even more so when they are inter-connected by means of an Information System, namely, a network of devices for the acquisition, organization, storage, and communication of information that is developed and managed by the host institution. The users' variety forces databanks' designers to conceive the organisation of information in such a way as to push the boundaries of local use. The needs and interests of local users of a catalog, e.g., the members of a museum's scientific community, are more easily identified and narrower than the needs and interests of external users. Taking the latter into account forces databanks' designers to introduce new possibilities of investigation, e.g., new query types or combinations. This is a very difficult task indeed, as emphasized by Leonelli in the case of biomedical databanks: "[i]ncorporating a large variety of possible viewpoints and prospective queries has been, and continues to be, the most complex and labour-intensive task involved in the development of [the databanks]" (Leonelli 2013a). It amounts to try to guess what the new directions of research may be in order to make the databank usable and useful in the future. This anticipatory task can be said central in the conception and design of biodiversity databanks, whose role in the development of biodiversity knowledge will increase. In Sect. 3.3.3, we further explore the implications of the way information is organised within a databank by showing how their underlying organisational principles may foster the evolution of scientific knowledge.

3.3.3   Databanks' Organization and the Dynamics of Biodiversity Knowledge

The principles and functioning of databanks are too often ignored by philosophers of science, who tend to view them as black boxes, whereas looking inside allows one to discover valuable information about the way scientific knowledge is produced on a daily basis. This is why we aim at opening these black boxes and describe their internal functioning in order to show how information is acquired and transformed within them. We will illustrate our claims about the way data are typified and organised by a series of examples.

Each databank's organization obeys a dominant organisational law that is based on a specific type of information. For instance, as mentioned above, some databanks are centered on taxa whereas others are centered on geographical areas: these are examples of *types of information*. Other types of information include pictures, taxonomic papers presenting taxonomic descriptions, papers presenting taxonomic revisions, or genetic sequences. It is worth emphasizing that the notion of "type of information" we introduce here is defined with respect to the organisation of biodiversity databanks. For sure, pictures of specimens and taxonomic papers may contain information on the same organisms; however, the information contained in a picture will not play the same role in the process of knowledge-production as the information contained in a taxonomic paper. The databank user will not use a picture in the same way she uses a species description. This is why our notion of type of information is not defined with respect to the item in the world that the information is about, but with respect to the way the information is used by the databank's user.

Our notion of type of information allows us to establish a classification of biodiversity databanks into taxonomy-oriented, geography-oriented, picture-oriented, revision-oriented, etc. databanks. The type of information that associates a given databank to a class governs what we call the "organisational law" that defines a specific kind of links, within the databank, between the type of information that is the most important within it and other types for information. For instance, the organisational law of a taxonomy-centered databank relates taxonomic data with geographical data, data about endemism, genetic data, etc. The organisational law thus sets apart a center and a periphery within the databank. Center and periphery are defined relative to types of information, as defined above.

Let us make clear how this center-periphery organisation is implemented in various examples.

(i) We begin with WoRMS, the World Register of Marine Species (http://www. marinespecies.org/, already mentioned above). The organisational law for this databank connects original and current species names. As taxonomic revision is at the heart of taxonomic practice, it is highly important to keep track of revisions in order to avoid collective oversight, which would have devastating consequences. WoRMS contains, for each species name, the list of its revisions, dates, and names of associated taxonomists. The use of

WoRMS is name-oriented; it delivers information on the sequence of past revisions. It is an international databank. Peripheral information is, e.g., the geographical distribution of the species.

(ii) The OBIS geographical databank (Ocean Biogeographic Information System, http://www.iobis.org/) is sampling-event and geography-oriented, and specific to marine species; it also provides physical, chemical and topographic information on precise locations where specimens have been observed or sampled.

(iii) GBIF (Global Biodiversity Information Facility, http://www.gbif.org/) is a sampling-event databank mapping signalization of species (observation- or sample-event) all around the world.

(iv) By contrast with GBIF, let us mention INPN (Inventaire National du Patrimoine Naturel, https://inpn.mnhn.fr/accueil/index), which also maps signalization of species (observation- or sample-event) on the French territory, with an emphasis on landscape preservation.

(v) BoLD, already mentioned above, combines genetic sequences with information on individual specimens. It is important to emphasize the differences between BoLD and GenBank, another gene-centered databank (see (vi)). Contrary to BoLD, GenBank is centered on the link between genetic sequences and scientific publications. There is no link with actual specimens within GenBank. By contrast, the users of BoLD look for a genetic sequence and the associated species name and may check themselves, by performing standardized experiments, whether the stored genetic sequence (1) has been correctly established and (2) is associated with the right species name. The users of GenBank are not given this possibility because GenBank provides them with a link between a genetic sequence and a scientific paper; as a result, they cannot but trust the authors of the paper with respect to the validity of the gene sequence and associated species name. By contrast, the genetic sequences in BoLD may be not published (Ratnasingham and Hebert 2007). The difference between BoLD and GenBank illustrates an important division among biodiversity databanks: those that provide links to actual specimens, allowing data-checking by the users, and those that provide links to publications or sampling-events, which oblige the users to trust the original information providers.

(vi) The users of GenBank (https://www.ncbi.nlm.nih.gov/genbank/) are looking for comparisons between their own organisms of interest and model organisms whose genomes have been sequenced and included in the databank. It includes all the sequences that have been published in scientific papers.

(vii) Let us end this list with two examples from the Paris Muséum National d'Histoire Naturelle. COLLECTIONS is the databank of the Muséum's collections, whose functioning is explained here: http://collections.mnhn.fr/wiki/Wiki.jsp?page=Publication_Internet_en. Each museum has several collection databanks for fish, mollusks, crustaceans, etc. They are specimen-oriented: the users look for a specimen's number (called "voucher ID") or a name, in order to get information thereon (nomenclatural status: holotype, paratype, etc.; sample site; name).

(viii) BasExp: As mentioned above, this databank is expedition-oriented: the users look for information related with a given expedition. This databank allows for both non-taxa-oriented research and a global appraisal of an expedition in terms of sampling location, specimens, papers, and reports. It is the repository for geographical data related with expeditions.

Whereas each databank has its own organisational logic, many biodiversity databanks function collectively by means of a series of networks like GBIF. Networks of databanks obey their own laws that govern *data flow*. Usually, data flow runs from the most standardized and best validated data to lesser controlled databanks. Data flow is thus a powerful way of enhancing and homogenizing standards because downstream databanks, when they happen to become upstream relative to other, more local databanks, have to strengthen the reliability and accessibility of their data. Establishing a connection between two biodiversity databanks is indeed a common way to fill the gaps within each. It also contributes to improve data validation and warrant traceability: as emphasized by Costello and Vanden Berghe (2006), "[g]lobal databases that integrate information on species force the development of standard classifications".

Databank networking has a further important effect at a higher level of knowledge production: it makes clear that some areas of knowledge that had been considered independent beforehand actually entertain epistemic relationships that are now considered as pivotal. For instance, the links between gene sequencing and species identification, implemented in BoLD, has only been brought to light recently (compared with the long history of species identification and the less long, but still not so recent, history of gene sequencing). Realizing that gene sequencing may facilitate species identification has been a side-effect of developing genetic databanks and databanks networks. BoLD's strength, in this context, is to allow data-checking by providing links with actual specimens (which do not change over time), whereas other databanks depend on taxonomic descriptions, which are hypothetical and susceptible to be transformed.

Another important tendency in the evolution of the way biodiversity databanks get organised is the gradual expansion of the domain of relevant data. Many elements that were considered irrelevant for biodiversity knowledge are now emerging as constitutive, and thus worth collecting, taking care of, and connecting with other types of data. For instance, BasExp includes information on the extraction device of marine specimens, which it relates with sampling sites as well as with digital pictures illustrating the collected specimens in their substrate. The pictures shot during the expeditions have long been difficult to access by scientists who were not on board; now, their status as vehicles of knowledge has completely changed in the last few years as they became easily accessible within databanks and provide irreplaceable information about either the specimen's environment or the context of its extraction. BasExp thus nicely illustrates a trend that has been identified by Leonelli (2013a), who emphasizes that the databanks she examines (The Arabidopsis

Information Resource, TAIR and the cancer Biomedical Informatics Grid, caBIG) progressively include elements that were not considered important, like archives of data provenance (the methods and instruments originally used to generate data) and links to biological materials. We cannot but agree with Leonelli's statement that "setting up and updating these resources occupies much of curators' time and creative efforts".

A last point has to be made about the connection between the development of biodiversity databanks and the dynamics of biodiversity knowledge. It is about the pace of transformation. The different features of biodiversity databanks do not usually evolve at the same rate, which dictates a mandatory upgrading process. For instance, paper catalogs of large natural history museums have usually not been computerized at once, but rather step by step. Now, each step in this process of computerization took place within its own technological and scientific context, involving innovations that the next steps had to catch up with. Evolving biodiversity databanks is thus no linear process. Some events had large interfering consequences, like the establishment of the BoLD consortium (as described here: http://www.barcodeoflife.org/content/about/what-cbol). As soon as Barcoding of Life proved a useful and fruitful endeavour, each biodiversity databank had to take this program into account, which meant huge adaptive changes. Adapting databanks moreover involves satisfying basic requirements of cumulativeness, editability, and interoperability, as these requirements ensure homogeneous development.

In this section, we have presented and discussed the logic governing the organization of biodiversity databanks by means of examples. We have also emphasized the importance of their interconnection into networks and the interplay between their evolution and the changes within biodiversity knowledge. As transformation is pivotal in any reflection on databanks, we now turn briefly to the ways biodiversity databanks may improve in the future.

3.4   On the Properties of Useful Biodiversity Databanks: Concluding Remarks

Let us recall that, in this paper, we call a biodiversity databank "useful" when it provides scientists, managers of assessment programmes, and conservationists with the means to successfully achieve their aims. In this section, we present two types of requirements a biodiversity databank has to satisfy in order to be useful in this sense. The first type is more on the technical side (although it is not content-independent) whereas the second is linked to the distinguishing features of biodiversity knowledge.

As mentioned above, a few basic requirements have to be met in any databank that is meant to be used by scientists. First, data have to be standardized. Standardization involves defining different data types (i.e., building up a glossary) in order to optimize interoperability among databanks. We can now bring to light the peculiar difficulties raised by this operation by coming back to the example of the integration of non-taxonomic data in a biodiversity databank, namely, data coming from biological and ecological descriptions (cf. Sect. 3.2.2). In order to include biological and ecological traits within a databank for marine species, the main challenge is to identify which traits are (1) useful and (2) available to researchers and conservationists. This is the first step of data standardization, which obliges databank designers to struggle with linguistic subtleties:

> For example, "littoral" habitat can be the marine zone between the low and high tide marks, extend to the continental shelf and include coastal river catchments, and refer to the edge of freshwater lakes. The lack of standard use of terms can compromise the bringing together of this knowledge from different sources. (Costello et al. 2015).

We see in this example that standardization forces databank designers to formulate precise definitions of the terms they choose to use within the databank, as well as to determine measurement units. This is a crucial, non-trivial step in the elaboration of the databank. It is based on the previous identification of the set of terms that are used in other databanks and in the relevant literature. Standardization is not always possible, however: when the way data are produced remains unknown, i.e., when the databank designer does not know whether they are primarily found in scientific papers, reports, other databanks, or unpublished sources, they cannot possibly be standardized because standardization involves checking the validity of data. However, when data are from unknown origin, their validity cannot be properly checked. That is the reason why working groups for data standardization must involve scientists: standardization *is* knowledge production.

The most important requirement to build up a useful databank is that data have to be validated: they must be submitted to a process that provides scientists with the same kind of warrant as the peer review process so that they can trust the elements they retrieve from databanks. When checking the validity of data, users are assisted by the meta-data that provide them with contextual information allowing them to both evaluate reliability and quality of data. Meta-data are sets of data that accompany the constitutive data of a databank: for instance, in a taxonomy-oriented databank, the constitutive data may be species descriptions and associated meta-data geographical information, gene sequences, pictures, location of holo- and paratypes, etc. When links to actual specimens are provided, databank users can moreover check themselves the validity of the data they are interested in. These links are thus an important way to enhance the collective process of data validation, and, as a result, the overall scientific quality of biodiversity data and databanks.

The usefulness of a databank is conditioned by the basic requirements so far illustrated and by the definition of possible queries. By listing the words or expressions that constitute well-defined queries, databank designers identify and delineate the possible uses of the tool. As mentioned above, the databank designers' role is to guess the future uses of the databank and to anticipate the developmental paths of biodiversity knowledge. This can only be done by people in close contact with on-going research, assessment, and conservation practices: even though it may not always result in the publication of scientific papers, databank design and maintenance is genuine scientific work because it implies being well aware of the scientific state of the art at a given time, of how the relevant scientific community assesses the various scientific hypotheses at stake, and of the emerging links among different fields of research.

Even though biodiversity databanks already have a history, they are recent tools compared to the old practice of publication of papers and monographs. A major challenge for biodiversity databanks is to become a part in a set of older practices that themselves transform at a quick pace. This means that biodiversity databanks should connect their contents with other, older epistemic practices. More precisely, they should complement, rather than replace, more traditional reservoirs of biodiversity knowledge, like collections of natural history, because reliable biodiversity knowledge requires links with actual specimens. This is an important and difficult task for databanks designers and associated computer scientists and engineers.

References

Baetu, T. M. (2012). Genes after the human genome project. *Studies in History and Philosophy of Science Part C: Studies in History and Philosophy of Biological and Biomedical Sciences, 43*, 191–201.

Bastow, R., & Leonelli, S. (2010). Sustainable digital infrastructure. *EMBO Reports, 11*(10), 730–735.

Bouchet, P. (2006). The magnitude of marine biodiversity. In C. M. Duarte (Ed.), *The exploration of marine biodiversity. Scientific and technological challenges* (pp. 31–62). Bilbao: Fundacion BBVA.

Carlson, E. A. (1991). Defining the gene: An evolving concept. *The American Journal of Human Genetics, 49*, 475–487.

Costello, M. J., & Vanden Berghe, E. (2006). Ocean biodiversity informatics: A new era in marine biology research and management. *Marine Ecology Progress Series, 316*, 203–214.

Costello, M. J., Claus, S., Dekeyzer, S., Vandepitte, L., Tuama, É. Ó., Lear, D., & Tyler-Walters, H. (18 août 2015). Biological and ecological traits of marine species. *PeerJ, 3*, e1201. https://doi.org/10.7717/peerj.1201.

Covitz, P. A. (2004). *Cruising the cancer biomedical informatics grid caBIG: From village to city* (caBIG Workspace and Working Group Kickoff meeting, 2004).

Falk, R. (1986). What is a gene? *Studies in the History and Philsophy of Science, 17*, 133–173.

Fogle, T. (2000). The dissolution of protein coding genes in molecular biology. In P. Beurton, R. Falk, & H.-J. Rheinberger (Eds.), *The concept of the gene in development and evolution. Historical and epistemological perspectives* (pp. 3–25). Cambridge: Cambridge University Press.

Gerstein, M. B., Bruce, C., Rozowsky, J. S., Zheng, D., Du, J., Korbel, J. O., Emanuelsson, O., Zhang, Z. D., Weissman, S., & Snyder, M. (2007). What is a gene, post-ENCODE? History and updated definition. *Genome Research, 17*, 669–681.

Grehan, J. R. (1993). Conservation biogeography and the biodiversity crisis: A global problem in space/time. *Biodiversity Letters, 1*(5), 134–140.

Hubbell, S. P. (2001). *The unified neutral theory of biodiversity and biogeography.* Princeton: Princeton University Press.

James, S. W., Porco, D., Decaëns, T., Richard, B., & Rougerie, R. (2010). DNA barcoding reveals cryptic diversity in *Lumbricus terrestris* L., 1758 (Clitellata): Resurrection of L. herculeus (Savigny, 1826). *PLoS One, 5*(12), e15629. https://doi.org/10.1371/journal.pone.0015629.

Kitcher, P. (1982). Genes. *British Journal for the Philosophy of Science, 33*, 337–359.

Leonelli, S. (2009). The impure nature of biological knowledge. In H. de Regt, S. Leonelli, & K. Eigner (Eds.), *Scientific understanding: Philosophical perspectives*. Pittsburgh: Pittsburgh University Press.

Leonelli, S. (2010). The commodification of knowledge exchange: Governing the circulation of biological data. In H. Radder (Ed.), *The commodification of academic research*. Pittsburgh: University of Pittsburgh Press.

Leonelli, S. (2013a). Global data for local science: Assessing the scale of data infrastructures in biological and biomedical research. *BioSociety, 8*(4), 449–465.

Leonelli, S. (2013b). Integrating data to acquire new knowledge: Three modes of integration in plant science. *Studies in History and Philosophy of Biological and Biomedical Sciences, 44*(4 Pt A), 503–514. https://doi.org/10.1016/j.shpsc.2013.03.020.

Leonelli, S. (2016). *Data-centric biology. A philosophical study*. Chicago: The University of Chicago Press.

OECD. (1999). *OECD megascience working group – Biological informatics – Final report*. 74 pp. Organisation for Economic Co-operation and Development. Available online at http://www.oecd.org/dataoecd/24/32/2105199.pdf

Olson, D. M., Dinerstein, E., Powell, G. V. N., & Wikramanayake, E. D. (2002). Conservation biology for the biodiversity crisis. *Conservation Biology, 16*(1), 1–3.

Ratnasingham, S., & Hebert, P. D. N. (2007). BOLD: The barcode of life data system (http://www.barcodinglife.org). *Molecular Ecology Notes, 7*(3), 355–364.

Richard, B., Decaëns, T., Rougerie, R., James, S. W., & Porco, D. (2009). Re-integrating earthworm juveniles into soil biodiversity studies: Species identification through DNA barcoding. *Molecular Ecology Ressources, 10*, 606–614.

Sarkar, S. (2016). *Ecology. The stanford encyclopedia of philosophy*. In E. N. Zalta (Ed.). https://plato.stanford.edu/archives/win2016/entries/ecology/

Singh, J. S. (2002). The biodiversity crisis: A multifaceted review. *Current Science, 82*(6), 638–647.

Sterelny, K., & Kitcher, P. (1988). The return of the gene. *Journal of Philosophy, 85*, 339–360.

Takacs, D. (1996). *The idea of biodiversity: Philosophies of paradise*. Baltimore: Johns Hopkins University Press.

Waters, K. C. (1994). Genes made molecular. *Philosophy of Science, 61*, 163–185.

Western, D. (1992). The biodiversity crisis: A challenge for biology. *Oikos, 63*(1), 29–38.

Wieczorek, J., Bloom, D., Guralnick, R., Blum, S., Döring, M., Giovanni, R., Robertson, T., & Vieglais, D. (2012). Darwin core: An evolving community-developed biodiversity data standard. *PLoS One, 7*(1), e29715.

Chapter 4
Problems and Questions Posed by Cryptic Species. A Framework to Guide Future Studies

Anne Chenuil, Abigail E. Cahill, Numa Délémontey, Elrick Du Salliant du Luc, and Hadrien Fanton

Abstract Species are the currency of biology and important units of biodiversity, thus errors in species delimitations potentially have important consequences. During the last decades, owing to the use of genetic markers, many nominal species appeared to consist of several reproductively isolated entities called **cryptic species** (hereafter CS). In this chapter we explain why CS are important for practical reasons related to community and ecosystem monitoring, and for biological knowledge, particularly for understanding ecological and evolutionary processes. To find solutions to practical problems and to correct biological errors, a thorough analysis of the distinct types of CS reported in the literature is necessary and some general rules have to be identified. Here we explain how to identify CS, and we propose a rational and practical classification of CS (and putative CS), based on the crossing of distinct levels of genetic isolation with distinct levels of morphological differentiation. We also explain how to identify likely explanations for a given CS (either inherent to taxonomic processes or related to taxon biology, ecology and geography) and how to build a comprehensive database aimed at answering these practical and theoretical questions. Our pilot review of the literature in marine animals established that half of the reported cases are not CS *sensu stricto* (i.e. where morphology cannot distinguish the entities) and just need taxonomic revision. It also revealed significant associations between CS features, such as a higher proportion of diagnostic morphological differences in sympatric than in allopatric CS and more frequent ecologi-

A. Chenuil (✉) · H. Fanton
Aix Marseille Univ, Avignon Université, CNRS, IRD, IMBE, Marseille, France
e-mail: anne.chenuil@imbe.fr; hadrien.fanton@imbe.fr

A. E. Cahill
Aix Marseille Univ, Avignon Université, CNRS, IRD, IMBE, Marseille, France

Albion College, Albion, MI, USA
e-mail: acahill@albion.edu

N. Délémontey · E. D. Salliant du Luc
Aix-Marseille University, Marseille, France

E. Casetta et al. (eds.), *From Assessing to Conserving Biodiversity*,
History, Philosophy and Theory of the Life Sciences 24,
https://doi.org/10.1007/978-3-030-10991-2_4

cal differentiation between sympatric than allopatric CS, both observations supporting the competitive exclusion theory, thus suggesting that ignoring CS causes not only species diversity but also functional diversity underestimation.

4.1 Introduction

Quoting Agapow et al. (2004) "species are the currency of biology". Long before the term "biodiversity" was coined and became widespread, the category of species was used as a major unit or category, not only to classify living things, but also to study ecological interactions and to assess the composition, resilience, evolution and risk of collapse of ecosystems (Gotelli and Colwell 2001). Nearly all descriptors of community assemblages and ecosystems - and their derived ecosystem functioning descriptors - require counting and separating species. The data used may contain variable amounts of information: (i) species richness (i.e. simply the number of distinct species), (ii) abundances of each species, (iii) relatedness among species and/or (iv) functional traits of species (Beauchard et al. 2017). The formal system naming the distinct species, established by Linnaeus in the eighteenth century, is the binomial nomenclature. The entities in the binominal nomenclature are called "nominal species" and are identified by a pair of Latin names, the first one corresponding to the genus to which the species belongs (*e.g. Homo sapiens*). Nominal species were described and defined exclusively from morphological characters until very recently, and are therefore sometimes called "morphospecies". Nominal species (or groups of nominal species) were the entities considered in all the inventories of multicellular life until only a few years ago. During the last decades however, numerous nominal species appeared to be composed of separate entities which could not interbreed (Fig. 4.1), i.e. genetically isolated units. Genetic isolation for a group of individuals is the inability of its members to breed successfully with individuals from another group due to geographical, behavioral, physiological, or genetic barriers or differences. When genetic isolation is not the mere consequence of an external constraint such as geographic separation, but inherent to behavioral or genomic incompatibilities, such units constitute, by definition, distinct *biological species* (Mayr 1942). The expression "cryptic species" (hereafter CS) designates the distinct biological species that belong to one given nominal species and which were overlooked by the taxonomists who described the species initially (Knowlton 1993). This is generally, though not always, due to the absence of conspicuous diagnostic morphological differences (i.e. characters whose states allow unambiguous discrimination between species). In this chapter, "cryptic species *sensu lato*" correspond to distinct biological species within a nominal species, whatever the morphological differences or knowledge thereof. We define cryptic species *sensu stricto* as those CS where the absence of diagnostic morphological characters has been verified (and below we further explain the need to distinguish more categories of CS or putative CS). Similarly to CS (*ss* and *sl*), but less restrictedly, we define **cryptic genetically isolated units** (CGI) (*ss* and *sl*) as entities that appear to be reproductively isolated *in fact* but which may potentially interbreed following range extension or after the disappearance of a geographical barrier

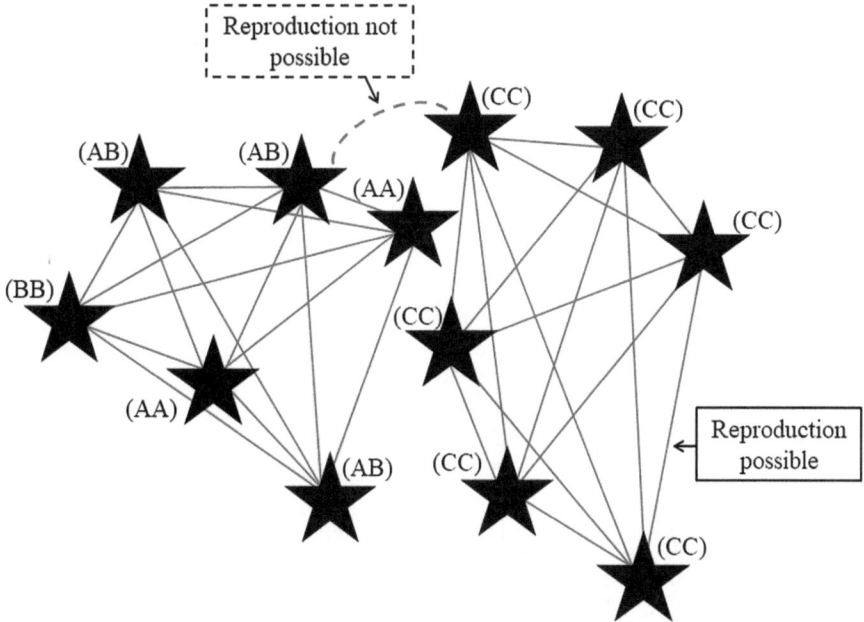

Fig. 4.1 One nominal species composed of two cryptic species: 12 individuals are represented by identical black stars (to illustrate their belonging to the same nominal species). Thin blue lines join all pairs of individuals that could potentially reproduce together and which thus belong to the same biological species. The curved dashed line joins two reproductively incompatible individuals (not all such cases are represented for clarity). Since there are two biological species, the nominal species is indeed a complex of two cryptic species (until a taxonomic revision eventually creates two nominal species). In parentheses are the individual genotypes at a codominant diagnostic locus (cf Sect. 4.3)

(Table 4.1). CS are particular cases of CGI but the problems and questions posed by CS and by the CGI that are only extrinsically isolated are essentially identical. Many reported CS (or putative CS) in the literature are indeed CGI (or *putative* CGI). We also consider *putative* CS and *putative* CGI, which are cases where the proof of genetic isolation is lacking although data suggest it may exist, because such cases are numerous and, generally, genetic isolation is confirmed when genetic information is supplemented with other types of data (cf. Sect. 4.3).

Putative CGI are being identified at an increasing rate owing to the development of genetic tools (Bickford et al. 2007; Fišer et al. 2018; Pfenninger and Schwenk 2007). Particularly in the marine realm, CS (*a fortiori* CGI) may be the rule rather than the exception ((Knowlton 1993); a seminal paper cited about 1000 times and (Nygren 2014)). One of the first marine species for which a whole genome was sequenced, the ascidian *Ciona intestinalis*, is indeed a complex of cryptic species (Nydam and Harrison 2011; Roux et al. 2013) that diverged particularly anciently (more than 10 Ma) and coexist in various regions of their distribution ranges. Interestingly, the fact that there were CS in this nominal species was ignored during the genome sequencing project and for many years despite the fact that this species was already the subject of numerous costly investigations. Our goal in this

Table 4.1 Classification of types of CGI (including putative cases) based on available knowledge and crossing the genetic isolation (GI) criteria (rows) and the morphological differentiation (MD) criteria (columns). The lower and isolated row does not belong to the classification itself but illustrates the possible causes of the origin of CGI. "BS" stands for biological species

MD \ GI		No morphological polymorphism	No morphological differentiation	Statistical morphological differentiation	Diagnostic morphological differences
		0	1	2	3
Biological Species	A	CS *sensu stricto*	CS *sensu stricto*	CS *sensu stricto*	CS *sensu lato* Taxonomic revision should define new species
Genetic Isolation	B	CGI *sensu stricto*	CGI *sensu stricto*	CGI *sensu stricto*	CGI *sensu lato* Should be considered at least as sub-species
Putative Genetic Isolation	C	Independent markers needed (genetic or not) to assess GI Many such cases in the "cryptic species" literature			
No genetic differentiation	D	(Not relevant)	(Not relevant)	(Not relevant)	(Not relevant)
Some possible causes (cf. section 4.4)		Stabilizing selection	Recently diverged entities	High effective sizes Advantageous morphological polymorphism	Poor taxonomy

chapter is not to participate in the debate about species concepts but to highlight problems (practical) and questions (theoretical) raised by the existence of CGI, with a particular effort to clarify the variety of causes generating CGI and CS and the features of CGI and CS that are useful to identify in order to explain their origins.

We will thus explain (i) why it is important to take CGI (and in particular CS) into account (identifying practical problems related to the assessment of biodiversity and ecosystem functioning, and theoretical problems for the understanding of community dynamics, biological evolution, etc.), (ii) how to detect CS or CGI (which is a dual task, implying both the distinction of biological species or genetically isolated entities and the characterization of morphological differentiation), (iii) how to correct inferences that are faulty due to CGI, and how to predict CGI occurrences and characteristics, which are similar questions that both require understanding of the factors responsible for the occurrence of CGI. These factors include human factors related to science history, and biological factors, such as the geographical distribution, habitat and life history traits of the species. Finally, we will present the results of a preliminary survey of the literature on marine species.

4.2 Why It Is Important to Recognize Cryptic Species

CGI and particularly CS challenge biodiversity estimations and, potentially, biodiversity management in several important ways. Figure 4.2 illustrates some of the consequences of (ignoring) CS on fundamental biodiversity parameters. What matters is that when these parameters are erroneous, the estimation of vulnerability (of

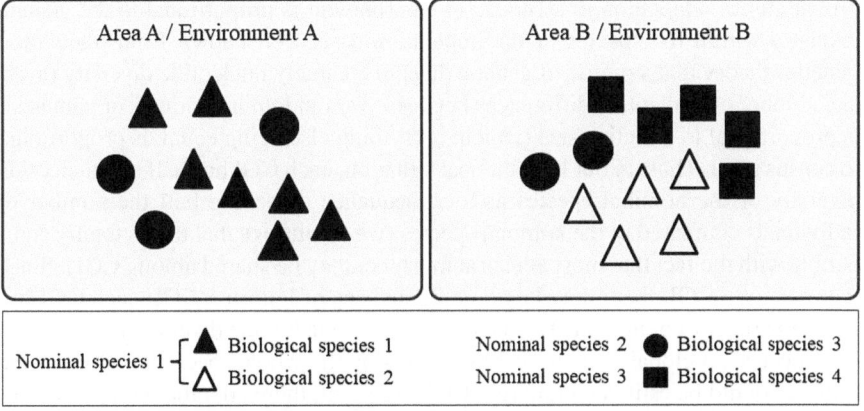

Parameter	Based on nominal species (blurred)	Based on biological species (true)
Species richness	3	4
Abundance (NS1)	12	7 and 5
Geographical range or Ecological niche * (NS1)	Wide	Narrow (for each biological sp.)

Fig. 4.2 Ignoring CGI has consequences on both assemblage parameters (e.g. species richness) and biological parameters (e.g. abundance, geographical range or ecological niche) defined for a given species. The figure represents hypothetical distributions and abundances of 3 nominal species, "nominal species 1" being a complex of two cryptic species (biological species 1 and 2). (*): The two separate zones (A and B) in which the individuals are distributed may represent either distinct geographic areas or distinct environments (i.e. habitats or ecological niches). We represented a situation where CGI have allopatric distributions or differentiated niches because these are the problematic cases, but there are situations where the CGI of a given species complex have the same geographic range or ecological niche. "NS" stands for nominal species

a species or an ecosystem) is wrong, and management measures based on these parameters may be inefficient or even deleterious.

The most conspicuous consequence of ignoring CGI is *an underestimation of species number* in a community or in an ecosystem because one nominal species is composed of several biological species. From a common biodiversity conservation point of view, this error would result in being more pessimistic than we should be about species richness in an area, species richness often being considered as a proxy of good ecological status or as a parameter to maximize. A direct corollary of the underestimation of species numbers is *the overestimation of the abundance for individual species* (by comparison to the nominal species abundance). In this case, the bias is toward undue optimism about a species' conservation status. If, instead of having one species with 2 N individuals, there are two separate entities of N individuals, the global risk of extinction at the level of the nominal species (i.e. pooling the two biological species) may change, depending on the vulnerability component considered (e.g. genetic diversity, or inbreeding rate), for the following reason. The probability of adaptation to a change in environment is proportional to the genetic diversity within the species or the population. It is well known from population genetics theory that a metric of genetic diversity, namely nucleotide diversity (average number of nucleotide differences between two random individuals or gametes), is proportional to effective size (which, everything else being equal, is proportional to census size). Thus, in our hypothetical situation, each CGI has half the nucleotide diversity of the nominal species as a consequence of having half the number of individuals compared to the nominal species (we emphasize that this is totally compatible with the fact that most alleles at most loci may be shared among CGI). Since there are two CGI, there may be no consequences of ignoring CGI: each CGI has twice the risk of going extinct by lack of adaptive nucleotide diversity but there are two species, so globally the probability of losing the whole species complex is the same as would be estimated ignoring CGI. However, there are other components of vulnerability where small population sizes are not compensated by the number of species, such as inbreeding. In hypothetical populations of N and 2 N individuals, the probabilities of self-reproduction are respectively $(1/N)^2$ and $(1/2 N)^2$, the latter equaling $\frac{1}{4} (1/N)^2$, which is a quarter of the former. Each CGI in this example therefore has a selfing probability four-fold higher than believed when ignoring that the nominal species is split in two, thus the vulnerability component is multiplied by four for each CGI which is not compensated by the presence of two (not four) CGI.

Another frequent consequence of ignoring CGI is an overestimation of the geographical range of a species: instead of a widespread (even cosmopolitan) species, there may be several geographically restricted species, allopatrically distributed or displaying partial sympatry (Egea et al. 2016; Eme et al. 2018). Again, this results in a systematic underestimation of the vulnerability of a species, particularly from a regional point of view because species with smaller geographical ranges are more vulnerable to environmental change and more threatened by extinction.

CGI may also lead to confounding numerous specialized species as a single generalist species (Morard et al. 2016), which is typically less sensitive to environmental change (Büchi and Vuilleumier 2014). More generally, functional diversity estimates may be affected depending on niche differentiation between CS: the competitive exclusion theory implies that sympatric CS may have diverged in the way they exploit limiting resources (otherwise one species would have eliminated the other by outcompeting it), with the consequence that the average niche widths of these CS may be overestimated (Van Campenhout et al. 2014) and as a result, vulnerability to perturbations would be underestimated. However, non-equilibrium situations, or more generally the neutralist theory of biodiversity, supported by many empirical studies (Hubbell 2001), prevent us from taking for granted that the ecological niches of all sympatric CS of a given complex have diverged. However, when CS share the same niche, there are also mistakes in assessing functional diversity because functional redundancy -the fact that several species ensure the same function in the ecosystem and may compensate one another in case one of them is going extinct- is underestimated when CS are ignored.

Another important element for bioconservation is the connectivity pattern of species' populations (i.e. the exchange of migrants able to reproduce with local individuals among distinct populations). The realized connectivity among populations, inferred by population genetics studies, is a key piece of information guiding the design of networks of protected areas. Inferred connectivity patterns may be erroneous when CS are ignored (Pante et al. 2015): for instance, if in two sympatric CS, samples from one area contain, by chance, only individuals of one species, and samples from another area individuals from the other species, genetic differentiation may appear very high, even if individuals migrate extensively and reproduce randomly among those areas (panmixia).

Thus far we have taken the viewpoint of community ecology, but biases induced by CGI also impact stock management of exploited species (population and range size overestimations, realized connectivity underestimations). Lastly, numerous parasites (including human parasites) are complexes of CS which may affect the efficiency of treatments (Tibayrenc 1996). CGI therefore strongly impact scientific data used by biodiversity managers and medicine.

Obviously, basic biological understanding also is challenged by CGI. Without accurate taxonomy, distributional and diversity patterns can become obscured (Paulay and Meyer 2006), and variation in taxonomic opinion can be an important source of confusion in diversification analyses (Faurby et al. 2016). For instance, ignored CGI may result in incorrectly indicating that rates of speciation have decreased toward the present (Cusimano and Renner 2010), causing false inferences of major ecological and evolutionary processes.

Beyond the erroneous inferences caused by CGI, numerous CGI are not taxonomical artefacts (i.e. morphological diagnostic differences among CGI are actually absent, not just overlooked) but they result from a significant decoupling of morphological and genetic divergence (cf below) which calls for an explanation involving evolutionary forces. Such CGI thus deserve to be studied as an important fundamental research question, not just for practical reasons (e.g. correcting biodi-versity estimates).

For all these reasons, it is necessary to undertake a thorough study of the phe-nomenon. Various factors may cause the presence of CGI, including human factors (e.g. the particular way in which taxonomists happened to describe and delimit the nominal species) and the habitat, biogeography and biological traits of the species. Understanding how these factors determine (i) the probability of having a CGI com-plex, (ii) the structure of morphological diversity in the species complex, (iii) the average number of CGI per nominal species, (iv) the probability that the CGI are ecologically differentiated or not and (v) their respective geographical ranges requires a compilation of case studies and their in-depth analysis. In Sect. 4.4, we will explain the role such factors may have in theory. Since different causes lead to different patterns of CGI, it is important to classify CGI in a relevant way. Furthermore, there are many cases of putative CGI in the literature but not as many confirmed cases; it is thus important to explain how to identify them reliably (Sect. 4.3: how to detect and classify CGI).

4.3 How to Detect and Classify Cryptic Species

There are two components in the notion of cryptic species. The first and most impor-tant component is that of genetic isolation, i.e. the presence, in a nominal species, of reproductively separated entities (though this isolation may be partial), which may correspond to distinct biological species *sensu* Mayr. The first part of this Sect. 4.3.1 presents the different levels of genetic isolation or levels of evidence of genetic isolation. In the absence of any degree of genetic isolation within a nominal species, there are no CGI, even in the wide sense (*sensu lato*). The second component is morphology (Sect. 4.3.2). Although CGI are sometimes defined as distinct biologi-cal species with similar morphology, we decided to consider as CGI (but *sensu lato*) the cases where biological species are indeed differentiated morphologically, while having the same Latin name. This choice was motivated by the fact that CGIsl as defined above pose many of the practical problems posed by CGI *sensu stricto* (where the distinct genetic entities have no diagnostic morphological differences). To avoid confusion about definitions, Table 4.1 displays our nomenclature in a 2-dimensional classification of CGI.

4.3.1 Identification of Genetic Isolation and Biological Species

The following explanations naturally only hold for taxa where "reproductive isola-tion" has a meaning (i.e. taxa in which there is sexual reproduction) and which also have a diploid life stage (with two copies for each marker/gene).

The most direct way to assess genetic isolation between two groups of individuals is to perform controlled crosses. However, in "non-model" species, in case of reproductive failure it is often impossible to determine whether genetic isolation or experimental conditions are responsible for the absence of offspring (or even mating). Moreover, when one does not know how to define the groups of individuals (typically the case of CGI*ss*, due to the lack of conspicuous morphological differences), the problem has no solution. This explains why CGI*ss* have always been discovered using genetic markers (characterized in a sufficient number of individuals).

Genetic markers may come from the nuclear genome. Since the nuclear genome is diploid, individuals have two copies for each nuclear marker, inherited from the two gametes that fused to form their first cell. There are also genetic markers that come from organellar genomes (chloroplastic or mitochondrial) which are transmitted to the (diploid) individual from a single gamete, generally the maternal gamete (oocyte) for animal mitochondrial genome, and often the paternal (pollen) for chloroplastic genomes.

When two groups are fully reproductively isolated, no genetic material is exchanged across groups (except viruses or mobile elements). There are necessarily some genetic differences among groups (otherwise they could exchange genes, if they were in contact). Diagnostic markers are those for which no allele is shared between group 1 and group 2 (yet there can be several alleles per group): if you know the allele, you can assign the organism to one of the two groups precisely. Semi-diagnostic markers are markers for which at least one allele is private to a group (absent from the other groups).

Two main types of genetic markers account for most CGI discoveries. Historically, the first type of markers which demonstrated genetic isolation within many nominal species were codominant markers, which are nuclear markers that reflect the state of both the maternal and the paternal allele of an individual. Most studies reported in the seminal review of Knowlton (1993) demonstrated CGI using such markers, in particular allozymes. By contrast, a dominant marker only provides two possible phenotypes (either presence or absence of the variant): when the variant is detected, which is often symbolized by [1], one cannot determine whether the genotype is homozygous (11) or heterozygous (10); when the phenotype is not observed [0], the genotype is necessarily (00).

A given diagnostic and codominant marker is a powerful tool to detect genetic isolation. For instance, imagine a scientist characterized 200 individuals with a marker with three alleles that are diagnostic of two biological species (alleles A and B for species 1, allele C for species 2). If the sample contains individuals from species 1 and 2, the scientist may find 4 genotypes, namely AA, AB, BB and CC. A possible distribution of the individual genotypes could be 25 individual (AA), 50 (AB), 25 (BB), and 100 (CC). Genotypes AC and BC do not exist because no genetic exchange is possible between species 1 and 2. Missing genotypes can only be explained by genetic isolation. However, to establish that the alleles are diagnostic in such a case, sample size and relative frequency among species (and also relative allele frequencies within species) matter: if only 10 individuals had been genotyped, the absence of AC and BC genotypes could have resulted by chance alone (as a result of random sampling). If species 2 was very rare in the global sample (say 7 individuals) the absence of AC and BC would not be considered evidence of genetic isolation. So conclusions are not always straightforward and require population genetic approaches where many individuals are genotyped and analyzed using relevant (basic) statistical tests. Note that semi-diagnostic markers also produce missing genotypes which may reveal the presence of genetic isolation, but they do not allow precise species delimitation based on genotypic data because some genotypes (those composed of shared alleles) can belong to both species. With dominant markers, it is not possible to identify missing genotypes (i.e. the absence of combination of some variants in a given individual from the whole population).

During the 1990's, the use of allozymes declined in favor of approaches based on DNA. These allow field collection without refrigeration and DNA characterization was greatly facilitated by the PCR technology (Avise 1994). However, at this time, current technology did not allow routine sequencing of both alleles of many diploid individuals and the commonest data produced thus became dominant markers or sequence data from a haploid genome (mitochondrial, chloroplastic) which is represented by a single gene per individual. The distribution of haploid genotypes such as (A), (B) or (C) among individuals does not reveal anything about isolated groups, in the absence of independent information, whatever their frequencies (and whatever the divergence among these alleles). In codominant markers (example just above), genetic isolation is simply deduced by the fact that some combinations of alleles are never found associated within the same individuals, which obviously cannot be assessed with haploid markers. Among such haploid markers, however, some contributed much to the detection of putative CGI. These are the markers in which the alleles were characterized by their DNA sequences, or more generally those for which it was possible to characterize distances among alleles. Imagine now that alleles A and B are very closely related DNA sequences (differing only by one out of 500 positions), and C is very different from A and B (by 20 positions) (Fig. 4.3). The temptation is great to infer that A and B belong to one species, and C to another one. In numerous studies, alternative explanations were not even considered and the presence of CGI was inferred by such patterns. But there are alternative explanations for the observation of highly divergent alleles within a single species, even when intermediate alleles are absent. For instance, a past bottleneck in the effective size of a species (high mortality events) can lead to loss of various alleles, with only a few divergent alleles remaining (for instance 2 alleles, which may differ by 10 nucleotide positions out of 500). Then, with time, new alleles arise by mutations, which differ from their parental allele by a single mutation, leading to the presence of various (e.g. 10–15) very closely related alleles (differing by a single mutation from their parental allele, because mutations rarely hit the same nucleotide position at short time intervals) for each of the two surviving ancient alleles. The typical pattern arising from this is shown in Fig. 4.3. Note that selective sweeps, i.e. the removal of genetic diversity due to spread in the species of an advantageous allele, within a single biological species can also produce similar patterns. It looks exactly the same as the result of divergence of distinct biological species. Therefore, when a pattern like Fig. 4.3 is observed, the confirmation that there are distinct bio-

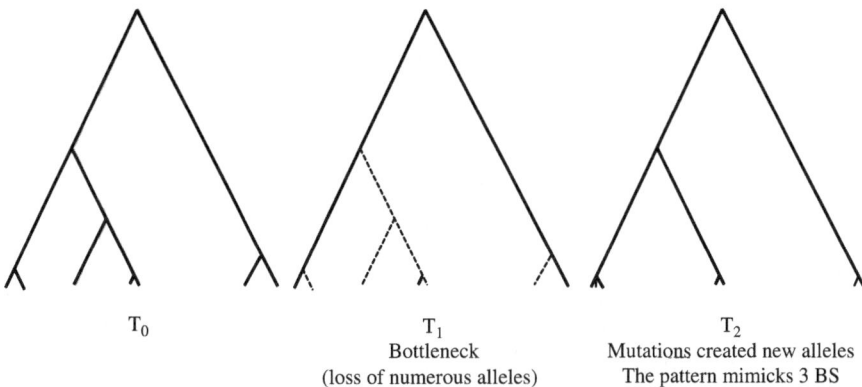

T_0

T_1
Bottleneck
(loss of numerous alleles)

T_2
Mutations created new alleles
The pattern mimicks 3 BS

Fig. 4.3 Phylogeny of alleles may erroneously suggest the presence of several biological species. Time T_0: Representation of a hypothetical allele phylogeny in a population of constant size, at mutation-drift equilibrium. At time T_1 a severe decrease in population size (bottleneck) causes the loss of many alleles (dashed lines). At T_2, the population has recovered its size and mutations created new alleles closely related to the survivor alleles. The allele phylogeny mimics a pattern with 3 distinct biological species

logical species requires obtaining independent evidence supporting the genetic partition displayed by the single haploid marker, i.e. a polymorphic trait whose states appear to be linked to the marker's states. This can come from any other genetic or phenotypic (in the widest sense) marker, provided this marker is not constrained by its nature to remain tightly linked to the first marker.

As an example with genetic markers if individuals with sequences A or B (at marker 1) always have the allele X (at marker 2), and individuals with sequences C (at marker 1) have the allele Y at the independent locus (at marker 2), and if the two markers are not physically linked in the genome (which means that at each reproduction event, these two loci segregate independently and their respective alleles do not remain linked), it establishes that genes are not exchanged among groups of individuals (the first group bearing alleles A, B, and X and the other group bearing C and Y). This situation (when applied to genetic markers) corresponds to an extreme case of linkage disequilibrium. Linkage disequilibrium is defined as the non-random association between alleles at distinct loci within individuals in a population. Linkage disequilibrium, even when it is not extreme (for instance when all possible allele combinations are observed) is useful because it can detect the presence of two genetic entities (such as CGI) in a sample even when there is hybridization between them. Indeed, there are many studies reporting occasional hybridizations among distinct biological species. If such hybrids were as fertile as "pure" individuals, the two species would fuse together and after a number of generations there would be a single species. However, in most cases after long term isolation between incipient species, some incompatibility has arisen and hybrids are either sterile or less fertile. In such cases, reproductive isolation is partial, but the presence of rare hybrids does not refute the presence of reproductively isolated entities that remain genetically distinct in the long term. Even in such cases, population genetics can reveal the presence of partially isolated populations (or hybridizing species) in a sample of individuals by the detection of linkage disequilibrium between loci that are physically unlinked.

Karyotypes (shape and numbers of chromosomes), ecological characters (habitats, phenology, diet... (Johannesson 2003)) and behavior are typical phenotypic traits which can distinguish reproductively isolated units. The great majority of putative CGI detected by DNA sequences in animals were detected by mitochondrial DNA markers (haploid); thus markers from the nuclear genome (which segregates independently from the mitochondrial genome) are ideal candidates to check whether the putative biological species are true biological species (Chenuil 2012; Chenuil et al. 2010; Egea et al. 2016) as well as any phenotype not determined by the mitochondrial genome (probably more than 99.9% of phenotypes). What we called putative CGI (and putative CS), being often identified by a single molecular marker, are similar to the "Primary Species Hypotheses" of previous authors (Castelin et al. 2016; Pante et al. 2015) that need to be confirmed by independent markers or by an integrative taxonomy approach.

Apart from direct methods that are clear cut and based on a small number of markers, there is a variety of recent methods to identify and validate species delimitations using information from several independent genetic markers. Some do not require codominant markers but use DNA sequence information (Yang and Rannala 2010). For their success, some alleles must have diverged between species as a result of mutations, not only genetic drift. Other methods do not use DNA sequences but codominant markers, and can have good results even when genetic markers are not diagnostic (i.e. some alleles are shared among CGI) (Huelsenbeck et al. 2011; Jombart et al. 2010). Although these clustering methods are rarely used to assess genetic isolation, they may be the only solution for recently diverged CGI that retain ancestral shared genetic polymorphism (Weber et al. 2019). Recent methods still account for a negligible number of CGI reports.

We have thus shown how to determine genetic isolation with genetic markers and other traits recorded in samples of sufficiently numerous individuals: either using codominant markers or using distinct markers (that may be dominant) that are not inherited in a linked manner, so that their statistical association (linkage) in individuals proves that they are genetically isolated.

Let us come back to the distinction between CGI and CS (CS being particular cases of CGI). Genetic isolation may be caused by geographical isolation among groups whose genomes remain intrinsically compatible: in such cases, if individuals were put into contact (for instance by human intervention), they may be able to produce fertile offspring (thus they belong to the same biological species). We thus considered as CGI all cases where genetic isolation was established but intrinsic incompatibility was not proven. Using genetic markers exclusively, it is not possible to know whether allopatric groups are still interfertile: such groups may display diagnostic markers as a result of genetic drift and mutation because they evolved separately for many generations. By contrast, in some (numerous) cases, genetically isolated groups detected by genetic markers are sympatric and completely inter-mixed in the field (Boissin et al. 2008a, b; Egea et al. 2016; Weber et al. 2014), so their reproductive incompatibility is not questioned and they deserve the status of cryptic (biological) species (CS). When the genetically isolated groups are allopatric, whether or not they kept the possibility to interbreed has few consequences for biodiversity characterization at the community level since most consequences highlighted in Sect. 4.2 still hold (e.g. range overestimation). However, the distinction is important for practical aspects of bio-conservation: in a case of strong bottlenecks endangering one geographical group, artificial introduction of individuals can be envisaged (to help restoring population size) from the other geographical group only when transplanted individual are able to reproduce with indigenous ones, thus not for actual CS.

To conclude, a practical way to classify the type of structuration within a nominal species according to genetic isolation is the following one:

Level A (biological species): True genetic isolation is shown by markers and intrinsic incompatibility is confirmed between entities (either by the observation of the genetically isolated entities in sympatry, or by controlled crosses).

Level B (genetic isolation, putative biological species): genetic isolation is confirmed (either established by a single codominant genetic marker or by an association of a genetic marker with another independent "marker", which could be genetic, morphological, ecological or behavioral) **but** it distinguishes groups that are in allopatry, so **the status of biological species** *sensu* (Mayr 1942) **requiring** *intrinsic* **incompatibility** (and see Wheeler and Meier (2000)) **cannot be confirmed**.

Level C (Putative genetic isolation): putative genetic isolation that needs confirmation. These cases correspond to a high divergence among alleles in haploid or dominant markers (cf. Fig. 4.3) which has not been confirmed by any independent marker.

Level D (No genetic isolation evidence): Absence of any significant genetic differentiation within the nominal species with available genetic markers (or phenotypic characters). This does not allow rejecting the hypothesis that there are some biological species within the nominal species; we simply have no indication that there are some which need to be delimited.

This classification is a practical one which reflects available knowledge on a given nominal species. For instance, a nominal species classified as level D for genetic isolation may indeed correspond to true biological species but we lack data to confirm it. This classification will be useful when reviewing literature published on CGI because many studies report "cryptic species" while evidence of genetic isolation does not go beyond level C (i.e. genetic isolation needs to be confirmed by an independent marker, genetic or not).

4.3.2 Morphological Differentiation

Independently of the level of genetic differentiation among some groups within a nominal species, their morphological variation can be studied using various types of characters: some studies consider only very conspicuous external characters, others focus on the characters traditionally used to diagnose the species in the genus or family to which the nominal species belongs, while other ones endeavor to seek any possible character in order to find some characters corroborating groups revealed by genetic markers. For a given sample of a given nominal species, morphological differentiation and polymorphism depend on the (set of) character(s) used.

For instance, in spatangoid sea urchins, species are described and diagnosed by morphological indices from the test (i.e. the skeleton). Egea et al. (2016) revealed CS in *Echinocardium cordatum* using morphological indices from test shape: they did not find a single diagnostic character (despite the fact that morphological differentiation among CS was highly significant statistically), although sperm morphology (requiring microscopic observations) would probably reveal diagnostic differences (Drozdov and Vinnikova 2010). For taxonomists, fidelity in considering a set of characters has some justification: for example, in sea urchins, using test shape permits analyses combining extant and fossil specimens. Sperm morphology cannot be used on fossils because sperm lack hard and fossilizable structures (and also because of their microscopic size).

We propose the following classification to characterize morphological variation and differentiation among groups in a nominal species. What we name "groups" are entities which were necessarily defined independently of morphology, generally from genetic markers. This classification considers both morphological variability within groups and morphological differentiation among groups because both are relevant to interpret the nature of the evolutionary forces impinging on the evolutionary trajectory followed by the nominal species under study. As for genetic markers, the notion of diagnosticity for a morphological marker is crucial. It is useful to distinguish a situation with statistically significant morphological differentiation among groups, in the absence of diagnostic characters. For example, multivariate analyses using a set of morphological characters correctly assign more than 97% of the specimens to their genetic CS in *E. cordatum*, yet for each of the 20 morphological indices, values overlap among CS (Egea et al. 2016).

Level 0: No morphological polymorphism for this character in the nominal species, thus no differentiation among groups.

Level 1: Presence of morphological polymorphism but no differentiation among groups (not even a statistical differentiation).

Level 2: Significant morphological differentiation among groups, but no diagnostic character among groups (e.g. character values overlap for quantitative characters).

Level 3: Diagnostic morphological differences among groups.

Here again, as for the genetic component, sample sizes are crucial: it is not possible to determine if a marker is diagnostic when it was characterized in too few individuals. Beyond sample size, sample variety is important; in fact, given that individuals from a field sample may be close relatives, it is desirable to collect several field samples from reasonably distant locations. For instance, a morphological character (radial shield) appeared diagnostic of two brittle-star CS in Crete and was supported by large sample sizes (Weber et al. 2014) although this was not the case in other regions (Stohr et al. 2009).

Crossing the genetic and the morphological differentiation components, using the levels defined above, we obtain a table which provides a bi-dimensional classification of nominal species regarding the phenomenon of "cryptic species" (or CGI) (Table 4.1). Further considerations based on the different cells (or ranges of cells) from Table 4.1 rely on the assumption that the morphological differentiation status reported corresponds to the most discriminating morphological marker available in the nominal species and that such characters were investigated seriously enough. This condition is very constraining when performing a review of the literature: as shown by our preliminary survey, many studies lack sufficient detail regarding which characters were looked at and many of them do not even name any morphological character, yet conclude the absence of morphological differences among species. Therefore, rigorously establishing the absence of morphological differentiation (or diagnostic differences) within a nominal species may be impossible in the *absolute*: it is rarely possible to rule out the objections that other characters (microscopic ones, or from transitory life stages) which could have revealed stronger differentiation were dismissed/overlooked. But what is relevant for an evolutionary biology understanding of morphological evolution is to establish that the ratio of "morphological differentiation/genetic differentiation" is significantly different in the studied species than in other closely related taxa. The ideal approach to establish the morphological differentiation status in a nominal species thus requires morphological analyses of both numerous specimens from the studied nominal species as well as that of some specimens from at least one other, closely related, nominal species. This was done in (Egea et al. 2016): genetic distances between CS of the sea urchin *E. cordatum* are greater than those observed between two nominal species of another spatangoid sea urchin genus, namely *Spatagus purpureus* and *S. multispinus*.

The right-hand column in Table 4.1 (MD_3) corresponds to cases with diagnostic morphological differences. When diagnostic morphological differences confirm biological species, the possibility of having CS *sensu stricto* is ruled out, but we call such cases CS *sensu lato* because there are biological species lacking the taxonomical status of nominal species. The nominal species and its component CS are thus in need of taxonomic revision. There can be no cases in category C3 (putative genetic isolation) because, as explained above, a morphological difference diagnostic of the genetic groups (assuming this morphological character is not encoded by genes linked to the genetic marker) automatically confirms genetic isolation: this corresponds to the B3 category, in which genetic isolation is established but genetic analyses were not performed on sympatric samples so that the possibility of interbreeding, if individuals were in contact, cannot be discarded. When genetic groups are in allopatry, B2 cases correspond to sub-species.

Columns MD_0, 1 and 2 are cases without diagnostic morphological differences: these cases, when the presence of distinct biological species is confirmed (i.e. in the first row, GI_A) correspond to CS *sensu stricto* because a traditional taxonomical diagnosis of morphological species is not possible, due to lack of diagnostic morphological characters. Lower rows may also be CSss but genetic evidence is lacking to establish the presence of biological species. GI_B cases (proven genetic isolation, possible biological species), in the absence of diagnostic morphological differentiation, can be called "cryptic genetically isolated entities" (category B0 or B1). For many questions regarding biological evolution, these cases are equivalent to established biological species and should be included in meta-analyses aimed at testing hypotheses regarding the coupling of morphological and genetic divergence. Like for (C3), there are no cases in category (C2) because significant morphological differentiation among genetic groups constitutes evidence of a certain degree of genetic isolation that may only be partial (as for instance when hybridization is possible and hybrids have a lower fitness).

Two cells with putative genetic isolation and no significant morphological differentiation (C0 and C1) may be CSss but are not confirmed. Since the literature on animal CS contains many such cases, mostly from mitochondrial DNA markers, and since, when independent markers are available in addition to mitochondrial markers, they confirm genetic isolation rather frequently, we consider that such cases are worth being reported and analyzed in meta-analyses, provided their lower level of evidence of genetic isolation is recorded.

When no polymorphism at all is observed within the nominal species for the morphological character considered (left column of Table 4.1) one may just consider that information is lacking and interpretations are not possible. However, when the morphological character(s) considered is typically one that usually displays a certain amount of variability within species or that differentiates species in other, closely related nominal species, the absence of polymorphism itself can be considered informative. This leads us to part 4, where we discuss possible causes generating CGI.

4.4   Identifying the Multiple Causes of Cryptic Species

The causes of the presence of CS or CGI may be related to our taxonomic activities or to the species themselves. In the first case, they are somehow inherent to the taxonomic process (i.e. the human process of delimiting nominal species, which however may in some cases be affected by features of the species or their habitats). In the second case either they correspond to recent (young) divergences or they reflect a slow-down in the accumulation of diagnostic differences or a slow-down in morphological divergence relative to genetic divergence. After explaining possible causes and explaining how biological or habitat factors may trigger such phenomena, we explain how to determine if each of these causes is likely to explain a CS or a CGI case. The different causes and their hierarchy are summarized in Box 4.1.

Box 4.1: Classification of the Main Causes of CS

1. *Taxonomic work is needed*

   1.1. *Formal description of new nominal species is needed (for CSsl only)*
   1.2. *Other taxonomic cause (character choice/availability, lack of samples)*

      1.2.1. *Technology available for observation when the nominal species was described*
      1.2.2. *Prevailing theories of nature and species origins when the nominal species was described*
      1.2.3. *Accessibility of habitats when nominal species was described*
      1.2.4. *Availability, quality and nature (natural selection targets / selectively neutral) of morphological characters in the group studied*

2. *Other causes than taxonomic process*

   2.1. *Recent divergence*

      2.1.1. *low dispersal*
      2.1.2. *fragmented habitat or active landscape dynamics*

   2.2. *True slow-down of ratio Morphological divergence/Genetic divergence*

      2.2.1. *natural selection*

         2.2.1.1. *stabilizing (in narrow niches)*
         2.2.1.2. *diversifying (in generalists, broadcasters...)*

      2.2.2. *selective neutrality of morphology (high Ne)*

4.4.1   Taxonomic Process

There are two distinct cases where taxonomic processes (i.e. the way species were delimited) are responsible for the presence of CGI. In the first case, cryptic species sensu lato (or CGIsl) do indeed display diagnostic morphological differences corresponding to biological species (or to units displaying genetic isolation). These cases are thus just in need of a formal description of the morphological biological species or an upgrade to the status of nominal species (or the status of sub-species, for CGI which are not CS). A second situation corresponds to cases where taxonomy failed to reveal diagnostic or differentiated morphological characters in true biological species or in CGI for various reasons discussed below which are inherent to: (1) technology available for observation when the nominal species was described, (2) prevailing theories of nature and species origins when the nominal species was described, (3) accessibility of habitats when nominal species was described and (4) availability, quality and nature (natural selection targets/selectively neutral) of morphological characters in the group studied.

1. Technology available for observation at time of description may explain many CGI cases. Species that were described in times when (or in countries where) microscopes were not available may not have the same range of characters at their disposal to delimit morphological species. Indeed, the year in which a species was described represents a rich source of information to investigate the effects of science history in general on the presence of CGI (e.g. (Strand and Panova 2015)).

2. Nominal species of multicellular organisms correspond to the so-called "morphological species" or "morphospecies" (in more than 99.99% of nominal species) and morphological species may not correspond to biological species. Such discrepancies may lead to the presence of cryptic species *sensu stricto* but also to the opposite phenomenon (e.g. males and females, or young stages and adults, have been erroneously described as distinct species in various groups (Johnson et al. 2009)). Indeed, different species concepts may delimit species in different ways (Agapow et al. 2004). Depending on the groups, the morphological characters used to diagnose the species (and define species boundaries) may have benefitted from a cladistics approach (Hennig 1950), in which case they are more likely to reflect phylogenetic species (and also, to a lesser extent, biological species). Although the "phylogenetic species concept" includes a wide spectrum of definitions (Agapow et al. 2004; Wheeler and Meier 2000), in practice, it is often invoked (explicitly or not) to claim the presence of (cryptic) species on the basis of a phylogenetic tree inferred from a single molecular marker. Single-marker-phylogenetic-species boundaries may not delimit genetically isolated entities (cf 3–1 and Fig. 4.3), thus disagreeing with the "biological species concept". In our Fig. 4.3 example, some widely used automatic methods of species delimitation such as the ABGD (Puillandre et al. 2012) may erroneously indicate the presence of 3 putative species. However, the formal/official description of nominal species based on molecular markers is very rare in multicellular organisms and in such cases, care is taken to use several markers (Meyer-Wachsmuth et al. 2014). Indeed, using single marker phylogenies potentially causes false reports of CGI.

3. Accessibility to an environment might limit the number of samples available for morphological analyses or cause specimen damage. Such accessibility limitations may contribute to the abundance of CGI in some environments (e.g. deep sea organism destruction by strong decrease in pressure when collected (Vacelet, 2006). This may help to explain the high frequency of CGI in the marine environment (Barberousse and Bary 2015; Luttikhuizen et al. 2011).

4. Depending on the taxon under consideration, the morphological characters used for species diagnosis are more or less reliable. For instance, some characters may be the targets of natural selection, thus may fail to distinguish entities that have a similar niche component as a result of evolutionary convergence or stabi-

lizing selection: beak shapes in a group of birds having a similar diet may not allow species distinction, because natural selection constrains beak shapes to remain adapted to collect and grind their food. Because humans use visual information for nominal species delimitation, animals that use visual cues for mate recognition (such as vertebrates) are also much less likely to form CGI than animals that rely entirely on chemical cues for mating, such as marine invertebrates (e.g. spawning is generally triggered by chemical signals, and gametes from both sexes themselves are attracted by chemical signals (Weber et al. 2017)). Tiny organisms provide fewer characters that can be used for diagnosis, parasites often have lost many morphological characters with respect to their free-living relatives, because their bodies are simplified, having lost some major functions, etc.

4.4.2 Other Causes Besides the Taxonomic Process

Some CS or CGI are not explained by weaknesses of the taxonomic process. These are necessarily CSss or CGIss, where diagnostic characters are lacking to distinguish completely or partially genetically isolated entities.

4.4.2.1 Recent Divergence

One possible explanation for the existence of CSss (or CGIss) is the young age of divergence. Recently diverged species are more likely when speciation rates are high. Thus, factors promoting allopatric speciation may be frequently associated with CSss and more generally CGIss. Low dispersal as well as habitat fragmentation are the most conspicuous candidate factors. Thus, a review of CGIss may report dispersal ability as well as the habitat fragmentation for all cases.

4.4.2.2 Deceleration in the Accumulation of Diagnostic Morphological Differences or in Morphological Divergence Relative to Genetic Divergence

When divergence is not recent and a poor taxonomy is not involved, CSss or CGIss thus reflect an actual slow-down in the ratio of morphological over genetic diversity or divergence that persisted long enough to produce the observed pattern.

Natural selection on morphological characters may be responsible for the absence of diagnostic characters among species. (1) The cause which is most often invoked to explain such cases is stabilizing selection (Charlesworth et al. 1982; Lee and Frost 2002). When morphology is strongly constrained by natural selection, morphological variation is very low within species and following speciation, daughter species are

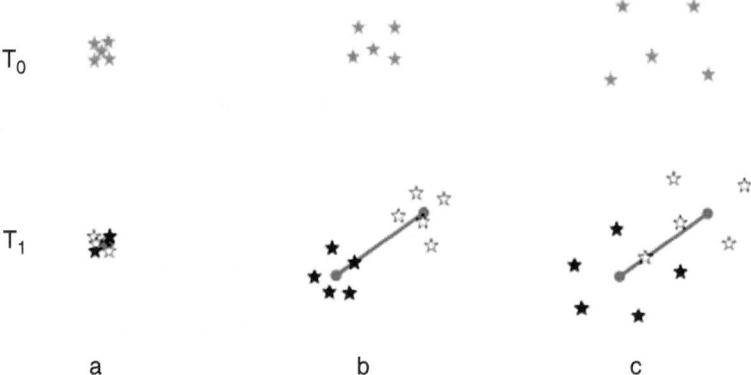

Fig. 4.4 Different patterns of morphological differentiation between genetically isolated entities. Individuals from the two entities are represented by filled or empty stars. Their relative position in the plane reflects their morphological similarity (e.g. horizontal and vertical coordinates may represent values for two continuous morphological characters). For a given divergence time (from T_0 before divergence to T_1 after several generations), the distribution of morphological variation within and among the two genetically isolated entities may correspond to one of three main patterns resulting from four main processes. (**a**) The absence of (or negligible) morphological diversity within species, probably resulting from stabilizing selection, impedes their divergence. (**b**) A standard situation without CGI *sensu stricto*. There is morphological variation within species and the genetically isolated entities diverged morphologically so there is no overlap between them (the character represented by the horizontal coordinate is diagnostic) that are therefore not CGI *sensu stricto*. (**c**) There is a higher level of diversity within species compared to case **b**, so despite identical divergence times compared to case b (represented by the same distance separating the barycenters represented by red dots as in case **b**), the morphological spaces of both species overlap and thus no diagnostic character distinguishes the species. Diversity may result from high effective sizes, or from natural selection favouring high morphological diversity (see text). We emphasize that pattern (**a**) could not be caused by low effective sizes: in such a case genetic drift would be high and lead to divergence at T1 (length between the two barycenters would be higher than in **b**, instead of null)

similarly monomorphic and do not diverge in their shapes (Fig. 4.4). (2) Paradoxically, an opposite pattern of morphological diversity may also lead to CS. This pattern is that of a high morphological polymorphism within species, which is selectively advantageous for species submitted to strong spatio-temporal fluctuations (Egea et al. 2016). Morphological polymorphism may be achieved by two different mechanisms: environmental phenotypic plasticity (a single genotype may lead to a variety of morphologies) or presence of a variety of genes determining morphology (i.e. presence in the species of distinct alleles, or genetic variants, also called genetic polymorphism). Both mechanisms prevent the appearance of diagnostic morphological differences between species because, as a result of character polymorphism, the range of possible character states overlaps between sister-species (Fig. 4.4).

Recently, it has been suggested that neutral (i.e., non-adaptive) processes, may also lead to absence of diagnostic morphological differences among genetically isolated entities (Egea et al. 2016). Higher polymorphism at neutral loci is expected for taxa with larger effective population sizes. When such taxa speciate, ancestral polymorphism remains shared among daughter species for a higher number of generations than in taxa with lower effective sizes. When the phenotypic traits used to diagnose species are selectively neutral this leads to an absence of diagnostic characters for longer temporal periods in the taxon with higher effective sizes, making the occurrence of CS more likely (because the taxonomists delimiting species cannot identify any diagnostic character). This novel neutral theory of morphological evolution provides a null model for the existence of CS, and may help to explain the abundance of marine CS because in the marine realm many species have high fecundities, abundances and range sizes.

Figure 4.4 illustrates the distribution of morphological diversity between two sister biological species corresponding to the above cases and compared to a species pair displaying diagnostic characters.

To summarize this section, five major types of causes correspond to the distinct levels of morphological differentiation of our classification (i.e. Table 4.1 columns): stabilizing selection for MD0, recent divergence for MD1, high effective sizes or advantageous morphological polymorphism for MD2, and poor taxonomy for MD3 (not excluding that various factors may interact).

4.4.3   How to Determine If a Cause Is Likely to Explain a CGI Case

Not all causes are possible for a given category of putative CGI (i.e. for a given cell or cell range in Table 4.1). The possible causes identified above are compiled in Table 4.2. For each cause, the "cell range" column displays the putative CGI category that can be explained by this cause, and which traits or factors are useful to assess the validity of the cause. Most causes can be assessed at two levels: for individual putative CGI or at a global level, in a higher order taxon. For instance, one may test whether the cryptic species observed in the species complex *Echinocardium cordatum* can be explained by stabilizing selection or not (Egea et al. 2016), but also whether CGI in the phylum Echinodermata are explained by stabilizing selection more often than expected at random. Testing the importance of a possible process globally (i.e. in generating CGI in a given higher-rank taxon) requires including in the (meta-) analysis not only the taxa for which CGI have been reported or suggested in the literature, but also all nominal species of the taxon for which genetic data have been published.

At this step of the analysis, we can list the different data fields that appear useful to include in a database aimed at studying the CS phenomenon. They should include both information enabling CGI characterization (both GI and morphological differentiation levels; Table 4.1) and information useful to determine the possible causes of the CS (Table 4.2; acknowledging the fact that most cases lack information in some fields). Potentially useful data fields include: (1) genetic marker type (haploid/diploid, codominant or not, number of markers), genetic structure (sample sizes, significant differentiation among groups, genetic diversity within populations/

Table 4.2 Possible causes (column 1) for different types of (putative) CGI (column 2) and traits or factors to check (column 3) to evaluate the validity of the hypothetical cause (rather than an exhaustive list, we proposed examples of the most relevant ones)

Cause or hypothetical process	Type of CGI (cell range in Table 4.1)	Traits or factors to check
T1 available technology	A-C x 0–3	Year (+place) of NS description
		Material needed for diagnosis (microscope, …)
T2 history of science	A-C x 0–3	Year (+place) of NS description
		Higher order taxon name
T3 accessibility	A-C x 0–3	Habitat of species
T4 morphological character	A-C x 0–3	Organism size
		Mode of life (endosymbiotic)
		Selective neutrality of character
		Variability of character in higher rank taxon
Recent divergence	A-C x 0–2	Genetic divergence (CS of the NS + at least 1 pair of closely related sister species)
		Biogeography (CS sympatry/allopatry)
Low dispersal	A-C x 0–2	Life traits related to dispersal ability (but also to effective size: fecundity, reproduction mode)
High fragmentation	A-C x 0–2	Habitat of species
Stabilizing selection	A-C x 0	Morphological variability within BS
		Spatio-temporal variability of environment
Diversifying selection	A-C x 1–2	Same as above
		Check knowledge on species plasticity
Neutral (high Ne)	A-C x (1)-2	Morphological variability within BS
		Genetic diversity within populations/BS
		Life history traits (fecundity, reproduction variance)
		Size of geographic species range

"NS" stands for "nominal species" and "BS" stands for "biological species"

species and comparison with closely related taxa external to the nominal species if possible), (2) reproductive isolation among groups if tested by crosses, (3) ecological differentiation among groups, (4) any phenotypic differentiation (in the wide sense) that corresponds to genetic differentiation to confirm GI, (5) morphological variability within and among groups (and sample sizes), and also, when possible, in closely related pairs of sister species, (6) year and place of nominal species description, (7) nature of morphological characters analyzed, (8) habitat (physical fragmentation, accessibility), (9) biogeographical distribution (allopatry, sympatry among CGI, size of species range) and (10) life history and other biological traits (dispersal ability, fecundity, reproductive success variance, parasite or not, use of visual cues for mating).

4.5 Preliminary Results

A pilot study by undergraduate students (Délémontey et al. 2014) compiled articles reporting cryptic species in the marine realm and recorded information relative to some of these fields. This study collected useful data about the relative proportions of different cases of CS in the literature and revealed some associations among CS features, phyla, habitat and biological traits. For the pilot study, successive groups of search terms were used in Web of Science. We detail the different steps of the first search. The first step using « cryptic species » OR « sibling species » provided 11,416 papers (this was done in 2014). After adding «morpho* OR phenotyp*» (second step) 4417 articles remained, after adding «genetic OR molecular OR mitochondrial» (third step) we had 3055 papers, and with «marine OR sea OR ocean» (fourth step) 647 articles. To limit the number of papers while increasing the proportion of cases corresponding to validated CGI, we added the terms «nuclear marker* OR microsatellite* OR allozyme* OR intron OR ribosomal» (fifth step) to favor studies combining several molecular markers. This resulted in 222 articles. We carried out a second search identical to this one except that we replaced the fourth step (marine or sea or ocean) by the title of scientific journals dealing with marine biology («ANNU REV MAR SCI» OR «DEEP SEA RES» OR «ESTUAR COAST SHELF S» OR «HELGOLAND MAR RES» OR «ICES J MAR SCI» OR «J OCEANOGR» OR «J PLANKTON RES» OR «LIMNOL OCEANOGR» OR «MAR ECOL PROG SER» OR «OCEANOGR MAR BIOL» OR «CORAL REEFS» OR «MAR ECOL-EVOL PERSP» OR «MAR BIOL» OR «CAN J FISH AQUAT SCI» OR «J EXP MAR BIOL ECOL» OR « J FISH BIOL»). This second search provided 41 papers. For the last search we changed, again, the fourth step to select taxon names («Echinoderm*» OR « Echinoid*» OR «Asteroid*» OR «ophiuroid*» OR «bivalv*» OR «mollus*» OR «fish*» OR «sponge*» OR «porifera*» OR «cnidaria*» OR «coral*» OR «bryozoan*» OR « ascidia*» OR «mysidac*» OR «nematod*» OR «gastropod*» OR «copepod*» OR «amphipod*») which led to 264 articles. The fusion of the three searches provided 402 different articles. After abstract reading, we discarded papers that dealt with plants and algae, terrestrial and freshwater animals, endoparasites, protists and foraminiferans, and papers reporting new species but not CGI. These studies corresponded to 126 nominal species (556 CGI) from 86 families, 55 orders, 25 classes and 11 phyla. For all nominal species, putative CGI were defined based on genetic markers; for three nominal species controlled crosses were performed to determine CS; in 14 nominal species, there were some differences between CGI for at least one factor among ecology, reproduction, nutrition, hosts, gamete morphology, color and in 3 nominal species, CGI had distinct karyotypes. This preliminary survey confirmed that CS were present in a diversity of animal phyla and established an average of 4.41 CGI per nominal species (Fig. 4.5).

Out of 126 nominal species cases, 70 (56%) had been the subject of a morphological study: 37 of these display diagnostic differences among CGI (53%, thus they are not CGI *sensu stricto*), 16 display statistical morphological differences among CGI (23%), and 17 do not display morphological differences among CGI (24%).

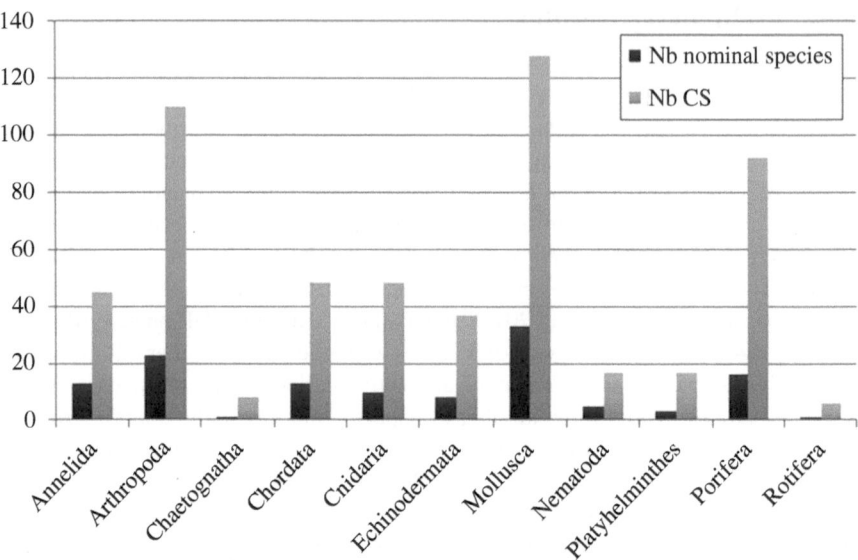

Fig. 4.5 Number of nominal species and CGI per phylum in our pilot study of 402 articles

This highlights that among reported CGI complexes, about half are just in need of taxonomic revision and may not correspond to any phenomenon of deceleration of morphological evolution. Among the 33 CGI complexes that may be CGIss, half display statistical differences in morphology and half do not display any morphological differentiation among genetic entities. These proportions are helpful to plan studies aimed at testing various hypotheses regarding the CS phenomenon. Among the hundreds of studies reporting CGI, about half may just need taxonomic revision, a quarter may be good candidates for testing hypotheses regarding natural selection, effective sizes, etc. Indeed, the categories of our classification based on crossed genetic isolation and morphological differentiation levels (Table 4.1) seem relatively well balanced. However, proportions of "diagnostic/statistically significant/ not significant" morphological differences among CGI vary among phyla (these differences are statistically very significant) (Fig. 4.6).

We investigated the relative geographical distribution of CGI and their ecological differentiation and found that (i) 50% of cases have exclusively allopatric sibling-species, (ii) the ratio of cases displaying "strict allopatry" versus "sympatry" varies among phyla (this result is statistically significant), (iii) there is a higher proportion of diagnostic morphological differences in "sympatric" than in "strictly allopatric" CGI (statistically significant result), (iv) ecological differentiation within CGI is more frequent in sympatric than in allopatric CGI, supporting the competitive exclusion theory (highly significant result) which stipulates that sympatric species cannot coexist stably if they have the same niche: either they evolve distinct niches or one eliminates the other. Returning to our first section on the practical importance of CGI, *this suggests that ignoring CGI leads to underestimating not only species diversity but also local functional diversity.*

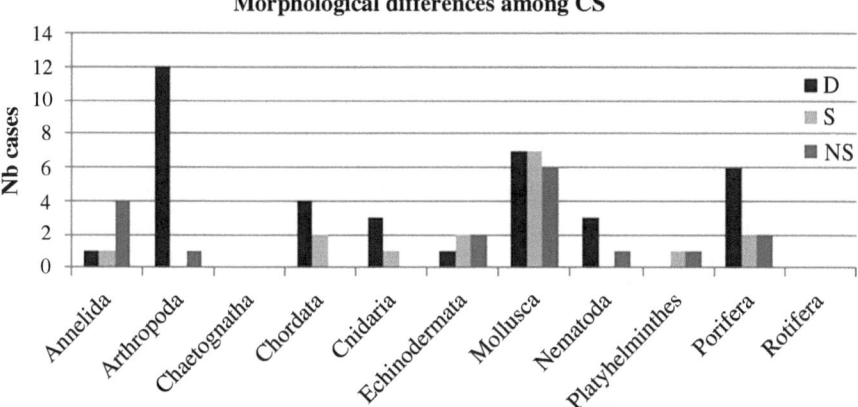

Fig. 4.6 Distribution of studies reporting CGI per phylum according to the status of morphological differentiation among CGI. Abbreviations correspondence: *D* diagnostic differences, *S* statistically significant differences, *NS* non-significant differences

To rapidly infer the ratio of morphological to genetic divergence (indirectly) we looked at (or computed) molecular phylogenies and divergences; we found that: (i) sibling species diverged more than some nominal species of the same group *in 2/3 of the cases, ruling out a "recent speciation" explanation for morphological similarity and confirming decoupling between morphological and genetic divergence* for these CGI, (ii) molecular divergence within CGI was higher for wider habitat ranges (statistically significant), and (iii) there were more diagnostic morphological differences in high dispersal taxa (statistically significant). No straightforward explanations were found for the former results. A much greater survey, also limited to marine metazoans and excluding parasites, has been carried out and its thorough analysis is ongoing (Cahill and Chenuil, unpublished). It selected 1209 studies compiled from more than 4000 titles, of which 55% report CGI, from which another 55% have morphological data, and 12% report ecological comparisons among CGI. As many studies are expected for macrophytes, perhaps more from parasites, and many additional ones would be found in terrestrial taxa. Based on these proportions, there is no doubt that scientists will be able to test many of the hypotheses raised above about factors favoring the presence of CGI in numerous phyla.

4.6   Concluding Remarks on the Use of Morphospecies for Biodiversity Assessment

Since the task is huge, one may argue that it would be more efficient to consider alternative approaches to replace the morphological identification of species in future studies of biological communities, ecosystem monitoring and conservation actions. Taxonomic sufficiency approaches, focusing on higher taxa (instead of the species level), may appear less affected by CGI. However, by lumping related species together they often lose or bias the functionality signal (Thiault et al. 2015) which consists of the variability of ecological functions, because even closely related species frequently have distinct functional traits. Parataxonomy is another approach that eliminates the requirement of rigorous taxonomic identification: it consists of sorting samples to recognizable taxonomic units (RTU). However, the error in this approach is not predictable and depends on the sorter (Krell 2004), precluding comparisons of datasets processed by distinct persons, a big problem for monitoring programs. Neither taxonomic sufficiency nor parataxonomy allow using putative functional knowledge we may have on the entities (not necessarily "species") recorded.

Barcoding and its derived method, metabarcoding, enable the automatic identification of species based on their DNA sequence at a given marker for which there is a huge database containing species names and their corresponding DNA sequence. Diversity estimates based on barcoding are less sensitive to CGI but have other drawbacks (Bucklin et al. 2011; Krishnamurthy and Francis 2012). Typical barcoding or metabarcoding was based on a single marker until now. The largest database is probably the 18S rDNA (and its homologous database, the 16S rDNA, for prokaryotes), which can be used in virtually all eukaryote phyla, but which sequences are not variable enough to distinguish related species within a genus and often within a family. For animals, the well-recognized "barcoding molecule" COI is much more useful than 18S due to its high variability (Chenuil 2006). Fungi and plants also have their own barcoding databases in BOLDSYSTEM (barcoding of life data system) (Ratnasingham and Hebert 2013). As explained above (Sect. 4.3), single marker data cannot establish genetic isolation. When at least another marker will have a sufficiently large database to be used in conjunction with the marker currently used for barcoding in the three main groups of living things, the identification of biological species (or GI entities) not requiring morphological identification will be possible. Another limitation of metabarcoding is its very poor representativeness of species biomass or abundances which may not be completely overcome by the use of various markers. But even with improved barcoding, understanding the discrepancy between morphological, phylogenetic, and biological species will remain necessary to validate fossil data and properly analyze the consequences of past environmental changes. This is particularly important because inferring past changes may help to predict future biodiversity responses to climate change (Condamine et al. 2013).

Once a database compiling putative CGI and containing information on GI levels, morphological differentiation, life history traits, biogeographical distribution and habitat is available, several practical questions related to bioconservation may be answered. (1) Is the error on biodiversity estimators caused by ignored CGI important or do the different errors and biases compensate each other? (2) Do barcoding approaches based on a single sequence marker represent a good solution to correct the CGI problem in common biodiversity estimates? (3) Would barcoding approaches based on *two* independent sequence markers (or more) improve biodiversity estimates? (4) Can we propose correction equations (based on meta-analysis) to solve the problem?

This study provides a robust framework to tackle the very complex question of CGI, by providing a bi-dimensional classification system, and identifying fields to be filled in a database reporting CGI cases. Our application of such a method on a pilot dataset provided promising results since the proportions of the distinct types of CGI appeared well balanced, potentially allowing the testing of all hypotheses raised in this study. Furthermore, it revealed meaningful significant associations among CGI features.

Acknowledgements A research proposal on cryptic species was selected after peer-review to be funded by the CESAB (French CEntre de synthèse et d'analyse sur la biodiversité) which belongs to the FRB (Fondation pour la Recherche sur la Biodiversité). Unfortunately, no funding was provided to most laureates of this call. We nevertheless thank some of the scientists of the consortium whose interest and participation in the CRYSPIM proposal motivated us to write this chapter: Philippe Borsa, Elena Casetta, Julien Claude, Fabien Condamine, Nancy Knowlton, Pieternella Luttikhuizen, Marina Panova.

References

Agapow, P.-M., Bininda-Emonds, O. R., Crandall, K. A., Gittleman, J. L., Mace, G. M., Marshall, J. C., & Purvis, A. (2004). The impact of species concept on biodiversity studies. *The Quarterly Review of Biology, 79*, 161–179.

Avise, J.C. (1994). Molecular markers, natural history and evolution. Springer Science & Business Media.

Barberousse, A., & Bary, S. (2015). Ideal and actual inventories of biodiversity. *Riv Estet, 59*(2), 14–31.

Beauchard, O., Veríssimo, H., Queirós, A. M., & Herman, P. M. J. (2017). The use of multiple biological traits in marine community ecology and its potential in ecological indicator development. *Ecological Indicators, 76*, 81–96.

Bickford, D., Lohman, D. J., Sodhi, N. S., Ng, P. K., Meier, R., Winker, K., Ingram, K. K., & Das, I. (2007). Cryptic species as a window on diversity and conservation. *Trends in Ecology & Evolution, 22*, 148–155.

Boissin, E., Feral, J. P., & Chenuil, A. (2008a). Defining reproductively isolated units in a cryptic and syntopic species complex using mitochondrial and nuclear markers: The brooding brittle star, *Amphipholis squamata* (Ophiuroidea). *Molecular Ecology, 17*, 1732–1744. https://doi.org/10.1111/j.1365-294X.2007.03652.x.

Boissin, E., Hoareau, T. B., Feral, J. P., & Chenuil, A. (2008b). Extreme selfing rates in the cosmopolitan brittle star species complex Amphipholis squamata: Data from progeny-array and heterozygote deficiency. *Marine Ecology Progress Series, 361*, 151–159. https://doi.org/10.3354/meps07411.

Büchi, L., & Vuilleumier, S. (2014). Coexistence of specialist and generalist species is shaped by dispersal and environmental factors. *The American Naturalist, 183*, 612–624. https://doi.org/10.1086/675756.

Bucklin, A., Steinke, D., & Blanco-Bercial, L. (2011). DNA barcoding of marine metazoa. *Annual Review of Marine Science, 3*, 471–508. https://doi.org/10.1146/annurev-marine-120308-080950.

Castelin, M., Van Steenkiste, N., Pante, E., Harbo, R., Lowe, G., Gilmore, S. R., Therriault, T. W., & Abbott, C. L. (2016). A new integrative framework for large-scale assessments of biodiversity and community dynamics, using littoral gastropods and crabs of British Columbia, Canada. *Molecular Ecology Resources, 16*, 1322–1339.

Charlesworth, B., Lande, R., & Slatkin, M. (1982). A neo-darwinian commentary on macroevolution. *Evolution, 36*, 474–498. https://doi.org/10.2307/2408095.

Chenuil, A. (2012). How to infer reliable diploid genotypes from NGS or traditional sequence
data: From basic probability to experimental optimization. *Journal of Evolutionary Biology,
25*, 949–960. https://doi.org/10.1111/j.1420-9101.2012.02488.x.

Chenuil, A. (2006). Choosing the right molecular genetic markers for studying biodiversity: From
molecular evolution to practical aspects. *Genetica, 127*, 101–120. https://doi.org/10.1007/
s10709-005-2485-1.

Chenuil, A., Hoareau, T. B., Egea, E., Penant, G., Rocher, C., Aurelle, D., Mokhtar-Jamai, K.,
Bishop, J. D. D., Boissin, E., Diaz, A., Krakau, M., Luttikhuizen, P. C., Patti, F. P., Blavet,
N., & Mousset, S. (2010). An efficient method to find potentially universal population
genetic markers, applied to metazoans. *BMC Evolutionary Biology, 10*, 276. https://doi.
org/10.1186/1471-2148-10-276.

Condamine, F. L., Rolland, J., & Morlon, H. (2013). Macroevolutionary perspectives to environmental change. *Ecology Letters, 16*, 72–85. https://doi.org/10.1111/ele.12062.

Cusimano, N., & Renner, S. S. (2010). Slowdowns in diversification rates from real phylogenies
may not be real. *Systematic Biology, 59*, 458–464.

Délémontey, N., Du Salliant du Luc, E., & Fanton, H. (2014). Recherche de facteurs associés à
la présence d'espèces cryptiques en mer par analyse de données bibliographiques (encadrant:
Chenuil).

Drozdov, A. L., & Vinnikova, V. V. (2010). Morphology of gametes in sea urchins from Peter the
Great Bay, Sea of Japan. *Russian Journal of Developmental Biology, 41*, 37–45.

Egea, E., David, B., Chone, T., Laurin, B., Feral, J. P., & Chenuil, A. (2016). Morphological and
genetic analyses reveal a cryptic species complex in the echinoid *Echinocardium cordatum*
and rule out a stabilizing selection explanation. *Molecular Phylogenetics and Evolution, 94*,
207–220. https://doi.org/10.1016/j.ympev.2015.07.023.

Eme, D., Zagmajster, M., Delic, T., Fiser, C., Flot, J.-F., Konecny-Dupre, L., Palsson, S., Stoch,
F., Zaksek, V., Douady, C. J., & Malard, F. (2018). Do cryptic species matter in macroecology?
Sequencing European groundwater crustaceans yields smaller ranges but does not challenge
biodiversity determinants. *Ecography, 41*, 424–436. https://doi.org/10.1111/ecog.02683.

Faurby, S., Eiserhardt, W. L., & Svenning, J.-C. (2016). Strong effects of variation in taxonomic
opinion on diversification analyses. *Methods in Ecology and Evolution, 7*, 4–13.

Fišer, C., Robinson, C. T., & Malard, F. (2018). Cryptic species as a window into the paradigm shift
of the species concept. *Molecular Ecology, 27*, 613–635. https://doi.org/10.1111/mec.14486.

Gotelli, N. J., & Colwell, R. K. (2001). Quantifying biodiversity: Procedures and pitfalls in the
measurement and comparison of species richness. *Ecology Letters, 4*, 379–391. https://doi.
org/10.1046/j.1461-0248.2001.00230.x.

Hennig, W. (1950). *Grundzüge einer Theorie der phylogenetischen Systematik*. Berlin: Deutscher
Zentralverlag.

Hubbell, S. P. (2001). *The unified neutral theory of biodiversity and biogeography*. Princeton:
Princeton University Press.

Huelsenbeck, J. P., Andolfatto, P., & Huelsenbeck, E. T. (2011). Structurama: Bayesian inference
of population structure. *Evolutionary Bioinformatics, 55*. https://doi.org/10.4137/EBO.S6761.

Johannesson, K. (2003). Evolution in *Littorina*: ecology matters. *Journal of Sea Research, 49*,
107–117. https://doi.org/10.1016/S1385-1101(02)00218-6.

Johnson, G. D., Paxton, J. R., Sutton, T. T., Satoh, T. P., Sado, T., Nishida, M., & Miya, M. (2009).
Deep-sea mystery solved: Astonishing larval transformations and extreme sexual dimorphism
unite three fish families. *Biology Letters, 5*, 235–239. https://doi.org/10.1098/rsbl.2008.0722.

Jombart, T., Devillard, S., & Balloux, F. (2010). Discriminant analysis of principal components:
a new method for the analysis of genetically structured populations. *BMC Genetics, 11*, 94.
https://doi.org/10.1186/1471-2156-11-94.

Knowlton, N. (1993). Sibling species in the sea. *Annual Review of Ecology and Systematics, 24*,
189–216.

Krell, F.-T. (2004). Parataxonomy vs. taxonomy in biodiversity studies – pitfalls and applicability of 'morphospecies' sorting. *Biodiversity and Conservation, 13*, 795–812. https://doi.org/10.1023/B:BIOC.0000011727.53780.63.

Krishnamurthy, P., & Francis, R. A. (2012). A critical review on the utility of DNA barcoding in biodiversity conservation. *Biodiversity and Conservation, 21*, 1901–1919. https://doi.org/10.1007/s10531-012-0306-2.

Lee, C. E., & Frost, B. W. (2002). Morphological stasis in the Eurytemora affinis species complex (Copepoda: Temoridae). *Hydrobiologia, 480*, 111–128. https://doi.org/10.1023/A:1021293203512.

Luttikhuizen, P. C., Bol, A., Cardoso, J. F., & Dekker, R. (2011). Overlapping distributions of cryptic *Scoloplos cf. armiger* species in the western Wadden Sea. *Journal of Sea Research, 66*, 231–237.

Mayr, E. (1942). *Systematics and the origin of species*. New York: Columbia University Press.

Meyer-Wachsmuth, I., Curini Galletti, M., & Jondelius, U. (2014). Hyper-cryptic marine meiofauna: Species complexes in nemertodermatida. *PLoS One, 9*, e107688. https://doi.org/10.1371/journal.pone.0107688.

Morard, R., Escarguel, G., Weiner, A. K. M., Andre, A., Douady, C. J., Wade, C. M., Darling, K. F., Ujiie, Y., Seears, H. A., Quillevere, F., de Garidel-Thoron, T., de Vargas, C., & Kucera, M. (2016). Nomenclature for the nameless: A proposal for an integrative molecular taxonomy of cryptic diversity exemplified by planktonic foraminifera. *Systematic Biology, 65*, 925–940. https://doi.org/10.1093/sysbio/syw031.

Nydam, M. L., & Harrison, R. G. (2011). Introgression despite substantial divergence in a broadcast spawning marine invertebrate. *Evolution, 65*, 429–442.

Nygren, A. (2014). Cryptic polychaete diversity: a review. *Zoologica Scripta, 43*, 172–183.

Pante, E., Puillandre, N., Viricel, A., Arnaud-Haond, S., Aurelle, D., Castelin, M., Chenuil, A., Destombe, C., Forcioli, D., Valero, M., Viard, F., & Samadi, S. (2015). Species are hypotheses: Avoid connectivity assessments based on pillars of sand. *Molecular Ecology, 24*, 525–544. https://doi.org/10.1111/mec.13048.

Paulay, G., & Meyer, C. (2006). Dispersal and divergence across the greatest ocean region: do larvae matter? *Integrative and Comparative Biology, 46*, 269–281.

Pfenninger, M., & Schwenk, K. (2007). Cryptic animal species are homogeneously distributed among taxa and biogeographical regions. *BMC Evolutionary Biology, 7*, 121.

Puillandre, N., Lambert, A., Brouillet, S., & Achaz, G. (2012). ABGD, Automatic Barcode Gap Discovery for primary species delimitation. *Molecular Ecology, 21*, 1864–1877.

Ratnasingham, S., & Hebert, P. D. N. (2013). A DNA-Based registry for all animal species: The barcode index number (BIN) system. *PLoS One, 8*, e66213. https://doi.org/10.1371/journal.pone.0066213.

Roux, C., Tsagkogeorga, G., Bierne, N., & Galtier, N. (2013). Crossing the species barrier: Genomic hotspots of introgression between two highly divergent *Ciona intestinalis* species. *Molecular Biology and Evolution, 30*, 1574–1587.

Stohr, S., Boissin, E., & Chenuil, A. (2009). Potential cryptic speciation in Mediterranean populations of *Ophioderma* (Echinodermata: Ophiuroidea). *Zootaxa, 1*(20).

Strand, M., & Panova, M. (2015). Size of genera–biology or taxonomy? *Zoologica Scripta, 44*, 106–116.

Thiault, L., Bevilacqua, S., Terlizzi, A., & Claudet, J. (2015). Taxonomic relatedness does not reflect coherent ecological response of fish to protection. *Biological Conservation, 190*, 98–106. https://doi.org/10.1016/j.biocon.2015.06.002.

Tibayrenc, M. (1996). Towards a unified evolutionary genetics of microorganisms. *Annual Review of Microbiology, 50*, 401–429. https://doi.org/10.1146/annurev.micro.50.1.401.

Vacelet, J. (2006). New carnivorous sponges (Porifera, Poecilosclerida) collected from manned submersibles in the deep Pacific. *Zoological Journal of the Linnean Society, 148*, 553–584.

Van Campenhout, J., Derycke, S., Moens, T., & Vanreusel, A. (2014). Differences in life-histories refute ecological equivalence of cryptic species and provide clues to the origin of bathyal Halomonhystera (Nematoda). *PLoS One, 9*, e111889.

Weber, A. A.-T., Stohr, S., & Chenuil, A. (2014). Genetic data, reproduction season and repro-
    ductive strategy data support the existence of biological species in *Ophioderma longicauda*.
    Comptes Rendus Biologies, 337, 553–560. https://doi.org/10.1016/j.crvi.2014.07.007.

Weber, A. A.-T., Stöhr, S., & Chenuil, A. (2019). Species delimitation in the presence of strong
    incomplete lineage sorting and hybridization: Lessons from *Ophioderma* (Ophiuroidea:
    Echinodermata). *Molecular Phylogenetics and Evolution, 131*, 138–148.

Weber, A. A.-T., Abi-Rached, L., Galtier, N., Bernard, A., Montoya-Burgos, J. I., & Chenuil,
    A. (2017). Positive selection on sperm ion channels in a brooding brittle star: Consequence
    of life-history traits evolution. *Molecular Ecology, 26*, 3744–3759. https://doi.org/10.1111/
    mec.14024.

Wheeler, Q., & Meier, R. (2000). *Species concepts and phylogenetic theory: A debate*. New York:
    Columbia University Press.

Yang, Z., & Rannala, B. (2010). Bayesian species delimitation using multilocus sequence data.
    Proceedings of the National Academy of Sciences, 107, 9264–9269.

Chapter 5
The Importance of Scaling in Biodiversity

Luís Borda-de-Água

Abstract Our main tenet is that biodiversity should be studied as a function of scale. The epitome of a similar approach was that of Mandelbrot in his studies on fractals. Although biodiversity patterns may not necessarily follow the mathematical description of fractals, we argue that much can be learnt if we adopt the perspective of studying biodiversity across scales. A case where the concept of scaling is routinely applied in ecology is the species-area relationship, a relationship describing how the number of species (species richness) changes, i.e. scales, as a function of area. However, the importance of scaling is often neglected in ecology. For instance, it is seldomly applied to another component of diversity, the relative abundances of species, being the latter often described using the proportion of individuals of each species. We exemplify the application of scaling to the species relative abundance with our own work. One of the advantages of studying biodiversity under the framework of scaling is that patterns tend to emerge. These patterns emerge from a myriad of processes and their respective interactions. However, understanding the role of each process individually, or quantifying its role in the community functioning, may be empirically impossible. Thus, we argue from theoretical and practical perspectives, including approaches to conservation problems, that we should concentrate our endeavours on the quantitative description of known patterns, as it is often done in other basic and applied sciences, even if that implies temporarily relegating to a secondary position the detailed analyses of the underlying mechanisms.

Keywords Biodiversity · Patterns · Scaling · Species abundance distributions · Species-area relationship

L. Borda-de-Água (✉)
CIBIO/InBio, Centro de Investigação em Biodiversidade e Recursos Genéticos,
Laboratório Associado, Universidade do Porto, Vairão, Portugal

CIBIO/InBio, Centro de Investigação em Biodiversidade e Recursos Genéticos, Laboratório
Associado, Instituto Superior de Agronomia, Universidade de Lisboa, Lisbon, Portugal

© The Author(s) 2019   107
E. Casetta et al. (eds.), *From Assessing to Conserving Biodiversity*,
History, Philosophy and Theory of the Life Sciences 24,
https://doi.org/10.1007/978-3-030-10991-2_5

5.1   Introduction

Among the several environmental challenges faced by humankind, biodiversity loss looms in the background, with the magnitude of its consequences wrapped in uncertainty. It is then natural that society looks for scientific basis, guidance and solutions in ecology to the present biodiversity crisis. However, ecology is a recently born science (e.g. Worster 1994) and the tools to understand biodiversity at a planetary scale are only now becoming available. In fact, despite the important strides in the nineteenth century by biogeographers such as Humboldt, Darwin or Wallace, the description of biodiversity and biogeography is still very much a piecemeal approach, mainly consisting of data sets of "small" size; "small" is in quotation marks because, though species data sets are indeed small samples of larger communities, they tend to require a tremendous sampling effort in terms of labour and financial costs, as is the case with the censuses of 50 ha plots on tropical tree species regularly undertaken by the Smithsonian Institution (Condit 1998).

Only now one starts to glimpse the possibility of using methodologies enabling sampling at large scales, such as remote sensing (e.g. Asner et al. 2017). But whether these new methods will be sufficiently developed in time to help solve the biodiversity crisis remains doubtful. Moreover, it is not clear if humankind, as a whole, is determined to put a concerted effort to understand biodiversity at a global scale. As S. P. Hubbell (personal communication) has put it, it seems to be easier to raise money to fund astronomy projects than to finance ecology ones, though the latter deal with problems closer to our daily concerns.

Besides the practical aspects faced by ecologists, there are also entrenched attitudes that may hinder progress in ecology. In particular, the tendency to scorn at the "mere" mathematical handling of quantitative patterns when the underlying mechanisms are not fully understood. I contend that this attitude is to be avoided and that much can be learnt from a proper mathematical description of patterns and their interrelationships. Of course, we should not ignore the need to explore the underlying mechanisms, though practical considerations may sometimes require postponing such endeavours.[1]

I devote this chapter to the importance of scaling in ecology, and when dealing with scaling patterns often emerge. In particular, we will be interested in biodiversity patterns across spatial scales. When one mentions biodiversity, we tend to think solely of the number of species. Such vision, however, does not fully describe the diversity of life on Earth. Indeed, the notion of biodiversity is far more encompassing. For instance, a common definition by the United Nations Environment Programme Convention on Biological Diversity (UNEP 1992) states that biodiversity

> means the variability among living organisms from all sources including, *inter alia,* terrestrial, marine and other aquatic ecosystems and the ecological complexes of which they are part: this includes diversity within species, between species and of ecosystems.

---

[1] This point is also emphasised in Casetta et al., Chap. 1, in this volume.

Similarly, but specifying more clearly the different levels of diversity within a species, it is the definition given by Wilcox (1984) and adapted by the IUCN:

> the variety of life forms, the ecological roles they perform and the genetic diversity they contain... at all levels of biological systems (i.e., molecular, organismic, population, species and ecosystem)...

Notice that the latter definition includes also the roles performed by the species in the community, which I identify by what is now referred as functional diversity (e.g. Cernansky 2017). As we will see, when studying communities, ecologists also define diversity not only in terms of number of species (species richness), but also in terms of their relative abundance (e.g. Magurran 1988). This distinction is important. Species richness takes into account only the presence of species; thus, each time an individual of a species not yet recorded is found, the number of species is raised by one, but subsequent individuals of the same species are ignored since that species has already been identified. On the other hand, when collecting information on species abundance all individuals (or some equivalent measure) of a given species have to be recorded since we want to know not simply which species are present but also their relative abundance.

The above definitions imply that we have information on all organisms on Earth. Even at the species level, and disregarding the difficulties of defining "species" (e.g., Coyne and Orr 2004 or Futuyma 2005), quantifying the total number of species is manifestly impossible with our present means. Perhaps for some groups, such as the cetaceans or primates, which have a relatively small number of species and tend to attract a large public and academic interest, it may be possible to identify all the species, but for other groups, such as insects, this is at present completely impossible. This impossibility arises mainly from practical reasons, such as the required labour intensive procedures that it entails or the economic costs involved, not to mention the negative impacts that gathering the data could have on the habitats. Naturally, ecologists have for long devised procedures to circumvent these problems by resorting to carefully devised sampling schemes and by taking advantage of well known patterns (e.g. May 1990). Patterns play a particularly important role in this chapter, as we will now see.

Patterns, by which we mean "regularities in what we observe in nature; that is, [...] 'widely observable tendencies'" (Lawton 1999), are essential to estimating biodiversity, especially those patterns concerned with how biodiversity *scales* as a function of a given variable. Because the notion of "scaling" is essential to us, we should try to give it a more precise meaning. In general, I will refer to the properties of how a quantity changes as its *scaling properties* or, simply, "*scaling*". A similar definition is that of Storch et al. (2007) according to whom scaling means "the effort to discover and explain how some state variable or dynamic parameter changes with some other variable". The relationship between the number of species and area, called the species-area relationship, a mathematical expression that relates how the number of species changes as a function of the size of the area sampled, is one of the best known scaling relationships in ecology (Rosenzweig 1995). Although the exact quantitative formulation of the scaling properties is often disputed (e.g., see

Rosenzweig 1995), my main tenet is that their existence is a fundamental component of studies on biodiversity. Therefore, I argue that the characterization of biodiversity should include not only the quantification of certain variables at one given scale, e.g. the number of species, but also the scaling properties associated with them. In other words, the scaling properties of biodiversity should be seen as an integral part of the biodiversity patterns.

In the previous paragraph, we used the species-area relationship, "one of the boldest and most robust patterns in ecology" as Lawton (1996) put it, as a typical example of a biodiversity scaling pattern. In fact, its application is common practice in theoretical and applied ecological studies. However, most other components of diversity tend to be studied at one single scale, for instance, the relative abundance of species (McGill 2003; Volkov et al. 2003), hence the ideas (and advantages) underlying the concept of scaling are not always explored. This chapter discusses a case study based on my own work concerning the relative abundance of species. Although there are several ways of measuring species abundances, for simplicity we assume that it is measured by the number of individuals.

To realize why it is important to take into account the relative abundance of species, imagine two communities with exactly the same number of species but with very different relative abundances. In one community the individuals are evenly distributed among species (a circumstance that is unlikely to be observed in nature) while in the other community individuals are unevenly distributed among species, with a few species having a large number of individuals and the majority being rare (a circumstance often observed in nature). Given the way individuals are distributed among species, these imagined communities are likely to exhibit very different temporal population dynamics, including the local risk of species extinction, and pose considerably different problems from a conservation perspective.

The relative abundance of individuals is often depicted using histograms called "species abundance distributions". However, while studies on the species-area relationship emphasize the rate of accumulation of species as area increases, studies on species abundance distributions usually concentrate on a single spatial scale, and efforts are directed towards determining which theoretical statistical distribution provides the best fitting to the empirically distribution. Implicit in this approach is that finding a theoretical distribution that would fit a large number, if not all, of the observed species abundance distributions will provide a pattern that could be used for evaluating theories of species diversity. Although fitting the abundance data at one spatial scale is a worthy endeavour, what has become clear from our studies is that the shape of the distributions depends on the scale at which it is sampled. Here too, as we will see, the concept of scaling may provide a unifying way of describing how the distributions change and reveal possible patterns.

In this chapter I shall argue for the importance of considering questions associated with scaling when dealing with biodiversity issues and associated patterns. But the chapter is in a sense anecdotal, because it deals with my own work on the scaling properties of species abundance distributions, which I use mainly to provide an instance of scaling in biodiversity, and in no way exhausts the topic; for a more complete role of scaling in biodiversity see Storch et al. (2007). But before dwelling into biodiversity proper, we discuss an example of scaling that has become a classic in fractal studies and that enshrines the basic ideas underlying our approach to the study of biodiversity.

5.2   An Example from Fractals

Richardson (1961) asked what the lengths of the borders between several countries were. Surprisingly, he found that the lengths were considerably different depending on the source of information he looked at. For example, Portuguese sources claimed that the border between Portugal and Spain was 1214 km while Spanish ones specified only a length of 987 km, a 23% difference! A likely explanation for such discrepancy was the size of the "ruler", that is, the *scale*, used to estimate the length of the border, as pictured in Fig 5.1a. For reasons that we will not speculate about, the Portuguese used a ruler that was smaller, probably half the size, of that used by the Spanish. As we can see from Fig. 5.1a, if we use the "blue" ruler we will obtain an estimate, if we use the "red" ruler, and because its smaller size allows it to follow better the shape of the border, we are likely to obtain a larger estimate. Naturally, we can imagine an even smaller ruler, that will give an even better fit to the border. Since smaller rulers give a better fit, one may be tempted to keep reducing the size of the ruler in the hope that the length would converge to its true value. Such strategy, however, would be self-defeating because the "roughness" of the border, with all its creases, leads to an estimate that keeps changing with the size of the ruler, as we exemplify in Fig. 5.1. Of course, at some subatomic level we would eventually find the true size of the border, but for all practical purposes the border's length is undefined since it is a function of the ruler's size (Mandelbrot 1967, 1982). In other words, for borders, and other objects that are fractals, the size is only meaningful when the scale at which it was measured (the size of the ruler, in our example) is also specified. Therefore, it would also not be appropriate to say that the border's length obtained by the Portuguese is more "correct" than that obtained by the Spanish, since the border does not have a defined length.

For those who seek concrete answers, the question of finding the border's size may look helpless because there is no numerical invariant characteristic associated with the length; in other words, the length keeps changing. However, if we draw in a double logarithmic plot the length of the border as a function of the ruler's size, we will see that the points fall approximately on a straight line, as shown in Fig. 5.1b. Because we will make use of logarithms often, a brief digression about its significance is in order. Logarithms are often used when one needs to depict a wide range of values in a single plot. For instance, while in an axis in a linear scale the distance between 1 and 2, 2 and 3, *et seq.*, are equal, in a logarithmic axis of base 10, the "distance" between 1 and 10 is equal to the "distance" between 10 and 100, and then between 100 and 1000, *et seq.* The latter happens because logarithms calculated the power at each a certain base needs to be raised to obtained the observed value. For instance, in base 10, the logarithm of 100 is 2 because $100 = 10^2$,

a

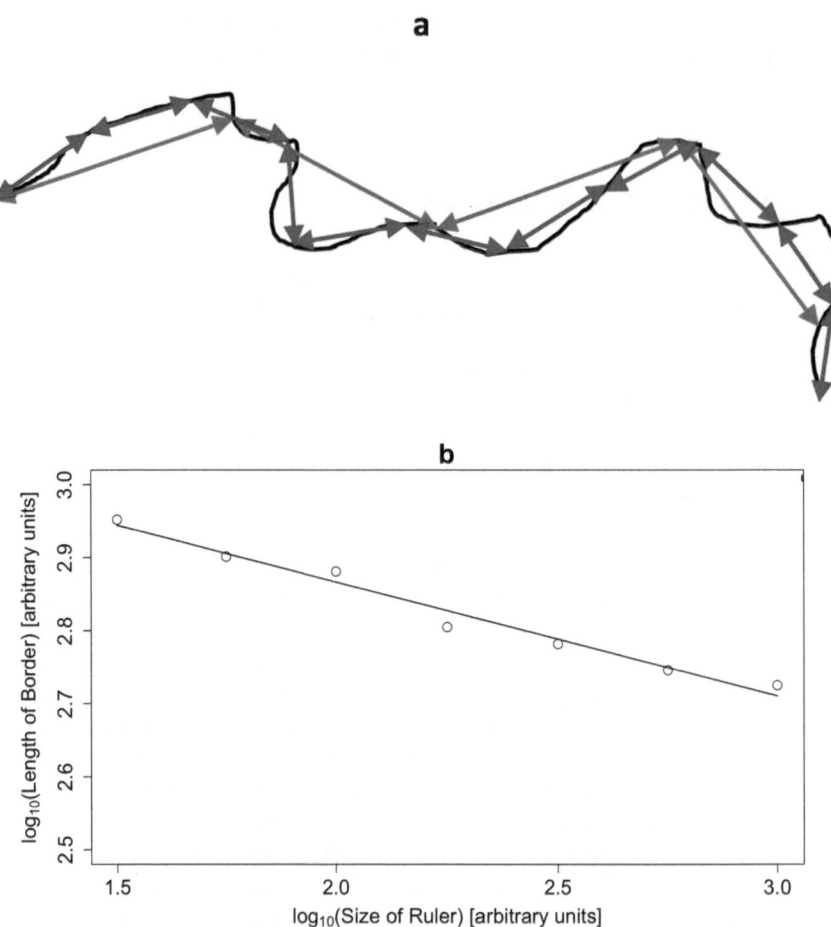

b

Fig. 5.1 A schematic representation of a "border" (black line) being fitted with rulers of two different sizes (blue and red "sticks"), plot a, and the hypothetical straight line that would have been obtained after fitting the "lengths" estimated with rulers of different sizes, plot b

and the one of 1000 is 3; or, in typical mathematical notation $\log_{10}(100) = 2$ and $\log_{10}(1000) = 3$. The implication is that logarithms compress the largest values in the axis of a graphic, with the advantage that the smallest ones are then visually better represented. Importantly, double logarithmic scales are used when one wants to identify power laws, and these play a very important role in ecology, as well as, in several scientific disciplines. Power laws are easily identified in double logarithmic plots because in these plots a power law becomes a straight line. To see this, take the general form of a power law, $y = ax^b$, where a and b are constants. If one takes the logarithms on both sides of this equation, we obtain $\log(y) = \log(a) + b\log(x)$. This means that if we plot $\log(y)$ as a function of $\log(x)$, which is what a plot with double logarithm scales does, we observe a straight line with slope b and intercept $\log(a)$.

Returning to the question of the length of the border, we can see from Fig. 5.1b that, although the length of the border changes as a function of the ruler's size, the estimated lengths define a straight line. Therefore, what is invariant, at least for a very large range of ruler's sizes (scales), is the slope of the (approximately) straight line defined by the points; incidentally, the slope is related to what Mandelbrot called the fractal dimension. Observe that, although we could not find any obvious patterns when we measured the borders with different ruler's sizes, a pattern consisting of a straight line emerges when we plot the measurements in a double logarithmic plot. It is now the slope of this line that provides an invariant characteristic that allows a succinct description of the length as a function of the ruler's size. This example shows that, first, it is not enough to study the length of the border at one single scale, since what is relevant is to study how the length changes as a function of the ruler's size, that is, its scaling properties, and, second, a *pattern emerges* when we display the measurements on a double logarithmic scale plot.

5.3   Scaling and the Species-Area Relationship

The above issues associated with the scaling of the borders' length are easily translated into biodiversity studies. Imagine we sample individuals of a given taxon in areas of different size using a nested sampling scheme, that is, a larger area includes smaller ones. When one plots the number of species as a function of the area size, a typical pattern emerges. At the beginning, for small area sizes, the number of species increases fast, but then the rate of accumulation of new species decreases until it reaches an apparent plateau. This is not surprising: first we observe a large number of new species but when the sample size increases, most individuals belong to species already identified, thus the rate of increase of new species progressively decreases. What is surprising is that if we plot the number of species as a function of the area in a double logarithmic plot, the points will fall on an almost straight line, not very differently from what we depicted in Fig. 5.1b but with a positive slope (e.g., Rosenzweig 1995). Because the points follow a linear relationship in a double logarithmic plot, this implies that the number of species scales with area as a power law, $S = cA^z$, where S is the number of species, A is the area, and c and z are constants. Interestingly, a power law species-area relationship is a rather robust pattern since it also arises under other sampling schemes. For instance, if we estimate the number of species in the islands of an archipelago and plot it in a double logarithmic plot as a function of the area of the islands, we will observe that the points fall approximately on a straight line.

The species-area relationship example is important because it represents a biodiversity pattern that exhibits a clear scaling property, with a clear quantitative relationship: a power law $S = cA^z$. Accordingly, ecologist have concentrated on studying the range of the parameter z, the one that controls the rate of increase of the number of species, and how it can be interpreted (e.g. Rosenzweig 1995). In fact, understanding how species change as a function of area has played an essential role in conservation biology because it (or related relationships) can be used to estimate species extinction due to habitat loss. The basic idea is that if we know how the number of species varies with area, then if part of the original habitat decreases, due to some natural or man-made destructive event, the number of species should also be reduced in accordance with the formula $S = cA^z$. Although simple, this method is not without problems. For instance, Pereira and Daily (2006) pointed out that destroying a habitat and converting it into another, for example, forests into pastures, does not necessarily leads to the extinction of all the species present in the original habitat, since some may still survive in the new one. On the other hand, He and Hubbell (2011) pointed out that the methods to assemble a species-area curve depend on finding the first individual of a new species, while extinction requires finding the very last individual, procedures that are not always equivalent (see also Pereira et al. 2012). In both cases, the consequence is that the number of predicted extinctions is larger than the ones really observed.

5.4 Scaling and Species Abundance Distributions

Species abundance distributions describe not only how many species can be found in a sample but also how the individuals (or other measure of abundance) are distributed among species. Typically, a species abundance distribution is depicted as a histogram where in the y-axis is the number of species and in the x-axis is the number of individuals, usually in a logarithmic scale of base 2. A logarithmic scale of base 2 is used because it is integer base that allows the best discrimination of the species abundances, in particular, it ensures that species with a small number of individuals are well represented in the histograms. Typically, the x-axis corresponds to classes of the logarithm of the number of individuals as follows: 1 individual, 2–3 individuals, 4–7 individuals, *et seq.*, that is, the bins are obtaining by doubling the number of individuals, hence it is called often an octave scale. Accordingly, these histograms contain information on the number of species with a given number of individuals; see Fig. 5.2, for some examples. Although species abundance and species richness are related concepts, the way species abundance distributions have been studied contrasts with that of the species-area relationship. In fact, as we mentioned before, while studies on species area relationships emphasize how species richness change as a function of area, studies on species abundance distributions are often carried at one single spatial scale.

This was not always the case. Some of the original studies on species abundance distributions did focus on how they changed as a function of sample size (Preston 1948, 1962); notice that very often the size of a sample is proportional to the size of the area sampled, therefore in the remainder of this chapter I will use "area size" or "sample size" interchangeably. Preston introduced the concept of *the veil line*, a line that would move to the left of the distribution as more data are gathered, progressively revealing more species and the full shape of the distribution. In particular, using data on birds, Preston showed that, once enough data had been collected, a maximum

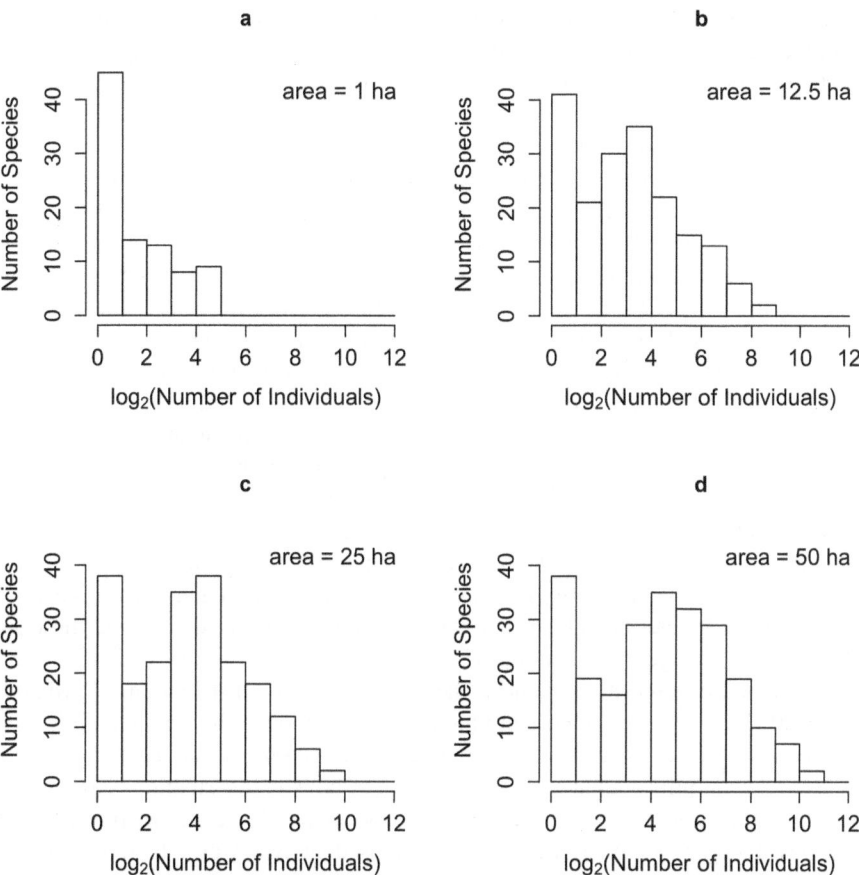

Fig. 5.2 The species abundance distribution for four different sizes of area within the BCI 50 ha plot. The x-axis corresponds to classes of the logarithm of the number of individuals as follows: 1 individual, 2–3 individuals, 4–7 individuals, *et seq.*

appeared for intermediate abundance classes (like in Fig. 5.2b–d), in contrast with the distribution observed for small sample sizes that were monotonically decreasing functions (like in Fig. 5.2a). The latter shape had been previously reported by Fisher et al. (1943) for data on Lepidoptera (moths and butterflies) who fitted the distribution using a logseries distribution. Because the histograms observed by Preston for large samples had a bell shape when plotted in a logarithmic scale, Preston suggested that a lognormal distribution rightly described the data. We now know that the concept of the veil line is not appropriate (Dewdney 1998; Green and Plotkin 2007) and computer simulations and empirical work suggest that the lognormal is not the distribution for very large sample sizes (ter Steege et al. 2006; Borda-de-Água et al. 2007).

While Preston did pay attention to scale, some of the most recent studies on species abundance distributions have focused on determining which theoretical distribution gives the best fit to an empirical observed distribution at a specific scale (Hubbell 2001; McGill 2003; Volkov et al. 2003). (I suspect this approach may partially reflect Preston's attempt to describe species abundances by assuming the lognormal to be the limiting distribution.) The problem with this approach is that the shape of the distributions keeps changing for a wide range of scales (cf. Fig. 5.2), and reaching a sample size that is representative of the community may not be practical or economically feasible; for example, collecting data for the first census on tree species in a 50 ha plot in Barro Colorado Island, Panama, (the data used to draw Fig. 5.2) took approximately 2 years (S. P. Hubbell, personal communication) but of course this is a very small part of the entire forest of which the plot is part of. Moreover, not only the shape of the distributions changes considerably, even for shapes that are visually similar, given the stochasticity inherent to any natural process and sampling procedure, it is very likely that a theoretical distribution that gives the best fitting at one scale may not give the best fitting at another.

Although it is undeniable that for a ecological community a specific instant in time there is a species abundance distribution, our experience based on empirical and computer simulations has shown that from a practical perspective all we can aim for is to understand how the species abundance distribution changes as a function of area. Therefore, we have developed an approach that, instead of analysing the species abundance distribution at one single scale, describes how the distributions change as a function of a given scale (typically the area size) using non-parametric descriptors, namely, the moments of the distributions (Borda-de-Água 2012, 2017).[2] The important finding was that, though the original motivation was partially due to impracticalities of obtaining large samples on large areas, there is also relevant information on the description of how the distributions change as a function of a scale, that is, on the patterns associated with the scaling properties of the distributions. Such information can be used in two ways, first, it provides patterns that can be used to checked whether theoretical models of species diversity predict the scaling for the species abundance distributions, as those empirically observed, and, second, the same patterns can be used to predict the species abundance distributions for larger scales.

The technical aspects of describing the change in the distributions by using their moments are easy. If a species i has X_i individuals, and we call the logarithm (usually of base 2) of this quantity $x_i = \log_2(X_i)$, then the nth moment, M_n, is calculated using the formula $M_n = \dfrac{1}{S} \sum_{i=1}^{S} x_i^n$, where S is the number of species; when $n = 1$, the previous expression is simply the formula of the mean. Thus, in order to describe the change in the distributions, we calculate several moments (e.g. the first 10) for each area and then plot each of the moments as a function of area. For instance, assume that we have information on the location of the individuals of all tree species in a given area, then we can apply the moments' formula for several sub-areas of different sizes and plot the result as a function of area. This is what was done to obtain Fig. 5.3, that shows the evolution of the moments as a function of area size for the

---

[2] For a complete description of diversity indices see Crupi, Chap. 6, in this volume.

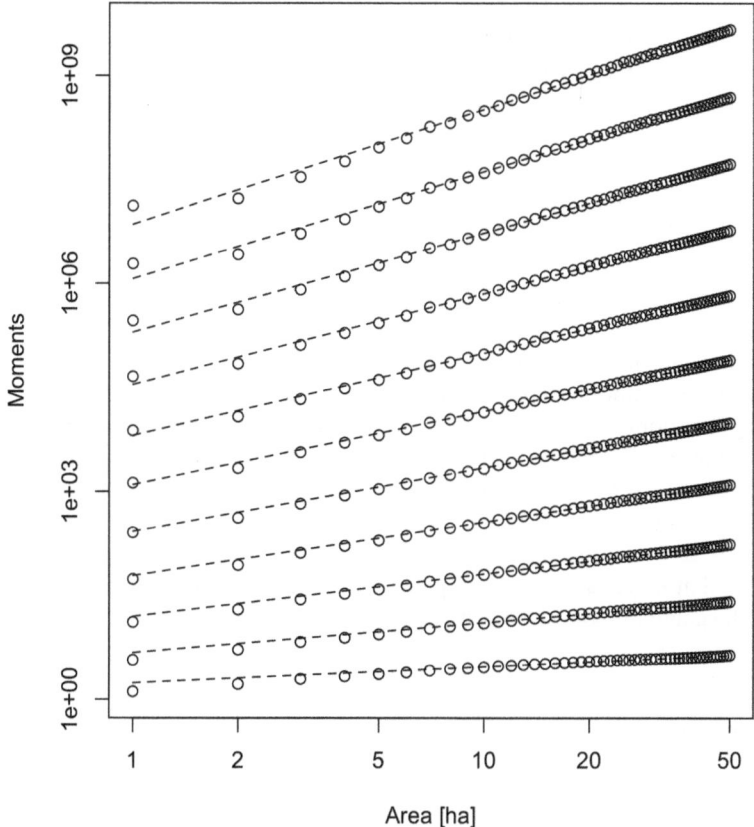

Fig. 5.3 The logarithm of moments from order 1 up to order 11 as a function of the logarithm of areas between 1 and 50 ha using data on tropical tree species from the 50ha plot in Barro Colorado Island, Panama, using all stems with diameter at breast height larger or equal to 10 cm. The order of the moments increases when we go from the bottom to the top lines. The dashed lines added to each moment are the curves of best fit obtained from linear regressions of the logarithm of the moments versus the logarithm of the area between 8 and 50 ha

tree species in a 50 ha plot in Barro Colorado Island, Panama. An important result of this exercise is that the moments are approximately linear in a double logarithmic plot; a point to which we will return soon.

Our approach to deal with the scaling of species abundance distributions is similar to the one underlying the species-area relationship. The species-area relationship describes how the number of species change as a function of the area, and, accordingly, it is a single curve. On the other hand, being a distribution, the information on the species abundance cannot be summarized in a single number. Although sometimes we use a single descriptor, such as the mean, the variance, the skewness or the kurtosis (which are in fact related to the moments), each of these reveals only one aspect of the distribution (e.g. Press et al. 1996). To fully describe a distribution we need all these descriptors (and in fact more). In the same way, we need several moments of the species abundance distribution. As a consequence, for a given area, instead of a single point (the number of species), we have several points (the moments). Therefore, to describe the scaling properties of the species abundance we need several curves, one for each moment. Nevertheless, besides the multiplicity of curves, this approach is essentially the same as that of the species-area relationship.

One of the advantages of using the moments of the distributions is the possibility of extrapolating the species abundance distributions for larger areas (Borda-de-Água et al. 2012). Naturally, to extrapolate the species abundance distributions several assumptions need to be met, in particular, it is implicit that the extrapolation is being performed to a habitat that is not too dissimilar from the habitat where the data were collected. However, instead of dwelling into technical details, I prefer to emphasise the importance of having observed what seems to be a general pattern relating the moments with area size. In fact, the linear behaviour of the moments observed in Fig. 5.3 for tropical tree species was also observed for arthropods from the Azorean archipelago (Borda-de-Água et al. 2017). This observation suggests that the scaling properties of the moments of the species abundance distributions may be a general property for several taxa and, that being the case, there is the prospect that this is a biodiversity pattern. Of course, more studies are required to assert the generality of this observation.

5.5   Final Remarks

The existence of unchanging properties is not given much weight by most statisticians. But they are beloved of physicists and mathematicians, like myself, who call them invariances and are happiest with models that present an attractive invariance property (Mandelbrot 1999).

I do not think it is fair to say that biologists fit the above quote (or statisticians, for that matter). After all, biology has one of the great examples of generalizations in science: Darwin and Wallace's theory of evolution by natural selection. However, I personally find easy to marvel over the diversity of species, and study diversity for its own sake, without looking for the underlying patterns or processes. As Lawton (1999) put it, there is "an almost suicidal tendency for many ecologists to celebrate complexity and detail at the expense of bold, first-order phenomena." I also agree with Lawton (1996) when he stated "[w]ithout bold, regular patterns in nature, ecologists do not have anything interesting to explain." However, it is my impression that often patterns, or attempts to identify them, are not given their due importance (see also Lawton 1992). For instance, we often hear that patterns are important but explaining the underlying mechanisms is what really matters. I am sympathetic with this view; in fact, it merely expresses natural curiosity and the desire to have a deeper understanding of the observed phenomena. However, patterns may emerge from a myriad of causes and have several equally valid explanations. Often, all these causes translate into a simple formula, like the species-area relationship, that combines the action of multiple inextricable causes. These causes occur at different levels of organization in the community, whose effects we can only hope to describe (at least at the moment) from an upper (macro) level of description, and by ignoring the subtle (micro) details occurring at the lower levels.

The existence of theories at different levels of complexity is common in other fields, such as physics, as it is illustrated by thermodynamics and statistical mechanics. The former deals only with macroscopic variables, such as pressure, temperature and volume, while the latter explicitly acknowledges the existence of atoms and tries to derive the laws of thermodynamics from considerations of the statistical properties arriving from a large number of particles (e.g. Reif 1983). For instance, Boyle's law merely states that the product of the pressure and volume of a gas, at a given temperature and (low) density, is constant. This law was obtained based solely on experimental observations, and in itself did not provide any explanation for why that should be the case. Later on, the kinetic theory of ideal gases provided an explanation based on the idealization of a gas as a set of particles with specific (idealized) properties (e.g. Resnick et al. 2001). However, this new development did not remove the practical importance of Boyle's law, but it helped understand its underlying mechanisms and, importantly, its limitations. From a historical perspective we can now realize that having ignored Boyle's law, because it lacked an underlying explanation, would have probably been rather detrimental by hindering a pattern and, hence, the motivation to develop the kinetic theory of gases, and could have delayed the development of practical applications. Furthermore, although we may have the mechanisms that connect the different scales from the microscopic (statistical mechanics) to the macroscopic level (thermodynamics), in some applications researchers and engineers may operate solely within the realm of thermodynamics without need to resort to a more complicated description based on statistical mechanics.

I would like to reiterate with another example the importance of being practical, that is, using what has been firmly experimentally established even if there is not a deeper knowledge of the concerned phenomena. The development of the electromagnetic theory may be the epitome of this approach (e.g. Kraus 1984). Indeed, Maxwell's equations, that still provide the theoretical foundations for electromagnetism, were introduced in the second half of nineteenth century when there was not even the concept of the electron. Still, relevant technological developments in electrical engineering, such as those by Edison and Tesla on the production and transmission of DC and AC currents, were largely independent of the developments in physics that led to the discovery of the electron (for an interesting account of the developing of electrical engineering see Bodanis (2006) or Meyer (1971)). Physicists did not stop at Maxwell's equation and further discoveries on more basic aspects of the structure of matter led to the quantum theory that now provides a model for how electric currents operate. But notice that these latter achievements in physics were often obtained thanks to the parallel developments in electrical engineering that experimentalists used to construct their apparatus. Interestingly, and as a final note, the difference between the models first idealized by physicists and engineers and the physical reality that was later uncovered led to a peculiar convention still in use today. Modelling the electric current as due to the displacement of electrons with their negative charge within a metal was not something that the first physicists and engineers knew, therefore it is no wonder that current was modelled as a fluid moving from the positive to the negative potentials, which is a perfectly natural model if one thinks of the electric current as the flow of a liquid, such as water, that moves from regions of higher to lower potentials. The convention of the electric current flowing from the positive to the negative potentials is the one still in use, because from a practical perspective it does not really matter, but, I suspect, it is unlikely to have been adopted had the researchers been aware of the negative charge of the electrons.

The important point is: we should not think less of a science because it lacks at present a deeper knowledge of its established patterns, nor shall we ignore these patterns because we do not have a clear understanding of their mechanisms. Ignoring, as a first approximation, the details of all ecological interactions was our approach to the study of species abundance distributions. In fact, we did not dwell into the explanations of the patterns observed, although in some cases they may have a simple interpretation in terms of dispersal ability or habitat diversity. Instead, we focused on strategies to obtain the patterns, and the advantages that may result from their identification; for instance, the possibility of forecasting species abundance distributions to hitherto unsampled areas.

Finally, our approach may have seemed unambitious. The reader did not find explanations or attempts to develop explanatory theories. Instead, I discussed only patterns that arose from our studies. This is because I think that patterns are the building blocks of the natural sciences and, as such, I believe that revealing patterns is a first step towards developing or testing theories, as in the previous examples from physics. The analysis of the scaling properties of the species abundance distribution described in this chapter fits in this general scheme, by suggesting a new biodiversity pattern associated with the scaling properties of the moments of species abundance distributions. In this regard two consequences of having detected a scaling pattern are important. First, and from a theoretical perspective, having detected a scaling pattern can be used as a benchmark criterion for evaluating the performance of the models that attempt to recreate patterns of relative species abundance. Second, and from a practical perspective, the pattern associated with the moments can be used to forecast species abundance distributions for areas that are too large to be sampled with our present technology; with obvious applications whenever the knowledge of the species abundance is required, such as in conservation studies. In summary, uncovering another pattern is important because enable us to take action in situations that cannot be postponed, even if the "microscopic" details are not fully understood. Such approach has worked well in physics and engineering, as the examples above illustrated; it remains to be seen if it works equally well in ecology and its practical applications.

Acknowledgements I would like to thank all my mentors, colleagues and friends on discussions on a variety of questions in ecology. In particular, I thank Stephen P. Hubbell, Henrique M. Pereira, Saeid Alirezazadeh, Paulo Borges, Pedro Cardoso, Filipe Dias, Francisco Dionísio, Rosalina Gabriel and Patricia Rodríguez. This work was funded by the FCT project MOMENTOS (PTDC/ BIA-BIC/5558/2014). I thank Davide Vecchi and Elena Casetta for comments that greatly improved the text.

References

Asner, G. P., Martin, R. E., Knapp, D. E., Tupayachi, R., Anderson, C. B., Sinca, F., Vaughn, N. R., & Llactayo, W. (2017). Airborne laser-guided imaging spectroscopy to map forest trait diversity and guide conservation. *Ecography, 355*, 385–389. https://doi.org/10.1126/science.aaj1987.

Bodanis, D. (2006). *Electric universe. The shocking true story of electricity.* New York: Crown Publishers.

Borda-de-Água, L., Hubbell, S. P., & He, F. (2007). Scaling biodiversity under neutrality. In D. Storch, P. A. Marquet, & J. H. Brown (Eds.), *Scaling biodiversity* (pp. 347–375). Cambridge: Cambridge University Press.

Borda-de-Água, L., Borges, P. A., Hubbell, S. P., & Pereira, H. M. (2012). Spatial scaling of species abundance distributions. *Ecography, 35*, 549–556. https://doi.org/10.1111/j.1600-0587.2011.07128.x.

Borda-de-Água, L., Whittaker, R. J., Cardoso, P., Rigal, F., Santos, A. M., Amorim, I. R., Parmakelis, A., Triantis, K. A., Pereira, H. M., & Borges, P. A. (2017). Dispersal ability determines the scaling properties of species abundance distributions: A case study using arthropods from the Azores. *Scientific Reports, 7*(3899). https://doi.org/10.1038/s41598-017-04126-5.

Cernansky, R. (2017). The biodiversity revolution. *Nature, 546*, 22–24. https://doi.org/10.1038/546022a.

Condit, R. (1998). *Tropical forest census plots: Methods and results from Barro Colorado Island, Panama and a comparison with other plots.* Berlin: Springer.

Coyne, J. A., & Orr, H. A. (2004). *Speciation.* Sunderland: Sinauer Associates.

Dewdney, A. K. (1998). A general theory of the sampling process with applications to the "veil line". *Theoretical Population Biology, 54*, 294–302. https://doi.org/10.1006/tpbi.1997.1370.

Fisher, R. A., Corbet, A. S., & Williams, C. B. (1943). The relation between the number of species and the number of individuals in a random sample of an animal population. *The Journal of Animal Ecology, 12*, 42–58. https://doi.org/10.2307/1411.

Futuyma, D. (2005). *Evolution.* Sunderland: Sinauer Associates.

Green, J. L., & Plotkin, J. B. (2007). A statistical theory for sampling species abundances. *Ecology Letters, 10*, 1037–1045. https://doi.org/10.1111/j.1461-0248.2007.01101.x.

He, F., & Hubbell, S. P. (2011). Species-area relationships always overestimate extinction rates from habitat loss. *Nature, 473*, 368–371. https://doi.org/10.1038/nature09985.

Hubbell, S. P. (2001). *The unified neutral theory of biodiversity and biogeography.* Princeton: Princeton University Press.

Kraus, J. D. (1984). *Electromagnetics.* Tokyo: McGraw Hill.

Lawton, J. H. (1992). There are not 10 million kinds of population dynamics. *Oikos, 63*, 337–338.

Lawton, J. H. (1996). Patterns in ecology. *Oikos, 75*, 145–147.

Lawton, J. H. (1999). Are there general laws in ecology? *Oikos, 84*, 177–192.

Magurran, A. E. (1988). *Ecological diversity and its measurement.* Princeton: Princeton University Press.

Mandelbrot, B. B. (1967). How long is the coast of Britain? Statistical self-similarity and fractional dimension. *Science, 156*, 636–638. https://doi.org/10.1126/science.156.3775.636.

Mandelbrot, B. B. (1982). *The fractal geometry of nature.* New York: W. H. Freeman.

Mandelbrot, B. B. (1999, February). A multifractal walk down Wall Street. *Scientific American, 280*, 70–73.

May, R. M. (1990). How many species? *Philosophical Transactions of the Royal Society of London B: Biological Sciences, 330*, 293–304. https://doi.org/10.1098/rstb.1990.0200.

McGill, B. J. (2003). A test of the unified neutral theory of biodiversity. *Nature, 422*, 881–885. https://doi.org/10.1038/nature01583.

Meyer, H. W. (1971). *History of electricity and magnetism.* Cambridge: MIT Press.

Pereira, H. M., & Daily, G. C. (2006). Modeling biodiversity dynamics in countryside landscapes. *Ecology, 87*, 1877–1885. https://doi.org/10.1890/0012-9658.

Pereira, H. M., Borda-de-Água, L., & Martins, I. S. (2012). Geometry and scale in species-area relationships. *Nature, 482*, E3–E4. https://doi.org/10.1038/nature10857.

Press, W. H., Teukolsky, S. A., Vetterling, W. T., & Flannery, B. P. (1996). *Numerical recipes in C*. Cambridge: Cambridge University Press.

Preston, F. W. (1948). The commonness, and rarity, of species. *Ecology, 29*, 254–283. https://doi.org/10.2307/1930989.

Preston, F. W. (1962). The canonical distribution of commonness and rarity: Part II. *Ecology, 43*, 410–432. https://doi.org/10.2307/1933371.

Reif, F. (1983). *Fundamental of statistical and thermal physics*. Tokyo: McGraw-Hill.

Resnick, R., Halliday, D., & Krane, K. S. (2001). *Physics* (Vol. 1, 5th ed.). New York: Wiley.

Richardson, L. F. (1961). The problem of contiguity: An appendix to statistics of deadly quarrels. *General Systems Yearbook, 6*, 139–187.

Rosenzweig, M. L. (1995). *Species diversity in space and time*. Cambridge: Cambridge University Press.

Storch, D., Marquet, P. A., & Brown, J. H. (2007). Introduction: Scaling biodiversity–what is the problem? In D. Storch, P. A. Marquet, & J. H. Brown (Eds.), *Scaling biodiversity* (pp. 1–11). Cambridge: Cambridge University Press.

ter Steege, H., Pitman, N. C., Phillips, O. L., Chave, J., Sabatier, D., Duque, A., Molino, J. F., Prevost, M. F., Spichiger, R., Dastellanos, H., Von Hildebrand, P., & Vásquez, R. (2006). Continental-scale patterns of canopy tree composition and function across Amazonia. *Nature, 443*, 444–447. https://doi.org/10.1038/nature05134.

UNEP [United Nations Environment Programme]. (1992). *Convention on biological diversity*. United Nations Environment Programme, Nairobi, Kenya.

Volkov, I., Banavar, J. R., Hubbell, S. P., & Maritan, A. (2003). Neutral theory and relative species abundance in ecology. *Nature, 424*, 1035–1037. https://doi.org/10.1038/nature01883.

Wilcox, B. A. (1984). In situ conservation of genetic resources: Determinants of minimum area requirements. In *National parks, conservation, and development: The role of protected areas in sustaining society*. Washington D.C.: Smithsonian Institution Press.

Worster, D. (1994). *Nature's economy: A history of ecological ideas*. Cambridge: Cambridge University Press.

Chapter 6
Measures of Biological Diversity: Overview and Unified Framework

Vincenzo Crupi

Abstract A variety of statistical measures of diversity have been employed across biology and ecology, including Shannon entropy, the Gini-Simpson index, so-called effective numbers of species (aka Hill's measures), and more besides. I will review several major options and then present a comprehensive formalism in which all these can be embedded as special cases, depending on the setting of two parameters, labelled *degree* and *order*. This mathematical framework is adapted from generalized information theory. A discussion of the theoretical meaning of the parameters in biological applications provides insight into the conceptual features and limitations of current approaches. The unified framework described also allows for the development of a tailored solution for the measurement of biological diversity that jointly satisfies otherwise divergent desiderata put forward in the literature.

Keywords Diversity · Richness · Evenness · Entropy · Information theory

Suppose that four different species, X, Y, W, and Z, are present in a given environment at a certain time, counting exactly 50, 25, 15, and 10 organisms each, respectively. At the same time in a different location (alternatively: at a later moment in the same area) the numbers are 40, 30, 30, and 0, respectively. *In which of these two situations one deals with a more diverse community*?

This example, although drastically simplified, illustrates a rather general problem. With minor variations, X, Y, W, and Z might just as well be the firms operating in a sector of the economy (see, e.g., Chakravarty and Eichhorn 1991), the languages spoken in a region (see, e.g., Greenberg 1956), the types of television channels in a country (see, e.g., Aslama et al. 2004), or the parties in a political system (see, e.g., Golosov 2010) characterized by their shares of market, speakers, overall

V. Crupi (✉)
Center for Logic, Language, and Cognition, Department of Philosophy and Education, University of Turin, Turin, Italy
e-mail: vincenzo.crupi@unito.it

© The Author(s) 2019

E. Casetta et al. (eds.), *From Assessing to Conserving Biodiversity*,
History, Philosophy and Theory of the Life Sciences 24,
https://doi.org/10.1007/978-3-030-10991-2_6

123

broadcasting, or parliamentary seats. In each of these domains, and still others, measuring *diversity* (or, conversely, *concentration*) has been a significant scientific issue. In biology and ecology, tracking diversity over space and time is of course a key topic for environmental concerns, but biological diversity also plays a relevant theoretical role for its connections with other variables of interest, such as stability, predation pressure, and so on.

In this chapter, I will review a variety of measures of biological diversity that have been employed and discussed across the scientific literature. Relying on a few intuitive illustrations, I will carry out an assessment of the strengths and limitations of some major options, including Shannon entropy, the Gini-Simpson index, so-called effective numbers of species, and more besides. I will also highlight the appeal of one specific measure, which might have not received adequate attention so far. Finally, I will describe a comprehensive formalism and point out how all diversity measures previously considered can be conveniently embedded in it as special cases, depending on the setting of two parameters, labelled *order* and *degree*. This unified mathematical framework is adapted from generalized information theory (Aczél 1984). As we will see, it provides insight into the conceptual features of current approaches and can allow for tailored technical solutions for the measurement of biological diversity.

6.1 Richness

For our current purposes, measuring diversity has to do with how a given quantity is distributed among some well defined categories.[1] In biological applications, it is typical (although by no means necessary) that such categories amount to distinct species characterized by their relative abundance. The latter quantity, in turn, is often simply measured by the proportion of organisms of that species in the overall target population (but biomass can be employed as well, for instance). In what follows, we will denote diversity as $D(p_1, ..., p_n)$, where n species—$s_1, ..., s_n$—are involved and p_i is the relative abundance of the i-th species, s_i. With this bit of formal notation, we can thus represent our initial example with categories X, Y, W, and Z as concerning the comparison between $D(0.5, 0.25, 0.15, 0.10)$ and $D(0.40, 0.30, 0.30, 0)$. Note that, as a direct consequence of our definition, $p_1, ..., p_n$ will always be positive numbers summing to 1, so that $(p_1, ..., p_n)$ actually represents a probability distribution. In our canonical

---

[1] Our use of the term *category* here is very general: essentially, categories in our current sense are the elements of any partition of interest. This terminology is thus *not* constrained by the more technical and specific distinction between "species category" and "species taxon" (e.g., Bock 2004). In particular, a set of different taxa can be treated as a partition of categories in our terms.

interpretation, p_i equals the probability that a randomly selected individual from the target population belongs to species s_i.

The simplest way to measure diversity, and a useful starting point for discussion, is to just count out the number of species with non-zero relative abundance. This straightforward approach relies on what is usually labelled *richness*, namely, how many different species are represented in an environment. In our formalism, it can be computed as follows (with the convenient convention that $0^0 = 0$):

$$Richness(p_1,...,p_n) = \sum_{i=1}^{n} p_i^0$$

As $p_i^0 = 1$ whenever $p_i > 0$, *Richness* always takes an integer value corresponding to how many ps are strictly positive, i.e., how many species are effectively instantiated by some organism. In order to satisfy the appealing constraint that diversity is null (rather than 1) in the extreme case when *only one* species is present, the following minor variation is sometimes employed (Patil and Taille 1982, p. 551):

$$Richness^*(p_1,...,p_n) = \sum_{i=1}^{n} p_i^0 - 1$$

As austere as it may seem as a measure of diversity, *Richness* yields "an intuitive property that is implicit in much biological reasoning about diversity" (Chao and Jost 2012, p. 204). A basic illustration of such property (aka the *replication principle*, see Jost 2009, p. 927) is conveyed by the following example.

Test Case 1 (Jost 2006, p. 363). Let us consider communities consisting of n equally common species, like (1/5, ..., 1/5) (with $n = 5$) and (1/10, ..., 1/10) (with $n = 10$). Arguably, diversity in the latter case should just be twice as in the former. A compelling measure of diversity should recover such assessment. *Richness* clearly does, because *Richness*(1/10, ..., 1/10) = 10 and *Richness*(1/5, ..., 1/5) = 5.

The richness measure is of course completely transparent in its interpretation, but it is also simplistic in a fairly obvious way: it is entirely insensitive to how even/uneven the distribution is. Here is a second test case to clarify the point.

Test Case 2 (*evenness sensitivity*, see Pielou 1975, p. 7). For a given number of species n, a compelling measure of diversity should assign maximum value to a completely even distribution (with $p_1 = ... = p_n = 1/n$), and a strictly lower value to a distribution which is much more skewed. *Richness*, however, clearly fails this condition. For instance, one has *Richness*(0.25, 0.25, 0.25, 0.25) = 4 = *Richness*(0.97, 0.01, 0.01, 0.01).

6.2 Entropies and Diversity

One traditional approach to meet the requirement underlying *Test case 2* is to analyze biological diversity on the basis of *entropy* measures developed in information
theory (Csizár 2008). By far the most widely known such measure is Shannon's
(Shannon 1948). In our current notation, it amounts to the following[2]:

$$D_{Shannon}\left(p_1,\ldots,p_n\right)=\sum_{i=1}^{n}p_i\log\left(\frac{1}{p_i}\right)$$

How can this measure be interpreted in the biological context? The quantity
$\log\left(\frac{1}{p_i}\right)$ can be seen as representing potential *surprise*, to wit, how surprising it
would be to find out that a randomly selected individual from the target population
belongs to species s_i. In fact, such index of surprise is null in the extreme case when
the outcome is already known for sure (so that $p_i = 1$, and $log(1) = 0$) and it is
increasingly and indefinetely large as p_i approches 0. As a consequence,
$D_{Shannon}(p_1, \ldots , p_n)$ quantifies the average (expected) surprise should one get to
know the species to which a randomly sampled element will belong. Appropriately,
such expected surprise will be low when a very uneven distribution—such as (0.97,
0.01, 0.01, 0.01)—implies a low level of uncertainty about the outcome, because
one species is (or few of them are) very likely to be instantiated in a random draw.
On the other hand, expected surprise gets its maximum value (for given n) when a
completely even distribution—namely, (0.25, 0.25, 0.25, 0.25)—implies the highest level of uncertainty, because each species is equally likely to be instantiated in
a random draw. In fact, the Shannon index of diversity gets *Test case 2* just right: in
particular, $D_{Shannon}(0.25, 0.25, 0.25, 0.25) = 1.386 > 0.168 = D_{Shannon}(0.97, 0.01,
0.01, 0.01)$.

Another very popular index of diversity which is appropriately sensitive to the
unevenness of the abundance distribution is *quadratic* entropy (Vajda and Zvárová
2007), also widely known as the Gini or the Gini-Simpson index (after Gini 1912
and Simpson 1949):

$$D_{Gini}\left(p_1,\ldots,p_n\right)=1-\sum_{i=1}^{n}p_i^2$$

D_{Gini}, too, can be given a convenient interpretation. It amounts to 1 minus the probability that two random draws (with replacement) from the background population
instantiate the same category (in fact, the latter probability is $p_i \times p_i$, for each species

---

[2] The choice of a base for the logarithm is a matter of conventionally setting a unit of measurement.
Usual options include 2, 10, and e. We will adopt the latter throughout our discussion, thus employing the natural logarithm in subsequent calculations.

s_i). For this reason, in biological and ecological applications, D_{Gini} is often seen as the *probability of interspecific encounter* (see Patil and Taille 1982, pp. 548–550), to wit, the probability that two random draws do *not* instantiate the same species. Also, $1—p_i$ can be taken as a natural measure of the *rarity* of species s_i in the target environment. Then, $D_{Gini}(p_1, \ldots, p_n)$ computes the average (expected) rarity of the species to which a (randomly sampled) individual would belong, as emphasized by the following equivalent rendition:

$$D_{Gini}\left(p_1,\ldots,p_n\right)=\sum_{i=1}^{n}p_i\left(1-p_i\right)$$

Expected rarity will be low when a very uneven distribution—such as (0.97, 0.01, 0.01, 0.01)—implies that one very common species is (or few of them are) likely to be instantiated in a random draw. On the other hand, expected rarity gets its maximum value (for given n) when a completely even distribution—namely, (0.25, 0.25, 0.25, 0.25)—implies that the species instantiated in a random draw will always have the same, and substantial, degree of rarity. Accordingly, the Gini index of diversity also gets *Test case 2* right: $D_{Gini}(0.25, 0.25, 0.25, 0.25) = 0.75 > 0.06 = D_{Gini}(0.97, 0.01, 0.01, 0.01)$.

One important and well-known feature of both the Shannon and the Gini index of diversity is that they are *concave* functions. The formal definition of concavity involves the notion of a *mixture* $M_{\alpha}^{P,P^*} =\left(\alpha p_1 +\left(1-\alpha\right)p_1^*,\ldots,\alpha p_n +\left(1-\alpha\right)p_n^*\right)$ of two distributions of relative abundance $P = (p_1, \ldots, p_n)$ and $P^* =\left(p_1^*,\ldots,p_n^*\right)$, where $\alpha \in [0,1]$ determines the relative weights of the distributions combined. As a plain illustration, if $P = (0.9, 0.1)$, $P^* = (0.7, 0.3)$, and $\alpha = 0.5$, then the mixture is $M_{\alpha}^{P,P^*} = (0.8, 0.2)$. A measure of diversity D is then said to be concave if and only if, for any P, P^*, and any $\alpha \in [0,1]$, one has $\alpha D\left(P\right)+\left(1-\alpha\right)D\left(P^*\right)\le D\left(M_{\alpha}^{P,P^*}\right)$. Concavity is sometimes advocated for measures of biological diversity as conveying the idea that, if one pools together two distinct populations X and Y composed by possibly different abundance distributions (P vs. P^*) of the same list of n species, then the aggregate should have a diversity that is at least as great as the average of the initial diversities of X and Y. Whether or not such condition is meant to be compelling in general, one significant implication of concavity is involved in the following example.

Test Case 3 Consider three subsequent moments in time t_1, t_2, and t_3, with corresponding relative abundace distributions of a population with $n = 10$, as follows.

t_1: (0.1,   0.1,   0.1,   0.1,   0.1,   0.1,   0.1,   0.1,   0.1,   0.1)

t_2: (0.5,   0.1,   0.05,   0.05,   0.05,   0.05,   0.05,   0.05,   0.05,   0.05)

t_3: (0.9,   0.1,   0,   0,   0,   0,   0,   0,   0,   0)

Quite clearly, a drop in diversity occurred from t_1 to t_2. But, arguably, the loss of diversity was even greater from t_2 to t_3, essentially because most of the species (in fact, 80% of them) disappeared altogether. A compelling measure of diversity should recover such assessment, and concave measures such as $D_{Shannon}$ and D_{Gini} do. $D_{Shannon}$ takes values 2.303 at t_1, 1.775 at t_2, and then drops to 0.325 at t_3. With D_{Gini}, we have 0.90, 0.72, and 0.18, respectively.

As we already know, because it lacks evenness sensitivity entirely, the *Richness* measure fails our *Test case 2* above. Concerning *Test case 3*, *Richness* gets the main point right, namely that a larger drop in diversity occurs from t_2 to t_3 (although, again because of evenness insensitivity, it fails to detect *any* change in diversity from t_1 to t_2). Noting that $D_{Shannon}$ and D_{Gini} do well in *both* cases 2 and 3, one might conclude that diversity should be quantified by these measures. And yet, the assessment of biological diversity on the basis of entropy measures such as $D_{Shannon}$ and D_{Gini} has been forcefully questioned because such measures do not fulfil the replication principle, and in fact fail as basic a benchmark as *Test case 1*. As it turns out, $D_{Shannon}(1/10, ..., 1/10) = 2.303$, which is definitely less than twice $D_{Shannon}(1/5, ..., 1/5) = 1.609$, and $D_{Gini}(1/10, ..., 1/10) = 0.9$, which is definitely less than twice $D_{Gini}(1/5, ..., 1/5) = 0.8$. To emphasize the troubling consequences of these failures, Jost (2009, p. 926) put forward a variation that is even more extreme (notation slightly adapted):

> Biologists frequently use measures of diversity to detect changes in the environment due to pollution, climate change, or other factors. [...] Suppose a continent has a million equally common species, and a meteor impact kills 999,900 of the species, leaving 100 species untouched. Any biologist, if asked, would say that this meteor impact caused a large absolute and relative drop in diversity. Yet D_{Gini} only decreases from 0.999999 to 0.99, a drop of less than 1%. Evidently, the metric of this measure does not match the intuitive concept of diversity as used by biologists, and ecologists relying on D_{Gini} will often misjudge the magnitude of ecosystem change. This same problem arises when $D_{Shannon}$ is equated with diversity.

6.3   Effective Numbers

Consider Jost's meteor illustration above. According to the replication principle (which Jost 2009 strongly advocates), diversity should be 10.000 times lower after the impact than it was before (1.000.000/100). Is it possible to define a measure such that (as it happens with richness but not with entropy) the replication principle is retained and at the same time (as it happens with entropy but not richness) the (un)evenness of the distribution is also integrated in the assessment of diversity?

A crucial step in this direction was made in classical work by MacArthur (1965) and Hill (1973). To introduce their proposal, consider an entropy measure such as D_{Gini}, and take one specific example such as $D_{Gini}(0.4, 0.3, 0.2, 0.1)$, which amounts to 0.7. We now ask: how many species should a *completely even* population include in order for its diversity to be just the same (i.e., 0.7) according to D_{Gini} itself? Note that, for a completely even population of n species, D_{Gini} equals $\sum_{i=1}^{n} \frac{1}{n}\left(1-\frac{1}{n}\right)=1-\frac{1}{n}$. For such a hypothetical population of equally abundant species to have D_{Gini}-diversity of 0.7, it should then hold that $1-\frac{1}{n} = 0.7$, by which we compute $n = \frac{1}{1-0.7}$, and thus $n = 3.333...$ So a hypothetical completely even population of 3.333... species would have the same D_{Gini}-diversity as our initial population with actual distribution (0.4, 0.3, 0.2, 0.1). Given a canonical diversity index such as D_{Gini}, this number of corresponding equally abundant categories is usually called the *effective number* of species (relative to the index at issue, D_{Gini} in this case). As our illustration shows, the effective number of species is a theoretical construct: often it will *not* be an integer. Generalizing from our computation above, one can see that the effective number corresponding to D_{Gini} is as follows:

$$D_{Gini-EN}\left(p_1,...,p_n\right)=\frac{1}{1-D_{Gini}\left(p_1,...,p_n\right)}=\frac{1}{\sum_{i=1}^{n} p_i^2}$$

In this case, too, in order to have null diversity (rather than 1) in the extreme case when *only one* species is present, a minor variation can be employed:

$$D_{Gini-EN}^{*}\left(p_1,...,p_n\right)=\frac{1}{\sum_{i=1}^{n} p_i^2}-1$$

An effective number measure like $D_{Gini-EN}$ meets the requirement of combining the replication principle and evenness sensitivity. Indeed, it can be shown that $D_{Gini-EN}$ $(p_1, ..., p_n) = Richness(p_1, ..., p_n)$ whenever the abundance distribution is uniform, so that, for instance, $D_{Gini-EN}(1/10, ..., 1/10) = 10$ and $D_{Gini-EN}(1/5, ..., 1/5) = 5$ (see *Test case 1* above). On the other hand, $D_{Gini-EN}$ is a smooth and strictly increasing function of D_{Gini}, thus it retains the evenness sensitivity of the latter when distributions are *not* uniform: for instance, $D_{Gini-EN}(0.25, 0.25, 0.25, 0.25) = 4$ while $D_{Gini-EN}(0.97, 0.01, 0.01, 0.01) = 1.064$ (see *Test case 2* above). To achieve the same results, one can alternatively generate an effective number measure from yet another evenness sensitive index of diversity, such as Shannon entropy. The general method is the same: take the actual value of $D_{Shannon}$ $(p_1, ... , p_n)$, equate it to $D_{Shannon}\left(\frac{1}{n},...,\frac{1}{n}\right)$, which amounts to $log(1/n)$, then solve for n. The resulting measure is $D_{Shannon-EN}\left(p_1,...,p_n\right)=e^{D_{Shannon}\left(p_1,...,p_n\right)}$ (see, e.g., Jost 2006, p. 364–365).

According to some authors, using an effective number measure is the one right way to quantify *true diversity* in biological and ecological applications (see Hoffmann

Table 6.1 Some properties of alternative ways to quantify biological diversity.

	Test case 1 (replication principle)	Test case 2 (evenness sensitivity)	Test case 3 (concavity)
Richness	Yes	No	Yes
Entropy (D_{Gini})	No	Yes	Yes
Effective number ($D_{Gini\text{-}EN}$)	Yes	Yes	No

Test cases are explained in the text, and associated to formally relevant mathematical conditions (in parenthesis). "Yes" / "no": the diversity measure (in row) yields/does not yield an intuitively adequate result in the test case at issue (in column)

and Hoffmann 2008, and again Jost 2009, for a debate). Following Hill (1973, pp. 429–430) and Ricotta (2003, pp. 191–192), one can highlight another attractive consequence of the replication principle, which is implied by all effective number measures. If a diversity measure $D(p_1, ..., p_n)$ satisfies the replication principle, then one can define a measure of evenness in a very natural way, as $Evenness(p_1, ..., p_n) = D(p_1, ..., p_n)/n$. A simple and compelling property of such a measure of evenness is that it equates a fixed maximum value of 1 (i.e., n/n) whenever the distribution P is uniform, regardless of the value of n. And a straightforward implication is that diversity can then be neatly factorized into richness and evenness as distinct and independent components, for instance as follows:

$$D_{Gini-EN}\left(p_1,...,p_n\right) = Richness\left(p_1,...,p_n\right) \times Evenness_{Gini-EN}\left(p_1,...,p_n\right)$$

One should note, however, that effective number measures do not retain the *concavity* of their generating indexes. For instance, the concavity of D_{Gini} is not retained in $D_{Gini-EN}$ (and the same applies to $D_{Shannon}$ and $D_{Shannon-EN}$). One disturbing consequence is that *Test case 3* above is not addressed in a convincing way. To illustrate, according to $D_{Gini-EN}$, diversity decreases from $D_{Gini-EN}(0.1, ..., 0.1) = 10$ to $D_{Gini-EN}(0.5, 0.1, 0.05, ..., 0.05) = 3.57$ between t_1 and t_2, but the drop is *smaller* from t_2 to t_3, with $D_{Gini-EN}(0.9, 0.1, 0, ..., 0) = 1.22$. As pointed out above, intuition clearly goes in the opposite direction, given that from time t_2 to t_3 eight out of ten species have disappeared entirely.

Table 6.1 summarizes our results so far. On inspection, it naturally suggests the question whether *there exist* a measure of diversity yielding a satisfactory response to *all* of our test cases above. As a final remark in our comparative discussion, I would like to point out that this can be done. One effective way is to adopt the following as a measure of diversity (see Arimoto 1971, p. 186, for an earlier occurrence in the information theory literature):

$$D_{Root}\left(p_1,...,p_n\right) = \left(\sum_{i=1}^{n}\sqrt{p_i}\right)^2$$

Once again, a − 1 correction can be employed to yield $D^*_{Root}(p_1,...,p_n)$, with null diversity (rather than 1) in the extreme case when only one species is present.[3] D_{Root} is demonstrably evenness sensitive (see Crupi et al. 2018). It is also concave and it satisfies the replication principle (see below for this). It addresses *Test case* 1 appropriately, because (according to the replication principle), $D_{Root}(1/10, ..., 1/10) = 10$ and $D_{Root}(1/5, ..., 1/5) = 5$. Moreover, it gets *Test case* 2 right, because (according to evenness sensitivity), $D_{Root}(0.25, 0.25, 0.25, 0.25) = 4 > 1.651 = D_{Root}(0.97, 0.01, 0.01, 0.01)$. And finally, in virtue of concavity, it also accommodates *Test case* 3, implying a moderate decrease in diversity between t_1 and t_2—from $D_{Root}(0.1, ..., 0.1) = 10$ to $D_{Root}(0.5, 0.1, 0.05, ..., 0.05) = 7.908$—and a much larger drop between t_2 and t_3—from $D_{Root}(0.5, 0.1., 0.05, ..., 0.05) = 7.908$ to $D_{Root}(0.9, 0.1, 0, ..., 0) = 1.6$.

6.4   Parametric Measures of Diversity

The discussion above suggests that statistical measure of diversity D_{Root} combines a number of appealing features, and it is good news, I submit, that such a measure exists. In general, however, the plurality of non-identical ways to quantify biological diversity needs not be a reason for concern or skepticism, like in Hurlbert's (1971, p. 585) complaint that "diversity *per se* does not exists". As noted by Patil and Taille (1982, p. 551), the plurality of measures is a very mundane phenomenon in various domains: in statistics, for instance, mean and median are non-equivalent measures of "central tendency"; variance, mean absolute variation, and range are non-equivalent measures of "spread", and so on. In fact, once their main distinctive properties become well understood, it is natural to think that different measures may be most useful relative to varying purposes or contexts. For this reason, several authors have put forward a comprehensive approach, based on *parametric families* of diversity measures. In the final part of this contribution, I would like to point out that *all* of the specific measures mentioned in the foregoing discussion can be embedded as special cases in a unified formalism taken from generalized information theory (Sharma and Mittal 1975; Hoffmann 2008), namely:

$$D^{(r,t)}_{Sharma-Mittal}(p_1,...,p_n) = \frac{1}{t-1}\left[1-\left(\sum_{i=1}^{n}p_i^r\right)^{\frac{t-1}{r-1}}\right]$$

Parameters r and t of the Sharma and Mittal (1975) family of measures are usually taken to be non-negative ($r,t \geq 0$), while for $r \to 1$ and $t \to 1$ $D^{(r,t)}_{Sharma-Mittal}(p_1,...,p_n)$ is known to yield the classical Shannon formula, $D_{Shannon}(p_1, ... , p_n)$ in our notation

---

[3] As pointed out by Arimoto (1971, p. 186), it also turns out that $D^*_{Root}(p_1,...,p_n) = \sum_{i=1}^{n}\sum_{j=1,j\neq i}^{n}\sqrt{p_i p_j}$.

Table 6.2 Some important special cases of the Sharma-Mittal framework for statistical measures of diversity

(r,t)-setting	Diversity measure
$r = 0$ $t = 0$	$D_{Sharma-Mittal}^{(0,0)}\left(p_1,\ldots,p_n\right) = \quad Richness^*\left(p_1,\ldots,p_n\right) = \sum_{i=1}^{n} p_i^0 - 1$
$r \to 1$ $t \to 1$	$D_{Sharma-Mittal}^{(1,1)}\left(p_1,\ldots,p_n\right) = \quad D_{Shannon}\left(p_1,\ldots,p_n\right) = \sum_{i=1}^{n} p_i \log\left(\frac{1}{p_i}\right)$
$r = 2$ $t = 2$	$D_{Sharma-Mittal}^{(2,2)}\left(p_1,\ldots,p_n\right) = \quad D_{Gini}\left(p_1,\ldots,p_n\right) = 1 - \sum_{i=1}^{n} p_i^2$
$r = 2$ $t = 0$	$D_{Sharma-Mittal}^{(2,0)}\left(p_1,\ldots,p_n\right) = \quad D_{Gini-EN}^*\left(p_1,\ldots,p_n\right) = \frac{1}{\sum_{i=1}^{n} p_i^2} - 1$
$r \to 1$ $t = 0$	$D_{Sharma-Mittal}^{(1,0)}\left(p_1,\ldots,p_n\right) = \quad D_{Shannon-EN}^*\left(p_1,\ldots,p_n\right) = e^{\sum_{i=1}^{n} p_i \log\left(\frac{1}{p_i}\right)} - 1$
$r = \frac{1}{2}$ $t = 0$	$D_{Sharma-Mittal}^{\left(\frac{1}{2},0\right)}\left(p_1,\ldots,p_n\right) = \quad D_{Root}^*\left(p_1,\ldots,p_n\right) = \left(\sum_{i=1}^{n} \sqrt{p_i}\right)^2 - 1$

(see Crupi et al. 2018). Accordingly, it is costumary to just posit $D_{Sharma-Mittal}^{(1,1)}\left(p_1,\ldots,p_n\right) = D_{Shannon}\left(p_1,\ldots,p_n\right)$. Other settings of parameters r, t (known as *order* and *degree*, respectively, in the information theory literature) generate all diversity measures mentioned above, as illustrated in Table 6.2.

What is the meaning of the order (r) and degree (t) parameters in the Sharma-Mittal formalism when employed in the measurement of biological diversity?

The order parameter r is an index of the *insensitivity to less abundant species*. In fact, as r increases, diversity gets closer and closer to a simple (decreasing) function of one single element p^* in the distribution $(p_1, \ldots, p_n)$, that is, the relative abundance of the most common species. As an illustration, on the basis of the limit for $r \to \infty$ when $t = 2$, one has $D_{Sharma-Mittal}^{(\infty,2)}\left(p_1,\ldots,p_n\right) = 1 - p^*$ (see Crupi et al. 2018). When $r = 0$, on the contrary, diversity becomes a (increasing) function of the plain number of species with non-null relative abundance. The simplest illustration here is just $D_{Sharma-Mittal}^{(0,0)}\left(p_1,\ldots,p_n\right) = Richness^*\left(p_1,\ldots,p_n\right)$ (see Table 6.2 and Fig. 6.1). This shows that the order parameter r indicates how much a diversity measure disregards relatively rare species. For order-0 measures, the actual distribution of relative abundance is neglected: non-zero abundance species are just counted, as if they were all equally important. For order-∞ measures, on the other hand, only the most abundant species matters, and all others are neglected altogether. The higher [lower] r is, the more [less] the common species are regarded and the rare species are discounted in the measurement of diversity.

Importantly, for extreme values of the order parameter, an otherwise natural idea of *continuity* fails in the measurement of diversity: when r goes to either zero or infinity, it is not the case that small (large) changes in the abundance distribution produce comparably small (large) changes in diversity. To illustrate, all order-0

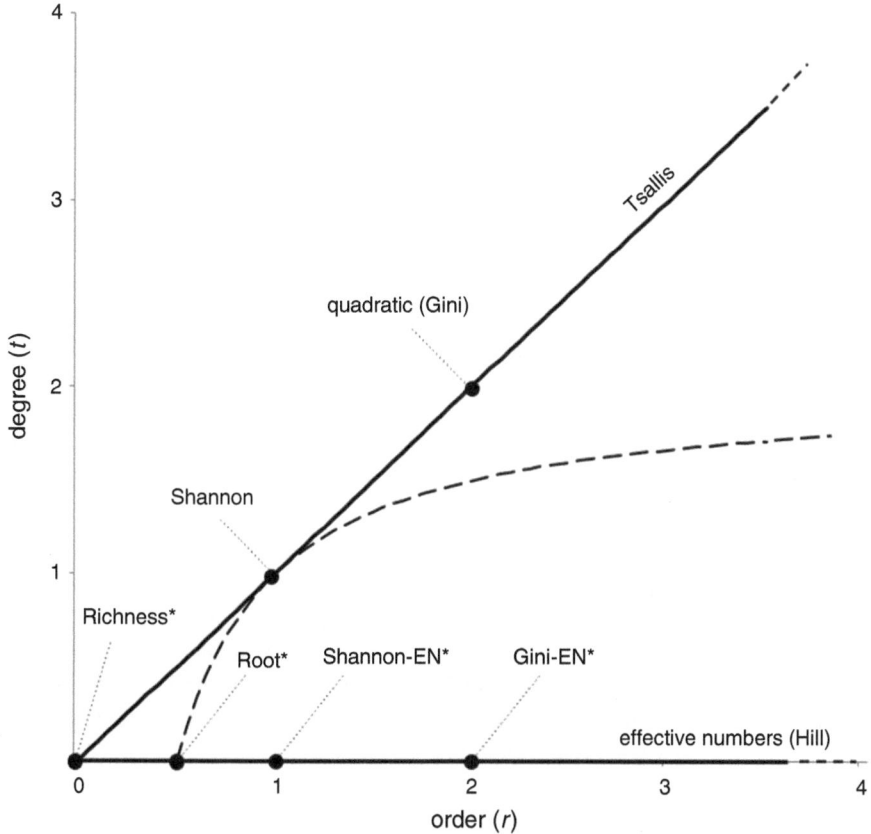

Fig. 6.1 The Sharma-Mittal family of diversity measures is represented in a Cartesian quadrant with values of the order parameter r and of the degree parameter t lying on the x– and y–axis, respectively. Each point in the quadrant corresponds to a specific measure. A line corresponds to a distinct one-parameter generalized diversity function. Several special cases are highlighted. A point in the plane represents a concave diversity measure unless it lies strictly below the dotted line where $t = 2-1/r$

entropies remain entirely invariant upon as large a change as that from, say, $(1/3, 1/3, 1/3)$ to $(0.98, 0.01, 0.01)$, while they yield clearly different values for as small a change as that from $(0.98, 0.01, 0.01)$ to $(0.99, 0.01, 0)$. Order-∞ entropies, in turn, remain entirely invariant upon as large a change as that from, say, $(0.50, 0.25, 0.25)$ to $(0.50, 0.50, 0)$, yet they still yield distinct values for as small a change as that from $(0.50, 0.25, 0.25)$ to $(0.52, 0.24, 0.24)$.

The role of the degree parameter t is somewhat more technical: it affects a few important metric properties. To appreciate this, it is useful to consider that all specific measures we considered earlier lie either (i) on the x-axis or (ii) on the diagonal line in Fig. 6.1. This is not by chance. Let us conclude our discussion by briefly considering cases (i) and (ii) in turn.

(i) Measures lying on the x-axis are obtained by positing $t = 0$, thus yielding:

$$D^{(r,0)}_{Sharma-Mittal}\left(p_1,\ldots,p_n\right) = \left(\sum_{i=1}^{n} p_i^r\right)^{\frac{1}{1-r}} - 1$$

For a diversity measure in the Sharma-Mittal family, having degree 0 ($t = 0$) is known to be a necessary and sufficient condition to satisfy the replication principle. In fact, this was a major reason for Hill (1973) to advocate this formalism as a one-parameter generalized approach to measure diversity. More precisely, the replication principle is satisfied by $D^{(r,0)}_{Sharma-Mittal}\left(p_1,\ldots,p_n\right) + 1$ (for any r) (see Hoffmann 2008, pp. 20–21), and that is equivant to the formula originally employed by Hill (1973, p. 428). The comprehensive approach presented here reveals one striking aspect of Hill's measures: for any Sharma-Mittal measure of a specified order r (regardless of the concurrent value of the degree parameter t!) $D^{(r,0)}_{Sharma-Mittal}\left(p_1,\ldots,p_n\right) + 1$ computes the corresponding effective number as defined earlier, i.e., the theoretical number of equally abundant categories that would be just as diverse as $(p_1, \ldots, p_n)$ is under that measure (see Crupi et al. 2018).

(ii) As we have seen with some special cases like $D^*_{Gini-EN}\left(p_1,\ldots,p_n\right)$, effective number measures of diversity may not be concave functions. Most Sharma-Mittal measures are concave, however: $D^{(r,t)}_{Sharma-Mittal}\left(p_1,\ldots,p_n\right)$ generates a concave function as long as $t \geq 2-1/r$ (see Hoffmann 2008 for a proof). This implies, in particular, the concavity of all measures lying on the diagonal line in Fig. 6.1, which are obtained by positing $r = t$, thus yielding:

$$D^{(t,t)}_{Sharma-Mittal}\left(p_1,\ldots,p_n\right) = \frac{1}{t-1}\left[1 - \sum_{i=1}^{n} p_i^t\right]$$

Measures of this kind are often labelled after Tsallis's (1988, 2004) work in generalized thermodynamics. Partly because of the concavity property, the Tsallis one-parameter continuum has been recently advocated as a compelling approach to the measurement of biological diversity by Keylock (2005).

The relevance of statistical measures of diversity is an open issue for the theoretical biologist and the philosopher addressing the investigation of biodiversity, and indeed a matter of much debate (see, e.g., Barrantes and Sandoval 2009 and Blandin 2015). In this chapter, no claim has been made to the resolution of divergences in this respect. Consideration of the variety and integration of diversity measures remains important, however, for the debate to be adequately informed. Advocates of the measurement of diversity should of course be aware of the tools at their disposal. Opponents and skeptics, on the other hand, should be careful to make sure that their legitimate doubts are not inflated by too narrow an outlook on the ways in which the notion of biological diversity can be formally unpacked and assessed.

References

Aczél, J. (1984). Measuring information beyond communication theory: Why some generalized information measures may be useful, others not. *Aequationes Mathematicae, 27*, 1–19.

Arimoto, S. (1971). Information-theoretical considerations on estimation problems. *Information and Control, 19*, 181–194.

Aslama, M., Hellman, H., & Sauri, T. (2004). Does market-entry regulation matter? *Gazette: The International Journal for Communication Studies, 66*, 113–132.

Barrantes, G., & Sandoval, L. (2009). Conceptual and statistical problems associated with the use of diversity indices in ecology. *International Journal of Tropical Biology, 57*, 451–460.

Blandin, P. (2015). La diversità del vivente prima e dopo la biodiversità. *Rivista di Estetica, 59*, 63–92.

Bock, W. J. (2004). Species: The concept, category, and taxon. *Journal of Zoological Systematics and Evolutionary Research, 42*, 178–190.

Chakravarty, S., & Eichhorn, W. (1991). An axiomatic characterization of a generalized index of concentration. *Journal of Productivity Analysis, 2*, 103–112.

Chao, A., & Jost, L. (2012). Diversity measures. In A. Hastings & L. J. Gross (Eds.), *Encyclopedia of theoretical ecology* (pp. 203–207). Berkeley: University of California Press.

Crupi, V., Nelson, J., Meder, B., Cevolani, G., & Tentori, K. (2018). Generalized information theory meets human cognition: Introducing a unified framework to model uncertainty and information search. *Cognitive Science, 42*, 1410–1456.

Csizár, I. (2008). Axiomatic characterizations of information measures. *Entropy, 10*, 261–273.

Gini, C. (1912). Variabilità e mutabilità. In *Memorie di metodologia statistica, I: Variabilità e concentrazione* (pp. 189–358). Milano: Giuffrè, 1939.

Golosov, G. V. (2010). The effective number of parties: A new approach. *Party Politics, 16*, 171–192.

Greenberg, J. H. (1956). The measurement of linguistic diversity. *Language, 32*, 109–115.

Hill, M. (1973). Diversity and evenness: A unifying notation and its consequences. *Ecology, 54*, 427–431.

Hoffmann, S. (2008). *Generalized distribution-based diversity measurement: Survey and unification*. Faculty of Economics and Management Magdeburg (Working Paper 23). http://www.ww.uni-magdeburg.de/fwwdeka/femm/a2008_Dateien/2008_23.pdf. Accessed 25 Sept 2018.

Hoffmann, S., & Hoffmann, A. (2008). Is there a "true" diversity? *Ecological Economics, 65*, 213–215.

Hurlbert, S. H. (1971). The nonconcept of species diversity: A critique and alternative parameters. *Ecology, 52*, 577–586.

Jost, L. (2006). Entropy and diversity. *Oikos, 113*, 363–375.

Jost, L. (2009). Mismeasuring biological diversity: Responses to Hoffmann and Hoffmann (2008). *Ecological Economics, 68*, 925–928.

Keylock, J. C. (2005). Simpson diversity and the Shannon-Wiener index as special cases of a generalized entropy. *Oikos, 109*, 203–207.

MacArthur, R. H. (1965). Patterns of species diversity. *Biological Reviews of the Cambridge Philosophical Society, 40*, 510–533.

Patil, G., & Taille, C. (1982). Diversity as a concept and its measurement. *Journal of the American Statistical Association, 77*, 548–561.

Pielou, E. C. (1975). *Ecological diversity*. New York: Wiley.

Ricotta, C. (2003). On parametric evenness measures. *Journal of Theoretical Biology, 222*, 189–197.

Shannon, C. E. (1948). A mathematical theory of communication. *Bell System Technical Journal, 27*, 379–423 and 623–656.

Sharma, B., & Mittal, D. (1975). New non-additive measures of entropy for discrete probability distributions. *Journal of Mathematical Sciences (Delhi), 10*, 28–40.

Simpson, E. H. (1949). Measurement of diversity. *Nature, 163*, 688.

Tsallis, C. (1988). Possible generalization of Boltzmann-Gibbs statistics. *Journal of Statistical Physics, 52*, 479–487.

Tsallis, C. (2004). What should a statistical mechanics satisfy to reflect nature? *Physica D, 193*, 3–34.

Vajda, I., & Zvárová, J. (2007). On generalized entropies, Bayesian decisions, and statistical diversity. *Kybernetika, 43*, 675–696.

Chapter 7
Essential Biodiversity Change Indicators for Evaluating the Effects of Anthropocene in Ecosystems at a Global Scale

Cristina Branquinho, Helena Cristina Serrano, Alice Nunes, Pedro Pinho, and Paula Matos

Abstract Understanding and predicting the impact of global change drivers on bio-diversity, the basis of the delivery of goods and services to humans, is a critical task in the *Anthropocene Era*. This has led to the development of international monitoring networks and frameworks to evaluate changes in biodiversity, the *Essential Biodiversity Variables*, though still somewhat ineffective. Biodiversity drivers have changed their relative importance in time and space, e.g. due to policies to combat air pollution, the increasing nitrogen pollution or climate change. Hence, to monitor their impact on biodiversity in space and time, we need appropriate *Biodiversity Change Indicators* and *Surrogates*, measured through distinct metrics. In this chapter, we propose a conceptual model to select the most cost-effective metrics of biodiversity-change based on both the type and intensity of the drivers that limit or impact biodiversity, and the nature of the *Essential Biodiversity Variables* which may be affected in each case. We propose ecophysiology-based metrics for low intensity limiting/impacting drivers, affecting organisms' individual performance; trait-based metrics for medium intensity drivers, affecting the ecological performance of sensitive species before tolerant ones, changing species abundance and community functional traits; taxonomic-based metrics for high driver intensities which may culminate in species loss. We further discuss the utility of remote sensing data to measure some of these indicators or surrogates, allowing to upscale and/or generalize spatial and temporal information.

Keywords Biodiversity · Driver · Surrogates · Metrics

C. Branquinho (✉) · H. C. Serrano · A. Nunes · P. Pinho · P. Matos
Centre for Ecology, Evolution and Environmental Changes, Faculdade de Ciências,
Universidade de Lisboa, Lisbon, Portugal
e-mail: cmbranquinho@fc.ul.pt; hcserrano@fc.ul.pt; amanunes@fc.ul.pt; ppinho@fc.ul.pt;
psmatos@fc.ul.pt

E. Casetta et al. (eds.), *From Assessing to Conserving Biodiversity*,
History, Philosophy and Theory of the Life Sciences 24,
https://doi.org/10.1007/978-3-030-10991-2_7

7.1 Introduction

7.1.1 The Need for Essential Biodiversity Variables

The establishment of the Convention on Biological Diversity, at the Rio Earth Summit in 1992, has put biodiversity at the centre stage. Since then, several agreements have been signed, such as Contracting Parties' agreement on the United Nations Strategic Plan for Biodiversity 2011–2020, and associated Aichi Biodiversity Targets. Despite these agreements, biodiversity continued to change globally (Butchart et al. 2010; Dornelas et al. 2014; Tittensor et al. 2014), with ongoing species loss and/or changes in communities (without species loss), and our knowledge is still too small to understand the exact consequences for human's wellbeing and ecosystems resilience, presently and in the future. Hence, monitoring biodiversity change is an essential task for the XXI century, not only to assure we meet the proposed political goals (Aichi Targets for 2020, Parties to the United Nations, Convention on Biological Diversity, and Sustainable Development Goals for 2030 of the United Nations), but also to guarantee that the provision of basic ecosystem services (MEA 2005) is maintained, securing our survival on Planet Earth.

To monitor biodiversity-change at the global scale, harmonized observation system and timely data are needed (Pereira et al. 2013; Matos et al. 2017). In an attempt to solve this problem and simultaneously trying to answer to the question "what to monitor?", the Group on Earth Observations – Biodiversity Observation Network, proposed the concept of "Essential Biodiversity Variables" (Pereira et al. 2013). This process intended to be the basis of monitoring programs worldwide, and was inspired by the Essential Climate Variables, that guided the implementation of the Global Climate Observing System by the Parties of the United Nations Framework Convention on Climate Change (WMO 2010). The Essential Biodiversity Variables aim to help biodiversity observation-communities: (i) to harmonize monitoring, by identifying how variables should be sampled and measured; (ii) to help prioritize, by defining a minimum set of essential measurements to capture major dimensions of biodiversity change, complementary to one another and to other environmental change observation initiatives; (iii) to facilitate data integration, by providing an intermediate abstraction layer between primary observations and indicators (Pereira et al. 2013). In short, the Essential Biodiversity Variables framework aims to identify a minimum set of variables that can be used to inform scientists, managers and the public on global biodiversity change.

The Essential Biodiversity Variables framework recognizes three distinct levels of biodiversity information: (i) Primary Observations (i.e., raw data); (ii) Essential Biodiversity Variables; and (iii) Biodiversity Indicators (Collen et al. 2009) (Fig. 7.1). In a first attempt, the minimum set of Essential Biodiversity Variables aggregated candidate variables into six classes: "genetic composition," "species populations," "species traits," "community composition," "ecosystem structure,"

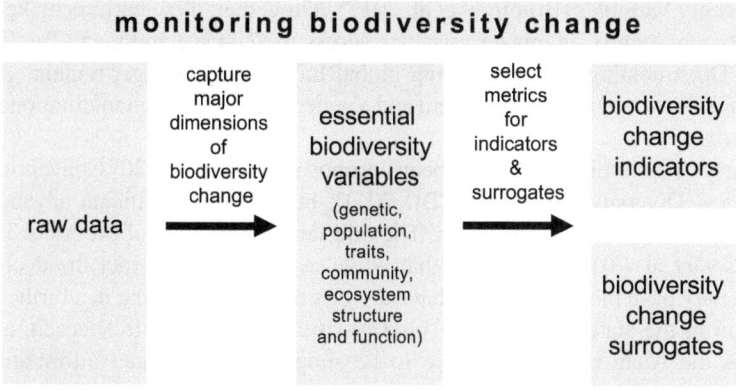

Fig. 7.1 Conceptual framework for "Monitoring Biodiversity Change" and for the selection of the metrics of indicators and surrogates that might reflect biodiversity change at the global scale

and "ecosystem function" (Pereira et al. 2013). The six classes proposed by Pereira et al. (2013) are now widely adopted as part of the essential biodiversity variables framework (Geijzendorffer et al. 2016; Chandler et al. 2017; Turak et al. 2017).

7.1.2 The Challenges of Biodiversity Change Indicators

Putting into practice a global monitoring network to track biodiversity change is far from being an easy endeavour. The Essential Biodiversity Variables framework will undoubtedly help global consistent reporting of changes in the state of biodiversity, but it is unlikely to contribute much to halt biodiversity decline, unless it can be effectively applied at scales relevant to decision-making regarding conservation (Henle et al. 2014). The complexity of biodiversity (different taxa, considerable species diversity within each, complex ecological interactions, numerous pressures interacting synergistically to impact multiple aspects of biodiversity, etc.) arises as a major obstacle, turning this purpose of tracking the trends and the state of biodiversity, in face of controllable and easily achievable conservation goals, a herculean task (Noss 1990; Brooks et al. 2014). In addition, major scientific challenges are faced when distilling biodiversity into a limited number of essential variables. These include: (i) identification of the taxa to be measured among the several that are important for biodiversity conservation and ecosystem services delivery; (ii) identification of a single variable for a critical aspect of biodiversity; (iii) the translation of information between different biological and geographical realms (e.g. terrestrial and marine); (iv) the heterogeneity of methods and data used for measuring and recording different components of biodiversity; and (v) the selection of appropriate metrics (units and scales) of measurement to ensure comparability between Essential

Biodiversity Variables (Brummitt et al. 2017). Altogether, these barriers make it difficult to consistently aggregate variables across time, space and taxa (Turak et al. 2017). Documenting and quantifying global biodiversity change, remains a huge challenge due to sparse or biased data and a general lack of agreed international data standards.

Nearly 100 indicators have been proposed for the 2020 Convention on Biological Diversity targets (UNCBD 2011), but without significant advances on the indicator's practical application. The mid-term assessment of the Aichi Targets (Tittensor et al. 2014), suggested that while actions to counteract the decline of biodiversity have increased, so too have driver's intensity, resulting in a further deterioration in the state and trends of biodiversity. Meaning that to succeed, actions towards the Aichi targets will have to be supported by updated information on regional and global patterns of biodiversity change, on drivers of biodiversity change, and on the effectiveness of conservation policies (Pereira and Cooper 2006; Scholes et al. 2012; Tittensor et al. 2014; Proença et al. 2017).

A successful example was attained with typical ecological indicators, the epiphytic lichens (Matos et al. 2017). Since the industrial revolution (sulphur dioxide) to the present (nitrogen, land use, climate change), lichens have been used to track the major drivers of global change. Yet, currently the challenge is to harmonize methodologies, so they can be used at the global scale. Two protocols are applied at the continental scale, the United States and the European (Fig. 7.2). This work developed a framework to help bridge existing long-term monitoring data sets, investigating the compatibility of the interpretation of their outcomes using broadly accepted biodiversity metrics. This work showed that both methodologies generate similar interpretation trends. The framework developed incorporates measures of species richness, community shits and functional trait metrics. This enabled the use of available data and new data together in jointly analysis under a biodiversity change perspective, giving information on the drivers of change and on the effectiveness of conservation policies at the global scale.

However, very few biodiversity data sets of sufficient quality, across broad taxonomic, temporal and spatial scales are available for official reporting, all of which result in a reduced ability to reliably detect biodiversity change. This leads to information gaps and geographical, temporal and taxonomic biases in reporting efforts worldwide; for example, most data come from less biodiverse areas such as North America and Europe rather than biodiversity-rich areas (Collen et al. 2008; Mora et al. 2008; Pereira et al. 2012). Similarly, vertebrates are much better covered than other taxa (Pereira et al. 2012). In global assessments such as the Global Biodiversity Outlook 4 (CBD 2014), this leads to the predominant use of bird data for many biodiversity indicators (Pereira et al. 2012), undermining the comprehensiveness of the effects of global change on biodiversity and biasing also policy responses based on these reports. Information gaps and biases can originate from the indicator set used or from a lack of robust and reliable data. These constitute practical barriers that require increased efforts before biodiversity data can be used in assessments. Concerning the gaps in data, mobilization of existing data and the collection of new data could help fill current information gaps (Kot et al. 2010). The potential for

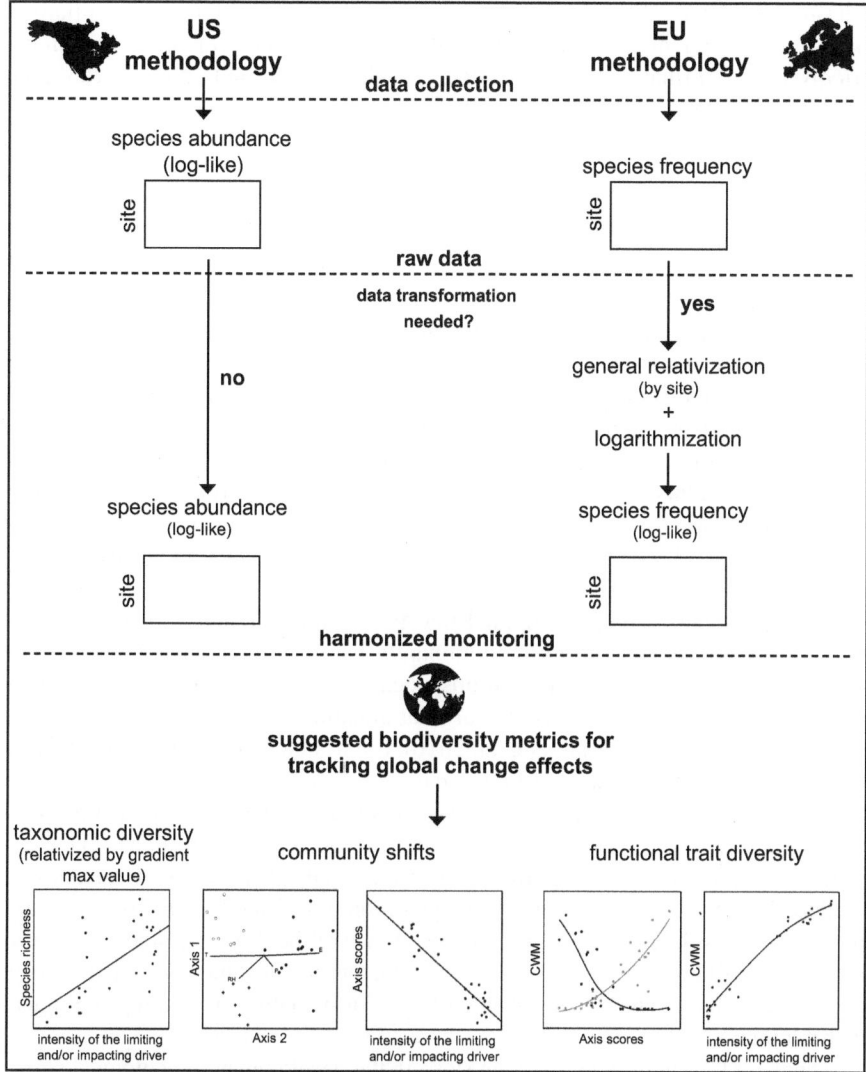

Fig. 7.2 Successful example of a harmonized monitoring framework to be included in the global monitoring network to track biodiversity change. This framework refers to the use of epiphytic lichen diversity collected with the most widely applied methodologies, the United States and European. (Adapted from Matos et al. 2017)

data mobilization is internationally recognized, and several long-term initiatives have focused on mobilizing biodiversity data and metadata (e.g. the Global Biodiversity Information Facility – GBIF, and the Long-Term Ecological Research Network – LTER).

7.1.3   The Need for Surrogates of Biodiversity Change

Most species have not yet been described and even for those that are known, data on spatial distributions are sparse and often unreliable. Given the inability to properly and comprehensively cover all taxa, planning for biodiversity conservation requires surrogates of biodiversity. In an environmental context, a "surrogate" is a component of the system of concern that one can more easily measure or manage than others, and that is used as an indicator of the attribute/trait/characteristic/quality of that system (Mellin et al. 2011). The use of surrogates is important and often necessary because resource constraints in monitoring and management require cost-effective yet useful ways to assess ecosystem responses and key ecological processes (Lindenmayer et al. 2015).

Surrogates can be roughly divided into taxonomic and environmental categories. Taxonomic surrogates are predominantly based on biological data and normally include known taxonomic groups, focal species, umbrella species, species assemblages, and various ecological classifications (Grantham et al. 2010; Lindenmayer et al. 2015; Hunter et al. 2016). Environmental surrogates are usually based on a mix of physical and biological data, subdivided into two types: those based on discrete classes (ecological classifications or land types) and surrogates where continuous data are analysed directly in the selection of areas. They can reflect drivers known to be important in determining the distribution of species and, modelled with species data, can be mapped more consistently, quickly, and inexpensively across large areas (Fig. 7.3). The choice of the drivers is determined also by data availability, spatial scale, choice of data merging techniques, biogeography, and perceptions about the importance of specific variables in shaping biological distributions.

There are four important common steps in the development of ecological surrogates: (i) identify well-developed goals for the use of ecological surrogates (McGeoch 1998; Collen and Nicholson 2014); (ii) develop a robust conceptual model of the system in question to then guide the identification of appropriate surrogates (Niemeijer and de Groot 2008); (iii) rigorously test the ecological surrogates (Bockstaller and Girardin 2003); and (iv) overcome widespread problems of translating the scientific knowledge on ecological surrogates in a way that effectively informs managers and decision-makers, or even the wider public (Halpern et al. 2012; Westgate et al. 2014) (Fig. 7.3).

Recently, Lindenmayer and co-workers (2015) developed a new conceptual Adaptive Surrogacy Framework to explicitly address five trade-offs: (i) whether it is better to employ surrogates or address (e.g. measure) an entity directly; (ii) the

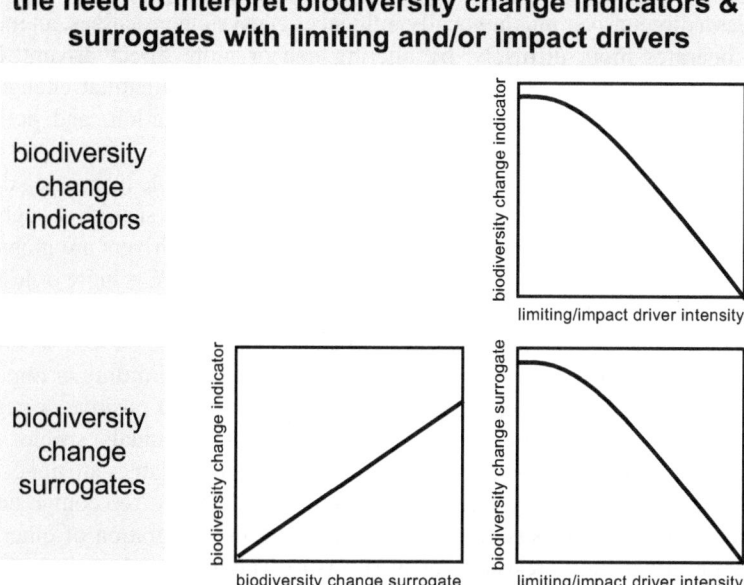

Fig. 7.3 Diagram representing the use of Indicators and Surrogates of Biodiversity Change. Before using Surrogates of Biodiversity Change (bottom-left), they need first to be modelled and validated in relation to the Biodiversity Change Indicator of interest (bottom-right) which must be related to changes in a limiting/impact driver (top-right)

accuracy versus generality of a surrogate; (iii) the temporal stability of a surrogate versus its ability to detect change over time; (iv) simple communication value versus communication complexity associated with caveats and details of methodology; and, (v) cost-effectiveness versus certainty.

7.1.4 The Importance of Drivers Limiting or Impacting Biodiversity Change

The preliminary analysis of biodiversity indicators showed that it requires additional information on non-biodiversity variables, i.e. the drivers limiting or impacting biodiversity in some way (Tittensor et al. 2014). These are essential to inform a specific objective of a policy target (e.g. progress in policy implementation, public awareness, and policy and management responses, such as the change in global surface temperature) and to provide an interpretation of the detected changes in biodiversity (e.g. driven by pollution or climate change), so that actions can be taken to address that problem. A driver is any natural or human induced pressure factor that directly or indirectly causes a change in an ecosystem. Whereas a direct driver unequivocally influences ecosystem processes, an indirect driver operates more diffusely, by altering one or more direct drivers. Some important anthropic direct drivers affecting biodiversity are habitat change, climate change, invasive species, eutrophication, overexploitation, and pollution (MEA 2005).

Changes in biodiversity are almost always caused by multiple interacting drivers working in space and over time, that can occur intermittently (such as droughts) or permanently (such as land-use) (MEA 2005). Though some drivers are global, the actual set of interactions that brings about an ecosystem change is more or less specific to a particular place (Ribeiro et al. 2013). The strength of a driver effect, however, is determined by a range of location-specific factors (MEA 2005). Therefore, it is crucial to identify and evaluate the intensity of the driver limiting or impacting biodiversity (recognized threats and pressures), and the actual positive or negative effects on physiological and ecological performance of individuals, species, communities, habitats and ecosystems. These non-biodiversity variables are not covered by the Essential Biodiversity Variables list (Pereira et al. 2013). Yet, comprehensive interpretation of biodiversity trends clearly requires the integration of other data, notably on drivers and pressures for biodiversity. This is crucial to decide which type of metric of biodiversity change indicator/surrogate should be selected for each Essential Biodiversity Variable.

7.1.5   The Nature and Intensity of the Drivers from the Past to the Future

Direct drivers vary in their importance within and among systems and in the extent to which they are increasing their impact. Historically, habitat and land use change have had the biggest impact on biodiversity across biomes. Overexploitation and invasive species have been important as well, and continue to be major drivers of change (MEA 2005). Pollution, in the past by metals and sulphur dioxide, and more recently the deposition of nitrogen and phosphorus, is expected to increase its impact, leading to declines in biodiversity across biomes. Climate change is projected to increasingly affect all aspects of biodiversity, from individual organisms, through populations and species, to ecosystem composition and function (MEA 2005).

Due to human activities and to climate change, ecosystems are experiencing changes from the local to the global scale (MEA 2005; Canadell et al. 2007). The impacts are of such magnitude that have led to the consideration of a new era – the Anthropocene (Zalasiewicz et al. 2010). Since the first warning by The Limits to Growth (Meadows et al. 1972), scientists have been trying to quantify the environmental, economic, and social limits to human activities on Earth. In a context of global change, Rockström et al. (2009), and more recently Steffen and co-workers

(2015) analysed the safety of nine planetary systems (with great importance for human habitability on Earth) and concluded that the rate of biodiversity loss, climate change and eutrophication have already crossed the safety borders of the terrestrial system. On the other hand, they show that chemical pollution, although undeniably important, had not yet been quantified. All these changes in the planetary systems impact the structure and functioning of ecosystems and the consequent delivery of goods and services they provide. Managing and understanding complex systems such as our society or ecosystems, requires simplification. It is therefore essential the construction of a simple image with a limited set of relevant factors: the "Indicators & Surrogates of Biodiversity Change".

7.2   Objective and Rationale

In the Era of Anthropocene and under the framework of the Essential Biodiversity Variables, the main aim of this work is to call attention for the need to develop Biodiversity Change Indicators and Surrogates to monitor biodiversity changes. These can only be interpreted and applied after knowing their relationship with the drivers that limit or impact biodiversity. We propose a conceptual model to select the most cost-effective metrics of biodiversity change, based on both the nature and the intensity of the drivers that limit or impact biodiversity. During the Anthropocene Era we expect most ecosystems to be affected by at least one anthropic driver (whether it is global or local). Some drivers, such as some pollutants (e.g. DDT) never existed in nature and others existed in much lower amounts or intensity than today (carbon dioxide or ammonia). Ecosystems are currently affected by drivers which limit or impact biodiversity with different intensities. In this work, we use the term 'Driver Intensity' to convey not only the amount but also the toxicity of the drivers affecting biodiversity. Additionally, driver's intensity changes over time. In the past, biodiversity was mostly affected by air pollution (e.g. sulphur dioxide and metals), while nowadays nitrogen pollution is the driver with stronger effects on biodiversity, particularly in rural areas. In the future, we expect changes in climate patterns due to climate change to become the most important driver. Sulphur dioxide in the atmosphere increased almost ten-fold during the industrial revolution (with its maximum in the 70s), whereas ammonia emissions increased more recently and with an intensity of approximately four-fold. Climate change effects on biodiversity are mostly related to changes in the deviation from average climatic variables, such as the case of global surface temperature (Fig. 7.4). Due to this, significant changes were only recently detected, and its intensity is still low in comparison with the magnitude of the previous drivers. Our conceptual model will be based on the selection of the most cost-effective biodiversity change metric having in mind the nature and intensity of drivers, and the nature of the Essential Biodiversity Variables.

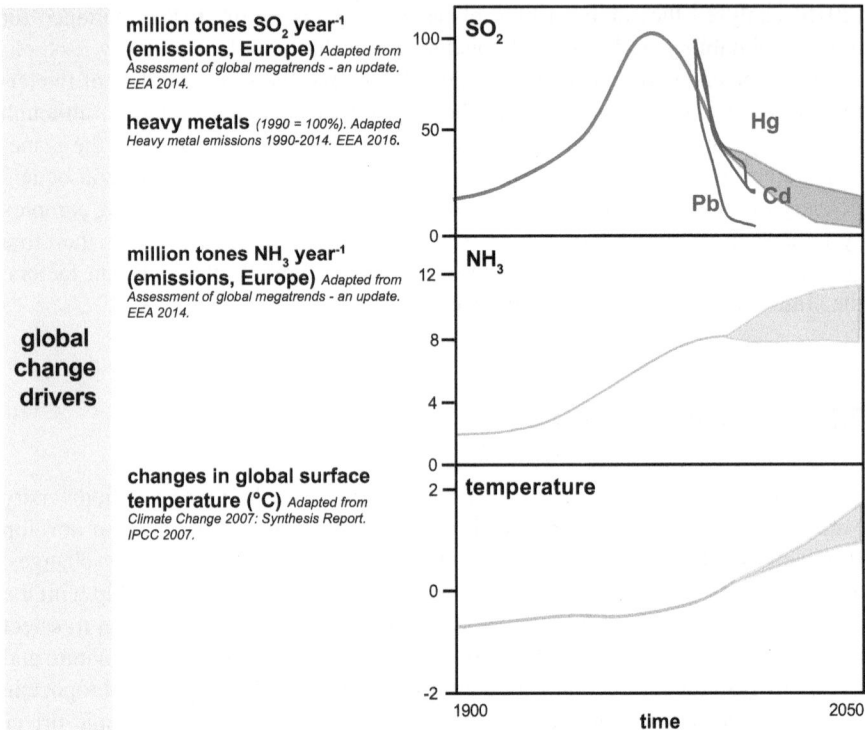

Fig. 7.4 Limiting and impacting drivers of biodiversity change over time. Emissions of gaseous (million tons of sulphur dioxide per year) and metal pollution (lead, mercury and cadmium emissions in relation to 1990), nitrogen pollution (ammonia emissions in tons per year) and climate change (global surface temperature change) since 1900 till present and predicted up till 2050. (Adapted from IPCC (2007) and EEA (2014, 2016))

7.3 How to Choose Biodiversity Change Metrics in Relation to Driver's Intensity

7.3.1 Low Intensity Drivers may Change Biodiversity Metrics from Genetic Composition to Species Populations

When the limiting or impacting drivers are of low intensity (due to both nature or magnitude), we expect them to interfere with individuals' performance, but not to the point of jeopardizing their existence. Individuals physiological performance might be different due to differences in their genetic pool and on acclimation conditions (plasticity). When a population is subject to limiting or impacting drivers, the most sensitive individuals tend to have a lower physiological performance (e.g. lower growth) in comparison with the most tolerant ones, that are not affected for

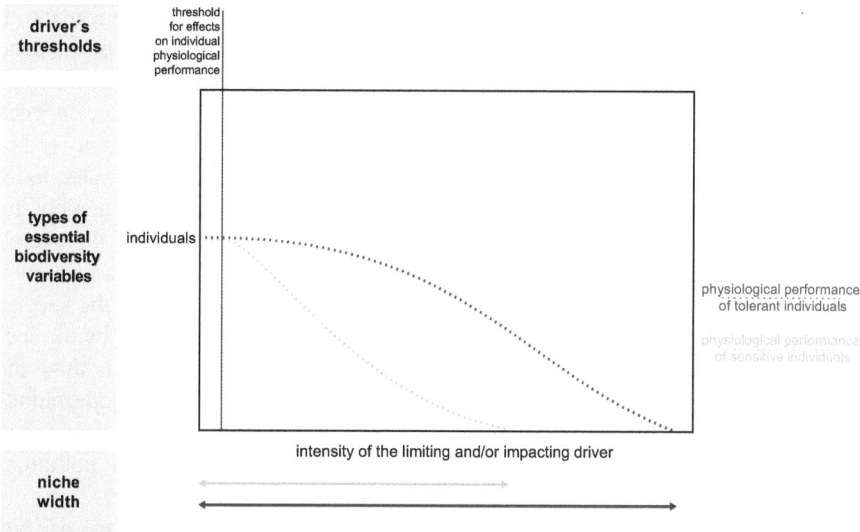

Fig. 7.5 Effect of the driver's intensity on Essential Biodiversity Variables, for individuals' performance. The decrease in physiological performance (dotted lines) of sensitive individuals is more abrupt than that of tolerant ones, though they show similar fundamental niches. Different starting points (genetic or phenotypic variability) of individual performance would also produce different ending points. The ecological performance (realised niche; double arrows) of the sensitive and tolerant individuals indicates their niche width and would be reflected in the biodiversity of the species population

the same driver intensity (Fig. 7.5). The amplitude of driver's intensity where sensitive and tolerant individuals grow determines their realised niche width (Fig. 7.5).

In the beginning there was the gene: one unit, more or less filled with variability (alleles) and prone to mutations that might expand biodiversity and drive evolution. An individual is a complex interaction of genes, with differences enough from the next to make it unique, but also limited within a common pool. Taxonomy studies these genetic pools, mostly by looking at the phenotypes, the physical expression of genes together with the environment, characterizing groups of similar individuals as a species, genus or phylum (e.g.). The more biodiverse a gene is (different alleles), the higher the chances for the outcome to be a more plastic phenotype, capable to withstand environmental variation (Fusco and Minelli 2010).

The fundamental niche concept is deeply connected with that of genetic pool: genes determine the limitations and plasticity ranges an individual can outstand. This concept of fundamental niche (theoretically, the widest range of abiotic conditions where a species can maintain a sustainable population (Hutchinson 1957)) can be considered one of the bases for biodiversity: different sets of abiotic conditions will encompass different performances for different individuals with different genetic backgrounds (Hutchinson 1991). Phylogeny intends to go a bit further than taxonomy, relating the species current taxonomic position to time and evolution.

Both taxonomy and phylogeny are measures of biodiversity that are dependent on the definition of species. With time and environmental drivers in action, the gene pool might be reduced or modified to a point of no return (speciation or extinction).

To add complexity to these interactions, single individual's performance depends not only on their genetic pool, but also on the interaction with the community (other individuals of the same and of different species) and with limiting (soil, water, light, temperature) or impacting drivers (pollution, eutrophication, etc.). The fitness of the several single individuals (their physiological performance) determines the overall performance of the population. Thus, the fitness of the several populations, which all together make the species (ecological) performance, is a proxy for the species realised niche, the part of the fundamental niche that is in fact occupied by the species in natural conditions (Colwell and Rangel 2009). An ecosystem is thus, the combination of a specific community, in a particular set of abiotic conditionings (limiting drivers), that make it unique. A change in community composition or in limiting (edaphoclimatic factors) or impact drivers (e.g. climate change; pollution) will eventually result in a change in biodiversity and the ecosystem status.

In water and nutrients rich ecosystems, and without other abiotic stress drivers, the main limiting drivers for the establishment of an individual should be its dispersion ability to more empty areas and tolerance to interactions with other species (e.g. pathogens, parasites, herbivores). As more individuals occupy those empty spaces, competition arises (intra or interspecific), mostly for light in the case of primary producers, for example. On the other extreme, in poor and hostile ecosystems, low resource availability implies a stronger competitive drive between nearby individuals. Biodiversity here, is shaped by the width of the realised niches, more or less reduced from the fundamental niche width, according to competitive ability to reach or occupy the areas with more resources (Colwell and Rangel 2009; Hortal et al. 2015).

In the case of strong competition (intra or interspecific), the strongest competitor will occupy the areas with more resources, while the weakest one will be pushed to areas with higher abiotic stress, where it doesn't perform so well physiologically (Colwell and Fuentes 1975). In this case, for the weak competitors, the areas of higher ecological performance (abundance) will not be the areas of higher physiological fitness, against what common sense would dictate (McGill 2012). Biodiversity in these communities can change dramatically, if the established ecosystem equilibrium is disrupted by strong drivers, like changes in community composition (e.g. introduction of exotic species) or abiotic factors (e.g. climate change), see Fig. 7.6 for an example. Therefore, at the community level (individuals or even species), the measure of physiological performance vs. ecological performance, would integrate the local biodiversity drivers that affect it, from genes, to individuals, to community, reflected in the niche width. Thus, individual physiological performance of the population can be a good Biodiversity Change Metric that respond to low intensity drivers.

Consider two drivers affecting a community: biotic – competition (intra or interspecific); and environmental – a particular stress. The response of the individuals in the community to those local drivers will determine its biodiversity (Fig. 7.6).

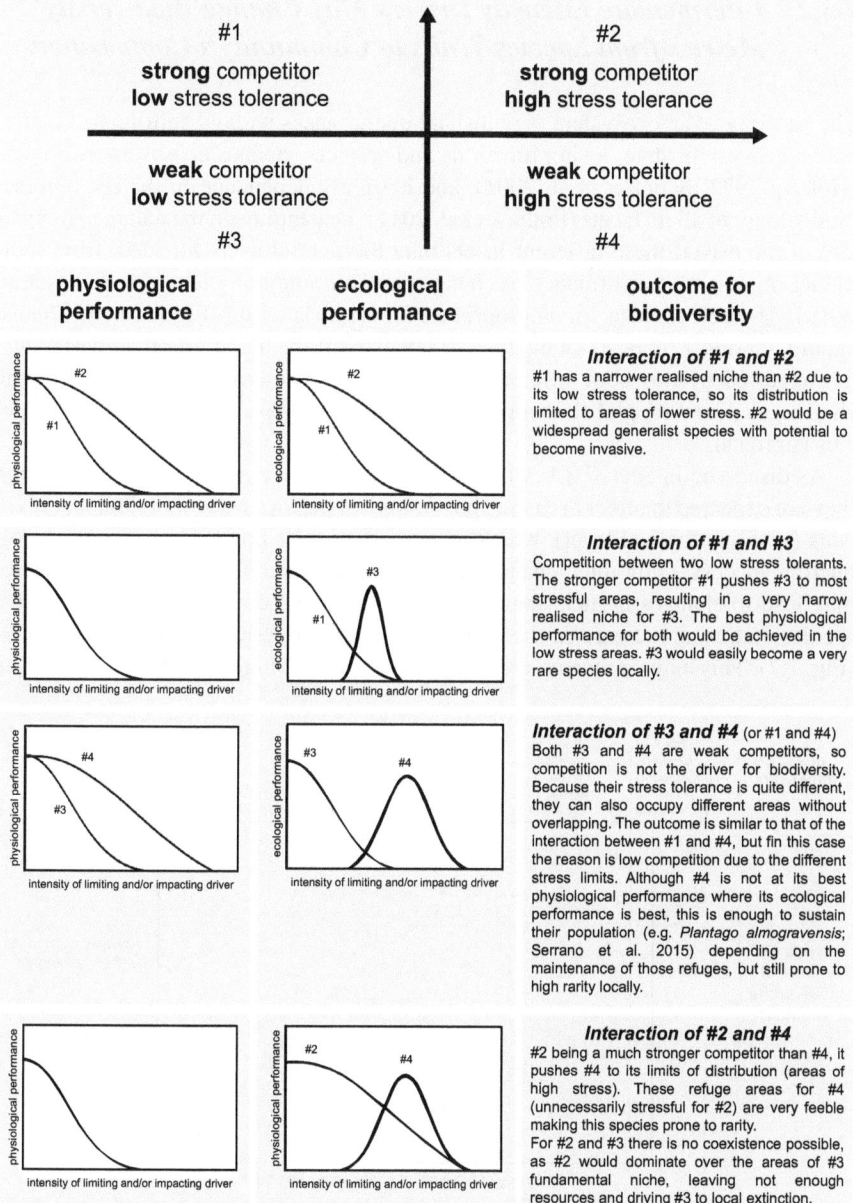

Fig. 7.6 Effect of the biodiversity change drivers in the physiological and ecological performance of two species with different characteristics

7.3.2 Intermediate Intensity Drivers May Change Biodiversity Metrics from Species Traits to Community's Composition

The capacity of an ecosystem to withstand disturbances without shifting to an alternative ecosystem state, losing functions and services, defines ecosystem resilience (Holling 1973; Scheffer et al. 2001), and is tightly dependent on the ecosystems' biodiversity in all its facets (Pollock et al. 2017). Depending on the nature and intensity of the driver, these different facets may be hierarchically affected, from individual organism's performance to changes at the community level (de Bello et al. 2013). The selection of the appropriate Metrics to measure Biodiversity Change should, therefore, depend on the type or intensity of the driver. For that reason, and as suggested by this book title, assessing and conserving biodiversity to face the challenges posed by global change implies moving beyond a mere individual species approach.

As discussed in Sect. 7.3.1, when the drivers act on ecosystems at low intensity, they are expected to affect firstly the physiological performance of biological organisms (at the individual level), which in turn affect their fitness, e.g. growth, reproduction success. Only then, changes in the performance of populations or species may lead to changes in their abundance, determining a decrease in more sensitive species and/or an increase in the most tolerant ones (Cornwell and Ackerly 2009) (Fig. 7.7). This may be seen as a response of species ecological performance that, as

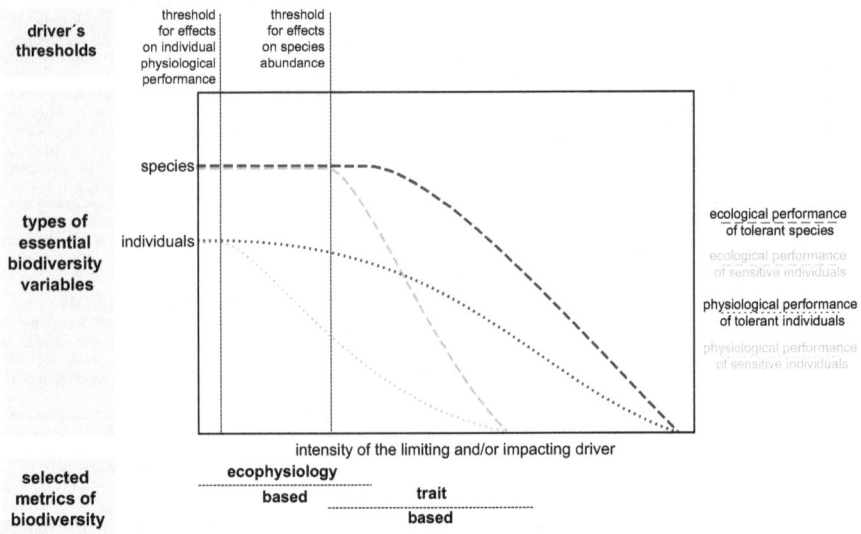

Fig. 7.7 Effect of the driver's intensity on Essential Biodiversity Variables at the individual and species performance level. A low intensity of the driver affects individual's performance, reflected by ecophysiology-based metrics. Increasing intensity of the driver will affect sensitive species before tolerant ones, leading to changes in species abundance and consequently to changes in community functional characteristics, reflected by trait-based metrics

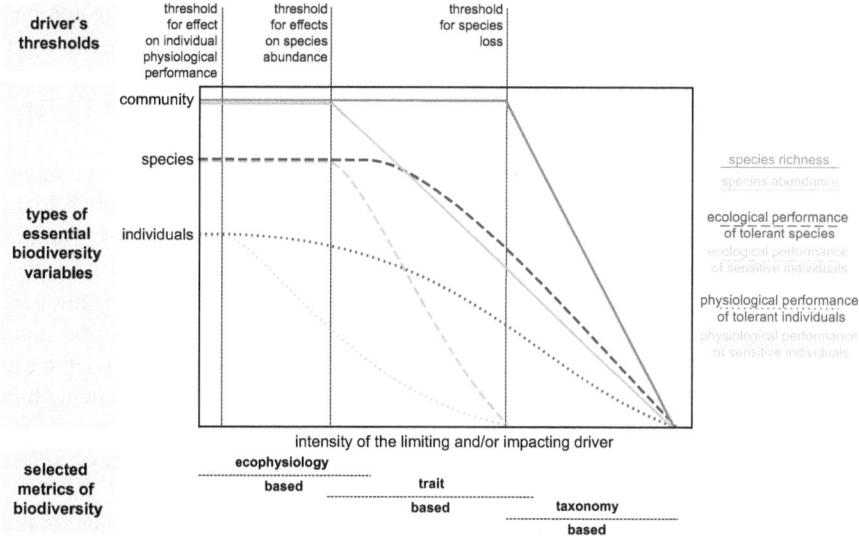

Fig. 7.8 Effect of the driver's intensity on Essential Biodiversity Variables, from individuals' performance to communities. Increasing intensity of the driver will affect sensitive species before tolerant ones, leading to changes in species abundances and consequently to changes in community functional characteristics, reflected by trait-based metrics. A higher intensity of the driver may culminate in species loss, reflected by taxonomic based metrics

mentioned in Sect. 7.3.1, may be due not only to the pressure drivers but also to biotic interactions (de Bello et al. 2012). If the pressure driver persists in time or increases its intensity, changes in ecological performance may culminate in species extinction or their substitution by others with specific traits that make them more tolerant to the pressure driver (Grime and Díaz 2006). Those less tolerant species may be filtered out of the community, affecting species composition and, ultimately, species richness (Fig. 7.8).

7.3.2.1 Intraspecific Trait Variation

It is known that species response to environmental drivers and their effect on ecosystem functioning depend on their characteristics, namely on their functional traits (Hooper et al. 2005; de Bello et al. 2010). For that reason, functional trait-based metrics are emerging as better indicators to quantify changes in ecosystems in response to global change drivers (Díaz and Cabido 1997; Díaz et al. 2007; Lavorel et al. 2011; Mouillot et al. 2013). Functional traits are species attributes, measurable at the individual level, that influence their responses to environmental conditions (by affecting their fitness), or determine their influence on ecosystem properties (Lavorel and Garnier 2002; Hooper et al. 2005). This approach can quantify compositional shifts accounting for species functional redundancy or uniqueness, and has the potential to be both universal and applicable at broad spatial scales (to compare very distinct communities), unlike some taxonomic diversity metrics, because it is not linked to species identity *per se*. Thus, in a case where compositional shifts occur, these metrics enable us to identify which group of species declines (i.e. sensitive species) and which remains (Fig. 7.7).

When studies are focused on the turnover of species (and their traits) in communities and compared across space or over time, the most frequent procedure is to use one mean trait value *per* species, ignoring trait variation within species. The trait value of each species (mean, range or categories) is then "weighted" by their abundance, to obtain trait-based metrics at the community level. However, some traits are more plastic than others, and may show a high intraspecific variation along environmental gradients (e.g. plant specific leaf area). In general, intraspecific trait variability should be also considered: (i) for more ubiquitous species showing high plasticity along environmental gradients; (ii) in communities with a low turnover in species composition; or (iii) in studies addressing trait variations at finer spatial scales.

Plant trait data may be obtained locally, using standard methodologies (Pérez-Harguindeguy et al. 2013), or retrieved from scientific literature or trait databases (Kattge et al. 2011). The first approach is considered crucial in the study of processes acting at the plot-scale (e.g. niche partitioning), while the use of database values is considered acceptable for studies at the site-level or at broader scales (Albert et al. 2011; Cordlandwehr et al. 2013; Shipley et al. 2016). Given that measuring species traits is often laborious and time consuming, and not always possible, trait databases are expected to support the change in paradigm from species to trait-based ecology at the global scale. However, trait data available in databases is still insufficient, and lacks large geographical coverage (Kattge et al. 2011). Trait data gaps are more pronounced, for instance, in northern and central Africa, parts of South America, southern and western Asia (Kattge et al. 2011). Thus, trait-based studies should contribute to fulfil trait data gaps as much as possible.

7.3.2.2   Functional Trait Metrics

Trait-based metrics may be described by several metrics comprising either the mean of functional traits, often called functional structure, or their range or dissimilarity, i.e. functional diversity (Díaz et al. 2007; Lavorel et al. 2008). Both were reported to respond to major environmental drivers or biotic interactions, and to affect major ecosystem processes like primary production or decomposition rates (de Bello et al. 2010; Mouillot et al. 2011, 2013; Valencia et al. 2015). The most widely used metric to measure functional structure at the community-level is the community-weighted mean (CWM) (Garnier et al. 2007). This metric reflects the dominant traits in a community, and derives from the "mass ratio hypothesis", according to which the effects of communities on ecosystem processes are largely determined by the traits of the dominant species (Grime 1998). CWM enables the quantification of community shifts in mean trait values due to environmental selection for certain traits, associated to changes in the abundance of dominant species (Mouillot et al. 2013). For instance, CWM metrics of epiphytic lichen traits and vascular plants were shown to be related to aridity, having the potential to work as indicators of its effects in drylands (Matos et al. 2015; Nunes et al. 2017).

Functional diversity (FD) shows the degree of functional dissimilarity within the community and can be expressed through various indices (Mason et al. 2005; Villeger et al. 2008; Laliberte and Legendre 2010). FD may be used to quantify the decrease or increase in trait dissimilarity in response to driver's intensity compared to a random expectation (i.e. trait convergence or divergence, respectively). For instance, some drivers may act as abiotic filters on communities, selecting for species with similar "more adapted" trait values, i.e., causing trait convergence. This has been found in vascular plant traits of Mediterranean dryland communities, as a response to increasing aridity (Nunes et al. 2017). Following the "niche complementarity hypothesis", a higher FD is thought to reflect an increase in complementarity in resource use between species, and thus an increase in ecosystem functioning (Tilman et al. 1997). Similarly to taxonomic diversity, FD may be divided into three components namely, functional richness, functional evenness, and functional dispersion (Mason et al. 2005). Functional dispersion, which considers trait abundance in addition to richness, has shown a better predictive ability than, for instance, functional richness (Schleuter et al. 2010; Mouillot et al. 2011). Functional diversity also allows the assessment of ecosystems' resilience towards disturbance drivers. The greater the presence of functionally similar species (higher functional redundancy), the higher the probability that disturbance-induced local extinctions of species will be compensated by the presence of similar species, ensuring higher ecosystem resilience (Pillar et al. 2013).

In short, functional trait-based metrics enable a universal and mechanistic understanding of species response to environmental drivers (Mason and de Bello 2013) and may improve predictions of the effect of global change drivers on biodiversity and on ecosystem functioning if drivers have an intermediate intensity (Suding et al. 2008).

7.3.2.3 Multi-trait Metrics

Over the last decade, several multi-trait metrics have been developed with the aim of resuming the functional diversity of multiple traits into a single FD value, estimating the "functional trait space" occupied by a community (Villeger et al. 2008; Laliberte and Legendre 2010; Schleuter et al. 2010). However, the integration of multiple traits into one metric has to take into account single-trait trends and their possible co-variation, to avoid misinterpretation (Butterfield and Suding 2013). Different traits may show divergent responses to environmental variation due, for instance, to trade-offs among species strategies. Another point to consider is that different traits may be redundant, i.e. convey the same information. Thus, joining them in a single FD value may overestimate their importance in relation to other traits. Two solutions may be adopted to avoid redundancy. The first one is to condense the information of single-traits FD into main axes of functional special-ization, with techniques that reduce trait data dimensionality in the most informa-tive variation axis, e.g. multivariate methods such as principal components analysis. Alternatively, and considering that correlated traits may not give exactly the same information, one may include all traits in a composite metric, giving a lower weight to correlated traits. However, in either case, it is important to start by checking for each particular context whether the considered traits co-vary or not, as species may show different combinations of traits under different environments, to maximize their performance (Maire et al. 2013; Volis and Bohrer 2013).

7.3.2.4 Taxonomic Diversity Metrics

Assessing biodiversity changes has largely relied on species richness metrics (i.e. a biodiversity-based taxonomy metrics) (Cadotte et al. 2011; Pereira et al. 2013). This was done firstly because this metric directly addresses biodiversity loss, enabling an assessment of the state of biodiversity change in relation to systematic loss. For several years, this metrics revealed highly effective in quantifying biodiversity changes in relation to drivers of great intensity (e.g. sulphur dioxide pollution), revealing trends of declining biodiversity (Cardinale et al. 2012; Hooper et al. 2012). In taxonomic diversity metrics, this is considered the α biodiversity compo-nent, which measures the diversity of spatially defined units (Magurran 2013).

Though undeniably useful to measure species loss, this metrics application has revealed some limitations. At the global scale, more than a sharp trend of systematic species loss over time, we seem to be witnessing consistent compositional shifts (Dornelas et al. 2014). These compositional shifts depict the β component of biodi-versity that measures spatial or temporal differences in composition between com-munities. The intensity and time of action of the drivers may be responsible for this. The most important drivers in the recent past (Fig. 7.4) have declined its intensity, while the new emerging ones have lower intensities at present, so only in a few years-time, with their continuous action, we will start to observe consistent declines (see Sects. 7.1.2 and 7.1.3). The adoption of measures to control industrial and urban pollutants emissions (e.g. sulphur dioxide or lead) has successfully reduced their levels in the atmosphere (Fenger 2009), since they peaked in late 1970s (Fig. 7.4). Since then, other drivers such as nitrogen pollution or even climate change, became more important (Fig. 7.4). Nonetheless, given their character, intensity, or still short time of action, they seem to trigger compositional shifts more than species loss (Balmford et al. 2003; Dornelas et al. 2014). This happens because species richness may not only be unresponsive, but it may also respond idiosyncrati-cally or peak at intermediate levels of disturbance, potentially showing no signal of change (Mouillot et al. 2013). However, contemplating only metrics of the β com-ponent of taxonomic diversity may also reveal insufficient from an ecosystem view-point. They are unable to account for species functional redundancy or uniqueness in the ecosystems when such compositional shifts occur, disregarding species func-tional role in the ecosystems (Petchey and Gaston 2006; Cadotte et al. 2011). For example, in a recent work, while a decreasing trend was found for functional diversity of several plant traits along a spatial climatic gradient in Mediterranean drylands, no pattern was observed for species richness (Nunes 2017; Nunes et al. 2017). Accordingly, other metrics that maximize tracking changes with a lowest effort should be considered in this context.

Thus, under intermediate driver's intensity a trait-based analysis should perform better, whereas under a strong driver's intensity a taxonomic analysis may be more cost-effective (Fig. 7.8).

7.3.3 Surrogates of Ecosystem Structure and Functioning Change from Remote Sensing

Making decisions regarding the conservation of biodiversity and the maintenance of ecosystem services, fostered the need to study indicators and/or surrogates that reflect the structure of ecosystems in a more in-depth way. Ecosystems are complex systems of energy and matter exchange, where a series of ecological processes operate at different scales and with different impacts on their functioning, dependent on their biodiversity. In turn, these processes respond to several drivers that can act simultaneously and interact, often in a non-linear way. This complicates how we can assess the integrity of ecosystems and the state of its processes.

Many of the indicators and/or surrogates used to measure the state of ecosystems do not consider the different effects that drivers have on the ecosystem they act upon, and on their degree of resilience. Measuring the impact on ecosystems is much more complex than measuring the drivers limiting or impacting them. Ecological indicators or surrogates are used to measure the effects of the drivers on the structure and functioning of ecosystems. They are used to communicate environmental data to stakeholders by helping describe them in a simpler and more concise way, more easily understood and used by non-scientists to make management decisions (Lindenmayer et al. 2015). Ecological indicators can be used to diagnose the cause of an environmental problem, to assess the effects of environmental conditions on ecosystem functioning or to provide an early warning signal in the event of changes in the environment. However, ecological indicators are expensive to use at the global scale, being commonly replaced by surrogates based on remote sensing. Before its use, the relationship between the ecological indicators and the surrogates must first be modelled (Fig. 7.3).

Of the six components of the Essential Biodiversity Variables, two can be comprehensively assessed by remote sensing, namely those related to ecosystem structure and ecosystem functioning (Pereira et al. 2013). Common examples include the use of NDVI – normalized difference vegetation index, as a surrogate of vegetation vigour (Gaitan et al. 2013), or the use of other wavelengths to infer measures of vegetation physiological status (Nestola et al. 2018). Remote sensing is also used to directly monitor species and populations, although this remains dependent on the size of such species and populations (Vihervaara et al. 2017). Remote sensing measures, offer a major advantage over ground measurements of species and populations in field assessments: they allow upscaling and/or generalizing spatial and temporal cover, which ground measures cannot provide. It also gives us the ability to look to ecosystems with a holistic perspective, for example allowing to look for critical thresholds resulting from multiple non-linear interactions (Scheffer et al. 2009), or to search for emergent properties, i.e., ecosystem properties that do not occur when ecosystems are considered isolated. The importance of remote sensing led to the development of the concept of the Satellite Remote Sensing – Essential Biodiversity Variables by the Group on Earth Observations – Biodiversity Observation Network (Pettorelli et al. 2016). Those are, within the Essential Biodiversity Variables, the ones that need remote sensing to be quantified.

Besides structure and functioning, the effects of many drivers related with biodiversity change can also be measured by remote sensing, including global change drivers like land-use change (Steffen et al. 2015). However, the link that relates ecosystem structure and functioning surrogates to its drivers is still missing. As in Sect. 7.3.2 dealing with species and communities, it is important to use the response to drivers to select the most appropriate remote sensing surrogates of ecosystem structure and functioning. This link would not only allow a better selection of the most appropriate surrogates of ecosystem structure and functioning, but also to upscale the knowledge obtained through a modelling approach.

Modelling approaches can be more direct (e.g. modelling changes in vegetation density by NDVI caused by frequency of wildfires) or more mechanistic (e.g. modelling the response of tree vitality to drought induced by climate change and mediated by insect's outbreaks). In either case, models' contribution would highly support the use of remote sensing in estimating ecosystem structure and functioning as surrogates of Essential Biodiversity Variables. The inclusion of explicit links to drivers in these models provides a much needed generalization capacity, and the possibility to track change over time and space at larger geographical scales. More specifically, that link would support the GEO-BON Global Biodiversity Change Indicators "Species distribution in habitats" and "Biodiversity loss by habitat degradation" (GEO BON 2015).

7.4   Final Remarks

There is a strong need for measuring biodiversity change and/or loss at the global scale and over time, to comply with all national and international conventions that protect species, habitats and ecosystems. Despite this, our main objective should be, at least, to live on Earth in a sustainable way, assuring the delivery of provision, cultural and regulation ecosystem services that allow our well-being. Measuring all forms of biodiversity everywhere and over time is an impossible task. That's why the framework of Essential Biodiversity Variables was developed, reducing the forms of biodiversity to six types ("genetic composition," "species populations," "species traits," "community composition," "ecosystem structure," and "ecosystem function" (Pereira et al. 2013)). However, even reducing it to six forms of biodiversity, it is not possible to measure all taxa relevant for the maintenance of the delivery of ecosystem services. Thus, we need to find good surrogates that can work as Biodiversity Change Metrics of the Essential Biodiversity Variables.

In this work, we expect biodiversity in the Anthropocene to be changed, in its different facets, by limiting and impacting drivers. We propose that a driver's intensity is determinant to the response of the different biodiversity change metrics. We expect low intensity drivers to influence the response of individual's physiological performance, and propose that Biodiversity Change Metrics should be ecophysiological-based in that case. At intermediate levels of driver's intensity, we expect the most cost-effective Biodiversity Change Metrics to be trait-based. Finally, in situations of strong driver's intensity we expect decreases in abundance and species loss, proposing a taxonomic-based approach as the most cost-effective to detect biodiversity changes. We further discuss the use of remote sensing data to measure changes in some of these indicators or surrogates of Essential Biodiversity Variables, particularly those reflecting ecosystem structure and functioning, allowing to upscale and/or generalize spatial and temporal information.

Acknowledgements We would like to acknowledge the financial support of the following projects: (i) NitroPortugal [European Union's Horizon 2020 research and innovation programme under grant agreement No 692331]; (ii) INMS, Towards the Establishment of an International Nitrogen Management System (INMS); (iii) ChangeTracker, financed by FCT, PTDC/ AAG-GLO/0045/2014.

References

Albert, C. H., Grassein, F., Schurr, F. M., Vieilledent, G., & Violle, C. (2011). When and how should intraspecific variability be considered in trait-based plant ecology? *Perspectives in Plant Ecology, Evolution and Systematics, 13*, 217–225. https://doi.org/10.1016/j. ppees.2011.04.003.

Balmford, A., Green, R. E., & Jenkins, M. (2003). Measuring the changing state of nature. *Trends in Ecology & Evolution, 18*, 326–330. https://doi.org/10.1016/S0169-5347(03)00067-3.

de Bello, F., Lavorel, S., Diaz, S., Harrington, R., Cornelissen, J. H. C., Bardgett, R. D., Berg, M. P., Cipriotti, P., Feld, C. K., Hering, D., da Silva, P. M., Potts, S. G., Sandin, L., Sousa, J. P., Storkey, J., Wardle, D. A., & Harrison, P. A. (2010). Towards an assessment of multiple ecosystem processes and services via functional traits. *Biodiversity and Conservation, 19*, 2873–2893. https://doi.org/10.1007/s10531-010-9850-9.

de Bello, F., Price, J. N., Muenkemueller, T., Liira, J., Zobel, M., Thuiller, W., Gerhold, P., Goetzenberger, L., Lavergne, S., Leps, J., Zobel, K., & Paertel, M. (2012). Functional species pool framework to test for biotic effects on community assembly. *Ecology, 93*, 2263–2273. https://doi.org/10.1890/11-1394.1.

de Bello, F., Lavorel, S., Lavergne, S., Albert, C. H., Boulangeat, I., Mazel, F., & Thuiller, W. (2013). Hierarchical effects of environmental filters on the functional structure of plant communities: A case study in the French Alps. *Ecography, 36*, 393–402. https://doi. org/10.1111/j.1600-0587.2012.07438.x.

Bockstaller, C., & Girardin, P. (2003). How to validate environmental indicators. *Agricultural Systems, 76*, 639–653. https://doi.org/10.1016/S0308-521X(02)00053-7.

Brooks, T. M., Lamoreux, J. F., & Soberón, J. (2014). Ipbes≠ ipcc. *Trends in Ecology & Evolution, 29*, 543–545. https://doi.org/10.1016/j.tree.2014.08.004.

Brummitt, N., Regan, E. C., Weatherdon, L. V., Martin, C. S., Geijzendorffer, I. R., Rocchini, D., Gavish, Y., Haase, P., Marsh, C. J., & Schmeller, D. S. (2017). Taking stock of nature: Essential biodiversity variables explained. *Biological Conservation, 213*, 252–255. https://doi.org/10.1016/j.biocon.2016.09.006.

Butchart, S. H., Walpole, M., Collen, B., Van Strien, A., Scharlemann, J. P., Almond, R. E., Baillie, J. E., Bomhard, B., Brown, C., & Bruno, J. (2010). Global biodiversity: Indicators of recent declines. *Science, 328*, 1164–1168. https://doi.org/10.1126/science.1187512.

Butterfield, B. J., & Suding, K. N. (2013). Single-trait functional indices outperform multi-trait indices in linking environmental gradients and ecosystem services in a complex landscape. *Journal of Ecology, 101*, 9–17. https://doi.org/10.1111/1365-2745.12013.

Cadotte, M. W., Carscadden, K., & Mirotchnick, N. (2011). Beyond species: Functional diversity and the maintenance of ecological processes and services. *Journal of Applied Ecology, 48*, 1079–1087. https://doi.org/10.1111/j.1365-2664.2011.02048.x.

Canadell, J. G., Le Quéré, C., Raupach, M. R., Field, C. B., Buitenhuis, E. T., Ciais, P., Conway, T. J., Gillett, N. P., Houghton, R., & Marland, G. (2007). Contributions to accelerating atmospheric CO2 growth from economic activity, carbon intensity, and efficiency of natural sinks. *Proceedings of the National Academy of Sciences, 104*, 18866–18870. https://doi.org/10.1073/pnas.0702737104.

Cardinale, B. J., Duffy, J. E., Gonzalez, A., Hooper, D. U., Perrings, C., Venail, P., Narwani, A., Mace, G. M., Tilman, D., Wardle, D. A., Kinzig, A. P., Daily, G. C., Loreau, M., Grace, J. B., Larigauderie, A., Srivastava, D. S., & Naeem, S. (2012). Biodiversity loss and its impact on humanity. *Nature, 486*, 59–67. https://doi.org/10.1038/nature11148.

CBD. (2014). *Convention on biological diversity*. Global Biodiversity Outlook 4. Montréal, 155 pages. https://www.cbd.int/gbo4/. Accessed 5 Nov 2017.

Chandler, M., See, L., Copas, K., Bonde, A. M., López, B. C., Danielsen, F., Legind, J. K., Masinde, S., Miller-Rushing, A. J., & Newman, G. (2017). Contribution of citizen science towards international biodiversity monitoring. *Biological Conservation, 213*, 280–294. https://doi.org/10.1016/j.biocon.2016.09.004.

Collen, B., & Nicholson, E. (2014). Taking the measure of change. *Science, 346*, 166–167. https://doi.org/10.1126/science.1255772.

Collen, B., Ram, M., Zamin, T., & McRae, L. (2008). The tropical biodiversity data gap: Addressing disparity in global monitoring. *Tropical Conservation Science, 1*, 75–88. https://doi.org/10.1177/194008290800100202.

Collen, B., Loh, J., Whitmee, S., McRAE, L., Amin, R., & Baillie, J. E. (2009). Monitoring change in vertebrate abundance: The living planet index. *Conservation Biology, 23*, 317–327. https://doi.org/10.1111/j.1523-1739.2008.01117.x.

Colwell, R. K., & Fuentes, E. R. (1975). Experimental studies of the niche. *Annual Review of Ecology and Systematics, 6*, 281–310. https://doi.org/10.1146/annurev.es.06.110175.001433.

Colwell, R. K., & Rangel, T. F. (2009). Hutchinson's duality: The once and future niche. *Proceedings of the National Academy of Sciences, 106*, 19651–19658. https://doi.org/10.1073/pnas.0901650106.

Cordlandwehr, V., Meredith, R. L., Ozinga, W. A., Bekker, R. M., Groenendael, J. M., & Bakker, J. P. (2013). Do plant traits retrieved from a database accurately predict on-site measurements? *Journal of Ecology, 101*, 662–670. https://doi.org/10.1111/1365-2745.12091.

Cornwell, W. K., & Ackerly, D. D. (2009). Community assembly and shifts in plant trait distributions across an environmental gradient in coastal California. *Ecological Monographs, 79*, 109–126. https://doi.org/10.1890/07-1134.1.

Díaz, S., & Cabido, M. (1997). Plant functional types and ecosystem function in relation to global change. *Journal of Vegetation Science, 8*, 463–474. https://www.jstor.org/stable/3237198.

Díaz, S., Lavorel, S., Chapin, F. S., III, Tecco, P. A., Gurvich, D. E., & Grigulis, K. (2007). Functional diversity – At the crossroads between ecosystem functioning and environmental filters. In J. G. Canadell, D. E. Pataki, & L. F. Pitelka (Eds.), *Terrestrial ecosystems in a changing world* (Global change – The IGBP series) (pp. 81–91). Berlin/Heidelberg: Springer.

Dornelas, M., Gotelli, N. J., McGill, B., Shimadzu, H., Moyes, F., Sievers, C., & Magurran, A. E. (2014). Assemblage time series reveal biodiversity change but not systematic loss. *Science, 344*, 296–299. https://doi.org/10.1126/science.1248484.

EEA. (2014). *European environment agency. Air quality in Europe – 2014 report* (EEA technical report No. 5/2014). https://doi.org/10.2800/22775.

EEA. (2016). *European environment agency. Air quality in Europe – 2016 report* (EEA technical report). https://doi.org/10.2800/80982.

Fenger, J. (2009). Air pollution in the last 50 years–From local to global. *Atmospheric Environment, 43*, 13–22. https://doi.org/10.1016/j.atmosenv.2008.09.061.

Fusco, G., & Minelli, A. (2010). Phenotypic plasticity in development and evolution: facts and concepts. *Philosophical Transactions of the Royal Society B, 365*, 547–556. https://doi.org/10.1098/rstb.2009.0267.

Gaitan, J. J., Bran, D., Oliva, G., Ciari, G., Nakamatsu, V., Salomone, J., Ferrante, D., Buono, G., Massara, V., Humano, G., Celdran, D., Opazo, W., & Maestre, F. T. (2013). Evaluating the performance of multiple remote sensing indices to predict the spatial variability of ecosystem structure and functioning in Patagonian steppes. *Ecological Indicators, 34*, 181–191. https://doi.org/10.1016/j.ecolind.2013.05.007.

Garnier, E., Lavorel, S., Ansquer, P., Castro, H., Cruz, P., Dolezal, J., Eriksson, O., Fortunel, C., Freitas, H., & Golodets, C. (2007). Assessing the effects of land-use change on plant traits, communities and ecosystem functioning in grasslands: a standardized methodology and lessons from an application to 11 European sites. *Annals of Botany, 99*, 967–985. https://doi.org/10.1093/aob/mcl215.

Geijzendorffer, I. R., Regan, E. C., Pereira, H. M., Brotons, L., Brummitt, N., Gavish, Y., Haase, P., Martin, C. S., Mihoub, J. B., & Secades, C. (2016). Bridging the gap between biodiversity data and policy reporting needs: An Essential Biodiversity Variables perspective. *Journal of Applied Ecology, 53*, 1341–1350. https://doi.org/10.1111/1365-2664.12417.

GEO BON. (2015). *The group on earth observations biodiversity observation network. Global biodiversity change indicators: Model-based integration of remote-sensing & in situ observations that enables dynamic updates and transparency at low cost.* http://orbit.dtu.dk/files/118107442/GBCI_Version1.2_low_Biodiversity_Index.pdf. Version 1.2, Nov 2015.

Grantham, H. S., Pressey, R. L., Wells, J. A., & Beattie, A. J. (2010). Effectiveness of biodiversity surrogates for conservation planning: Different measures of effectiveness generate a kaleidoscope of variation. *PLoS One, 5*, e11430. https://doi.org/10.1371/journal.pone.0011430.

Grime, J. P. (1998). Benefits of plant diversity to ecosystems: Immediate, filter and founder effects. *Journal of Ecology, 86*, 902–910. https://doi.org/10.1046/j.1365-2745.1998.00306.x.

Grime, J. P., & Díaz, S. (2006). Trait convergence and trait divergence in herbaceous plant communities: Mechanisms and consequences. *Journal of Vegetation Science, 17*, 255–260. https://doi.org/10.1111/j.1654-1103.2006.tb02444.x.

Halpern, B. S., Longo, C., Hardy, D., McLeod, K. L., Samhouri, J. F., Katona, S. K., Kleisner, K., Lester, S. E., O'Leary, J., & Ranelletti, M. (2012). An index to assess the health and benefits of the global ocean. *Nature, 488*, 615–620. https://doi.org/10.1038/nature11397.

Henle, K., Potts, S., Kunin, W., Matsinos, Y., Simila, J., Pantis, J., Grobelnik, V., Penev, L., & Settele, J. (2014). Scaling in ecology and biodiversity conservation. *Advanced Books, 1*, e1169. https://doi.org/10.3897/ab.e1169.

Holling, C. S. (1973). Resilience and stability of ecological systems. *Annual Review of Ecology and Systematics, 4*, 1–23. https://doi.org/10.1146/annurev.es.04.110173.000245.

Hooper, D. U., Chapin, F., Ewel, J., Hector, A., Inchausti, P., Lavorel, S., Lawton, J., Lodge, D., Loreau, M., & Naeem, S. (2005). Effects of biodiversity on ecosystem functioning: A consensus of current knowledge. *Ecological Monographs, 75*, 3–35. https://doi.org/10.1890/04-0922.

Hooper, D. U., Adair, E. C., Cardinale, B. J., Byrnes, J. E., Hungate, B. A., Matulich, K. L., Gonzalez, A., Duffy, J. E., Gamfeldt, L., & O'Connor, M. I. (2012). A global synthesis reveals biodiversity loss as a major driver of ecosystem change. *Nature, 486*, 105–108. https://doi. org/10.1038/nature11118.

Hortal, J., de Bello, F., Diniz-Filho, J. A. F., Lewinsohn, T. M., Lobo, J. M., & Ladle, R. J. (2015). Seven shortfalls that beset large-scale knowledge of biodiversity. *Annual Review of Ecology, Evolution, and Systematics, 46*, 523–549. https://doi.org/10.1146/ annurev-ecolsys-112414-054400.

Hunter, M., Westgate, M., Barton, P., Calhoun, A., Pierson, J., Tulloch, A., Beger, M., Branquinho, C., Caro, T., & Gross, J. (2016). Two roles for ecological surrogacy: Indicator surrogates and management surrogates. *Ecological Indicators, 63*, 121–125. https://doi.org/10.1016/j. ecolind.2015.11.049.

Hutchinson, G. E. (1957). Cold spring harbor symposium on quantitative biology. *Concluding Remarks, 22*, 415–427. https://doi.org/10.1101/SQB.1957.022.01.039.

Hutchinson, G. E. (1991). Population studies: Animal ecology and demography. *Bulletin of Mathematical Biology, 53*, 193–213. https://doi.org/10.1007/BF02464429.

IPCC. (2007). *Climate change 2007: Synthesis report. Contribution of working groups I, II and III to the fourth assessment report of the intergovernmental panel on climate change*. Geneva: IPCC.

Kattge, J., Diaz, S., Lavorel, S., Prentice, I. C., Leadley, P., Bönisch, G., Garnier, E., Westoby, M., Reich, P. B., & Wright, I. J. (2011). TRY–A global database of plant traits. *Global Change Biology, 17*, 2905–2935. https://doi.org/10.1111/j.1365-2486.2011.02451.x.

Kot, C. Y., Fujioka, E., Hazen, L. J., Best, B. D., Read, A. J., & Halpin, P. N. (2010). Spatio-temporal gap analysis of OBIS-SEAMAP project data: Assessment and way forward. *PLoS One, 5*, e12990. https://doi.org/10.1371/journal.pone.0012990.

Laliberte, E., & Legendre, P. (2010). A distance-based framework for measuring functional diversity from multiple traits. *Ecology, 91*, 299–305. https://doi.org/10.1890/08-2244.1.

Lavorel, S., & Garnier, E. (2002). Predicting changes in community composition and ecosystem functioning from plant traits: Revisiting the Holy Grail. *Functional Ecology, 16*, 545–556. https://doi.org/10.1046/j.1365-2435.2002.00664.x.

Lavorel, S., Grigulis, K., McIntyre, S., Williams, N. S. G., Garden, D., Dorrough, J., Berman, S., Quetier, F., Thebault, A., & Bonis, A. (2008). Assessing functional diversity in the field – Methodology matters! *Functional Ecology, 22*, 134–147. https://doi. org/10.1111/j.1365-2435.2007.01339.x.

Lavorel, S., de Bello, F., Grigulis, K., Lepš, J., Garnier, E., Castro, H., Dolezal, J., Godolets, C., Quétier, F., & Thébault, A. (2011). Response of herbaceous vegetation functional diversity to land use change across five sites in Europe and Israel. *Israel Journal of Ecology & Evolution, 57*, 53–72. https://doi.org/10.1560/IJEE.57.1-2.53.

Lindenmayer, D., Pierson, J., Barton, P., Beger, M., Branquinho, C., Calhoun, A., Caro, T., Greig, H., Gross, J., & Heino, J. (2015). A new framework for selecting environmental surrogates. *Science of the Total Environment, 538*, 1029–1038. https://doi.org/10.1016/j. scitotenv.2015.08.056.

Magurran, A. E. (2013). *Measuring biological diversity*. Oxford: John Wiley & Sons.

Maire, V., Gross, N., Hill, D., Martin, R., Wirth, C., Wright, I. J., & Soussana, J.-F. (2013). Disentangling coordination among functional traits using an individual-centred model: Impact on plant performance at intra-and inter-specific levels. *PLoS One, 8*, e77372. https://doi. org/10.1371/journal.pone.0077372.

Mason, N. W. H., & de Bello, F. (2013). Functional diversity: A tool for answering challenging ecological questions. *Journal of Vegetation Science, 24*, 777–780. https://doi.org/10.1111/ jvs.12097.

Mason, N. W. H., Mouillot, D., Lee, W. G., & Wilson, J. B. (2005). Functional richness, functional evenness and functional divergence: The primary components of functional diversity. *Oikos, 111*, 112–118. https://doi.org/10.1111/j.0030-1299.2005.13886.x.

Matos, P., Pinho, P., Aragón, G., Martínez, I., Nunes, A., Soares, A. M., & Branquinho, C. (2015). Lichen traits responding to aridity. *Journal of Ecology, 103*, 451–458. https://doi.org/10.1111/1365-2745.12364.

Matos, P., Geiser, L., Hardman, A., Glavich, D., Pinho, P., Nunes, A., Soares, A. M., & Branquinho, C. (2017). Tracking global change using lichen diversity: Towards a global-scale ecological indicator. *Methods in Ecology and Evolution, 8*, 788–798. https://doi.org/10.1111/2041-210X.12712.

McGeoch, M. A. (1998). The selection, testing and application of terrestrial insects as bioindicators. *Biological Reviews, 73*, 181–201. https://doi.org/10.1111/j.1469-185X.1997.tb00029.x.

McGill, B. J. (2012). Trees are rarely most abundant where they grow best. *Journal of Plant Ecology, 5*, 46–51. https://doi.org/10.1093/jpe/rtr036.

MEA. (2005). *Milliennium ecosystem assessment. Ecosystems and human well-being: Synthesis.* Washington, DC: Island Press.

Meadows, D. H., Meadows, D. L., Randers, J., & Behrens, W. W. (1972). *The limits to growth.* New York: Universe Books.

Mellin, C., Delean, S., Caley, J., Edgar, G., Meekan, M., Pitcher, R., Przeslawski, R., Williams, A., & Bradshaw, C. (2011). Effectiveness of biological surrogates for predicting patterns of marine biodiversity: A global meta-analysis. *PLoS One, 6*, e20141. https://doi.org/10.1371/journal.pone.0020141.

Mora, C., Tittensor, D. P., & Myers, R. A. (2008). The completeness of taxonomic inventories for describing the global diversity and distribution of marine fishes. *Proceedings of the Royal Society of London B: Biological Sciences, 275*, 149–155. https://doi.org/10.1098/rspb.2007.1315.

Mouillot, D., Villeger, S., Scherer-Lorenzen, M., & Mason, N. W. H. (2011). Functional structure of biological communities predicts ecosystem multifunctionality. *PLoS One, 6*, e17476. https://doi.org/10.1371/journal.pone.0017476.

Mouillot, D., Graham, N. A. J., Villeger, S., Mason, N. W. H., & Bellwood, D. R. (2013). A functional approach reveals community responses to disturbances. *Trends in Ecology & Evolution, 28*, 167–177. https://doi.org/10.1016/j.tree.2012.10.004.

Nestola, E., Scartazza, A., Di Baccio, D., Castagna, A., Ranieri, A., Cammarano, M., Mazzenga, F., Matteucci, G., & Calfapietra, C. (2018). Are optical indices good proxies of seasonal changes in carbon fluxes and stress-related physiological status in a beech forest? *Science of the Total Environment, 612*, 1030–1041. https://doi.org/10.1016/j.scitotenv.2017.08.167.

Niemeijer, D., & de Groot, R. S. (2008). A conceptual framework for selecting environmental indicator sets. *Ecological Indicators, 8*, 14–25. https://doi.org/10.1016/j.ecolind.2006.11.012.

Noss, R. F. (1990). Indicators for monitoring biodiversity: A hierarchical approach. *Conservation Biology, 4*, 355–364. https://doi.org/10.1111/j.1523-1739.1990.tb00309.x.

Nunes, A. (2017). *Plant functional response to desertification and land degradation – Contribution to restoration strategies.* Aveiro: Universidade de Aveiro. http://hdl.handle.net/10773/18814.

Nunes, A., Köbel, M., Pinho, P., Matos, P., de Bello, F., Correia, O., & Branquinho, C. (2017). Which plant traits respond to aridity? A critical step to assess functional diversity in Mediterranean drylands. *Agricultural and Forest Meteorology, 239*, 176–184. https://doi.org/10.1016/j.agrformet.2017.03.007.

Pereira, H. M., & Cooper, H. D. (2006). Towards the global monitoring of biodiversity change. *Trends in Ecology & Evolution, 21*, 123–129. https://doi.org/10.1016/j.tree.2005.10.015.

Pereira, H. M., Navarro, L. M., & Martins, I. S. (2012). Global biodiversity change: The bad, the good, and the unknown. *Annual Review of Environment and Resources, 37*, 25–50. https://doi.org/10.1146/annurev-environ-042911-093511.

Pereira, H. M., Ferrier, S., Walters, M., Geller, G. N., Jongman, R. H. G., Scholes, R. J., Bruford, M. W., Brummitt, N., Butchart, S. H. M., Cardoso, A. C., Coops, N. C., Dulloo, E., Faith, D. P., Freyhof, J., Gregory, R. D., Heip, C., Hoeft, R., Hurtt, G., Jetz, W., Karp, D. S., McGeoch, M. A., Obura, D., Onoda, Y., Pettorelli, N., Reyers, B., Sayre, R., Scharlemann, J. P. W., Stuart, S. N., Turak, E., Walpole, M., & Wegmann, M. (2013). Essential biodiversity variables. *Science, 339*, 277–278. https://doi.org/10.1126/science.1229931.

Pérez-Harguindeguy, N., Díaz, S., Garnier, E., Lavorel, S., Poorter, H., Jaureguiberry, P., Bret-Harte, M., Cornwell, W., Craine, J., & Gurvich, D. (2013). New handbook for standardised measurement of plant functional traits worldwide. *Australian Journal of Botany, 61*, 167–234. https://doi.org/10.1071/BT12225.

Petchey, O. L., & Gaston, K. J. (2006). Functional diversity: Back to basics and looking forward. *Ecology Letters, 9*, 741–758. https://doi.org/10.1111/j.1461-0248.2006.00924.x.

Pettorelli, N., Wegmann, M., Skidmore, A., Mücher, S., Dawson, T. P., Fernandez, M., Lucas, R., Schaepman, M. E., Wang, T., & O'Connor, B. (2016). Framing the concept of satellite remote sensing essential biodiversity variables: Challenges and future directions. *Remote Sensing in Ecology and Conservation, 2*, 122–131. https://doi.org/10.1002/rse2.15.

Pillar, V. D., Blanco, C. C., Mueller, S. C., Sosinski, E. E., Joner, F., & Duarte, L. D. S. (2013). Functional redundancy and stability in plant communities. *Journal of Vegetation Science, 24*, 963–974. https://doi.org/10.1111/jvs.12047.

Pollock, L. J., Thuiller, W., & Jetz, W. (2017). Large conservation gains possible for global biodiversity facets. *Nature, 546*, 141–144. https://doi.org/10.1038/nature22368.

Proença, V., Martin, L. J., Pereira, H. M., Fernandez, M., McRae, L., Belnap, J., Böhm, M., Brummitt, N., García-Moreno, J., & Gregory, R. D. (2017). Global biodiversity monitoring: From data sources to essential biodiversity variables. *Biological Conservation, 213*, 256–263. https://doi.org/10.1016/j.biocon.2016.07.014.

Ribeiro, M. C., Pinho, P., Llop, E., Branquinho, C., Sousa, A. J., & Pereira, M. J. (2013). Multivariate geostatistical methods for analysis of relationships between ecological indicators and environmental factors at multiple spatial scales. *Ecological Indicators, 29*, 339–347. https://doi.org/10.1016/j.ecolind.2013.01.011.

Rockström, J., Steffen, W., Noone, K., Persson, Å., Chapin, F. S., Lambin, E. F., Lenton, T. M., Scheffer, M., Folke, C., & Schellnhuber, H. J. (2009). A safe operating space for humanity. *Nature, 461*, 472–475. https://doi.org/10.1038/461472a.

Scheffer, M., Carpenter, S., Foley, J. A., Folke, C., & Walker, B. (2001). Catastrophic shifts in ecosystems. *Nature, 413*, 591–596. https://doi.org/10.1038/35098000.

Scheffer, M., Bascompte, J., Brock, W. A., Brovkin, V., Carpenter, S. R., Dakos, V., Held, H., van Nes, E. H., Rietkerk, M., & Sugihara, G. (2009). Early-warning signals for critical transitions. *Nature, 461*, 53–59. https://doi.org/10.1038/nature08227.

Schleuter, D., Daufresne, M., Massol, F., & Argillier, C. (2010). A user's guide to functional diversity indices. *Ecological Monographs, 80*, 469–484. https://doi.org/10.1890/08-2225.1.

Scholes, R. J., Walters, M., Turak, E., Saarenmaa, H., Heip, C. H., Tuama, É. Ó., Faith, D. P., Mooney, H. A., Ferrier, S., & Jongman, R. H. (2012). Building a global observing system for biodiversity. *Current Opinion in Environmental Sustainability, 4*, 139–146. https://doi.org/10.1016/j.cosust.2011.12.005.

Serrano, H. C., Antunes, C., Pinto, M. J., Máguas, C., Martins-Loução, M. A., & Branquinho, C. (2015). The ecological performance of metallophyte plants thriving in geochemical islands is explained by the Inclusive Niche Hypothesis. *Journal of Plant Ecology, 8*, 41–50. https://doi.org/10.1093/jpe/rtu007.

Shipley, B., De Bello, F., Cornelissen, J. H. C., Laliberté, E., Laughlin, D. C., & Reich, P. B. (2016). Reinforcing loose foundation stones in trait-based plant ecology. *Oecologia, 180*, 923–931. https://doi.org/10.1007/s00442-016-3549-x.

Steffen, W., Richardson, K., Rockström, J., Cornell, S. E., Fetzer, I., Bennett, E. M., Biggs, R., Carpenter, S. R., de Vries, W., & de Wit, C. A. (2015). Planetary boundaries: Guiding human development on a changing planet. *Science, 347*, 1259855. https://doi.org/10.1126/science.1259855.

Suding, K. N., Lavorel, S., Chapin, F., Cornelissen, J. H., Diaz, S., Garnier, E., Goldberg, D., Hooper, D. U., Jackson, S. T., & Navas, M. L. (2008). Scaling environmental change through the community-level: A trait-based response-and-effect framework for plants. *Global Change Biology, 14*, 1125–1140. https://doi.org/10.1111/j.1365-2486.2008.01557.x.

Tilman, D., Knops, J., Wedin, D., Reich, P., Ritchie, M., & Siemann, E. (1997). The influence of functional diversity and composition on ecosystem processes. *Science, 277*, 1300–1302. https://doi.org/10.1126/science.277.5330.1300.

Tittensor, D. P., Walpole, M., Hill, S. L., Boyce, D. G., Britten, G. L., Burgess, N. D., Butchart, S. H., Leadley, P. W., Regan, E. C., & Alkemade, R. (2014). A mid-term analysis of progress toward international biodiversity targets. *Science, 346*, 241–244. https://doi.org/10.1126/science.1257484.

Turak, E., Brazill-Boast, J., Cooney, T., Drielsma, M., DelaCruz, J., Dunkerley, G., Fernandez, M., Ferrier, S., Gill, M., & Jones, H. (2017). Using the essential biodiversity variables framework to measure biodiversity change at national scale. *Biological Conservation, 213*, 264–271. https://doi.org/10.1016/j.biocon.2016.08.019.

UNCBD. (2011). *United Nations convention on biological diversity. Strategic plan for biodiversity 2011–2020 and the aichi biodiversity targets*. http://www.cbd.int/sp/. Accessed 5 Nov 2017.

Valencia, E., Maestre, F. T., Le Bagousse-Pinguet, Y., Luis Quero, J., Tamme, R., Boerger, L., Garcia-Gomez, M., & Gross, N. (2015). Functional diversity enhances the resistance of ecosystem multifunctionality to aridity in Mediterranean drylands. *New Phytologist, 206*, 660–671. https://doi.org/10.1111/nph.13268.

Vihervaara, P., Auvinen, A. P., Mononen, L., Torma, M., Ahlroth, P., Anttila, S., Bottcher, K., Forsius, M., Heino, J., Heliola, J., Koskelainen, M., Kuussaari, M., Meissner, K., Ojala, O., Tuominen, S., Viitasalo, M., & Virkkala, R. (2017). How essential biodiversity variables and remote sensing can help national biodiversity monitoring. *Global Ecology and Conservation, 10*, 43–59. https://doi.org/10.1016/j.gecco.2017.01.007.

Villeger, S., Mason, N. W. H., & Mouillot, D. (2008). New multidimensional functional diversity indices for a multifaceted framework in functional ecology. *Ecology, 89*, 2290–2301. https://doi.org/10.1890/07-1206.1.

Volis, S., & Bohrer, G. (2013). Joint evolution of seed traits along an aridity gradient: Seed size and dormancy are not two substitutable evolutionary traits in temporally heterogeneous environment. *New Phytologist, 197*, 655–667. https://doi.org/10.1111/nph.12024.

Westgate, M. J., Barton, P. S., Lane, P. W., & Lindenmayer, D. B. (2014). Global meta-analysis reveals low consistency of biodiversity congruence relationships. *Nature Communications, 5*, 3899. https://doi.org/10.1038/ncomms4899.

WMO. (2010). *GCOS – Global climate observing system. Implementation plan for the global observing system for climate in support of the UNFCCC (2010 Update)*. Geneva: WMO Publications.

Zalasiewicz, J., Williams, M., Steffen, W., & Crutzen, P. (2010). The new world of the Anthropocene 1. *Environmental Science & Technology, 44*, 2228–2231. https://doi.org/10.1021/es903118j.

Part II
Characterizing Biodiversity: Beyond the Species Approach

Chapter 8
Are Species Good Units for Biodiversity Studies and Conservation Efforts?

Thomas A. C. Reydon

Abstract While species have long been seen as the principal units of biodiversity, with prominent roles in biodiversity research and conservation practice, the long-standing debate on the nature of species deeply problematizes their suitability as such units. Not only do the metaphysical questions remain unresolved what kinds of things species *are*, and whether species are at all real, there also is considerable disagreement on how to define the notion of species for use in practice. Moreover, it seems that different organism groups are best classified using different definitions of 'species', such that species of organisms in very different domains of biodiversity are not generally comparable units. In this chapter I will defend and elaborate the claim that species are not good units of biodiversity, focusing in the issue of species realism. I will sketch a pragmatic notion of 'species' that can be used as an epistemic tool in the context of biodiversity studies, without however involving a view of species as basic units of biodiversity or as the focal, real entities in biodiversity conservation.

Keywords Species · Species concepts · Species problem · Units of biodiversity

8.1 Introduction

Biodiversity studies and conservation biology are rapidly growing fields of work, aimed, among other things, at mapping the diversity of life on our planet, studying its origins, assessing how humanity benefits from its presence, and achieving clarity about possible ways of preserving it (Ehrlich and Wilson 1991).[1] The growth of both

---

[1] Conservation biology is a well-defined discipline with its own textbooks, journals, scientific societies, and so on. Biodiversity studies, in contrast, is a loose collection of disciplines (or perhaps

---

T. A. C. Reydon (✉)
Institute of Philosophy & Centre for Ethics and Law in the Life Sciences (CELLS),
Leibniz Universität, Hanover, Germany
e-mail: reydon@ww.uni-hannover.de; http://www.reydon.info

© The Author(s) 2019
E. Casetta et al. (eds.), *From Assessing to Conserving Biodiversity*,
History, Philosophy and Theory of the Life Sciences 24,
https://doi.org/10.1007/978-3-030-10991-2_8

fields is fueled by a number of factors, one of which is the realization that still only a small fraction of the Earth's extant biological diversity is known, and that many organism groups – most importantly local populations and entire species – are threatened with extinction, such that there is an urgency to map out our planet's biological diversity and undertake appropriate conservation efforts before these have disappeared forever (e.g., Wheeler et al. 2012).

Studying and conserving biodiversity is not merely a matter of the disinterested acquisition of knowledge about the world we live in, though, nor is it a matter of simply wanting to preserve the world as we find it. The importance of mapping, studying and conserving biodiversity is related foremost to the availability of natural resources that are crucial to human survival and well-being, i.e., to the availability of ecosystem services (Costanza et al. 1997; Dirzo and Raven 2003; Mace 2015; Hunter and Gibbs 2007: 348–349; Mace et al. 2012). Ecosystem services include resources such as clean water, clean air, arable land, fuel, building materials, and so on, as well as natural environments that can be used for recreative and aesthetic purposes. In addition, potential resources that as yet unexplored organism groups could provide for the production of new medicinal products, better foodstuffs, etc. constitute potential ecosystem services that support the value of biodiversity studies and conservation efforts (Maclaurin and Sterelny 2008: 5–6). In sum, studying and conserving biodiversity is important because human survival and well-being depends on it.

Yet, notwithstanding the importance and rapid growth of biodiversity studies and conservation biology as academic fields as well as areas of activism and engagement, they are faced with profound challenges. As highlighted in the Introduction to this volume, while the many practical challenges (due, among other things, to political and economic interests making it difficult to achieve conservation targets) are widely recognized, conceptual challenges are much less widely discussed. As Casetta et al. rightly point out, "[w]hen biodiversity conservation is at issue, theoretical and practical matters go hand in hand, and practical challenges are intertwined with conceptual ones".[2] This chapter addresses such a conceptual challenge concerning the notion of 'species', and aims to resolve some conceptual difficulties to obtain a notion that is better suited for application in biodiversity studies and conservation biology. Considering that these fields are practice-oriented areas of work, it is surprisingly hard to pin down the meanings of many of their core concepts – a state of affairs that makes conceptual work a crucial prerequisite for practical interventions.

---

rather an interdisciplinary field of work) that are in some way or other concerned with earthbound biodiversity. The term 'biodiversity studies' is common in the literature (Ehrlich and Wilson 1991; Agapow et al. 2004; Raczkowski and Wenzel 2007), but is not used to denote a well-delimited set of disciplines. Ehrlich and Wilson (1991) conceive of biodiversity studies as combining elements of evolutionary biology, ecology, applied areas of biology (such as forestry, agriculture, and medicine) and public policy, and see conservation biology as a subdiscipline of biodiversity studies. For the purposes of the present chapter, I think the term does not need to be specified in more detail beyond the meaning of "any area of work that studies biodiversity".

[2] Casetta, Marques da Silva & Vecchi, this volume.

Consider for instance the notion of biodiversity itself. It is notoriously hard to define and is the topic of ongoing debate among philosophers of biology as well as biologists themselves (e.g., Purvis and Hector 2000; Hamilton 2005; Colwell 2009; Santana 2014, 2018; Faith 2016; Odenbaugh 2016; Burch-Brown and Archer 2017).[3] In its broadest sense, biodiversity simply means the diversity of living entities, of their parts, and of systems composed of them, at all levels of organization, from the genetic level all the way up to the ecosystem level (Hunter and Gibbs 2007: 22; Odenbaugh 2016: Section 1). A simple definition of the term 'biodiversity' thus would be "the variety of all forms of life, from genes to species, through to the broad scale of ecosystems" (Faith 2016). In addition to the various levels of organization in the living world, biodiversity is studied with respect to a number of different aspects of living systems, such as the roles organisms play as parts of ecosystems or food webs (functional diversity), the specific morphological or behavioral properties shared by organisms of particular groups (trait diversity), their lines of descent (phylogenetic diversity), and so on. Colwell (2009: 257), for example, explains the notion of biodiversity as encompassing "the variety of life, at all levels of organization, classified both by evolutionary (phylogenetic) and ecological (functional) criteria" and Dirzo and Raven (2003: 138) explain it as "the sum total of all of the plants, animals, fungi, and microorganisms on Earth; their genetic and phenotypic variation; and the communities and ecosystems of which they are a part." Accordingly, there are various ways of conceiving of biodiversity that differ from one another in regard to the specific levels of organization and the specific aspects of living systems one is interested in. Biodiversity can be thought of as the genetic diversity in a local population, the species richness of a local community or region, the diversity of functional groups (such as primary producers, herbivores, etc.) in a particular ecosystem, the diversity of habitats making up an ecosystem, and so on. As a result, researchers often focus on only one or a few aspects of biodiversity, rather than assessing the biodiversity of a particular area as such. Note that this multidimensionality of the concept is precisely the reason for which some authors are skeptical about the concept's scientific value: generally, it does not make much sense to talk about *the* biodiversity of a particular region or *the* biodiversity of planet Earth, because there are too many distinct aspects to consider and a region can be very diverse in some aspects and much less diverse in others (Santana 2014, 2018).

While 'biodiversity' thus is a multifaceted concept and the interests that guide work in biodiversity studies differ from those in conservation biology – the former being more aimed at producing scientific knowledge and the letter more oriented towards intervention and activism –,[4] both fields use largely the same basic units to individuate groups of organisms and living systems of interest. Many studies and

---

[3] See also, in this volume, Toepfer, Chap. 18; Sarkar, Chap. 18; and Santana, Chap. 19.

[4] Indeed, conservation biology is often conceived of as a "mission-driven discipline" and as "the applied science of maintaining the earth's biological diversity" (Hunter and Gibbs 2007: 14; see also Soulé 1985: 727; Meine et al. 2006: 631; Odenbaugh 2016). The mission of the discipline is sometimes framed in terms of "healing" an ailing patient (Casetta, Marques da Silva, and Vecchi, Chap. 1, in this volume).

interventions are aimed at a particular level of the taxonomic hierarchy, namely the species level. In the same way as species serve as the basic taxonomic units and as such constitute the basic kinds of organisms studied by biologists, they also count among the basic units of biodiversity and constitute targets of many conservation efforts. In this chapter, however, I want to argue that species are not good units of biodiversity and conservation, while at the same time trying to develop a pragmatic notion of 'species' that can be used in the context of biodiversity studies while avoiding the problems highlighted in this chapter. To do so, I will examine the main aspects of the role of species in biodiversity studies and conservation efforts from an epistemological and a metaphysical perspective. Section 8.2 sketches the central role of species in biodiversity studies and conservation efforts. In Sect. 8.3, I will look at relevant debates in the philosophy of biology as well as in biology itself to highlight epistemological and in particular metaphysical problems (connected to the reality of species) that confront the notion of species in these contexts. In Sect. 8.4, I will suggest an alternative interpretation of the notion of species based on Darwin's views that might better fit the role the notion of species plays in biodiversity studies and conservation biology. Section 8.5 concludes.

8.2   Species as the Units of Biodiversity and Conservation

While biodiversity can be studied at a variety of levels of organization and with respect to a variety of aspects of living systems, one or more units of biodiversity are needed as a basis for comparisons between different areas of work and different studies. Without a common "currency", the notion of biodiversity is meaningless. A number of biodiversity measures are currently in use, many of which (including species richness, species diversity, and species evenness) focus on species as basic units of biodiversity (Purvis and Hector 2000; Faith 2016: Section 2).[5] Hamilton (2005: 90), for example, writes:

> There are currently many definitions of biodiversity and most are vague, which probably reflects the uncertainty of the concept. Some consider it to be synonymous with species richness […], others see it as species diversity […], whereas many propound a much broader definition such as the 'full variety of life on Earth' […]. NRE [the Department of Natural Resources and Environment, State of Victoria, Australia] distinguish between native and introduced species, and others have put extra emphasis on threatened species […].

Sometimes species are even considered to be the most important or the most fundamental units of biodiversity (e.g., Claridge et al. 1997; Mace et al. 2012: 20; Hohenegger 2014). Given the centrality of species as the basic taxonomic units for the whole of the life sciences as well as in the context of public representations of organismal diversity (in natural history museums, science centers, zoos, botanical gardens, etc.), putting species central seems an appropriate choice.

---

[5] See also Crupi, Chap. 6, and Branquinho et al., Chap. 7, in this volume.

Closely connected to their role as units of biodiversity is the central position that species occupy in conservation efforts. Similar to biodiversity studies, conservation biology is concerned with living entities and systems on all levels of organization from the genetic level up to the ecosystem level, and conservation efforts can be directed at the genetic diversity of a particular area or taxon, at taxa themselves, at communities or ecosystems, and so on. Even though "conservation biology [...] does not yet have a general and coherent account of what should be conserved and why" (Maclaurin and Sterelny 2008: 26) and the field has since the 1990's been moving away somewhat from its earlier focus on species towards a focus on interactions between people and their environments (Mace 2015: 1558–1559), species are still among the principal entities in focus. In part this is for historical reasons. One of the main factors that contributed to both the establishment of conservation biology and the coining of the term 'biodiversity' in the 1980s was the realization that species extinction currently proceeds at a much higher rate than the natural extinction rate, and that this increased extinction rate is very probably due to the impact of the human population on the planet. Discussions of the causes of current biodiversity loss (often referred to as the "sixth mass extinction" – e.g., Barnosky et al. 2011; Ceballos et al. 2015, 2017) and of possible countermeasures tended to revolve around the extinction and conservation of local populations as representatives of a particular species in particular areas, as well as entire species (Soulé 1985; Meine et al. 2006: 637). Perhaps the best-known example of the ongoing centrality of species in conservation efforts is the *IUCN Red List of Threatened Species* that categorizes species according to the level at which they are threatened with extinction (Mace and Lande 1991; Mace et al. 2008; http://www.iucnredlist.org) and is widely used by researchers, NGOs, governments, politicians and the general public as a basis for conservation efforts. Another well-known example of the keystone role of species is the Convention on Biodiversity, which is aimed at facilitating biodiversity conservation and specifies that biodiversity "includes diversity *within* species, *between* species and of ecosystems" (see www.cbd.int, Article 2; emphasis added).

In addition to the historical focus on species in conservation biology, species also are sometimes highlighted in environmental ethics as morally relevant entities. In discussions on the normative aspects of human interactions with nature and the normative principles that could underwrite conservation efforts, several authors have argued that we have a moral obligation to preserve species, have attributed intrinsic value to species, or have argued that species should be counted among the entities toward which humans can have duties and obligations (e.g., Soulé 1985, and most famously Rolston 1975, 1985, 1995). Soulé (1985: 731), for example, writes: "Species have value in themselves, a value neither conferred nor revocable, but springing from a species' long evolutionary heritage and potential or even from the mere fact of its existence." Ignoring for the moment the question whether it makes sense at all to attribute intrinsic values to natural entities, such as individual organisms, species or ecosystems, it is clear that species are among the basic natural entities (along with individual organisms) that at least sometimes are seen as bearers of value in conservation contexts.[6]

Species, then, have long been among the principal units of biodiversity studies as well as one of the principal kinds of focal entities in conservation efforts. This role entails a number of epistemological and metaphysical requirements. For one, it seems that to be able to serve as units in biodiversity estimates of a particular area or as the subjects of biodiversity conservation efforts species must be real entities – at least it seems that for species to be entities that can become extinct or can be kept in existence by human efforts, or for them to be bearers of intrinsic value, they cannot be *purely* conventional or instrumental units and a minimal level of realism with respect to species is required. But if species are real entities, what exactly are they? This is the core issue in the long-standing debate on the nature of species, and it deeply problematizes the suitability of species as units of biodiversity.

In addition, not only do the questions what kinds of things species *are* and whether species are at all real entities remain highly controversial metaphysical issues, there also is considerable epistemological disagreement on how to best define the notion of species for use in practice. For practical purposes in biodiversity inventories and conservation contexts, species should possess an unequivocal and generally agreed upon epistemic status as well-delimited and recognizable groups of organisms that, by being groups at the same level of the biological hierarchy, are comparable throughout the whole of earthbound biodiversity. A species of fruit fly should be a similar kind of unit as a species of flowering plant, or a species of fungus – they all are species, after all, and not groups at higher or lower taxonomic levels. However, it seems that different organism groups are best classified using different definitions of 'species', such that species of organisms in very different domains of biodiversity are not generally comparable units.

The preceding issues suggest that species are not good units of biodiversity, that is, units that meet the various epistemological and metaphysical requirements for performing the roles assigned to them in biodiversity research and conservation practice. In what follows, I will explore this suggestion in more detail.

8.3   Why Species Are Not Good Units of Biodiversity and Conservation

A number of aspects of the role species play in biodiversity studies and conservation efforts are philosophically controversial. In this section, I will highlight four such aspects: the reality of species (and the connected issue of the naturalness of species), the role that species play as counting units, the idea that species should be targets of conservation efforts because they are repositories of genetic information, and the relation between species and ecosystem services.

---

[6] I am sceptical about using notions of intrinsic value in general and in the context of environmental ethics in particular. This is an issue for another paper, though.

A first issue that needs to be addressed is the reality and naturalness of species. If species are the sort of things that can come into existence in speciation events and go extinct, that can be attributed an intrinsic value, or can be counted in biodiversity surveys of particular areas, then surely they must be real things.[7] If species are purely conventional units without *any* basis in nature, then what exactly are we counting in biodiversity inventories?[8] The concept of species richness, for example, is the simple idea of the number of species in a particular area or system: "[E]cosystem A is easily recognized as more diverse than B or C because it has four species instead of three. This characteristic is called species richness or just richness, and it is a simple, commonly used measure of diversity" (Hunter and Gibbs 2007: 25).[9] For comparisons between ecosystems with respect to species richness to make sense, species counts must be counting real features of ecosystems – or at least features that are not *purely* conventional.

Accordingly, the view that species must be real entities is widespread among biologists. Cracraft, for example, writes:

> Unless species concepts are used to individuate *real, discrete entities in nature*, they will have little or no relevance for advancing our understanding of the structure and function of biological phenomena involving those things we call species. [...] If species are not considered to be discrete real entities [...] then it implies that evolutionary and systematic biology would be based largely on *units that are fictitious*, whose boundaries, if drawn, are done so arbitrarily. It would also mean that most, if not all, of the processes that we ascribe to species are concoctions of the mind and have no objective reality. Entities of postulated processes must be real and discrete if those processes are to have much meaning. [...] Unless a species concept can be used to individuate *real-world entities*, that concept will have limited utility for systematists getting on with their task of sorting out and understanding biological diversity. (Cracraft 1997: 327–328; emphasis added).

In a similar vein, Wilson writes:

> Not to have a *natural unit* such as the species would be to abandon a large part of biology into free fall, all the way from the ecosystem down to the organism. It would be to concede the idea of amorphous variation and *arbitrary limits* for such intuitively obvious entities as American elms (species: *Ulmus americana*), cabbage white butterflies (*Pieris rapae*), and human beings (*Homo sapiens*). Without natural species, ecosystems could be analyzed only in the broadest terms, using crude and shifting descriptions of the organisms that compose them. Biologists would find it difficult to *compare results* from one study to the next. How might we access, for example, the thousands of research papers on the fruit fly, which form much of the foundation of modern genetics, if no one could tell one kind of fruit fly from another? (Wilson 1992: 38; emphases added).

---

[7] Note that is not the case that only real things can be attributed a value. Here, however, species are thought to have *intrinsic* value, i.e., values that a species are supposed to have in and of themselves, and it is difficult to see how non-real things could be the bearers of such intrinsic values.

[8] An additional problem is the following. If there indeed are too many different aspects of biodiversity such that studying *the* biodiversity of a region does not make sense (see Sect. 8.1; Santana 2014, 2018), a question is what species as units of biodiversity are supposed to represent. They cannot represent biodiversity *as such*, but only one or a few aspects of it. So, species could not be units of biodiversity without further qualification. At most, they could be units of some aspect(s) of biodiversity.

[9] The authors make reference to a table in their book in which ecosystem A consists of four species, and ecosystems B and C three.

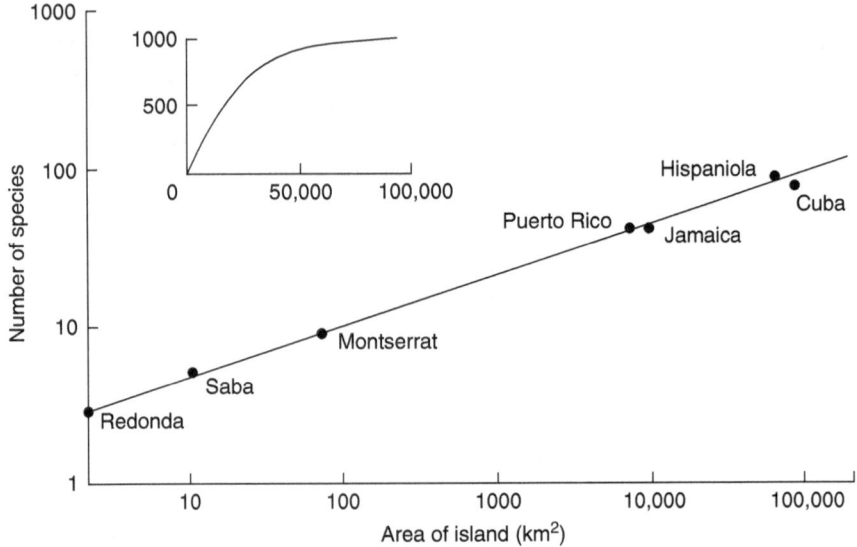

Fig. 8.1 The number of species of reptiles and amphibians of the West Indies. (From Lomolino et al. 2017, *Biodiversity (Fifth Edition)*, Fig. 13.5; reproduced by permission of Oxford University Press, USA)

And, as a final example, Mace expresses a strong view of the reality of species that involves the idea of objectivity: "the species rank is unique in the taxonomic hierarchy in that it has claim to *objective reality*, since gene flow is largely restricted within species […]. Almost all of the many variants on a species definition agree at least that *species are real and distinct entities in nature* […]" (Mace 2004: 711; emphasis added).

And there are empirical patterns that support this view. For example, the generality of the species-area curve, that plots species richness against area size, strengthens the suggestion that species are real things. The number of species in a particular area or ecosystem tends to be larger for larger areas or ecosystems. However, the number of species does not simply rise proportionally to the area, habitat or ecosystem size, but flattens out in a way that in island biogeography (and other contexts in which the areas are clearly delimited) is described by the equation $S = cA^z$ or $\log_{10}(S) = \log_{10}(c) + z\log_{10}(A)$, where S is the number of species and A the area.[10] The corresponding species-area curve has the general form pictured in Fig. 8.1 and Fig. 8.2: it is a curve that for S plotted against A rises sharply near the intersection of both axes and smoothly flattens out with increasing area, and a straight line for $\log S$ plotted against $\log A$. This is a semi-universal pattern – not one that holds up without exception and as a strict law of nature, but a pattern that

[10] The equation was first developed by Arrhenius (1921) and modified by Gleason (1922). See Connor and McCoy (1979: 794).

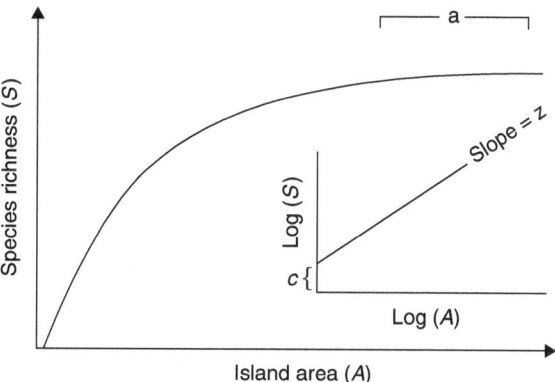

Fig. 8.2 The characteristic general species-area curve. (From Lomolino et al. 2017, *Biodiversity (Fifth Edition)*, Fig. 13.8; reproduced by permission of Oxford University Press, USA)

holds up pretty well nonetheless.[11] The species-area relationship "is often referred to as the closest thing to a rule in ecology" (Lomolino 2000: 17), and "one of the most general, best-documented patterns in nature" (Lomolino et al. 2017: 449). Some authors even call it "one of community ecology's few genuine laws" (Schoener, cited in McGuinness 1984: 423; Lomolino et al. 2017: 449).

For a pattern such as this to hold up so well, an epistemological requirement is that species must be uniquely countable (it must be possible to obtain clear species counts for areas and ecosystems) and comparable (species counts of different areas and ecosystems must have the same meaning).[12] Metaphysically this implies a minimal realism about species. With 'minimal realism' I mean the view that species are not *completely* artificial units, that is, units constructed by us without having any grounding in nature. Examples of such completely artificial units that are often mentioned in philosophical discussions are the group of all items that happen to be on my desk as I am writing this, or, for a biological example, the grouping of whales and fish together in one kind because they look very similar and live in water. Nothing prevents us from grouping things in this way or from referring to these groups in everyday or even scientific contexts if we feel this serves some purpose. But we cannot think of them as real groupings in nature – we could just as easily have grouped things differently (taking all items on the left half of my desk as one group and all items on the right half as another, or grouping whales with fish larger than a particular size and taking smaller fish as a distinct group), and in the case of whales and fish there are clear scientific reasons to deny the reality of the grouping.

---

[11] Although the pattern is strong, depending on sampling methods (e.g., how the sampling area is determined) and the model used to produce a species-area relationship on the basis of the data, different types of curves exist (see Scheiner 2003; Tjørve 2003, 2009).

[12] Note that this does not necessarily entail that species counts must be comparable for *all* species throughout the living world and for *all* kinds of areas, habitats and ecosystems. But at the very least, comparability must be guaranteed for counts of, say, mammalian species in forests or avian species on islands. Comparability is affected greatly by sampling methods and the possibility of uniquely delimiting the areas in which species are sampled (see footnote 5), but here I will only address the issue of species.

They are groups *by convention*, not because they represent some aspect of a natural order. For species-area curves to have scientific meaning, species cannot be artificial or conventional units in this sense, but must have at least *some* reality in nature.[13]

Indeed, the reality of species has long been a topic of discussion among philosophers of biology as well as biologists (e.g., Rolston 1995; Cracraft 1997; Mishler 1999, 2010; Bachmann 2001; Wilkins 2009; Claridge 2010; Richards 2010: Chapters 1 & 4; Kunz 2012; Slater 2013). As remarked above, there is a strong motivation for considering species to be real entities. Since the origins of biology as a science, species of organisms have been among the focal entities of research. They play an important role in nature conservation efforts. They are the basic constituents of the tree (or bush, or network) of life, as it is often presented in natural history museums. Species come into being in speciation events, participate in evolutionary processes, and go extinct. How could species not be real? But notwithstanding these strong motivations for assuming the reality of species, there are good reasons to doubt their reality. Perhaps the most important such reason is the persistent failure of biologists and philosophers of biology to agree what, exactly, species *are*. This is the metaphysical question at the heart of what has come to be known as "the species problem" (e.g., Mayr 1957; Stamos 2003; Reydon 2004, 2005; Wilkins 2009; Richards 2010). The problem has been a topic of debate at least since the publication of Darwin's *Origin of Species*, and Darwin himself addressed it in the book (Ereshefsky 2010, 2011, 2014).

The problem has turned out surprisingly difficult and at present there exists a massive volume of literature on the topic, most of which is concerned with the plurality of species concepts that have been proposed over the years and are currently used in biological science. There are around two dozen competing explanatory definitions of the term 'species' (so-called species concepts) available in the literature (Mayden 1997, 1999; Wilkins 2009; Mallet 2013).[14] Metaphysically, these definitions provide different views of what kind of things species are – populations or metapopulations, lineages, monophyletic groups of lineage segments, groups of morphologically and behaviorally similar organisms, groups of genetically similar organisms, groups of organisms that inhabit the same niche, or some combination of these characterizations – that roughly fall into the two main metaphysical categories of kinds and individuals. But none of the available definitions has gained acceptance as the one, correct definition and the nature of species remains elusive.

---

[13] This brings us to the highly difficult metaphysical issue of what it means for something to be real. For reasons of space, I have tried to avoid delving into this discussion and I am assuming that it is intuitively clear what it means for something to be minimally real in the sense discussed here: to be minimally real means to have some foundation in the world as it is independently of us.

[14] The exact number of species concepts (or definitions) proposed in the literature is unclear, and much depends on how one counts. Wilkins (2011) counts seven definitions and 27 variations and mixtures of those seven definitions, Mayden (1997, 1999) lists 22 definitions, and Lherminer and Solignac (2000) list 92 definitions. In the end, though, what matters is that there are multiple distinct and mutually incompatible definitions of the concept of species that are used side by side in biological science and yield mutually incompatible groupings of organisms into species (Reydon 2005).

The question thus remains what biodiversity inventories are counting and what conservation efforts – when successful – are preserving.

At present, the debate (that has been going on for more than 150 years) seems to have reached something of an impasse and it looks like we have run out of unexplored options that could break the deadlock between the available definitions. As a better metaphysical understanding of the nature of species as real entities in nature does not seem to be on the horizon, the suspicion is growing that the starting point of the quest – the assumption that species are real entities in nature – was wrongheaded all along and species should be considered conventional, instrumental units after all. This latter view – species antirealism – can take two forms. One is the denial that species names such as *Drosophila melanogaster* or *Quercus rubra* refer to real entities – they refer to artificial groupings made by us for particular purposes but do not represent real groupings in nature (for an overview of this position, see Stamos 2003: Chapter 2; Wilkins 2009: 221ff.). This version of species antirealism entails that in nature there is no such things as a species – species are our inventions. The other form is the denial that the species *category* is a real category, while taking species names such as *Drosophila melanogaster* or *Quercus rubra* as referring to real entities.[15] On this latter view, the entities referred to by means of species names may be real, but they are not members of one ontological category and thus are not comparable entities or units. Accordingly, there is nothing special about species: *Drosophila melanogaster* or *Quercus rubra* may be real entities in nature that are of interest in the sciences or in conservation efforts, but *not* because they are species.

The view that species are artificial groupings and in particular the view that the species category is not a real category (while species are real) may well strike many as odd. For the purposes of the present chapter, however, the precise ways in which one can be a species antirealist are less important than the point that realism about species is far from straightforward. The plurality of mutually incompatible explanatory definitions of the concept of species and the persistent problem of identifying one definition as the correct characterization of species strongly suggest that species are not real entities in nature (see Richards 2010: 10ff.). But the central role of species in biological science and the existence of patterns such as the species-area curve suggest otherwise. What can be made of this situation? At the very least, the doubts raised here regarding the reality of species should raise doubts regarding the suitability of species as focal units in biodiversity studies and conservation efforts.

These doubts are deepened when looking at practices of species counts. A number of recent studies have shown that species counts are strongly dependent on the definition of 'species' that is used, and that large discrepancies can exist between species counts based on different species concepts. The number of endemic bird species in Mexico, for example, was found to vary between 101 when using the Biological Species Concept and 249 when using a version of the Phylogenetic Species Concept (Peterson and Navarro-Sigüenza 1999). Similarly, depending on the species concept that is used, the whitefly species complex *Bemisia tabaci* can be seen as a single species (using the Morphological Species Concept) or as a complex

---

[15] Ereshefsky (2010, 2011, 2014) suggested that this was Darwin's view.

of 24–28 species when species are identified according to reproductive isolation and phylogeny (Liu et al. 2012). In a meta-analysis of 89 studies, Agapow et al. (2004) found that using versions of the Phylogenetic Species Concept on average led to an increase of more than 48% in species counts as compared to counts based on non-phylogenetic species definitions, such as the Biological Species Concept. These increases in species counts were accompanied by decreases in population sizes and ranges, thus not only yielding more species, but also resulting in those species being threatened to a higher degree. Mace summarizes the problem as follows:

> Without doubt, species need to be named and identified formally if they are to benefit from the conservationists' sets of legislative and planning tools. Unfortunately, all lists of species, and species richness measures generally, are extremely vulnerable to changes in species definitions. As the species concept becomes narrower, or species are split for whatever reason, the length of the list increases. The units making up the list can also alter radically. (Mace 2004: 713).

In the face of such practical difficulties we should be cautious when attributing reality to species in biodiversity studies and conservation contexts. It is crucial to ask what species counts mean, and why certain entities rather than others should be in focus of conservation efforts.

The problem is intensified by the fact that not only can different species definitions be used to partition organisms of one group into species, but it seems that different organism groups are best classified using different species concepts. In addition, microbes (which constitute the largest part of biomass on Earth) are currently classified into species on the basis of a pragmatic definition of 'species' based on a number of conventions guided by the availability of analytic technologies (such as a level of genetic similarity above which organisms should be counted as members of the same species), while it remains unclear to what extent (if at all) microbial species represent real groups in nature (Roselló-Mora and Amann (2001; Gevers et al. 2005; Doolittle and Zhaxybayeva 2009). This means that it is very difficult, if not impossible, to compare species groupings, species counts and the conservation status of species throughout the whole of biodiversity. If the same group of organisms can be classified into species in different ways, and species of birds are very different kinds of groupings from species of, say, flowering plants, then how can we be sure that we have assessed the conservation status of species correctly and how can we make good decisions on conservation priorities (which involve the comparison of species with respect to their conservation status, after all)? Moreover, how can we attribute – intrinsic or extrinsic – values to species and use these valuations as the basis for conservation efforts, if the species we pick out depend strongly on the definition we use and thus seem to lack reality as natural entities? While the practical consequences of using different species definitions for biodiversity studies and conservation efforts are widely acknowledged in the biological literature, authors have proposed diverging view of how the problem should be dealt with (e.g., Cracraft 1997; Agapow et al. 2004; Mace 2004; Dillon et al. 2005; George and Mayden 2005; Garnett and Christidis 2007; Frankham et al. 2012; Groves et al. 2017).

Taking a closer look at some of the reasons for which conservation value is attributed to natural entities is illuminating in this respect. One such reason is the view that species are repositories of genetic information, such that conserving species would be a way to conserve valuable genes and genotypes. For example, in the proceedings volume of the 1986 *National Forum on BioDiversity*, where the term 'biodiversity' was coined, E.O. Wilson asserted: "Each species is the repository of an immense amount of genetic information. The number of genes range from about 1,000 in bacteria and 10,000 in some fungi to 400,000 or more in many flowering plants and a few animals [...]. A typical mammal such as the house mouse (*Mus musculus*) has about 100,000 genes." (Wilson 1988: 7). For a number of reasons, this is a problematic basis for attributing value to species. For one, the view of species as genetic repositories involves the assumption that species are real entities in which a certain amount of genetic information is stored. Moreover, it involves the assumption that the genetic information stored in a species is specific for that species – i.e., that each species can be associated with a combination of genes that characterizes that species and distinguishes it from other species, such that in order to conserve specific genetic information we would need to conserve a particular species.

The quotation from Wilson suggests a comparatively simple view of a species' genome as consisting of a countable number of well-individuated units – genes – that together constitute the genetic information contained in the species. Recent developments in the philosophy of biology as well as in biology itself, however, have shown the picture to be much more complex. One aspect of the problem is the current debate on what, exactly, genes are. While the notion of 'gene' was explicitly introduced as a technical term in biology in 1909 by the Danish geneticist Wilhelm Johannsen, originally it was a term without any concrete material referent and the concept's meaning has been undergoing considerable change. Weber (2005: 227), for example, considered the notion of 'gene' a case of what he calls 'freely floating reference': biologists repeatedly began to use the term to refer to different kinds of DNA-segments as new molecular biological techniques became available, such that which entities were individuated as genes depended strongly on the investigative methods and techniques available, as well as on the specific interests of researchers. At present, it remains unclear what, exactly, genes are and how the notion of the gene is best conceptualized (Dietrich 2000; Griffiths and Stotz 2006, 2013). Thinking of the genetic information stored in a species in terms of the genes contained in the species' genome, then, is problematic. Thinking of genetic information in terms of whole-genome sequences does not constitute a better option. Even though colloquially references are often made to *the* human genome (International Human Genome Sequencing Consortium 2001; Venter et al. 2001) or the *Arabidopsis thaliana* genome (The Arabidopsis Genome Initiative 2000), it is clear that there is no unique, species-specific amount of genetic information stored in a species' genome. It is a fundamental fact of evolution that the organisms within a species always exhibit genetic variation, which is often considerable, while at the same time there are widespread genetic similarities between distinct species. This makes it unclear to what extent one can think of a species (or "its" genome) as a repository of genetic information. A better view might be to think of genetic information as not stored in species as such, but in small local populations in which genetic diversity is limited (but for the fundamental reason of intra-populational genetic variation this is problematic too).

Similarly problematic is a view of species as providers of ecosystem services – one other important reason to value natural entities. Ecosystem services, such as clean water, arable land, or recreational landscapes, are provided by entities located at higher levels of organization than species. Moreover, the ecosystems providing us with ecosystem services are not composed of species as such, but of local populations that are allocated to species. Local populations, then, seem better suited than species as units of biodiversity studies and conservation efforts. For a number of reasons, species – either conceived as real entities in nature or as conventional groupings – are not well suited as focal units in these contexts.

8.4   What to Do with the Species Concept?

How can the notion of species continue to play a role in biodiversity studies and conservation efforts in the face of the problems highlighted in the preceding section? To answer this question, I want to go back to early stages of the debate on the concept and draw clues from Darwin's work.

In the *Origin of Species*, Darwin seemed to defend a view of species as artificial units that did not represent natural groupings. As he famously writes:

> In short, we shall have to treat species in the same manner as those naturalists treat genera, who admit that genera are *merely artificial combinations made for convenience*. This may not be a cheering prospect; but we shall at least be freed from the vain search for the undiscovered and undiscoverable essence of the term species. (Darwin 1859: 485; emphasis added).

While Darwin indeed seems to propose that species are not real, Ereshefsky (2010, 2011, 2014) recently suggested that Darwin's view only was that the species *category* was not real while individual species could be thought of as real entities (see also the discussion above). But if this is correct, what should we make of Darwin's use of the term 'species' – why did he even use 'species' in the title of his book?

Ereshefsky (2010: 409; 2011: 71; 2014: 83) suggested that Darwin used the term for mere pragmatic reasons without attributing theoretical meaning to it. In contrast, but without entering into detailed Darwin exegesis, I want to suggest that for Darwin the notion of species did have a theoretical meaning. In the *Origin*, about 60 pages before the above quotation and in a part of the text in which he refers to the only figure in the book (the famous tree or bush of life in Chapter IV of the *Origin*, depicted here in Fig. 8.3), Darwin writes the following:

> I believe that the *arrangement* of the groups within each class [...] must be strictly genealogical in order to be natural; but that the *amount* of difference in the several branches or groups [...] may differ greatly, being due to the different degrees of modification which they have undergone; and this is expressed by the forms being ranked under different

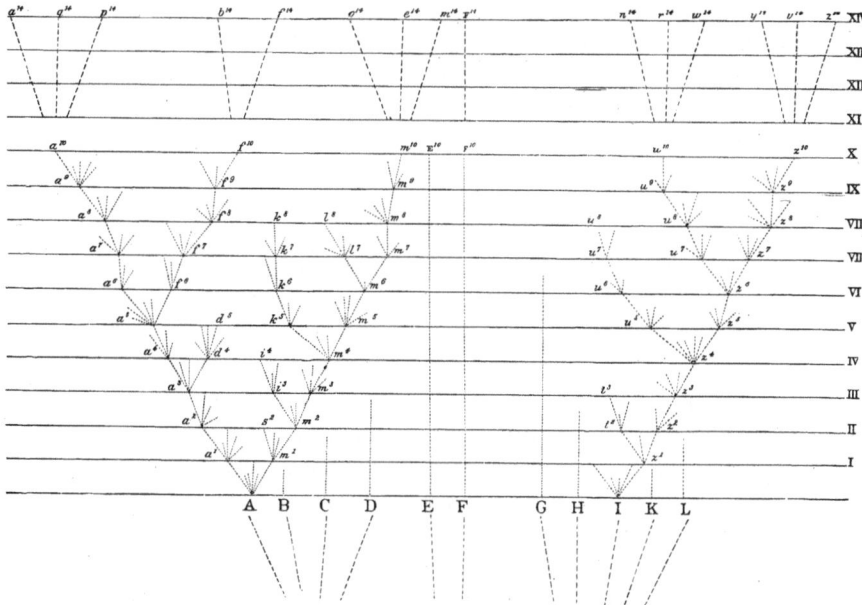

Fig. 8.3 The descent-with-modification diagram from the *Origin of Species*. (Darwin 1859)

genera, families, sections, or orders. [...] [T]he natural system is genealogical in its arrangement, like a pedigree; but the degrees of modification which the different groups have undergone, have to be expressed by ranking them under different so-called genera, sub-families, families, sections, orders, and classes. (Darwin 1859: 420 & 422; emphasis in original).

Here, Darwin talks about higher taxa and explains that the ordering of species into higher taxa represents degrees of modification that distinguish species of different higher taxa from each other. While Darwin does not say that the grouping of organisms into species can in a similar way be seen as representing degrees of modification, I want to suggest that such an interpretation fits Darwin's views and more generally would constitute a plausible way of treating the term 'species'.

In the figure in the *Origin*, to which Darwin is referring (Fig. 8.3 here), descent with modification is represented by dotted lines fanning out from common origins (common ancestor species) indicated with A, B, C, ... The span between two horizontal lines, indicated with Roman numerals, represents a distance of 1000 generations. Darwin explains:

After a thousand generations, species (A) is supposed to have produced two fairly well-marked varieties, namely a^1 and m^1. [...] After ten thousand generations, species (A) is supposed to have produced three forms, a^{10}, f^{10}, and m^{10}, which, from having diverged in character during the successive generations, will have come to differ largely, but perhaps unequally, from each other and from their common parent. If we suppose the amount of change between each horizontal line in our diagram to be excessively small, these three forms may still be only well-marked varieties; or they may have arrived at the doubtful category of sub-species; but we have only to suppose the steps in the process of modification to be more numerous or greater in amount, to convert these three forms into well-defined species: thus the diagram illustrates the steps by which the small differences distinguishing varieties are increased into the larger differences distinguishing species. (Darwin 1859: 117 & 120).

As is well known, Darwin did not make a fundamental distinction between species and varieties. As he stated in a famous passage in the *Origin*:

> I look at the term species, as one arbitrarily given for the sake of convenience to a set of individuals closely resembling each other, and that it does not essentially differ from the term variety, which is given to less distinct and more fluctuating forms. The term variety, again, in comparison with mere individual differences, is also applied arbitrarily, and for mere convenience sake. (Darwin 1859: 52).

So, Fig. 8.3 can be interpreted as showing very gradual divergence of forms, where smaller divergences are given the status of varieties, larger divergences the status of species, still larger divergences that status of genera, and so on. Being attributed the status of species, then, means that a group of organisms has achieved a particular level of modification in comparison to its ancestor group and to other groups in the same time-slice – a level of modification larger than that of a variety but smaller than that of a genus. This view of species – as a *status* that is attributed to a group of organisms on the basis of how far it has "evolved away" from its ancestor – fits the interpretation that Darwin thought the species category is not real. Species do not constitute a separate kind of things that are part of the furniture of the world, as for instance electrons or gold atoms, but rather the notion of species refers to a level of evolution that groups of organisms can achieve.

While this may be seen as a somewhat peculiar view of what the notion of species means – and as a view that is in conflict with much of what biologists say species are, as well as with both of the main metaphysical views of species as being either natural kinds or individuals – it is a view that can also be found in early work in the Modern Synthesis. Dobzhansky, in an early issue of the journal *Philosophy of Science*, puts it thus:

> Considered dynamically, the species represents that *stage of evolutionary divergence*, at which the once actually or potentially interbreeding array of forms becomes segregated into two or more separate arrays which are physiologically incapable of interbreeding. The fundamental importance of this stage is due to the fact that it is only the development of the isolating mechanisms that makes possible the coexistence in the same geographic area of different discrete groups of organisms. [...] [D]evelopment of isolating mechanisms renders the differences between groups relatively fixed and irreversible, and permits them to dwell side by side without losing their differentiating characteristics. This, in turn, opens the possibility for the organisms dwelling together to become adapted to different places in the general economy of nature. *The usage of the term "species" can and should be made to reflect the attainment by a group of organisms of this evolutionary stage.* (Dobzhansky 1935: 354; emphasis added).

As Dobzhansky explains in his book, *Genetics and the Origin of Species*: "[I]n the light of the evolution theories [...] such concepts as race, species, genus, family, etc., have come to be understood as connoting *nothing more than degrees of separation* in the process of a gradual phylogenetic divergence." (Dobzhansky 1937: 309; emphasis added). According to Dobzhansky, however, this does not mean that species are arbitrary or purely conventional units, that is, arbitrary divisions of a continuum into discrete units: as long as real discontinuities exist in the array of living forms, these can be interpreted as the natural boundaries of species as natural units (Dobzhansky 1937: 306–307).

The basic idea is that wherever we find stable discontinuities between groups of organisms, these natural boundaries can be taken to delimit species. Because in sexually reproducing organisms discontinuities between morphological groups are supported by reproductive isolation between populations, Dobzhansky suggested to define the notion of species in such a way as to link species status to the presence of isolating mechanisms. But the main idea is that the term 'species' indicates the achievement of a particular degree of separation by an evolving group, no matter by means of which mechanisms or other causal factors this is achieved. The development of reproductive isolating mechanisms by a group of organisms in Dobzhansky's view underlies the achievement of a level of evolutionary independence that allows us to individuate species by means of their natural boundaries: "The stage of the evolutionary process at which this fixation [of the discontinuity] takes place is fundamentally important, and the attainment of this stage by a group of organisms signifies the advent of species distinction" (Dobzhansky 1937: 312).[16] Note that achievement of this stage of evolution is not a yes-or-no matter, but a matter of degree: groups of organisms gradually develop isolating mechanisms, and may be more or less isolated from other groups. The term 'species' for Dobzhansky refers to this process stage and not to an ontological category of entities. As he emphasizes, his definition "lays emphasis on the dynamic nature of the species concept. Species is *a stage in a process, not a static unit.*" (Dobzhansky 1937: 312).

Drawing inspiration from Darwin and Dobzhansky, I want to suggest to take Dobzhansky literally and interpret the term 'species' as signifying a particular *stage of the evolutionary process* that evolving groups of organisms can achieve. The species stage is characterized by comparative evolutionary independence: attributing species status to a population or other group of organisms means that it has reached a stage at which it has become sufficiently independent from other parts of the system in which it exists for novel traits (or at least clearly different traits from those of other

---

[16] While Dobzhansky's definition of 'species' may seem the same as Mayr's widely used Biological Species Concept, there is an important difference between them. On the Biological Species Concept, species are a particular kind of breeding populations. That is, on this concept species are entities of a particular ontological category, i.e., the species category is a subcategory of the more general category of populations. On Dobzhansky's definition, however, 'species' denotes a status that may be attributed to populations that have achieved a particular stage in the evolutionary process: populations undergo a gradual process of modification in which at some point they may (or may not) achieve an evolutionary stage at which they exist as discrete groups next to other groups, marked by morphological and genetic discontinuities that are supported by reproductive isolation mechanisms. Accordingly, asexually reproducing and obligatory self-fertilizing organisms that do not develop reproductive isolating mechanisms also do not form species (Dobzhansky 1937: 321). But note that Dobzhansky did not define species as a kind of populations: any sort of group of organisms that reaches an evolutionary stage in which it exists as a discrete group can in principle be attributed species status (but Dobzhansky did emphasize reproductive isolating mechanisms of populations in this context).

groups) to emerge and to become fixated in the population. As such evolutionary independence is a matter of degree and independence never is *complete* independence, the corresponding attribution of species status is a matter of degree too.[17]

How exactly is this a different view of species from the ones reflected by available species concepts? While I do not have sufficient space in the present chapter to fully articulate the suggestion that a species is a stage in the evolutionary process, I want to point to some philosophical aspects of this suggestion by way of clarification. Being a species is an accidental property, i.e., a property that a group of organisms can come to have or lose without losing its identity. As was discussed above, the view of species as process stages involves a denial that the species category is a real category of entities. Much of the debate on the species problem has been fueled by the metaphysical question what species *are* – natural kinds, individuals, or something else (e.g., Ruse 1987). Accordingly, available species concepts tend to be explicative definitions of the metaphysics of species – they tell us what kind of entities species are, i.e., what kinds of entities constitute the ontological category of species. The two principal options in the debate are conceptions of species as a particular kind of natural kinds or a particular kind of individuals, but other options have been suggested too (for example, that species are processes – cf. Rieppel 2009). While one could perhaps think of process stages as real in some way, this would at least not be realism about entities of a particular kind. The *entities* that one would be a realist about, after all, are populations, metapopulations, lineages, clades, and other sorts of organism groups, i.e., entities of a number of different ontological categories. I would go further, though, and say that the attribution of species status to a group is an *epistemological* as well as *normative* attribution, and not a metaphysical one. To say that a group of organisms is a species is to say that it is a group – whatever its precise metaphysical nature – that has achieved a stage of evolution that is of importance to us in the light of evolutionary theory – it is a stage *we* highlight, because of its explanatory importance – and that we can value accordingly. The claim that a group of organisms is a species thus does not entail anything regarding its metaphysics; in particular, it does not entail that it is a natural kind, an individual, a historical entity, and so on.

Similar suggestions have been advanced in the recent philosophical literature on species and natural kinds. As philosophers of science have increasingly begun to examine scientific practices (in the context of what has been called the "practice turn" in the philosophy of science – Kendig 2016a: 3ff.), focus is increasingly placed on the question how scientific concepts are used in investigative practices, rather than on questions of what in nature these concepts represent. Kendig (2014, 2016a, b), for example, has proposed that classificatory notions, such as 'species', 'natural kind', and 'homologue', are best understood as denoting practices of

---

[17] Organisms always live interdependently in ecosystems, and often populations of organisms coevolve. In cases in which two populations coevolve – for instance a particular kind of plant and its specific pollinator – the two populations can be treated as independent in the sense of the present discussion, as they each evolve their own novelties. Still, as coevolving populations they of course are dependent on each other. The notion of evolutionary independence that is in play here simply means for a population to have the ability to evolve its own novelties, and should not be read in too strong a manner.

species-making and kind-making, that is, of the grouping of organisms into species, and of things more generally into natural kinds, on the basis of various epistemological and practical considerations. Accordingly, there are multiple ways of grouping organisms into species depending on which aspects of speciation, lineage-forming processes, inheritance processes, etc. one is interested in (Kendig 2014). Being a species, then, means *being used* by scientists in a particular context as a species. Similarly, the notion of natural kind can be understood as consisting principally in the making use of certain groups in investigative practices (Kendig 2016a: 6). And in a historical study of the notion of homology, Kendig (2016b) showed how the notion of homology is best understood as referring to a set of practices of comparing organisms and their traits, and how different biologists highlighted certain traits as homologous on the basis of different theoretical grounds and investigative interests. What happens, in sum, is that scientists make groups of entities, organisms, and traits, and attribute some of those groups the status of natural kind, species, or homologue on the basis of a variety of theoretical and practical grounds. Different grounds yield different, oftentimes incompatible groupings of the same entities, while no particular grouping can be said to yield *the* natural kinds, *the* species, or *the* homologues in a particular domain of nature.

In a similar fashion, Slater (2013, 2015) recently suggested that we should not try to develop a philosophical account of what natural kinds *are*, but rather should think of "natural kindness" as a kind of status that things can have on epistemological grounds (Slater 2013: 150; 2015: 378).[18] On Slater's view, rather than thinking of natural kinds as an ontological category of entities, we should think of "being a natural kind as a sort of status that things or pluralities of things (from various ontological categories) can have" (Slater 2015: 407), where the degree to which something can be attributed the status of natural kind corresponds to the degree to which it supports inferential practices in a particular context. The status of natural kindness, Slater emphasizes, is domain dependent: something may have that status in one domain of work but not (or to a lesser degree) in another (Slater 2013: 172). Slater thus takes the problem of natural kinds out of the domain of metaphysics and puts it squarely into the domain of epistemology: saying that something is a natural kind is attributing a particular epistemic status to it to a particular degree. In this case the epistemic status is that of being useful as a basis for inferences (where groups that are better suitable as bases for inferences have this status to a higher degree). The precise metaphysics underlying this status is secondary, and can take very different forms in different cases. Some things that we attribute the status of natural kind to belong to the ontological category of individuals, while others belong to other categories.[19]

---

[18] Slater (2013: 107 & 150) says that his account is a characterization of natural kindness as an adjective rather than of the nature of natural kinds as an ontological category.

[19] Note that here I am not endorsing Slater's account of natural kindness, nor am I claiming that it is an adequate interpretation of species as natural kinds. All I want to do is to highlight the similarity between Slater's approach to the notion of natural kinds and my approach to the notion of species.

The view of species proposed here fits taxonomic practice. Take for example the recent discovery of a new species of orangutan (Nater et al. 2017; Reese 2017). In the 1930s, explorers reported the existence of a small, isolated population of orangutans in the Tapanuli district in North Sumatra. Only after reading the reports in the mid-1990s did scientists start to look for the population again and eventually managed to find nests, the remains of a female orangutan and finally in 2013 one male that was killed by local inhabitants. On the basis of comparisons of the skull of the male specimen with 33 museum specimens of the two already described species of orangutans, *Pongo pygmaeus* (the Bornean orangutan) and *Pongo abelii* (the Sumatran orangutan), as well as genetic comparisons of 37 orangutan specimens, biologists identified the population as a new species, *Pongo tapanuliensis*. As the authors write, the species "encompasses a geographically and genetically isolated population found in the Batang Toru area at the southernmost range limit of extant Sumatran orangutans" (Nater et al. 2017: 3488). The central factor in the individuation of the local population as a new species, rather than a mere morphological variety of one of the extant species of the genus *Pongo* was its evolutionary independence from the other groups, as evidenced by morphology and genetics. As the authors write: "*P. tapanuliensis* and *P. abelii* have been on independent evolutionary trajectories at least since the late Pleistocene / early Holocene" (Nater et al. 2017: 3491–3492). While the researchers found morphological differences with the Sumatran orangutan, they also found that *P. tapanuliensis* is genetically more closely related to *P. pygmaeus*, from which it diverged much later, than to *P. abelii*. *P. tapanuliensis* thus has its own evolutionary identity in distinction of the group of orangutans living on the same island, Sumatra, and due to the geographical separation of the group of orangutans living on Borneo, it had its own evolutionary identity in distinction of that group too.[20]

How does the view of species as evolutionary process stages, and of the attribution of species status to groups as epistemological and normative attributions rather than metaphysical statements, affect biodiversity studies and conservation efforts? First, on the view of species suggested here species counts cannot be seen as counts of the real entities of a particular kind that exist in that area or ecosystem. Still, species counts are meaningful when thinking of them as counts of groups of organisms that share an important property, namely comparative evolutionary independence. When it comes to counting species in the context of making inventories of an area's or ecosystem's biodiversity it is crucial to count things that are comparable, and while on the view proposed here species cannot be seen as entities of the same kind, they still are comparable as entities that we highlight for the same epistemological reasons as the units in which evolutionary novelties can arise. For the same reason, evolutionary independence is an important basis for attributing conservation value

---

[20] The newly discovered species has immediately come in focus in conservation efforts. Because of its small population size of less than 800 individuals, biologists have expressed concerns for the survival of the population, implying an urgent need for conservation measures (Nater et al. 2017: 3493; Reese 2017).

to groups of organisms: conserving groups that have a comparative evolutionary independence means conserving evolutionary potential.[21]

In conservation biology, these ideas are sometimes embodied in the – ill-defined – concept of Evolutionarily Significant Unit (ESU). An ESU is a population or other group of organisms that stands at the focus of conservation efforts because of its evolutionary significance, where evolutionary significance is fleshed out in terms of evolutionary heritage (e.g., significant divergence from other groups, or representing an important aspect of a species' evolutionary legacy) and evolutionary potential (Moritz 1994; Crandall et al. 2000; Casacci et al. 2013). Waples (1995: 9) perhaps expressed the idea of evolutionary sugnificance most clearly in specifying that "[t]he evolutionary legacy of a species is the genetic variability that is a product of past evolutionary events and that represents the reservoir upon which future evolutionary potential depends." As usually conceived, ESUs are units *below* the species level, even though several authors have pointed out that in principle an ESU can coincide with a species (Moritz 1994: 374; Casacci et al. 2013: 183). Ryder, who introduced the ESU concept in the literature, for example suggested that rather than attempting to conserve all subspecies and varieties of a threatened species conservation focus in zoos should be placed on subspecies that represent significant adaptive variation and "zoos ought properly to address the conservation of evolutionarily significant units (ESUs) within species" (Ryder 1986: 9–10). Ryder noted that an alternative for the ESU concept could be the concept of evolutionarily significant population (ESP), thus highlighting that ESUs are not themselves species, but smaller units. Accordingly, conservation biologists may find that a species encompasses multiple ESUs that each should stand at the focus of conservation efforts. An example is the case of White Sands pupfish (*Cyprinidon tularosa*), where researchers have argued that two populations constitute distinct ESUs, where "[l]oss of either of the two ESUs of White Sands pupfish would result in a substantial loss of the evolutionary legacy of this species" (Stockwell et al. 1998: 219).

Dobzhansky's view of species as groups of organisms that have achieved a particular stage in the evolutionary process and the ESU concept express similar views of why some groups of organisms are highlighted as being of particular interest to us. First, species and ESUs do not exist independently of our interests, but rather the status of species or ESU is attributed by us to certain groups of organisms on the basis of theoretical considerations. Second, these considerations are fundamentally connected to evolutionary theory: species and ESUs are of particular importance because they represent important aspects of the evolutionary process. And third, both Dobzhansky's view and the ESU concept highlight that species and ESUs are important because of their evolutionary (that is, adaptive) potential. Dobzhansky's view and the ESU concept differ, however, by focusing on different taxonomic levels.

---

[21] Note that I am not suggesting that this is the only or even the most important basis for attributing a conservation value to a group of organisms. It merely is one such basis among many, as we may attach value to a group of organisms for a plethora of reasons.

On the view of species suggested here, then, populations and other groups of organisms should not be targets of conservation efforts because they would *represent* or *instantiate* a species that we want to retain in existence.[22] Nor should their species status be thought of as what we would want to conserve: in general, we do not want to retain the species in the evolutionary state in which it happens to be, but we want to conserve its potential for future evolution. Summarizing, on the view of the meaning of 'species' proposed here, what we count in species counts and focus on in conservation efforts are entities to which we attribute the same epistemological and normative status – the status of species –, even though that status may be underwritten metaphysically in very different ways for different kinds of organisms.

8.5   Concluding Remarks

I have started out this chapter by noting that species are usually seen as core units in biodiversity studies and conservation biology, and asking whether species indeed are good units in these contexts. My answer has been negative: the philosophical problems regarding the species concept are such that we cannot safely assume that species are real entities (entities existing independently of us in nature) or natural units. But the notion of species can still be used as a basis for biodiversity inventories and value attributions in conservation biology if the notion of species is interpreted differently.

I have suggested an interpretation of the species concept that would fit biodiversity studies and conservation biology better than views of species as constituting a particular category of real entities in nature. Contrary to recently advanced views but in line with Darwin's and Dobzhansky's views of species, I have suggested that the species concept does have theoretical significance, even though metaphysically one cannot say that the species category is real. Species as such are not real entities in nature, but grouping organisms into species is not a matter of mere convention either. The meaning of the species concept is connected to evolutionary reality in that attributions of species status to evolving populations reflects the achievement of populations of a degree of evolutionary independence that allows them to evolve their own adaptations and maintain their identity in distinction of other populations.

---

[22] Let me try to clarify this point. Often, local populations or other local groups of organisms are said to represent species ("Species *x* is represented in five countries by twelve populations.") or to instantiate species ("Species *x* is instantiated in twelve populations spread over five countries."). To be sure, this is an imprecise way of speaking. The point, though, is that usually species and populations are seen as two distinct entities between which there can be a relation such as representation, instantiation, etc. The species is seen as a larger entity (an abstract kind, a lineage, a collection of populations, or something else) than the population "out there" in the field. On the account proposed here, however, 'species' is a status attributed to a group of organisms such as a population, so there cannot be a relation of representation, instantiation, or otherwise between the group of organisms to which we attribute that status and "its" species.

Whether or not this can be seen as a (weak) realism about species is a question I want to leave open here, as my aim is not to defend a particular metaphysical position in the species debate. Also, it should be noted that my aim was not to devise an overall account of the species concept for the whole of the life sciences: my focus was on the meaning of 'species' for the purposes of biodiversity studies and conservation biology, and I leave open whether the view proposed here would be applicable throughout the whole of biology. And I have largely ignored core topics in the philosophy of biology regarding species, such as the monism-pluralism debate, the kinds-individuals debate, and other issues.

Suffice it to say that although there is no category of species as entities in the way there is a category of electrons or a category of cells, there are real entities (groups of organisms of various metaphysical categories) that can be attributed an epistemological and normative status on the basis of their having achieved evolutionary independence (that is, their having achieved some degree of "speciesness"). Putting those entities at the focus of biodiversity studies and conservation efforts (in a similar way as is suggested by the Evolutionary Significant Unit concept) means putting local populations and other groups of organisms as parts of landscapes and ecosystems in the foreground, moving away from the conservation of species as an aim of conservation efforts, but retaining the notion of species as a theoretically meaningful notion in biodiversity studies and conservation biology.[23]

References

Agapow, P. M., Bininda-Emonds, O. R. P., Crandall, K. A., et al. (2004). The impact of species concept on biodiversity studies. *The Quarterly Review of Biology, 79*, 161–179.

Arrhenius, O. (1921). Species and area. *Journal of Ecology, 9*, 95–99.

Bachmann, K. (2001). Species concepts: The continuing debate. *The New Phytologist, 149*, 367–368.

Barnosky, A. D., Matzke, N., Tomiya, S., et al. (2011). Has the Earth's sixth mass extinction already arrived? *Nature, 471*, 51–57.

Burch-Brown, J., & Archer, A. (2017). In defence of biodiversity. *Biology and Philosophy, 32*, 969–997.

Casacci, L. P., Barbero, F., & Balletto, E. (2013). The "evolutionarily significant unit" concept and its applicability in biological conservation. *The Italian Journal of Zoology, 81*, 182–193.

Ceballos, G., Ehrlich, P. R., Barnosky, A. D., et al. (2015). Accelerated modern human-induced species losses: Entering the sixth mass extinction. *Science Advances, 1*, e1400253.

Ceballos, G., Ehrlich, P. R., & Dirzo, R. (2017). Biological annihilation via the ongoing sixth mass extinction signaled by vertebrate population losses and declines. *Proceedings of the National Academy of Sciences of the United States of America, 114*, E6089–E6096.

Claridge, M. F. (2010). Species are real biological entities. In F. J. Ayala & R. Arp (Eds.), *Contemporary debates in philosophy of biology* (pp. 110–122). Chichester: Wiley.

Claridge, M. F., Dawah, H. A., & Wilson, M. R. (Eds.). (1997). *Species: The units of biodiversity.* London: Chapman & Hall.

---

[23] I am indebted to Philippe Huneman, Werner Kunz and an anonymous reviewer for helpful comments on earlier versions of this chapter.

Colwell, R. K. (2009). Biodiversity: Concepts, patterns, and measurement. In S. A. Levin (Ed.), *The Princeton guide to ecology* (pp. 257–263). Princeton: Princeton University Press.

Connor, E. F., & McCoy, E. D. (1979). The statistics and biology of the species-area relationship. *The American Naturalist, 113*, 791–833.

Costanza, R., d'Arge, R., de Groot, R., et al. (1997). The value of the world's ecosystem services and natural capital. *Nature, 387*, 253–260.

Cracraft, J. (1997). Species concepts in systematics and conservation biology – An ornithological viewpoint. In M. F. Claridge, H. A. Dawah, & M. R. Wilson (Eds.), *Species: The units of biodiversity* (pp. 325–339). London: Chapman & Hall.

Crandall, K. A., Bininda-Emonds, O. R. P., Mace, G. M., & Wayne, R. K. (2000). Considering evolutionary processes in conservation biology. *Trends in Ecology & Evolution, 15*, 290–295.

Darwin, C. R. (1859). *On the origin of species by means of natural selection, or, the preservation of favoured races in the struggle for life*. London: John Murray.

Dietrich, M. R. (2000). The problem of the gene. *Comptes Rendus de l'Académie des Sciences de Paris, Sciences de la Vie, 323*, 1139–1146.

Dillon, S., Fjeldså, J., & Kelt, D. (2005). The implications of different species concepts for describing biodiversity patterns and assessing conservation needs for African birds. *Ecography, 28*, 682–692.

Dirzo, R., & Raven, P. H. (2003). Global state of biodiversity and loss. *Annual Review of Environment and Resources, 28*, 137–167.

Dobzhansky, T. (1935). A critique of the species concept in biology. *Philosophy in Science, 2*, 344–355.

Dobzhansky, T. (1937). *Genetics and the origin of species*. New York: Columbia University Press.

Doolittle, W. F., & Zhaxybayeva, O. (2009). On the origin of prokaryotic species. *Genome Research, 19*, 744–756.

Ehrlich, P. R., & Wilson, E. O. (1991). Biodiversity studies: Science and policy. *Science, 253*, 758–762.

Ereshefsky, M. (2010). Darwin's solution to the species problem. *Synthese, 175*, 405–425.

Ereshefsky, M. (2011). Mystery of mysteries: Darwin and the species problem. *Cladistics, 27*, 67–79.

Ereshefsky, M. (2014). Consilience, historicity, and the species problem. In R. P. Thompson & D. M. Walsh (Eds.), *Evolutionary biology: Conceptual, ethical, and religious issues* (pp. 65–86). Cambridge: Cambridge University Press.

Faith, D. P. (2016). Biodiversity. In E. N. Zalta, (Eds.), *The Stanford Encyclopedia of philosophy* (Summer 2016 Edition), https://plato.stanford.edu/archives/sum2016/entries/biodiversity/

Frankham, R., Ballou, J. D., Dudash, M. R., et al. (2012). Implications of different species concepts for conserving biodiversity. *Biological Conservation, 153*, 25–31.

Garnett, S. T., & Christidis, L. (2007). Implications of changing species definitions for conservation purposes. *Bird Conservation International, 17*, 187–195.

George, A. L., & Mayden, R. L. (2005). Species concepts and the endangered species act: How a valid biological definition of species enhances the legal protection of biodiversity. *Natural Resources Journal, 45*, 369–407.

Gevers, D., Cohan, F. M., Lawrence, J. G., et al. (2005). Re-evaluating prokaryotic species. *Nature Reviews. Microbiology, 3*, 733–739.

Gleason, H. A. (1922). On the relation between species and area. *Ecology, 3*, 158–162.

Griffiths, P. E., & Stotz, K. (2006). Genes in the postgenomic era. *Theoretical Medicine and Bioethics, 27*, 499–521.

Griffiths, P. E., & Stotz, K. (2013). *Genetics and philosophy: An introduction*. Cambridge: Cambridge University Press.

Groves, C. P., Cotterill, F. P. D., Gippoliti, S., et al. (2017). Species definitions and conservation: A review and case studies from African mammals. *Conservation Genetics, 18*, 1247–1256.

Hamilton, A. J. (2005). Species diversity or biodiversity? *Journal of Environmental Management, 75*, 82–92.

Hohenegger, J. (2014). Species as the basic units in evolution and biodiversity: Recognition of species in the recent and geological past as exemplified by larger foraminifera. *Gondwana Research, 25*, 707–728.

Hunter, M. L., & Gibbs, J. P. (2007). *Fundamentals of conservation biology* (3rd ed.). Malden: Blackwell.

International Human Genome Sequencing Consortium. (2001). Initial sequencing and analysis of the human genome. *Nature, 409*, 860–921.

Kendig, C. E. (2014). Towards a multidimensional metaconception of species. *Ratio, 27*, 155–172.

Kendig, C. E. (2016a). S introduction: Activities of *kinding* in scientific practice. In C. E. Kendig (Ed.), *Natural kinds and classification in scientific practice* (pp. 1–13). Abingdon/New York: Routledge.

Kendig, C. E. (2016b). Homologizing as kinding. In C. E. Kendig (Ed.), *Natural kinds and classification in scientific practice* (pp. 106–125). Abingdon/New York: Routledge.

Kunz, W. (2012). *Do species exist? Principles of taxonomic classification.* Weinheim: Wiley-Blackwell.

Lherminer, P., & Solignac, M. (2000). L'espèce: Définitions d'auteurs. *Comptes Rendus de l' Académie des Sciences de Paris, Sciences de la Vie, 323*, 153–165.

Lomolino, M. V. (2000). Ecology's most general, yet protean pattern: The species-area relationship. *Journal of Biogeography, 27*, 17–26.

Lomolino, M. V., Riddle, B. R., & Brown, J. H. (2017). *Biogeography* (5th ed.). Sunderland: Sinauer Associates.

Liu, S.-S., Colvin, J., & De Barro, P. J. (2012). Species concepts as applied to the whitefly *Bemisia tabaci* systematics: How many species are there? *Journal of Integrative Agriculture, 11*, 176–186.

Mace, G. M. (2004). The role of taxonomy in species conservation. *Philosophical Transactions of the Royal Society of London B, 359*, 711–719.

Mace, G. M. (2015). Whose conservation? *Science, 345*, 1558–1560.

Mace, G. M., & Lande, R. (1991). Assessing extinction threats: Toward a reevaluation of IUCN threatened species categories. *Conservation Biology, 5*, 148–157.

Mace, G. M., Collar, N. J., Gaston, K. J., et al. (2008). Quantification of extinction risk: IUCN's system for classifying threatened species. *Conservation Biology, 22*, 1424–1442.

Mace, G. M., Norris, K., & Fitter, A. H. (2012). Biodiversity and ecosystem services: A multilayered relationship. *Trends in Ecology & Evolution, 27*, 19–26.

Maclaurin, J., & Sterelny, K. (2008). *What is biodiversity?* Chicago/London: University of Chicago Press.

Mallet, J. (2013). Species, concepts of. In S. A. Levin (Ed.), *Encyclopedia of biodiversity* (Vol. 6, 2nd ed., pp. 679–691). Amsterdam: Academic.

Mayden, R. L. (1997). A hierarchy of species concepts: The denouement in the saga of the species problem. In M. F. Claridge, H. A. Dawah, & M. R. Wilson (Eds.), *Species: The units of biodiversity* (pp. 381–424). London: Chapman & Hall.

Mayden, R. L. (1999). Consilience and a hierarchy of species concepts: Advances toward closure on the species puzzle. *Journal of Nematology, 31*, 95–116.

Mayr, E. (Ed.). (1957). *The species problem.* Washington, DC: American Association for the Advancement of Science.

McGuinness, K. A. (1984). Equations and explanations in the study of species-area curves. *Biological Reviews, 59*, 423–440.

Meine, C., Soulé, M., & Noss, R. F. (2006). "A mission-driven discipline": The growth of conservation biology. *Conservation Biology, 20*, 631–651.

Mishler, B. D. (1999). Getting rid of species? In R. A. Wilson (Ed.), *Species: New interdisciplinary essays* (pp. 307–315). Cambridge, MA: MIT Press.

Mishler, B. D. (2010). Species are not uniquely real biological entities. In F. J. Ayala & R. Arp (Eds.), *Contemporary debates in philosophy of biology* (pp. 91–109). Chichester: Wiley.

Moritz, C. (1994). Defining 'evolutionarily significant units' for conservation. *Trends in Ecology & Evolution, 9,* 373–375.

Nater, A., Mattle-Greminger, M. P., Nurcahyo, A., et al. (2017). Morphometric, behavioral, and genomic evidence for a new orangutan species. *Current Biology, 27,* 3487–3498. e10.

Odenbaugh, J. (2016). Conservation biology. In E. N. Zalta, (Ed.), *The Stanford Encyclopedia of philosophy* (Winter 2016 Edition), https://plato.stanford.edu/archives/win2016/entries/conservation-biology/

Peterson, A. T., & Navarro-Sigüenza, A. G. (1999). Alternate species concepts as bases for determining priority conservation areas. *Conservation Biology, 13,* 427–431.

Purvis, A., & Hector, A. (2000). Getting the measure of biodiversity. *Nature, 405,* 212–219.

Raczkowski, J. M., & Wenzel, J. W. (2007). Biodiversity studies and their foundation in taxonomic scholarship. *BioScience, 57,* 974–979.

Reese, A. (2017). New orangutan species identified. *Nature, 551,* 151.

Reydon, T. A. C. (2004). Why does the species problem still persist? *BioEssays, 26,* 300–305.

Reydon, T. A. C. (2005). On the nature of the species problem and the four meanings of 'species. *Studies in History and Philosophy of Biological and Biomedical Sciences, 36,* 135–158.

Richards, R. A. (2010). *The species problem: A philosophical analysis.* Cambridge: Cambridge University Press.

Rieppel, O. (2009). Species as a process. *Acta Biotheoretica, 57,* 33–49.

Rolston, H. (1975). Is there an ecological ethic? *Ethics, 85,* 93–109.

Rolston, H. (1985). Duties to endangered species. *BioScience, 35,* 718–726.

Rolston, H. (1995). Duties to endangered species. In W. A. Nierenberg (Ed.), *Encyclopedia of environmental biology* (Vol. 1, pp. 517–528). San Diego: Academic.

Roselló-Mora, R., & Amann, R. (2001). The species concept for prokaryotes. *FEMS Microbiology Reviews, 25,* 39–67.

Ruse, M. (1987). Biological species: Natural kinds, individuals, or what? *The British Journal for the Philosophy of Science, 38,* 225–242.

Ryder, O. A. (1986). Species conservation and systematics: The dilemma of subspecies. *Trends in Ecology & Evolution, 1,* 9–10.

Santana, C. (2014). Save the planet: Eliminate biodiversity. *Biology and Philosophy, 29,* 761–780.

Santana, C. (2018). Biodiversity is a chimera, and chimeras aren't real. *Biology and Philosophy, 33,* 15.

Scheiner, S. M. (2003). Six types of species-area curves. *Global Ecology and Biogeography, 12,* 441–447.

Slater, M. H. (2013). *Are species real? An essay on the metaphysics of species.* Basingstoke: Palgrave Macmillan.

Slater, M. H. (2015). Natural kindness. *The British Journal for the Philosophy of Science, 66,* 375–411.

Soulé, M. E. (1985). What is conservation biology? *BioScience, 35,* 727–734.

Stamos, D. N. (2003). *The species problem: Biological species, ontology, and the metaphysics of biology.* Lanham: Lexington Books.

Stockwell, C. A., Mulvey, M., & Jones, A. G. (1998). Genetic evidence for two evolutionarily significant units of White Sands pupfish. *Animal Conservation, 1,* 213–225.

The Arabidopsis Genome Initiative. (2000). Analysis of the genome sequence of the flowering plant *Arabidopsis thaliana. Nature, 408,* 796–815.

Tjørve, E. (2003). Shapes and functions of species-area curves: A review of possible models. *Journal of Biogeography, 30,* 827–835.

Tjørve, E. (2009). Shapes and functions of species-area curves (II): A review of new models and parametrizations. *Journal of Biogeography, 36,* 1435–1445.

Venter, J. C., Adams, M. D., Myers, E. W., et al. (2001). The sequence of the human genome. *Science, 291,* 1304–1351.

Waples, R. S. (1995). Evolutionarily significant units and the conservation of biological diversity under the endangered species act. *American Fisheries Society Symposium, 17,* 8–27.

Weber, M. (2005). *Philosophy of experimental biology*. Cambridge: Cambridge University Press.

Wheeler, Q. D., Knapp, S., Stevenson, D. W., et al. (2012). Mapping the biosphere: Exploring species to understand the origin, organization and sustainability of biodiversity. *Systematics and Biodiversity, 10*, 1–20.

Wilkins, J. S. (2009). *Species: A history of the idea*. Berkeley: University of California Press.

Wilkins, J. S. (2011). Philosophically speaking, how many species concepts are there? *Zootaxa, 2765*, 58–60.

Wilson, E. O. (1988). The current state of biological diversity. In E. O. Wilson (Ed.), *Biodiversity* (pp. 3–18). Washington, DC: National Academy Press.

Wilson, E. O. (1992). *The diversity of life*. Cambridge, MA: Harvard University Press.

Chapter 9
Why a Species-Based Approach to Biodiversity Is Not Enough. Lessons from Multispecies Biofilms

Jorge Marques da Silva and Elena Casetta

Abstract In recent years, we have assisted to an impressive effort to identify and catalogue biodiversity at the microbial level across a wide range of environments, human bodies included (e.g., skin, oral cavity, intestines). This effort, fostered by the decreasing cost of DNA sequencing, highlighted not only the vast diversity at the microbial level but also the importance of cells' social interactions, potentially leading to the emergence of novel diversity. In this contribution, we shall argue that entities other than species, and in particular multispecies biofilms, might play a crucial—and still underestimated—role in increasing biodiversity as well as in conserving it. In particular, after having discussed how microbial diversity impacts ecosystems (Sect. 9.1), we argue (Sect. 9.2) that multispecies biofilms may increase biodiversity at both the genetic and phenotypic level. In Sect. 9.3 we discuss the possibility that multispecies biofilms, both heterotrophic and autotrophic, are evolutionary individuals, i.e. units of selection. In the conclusion, we highlight a major limitation of the traditional species-based approach to biodiversity origination and conservation.

Keywords Microbial diversity · Biofilms · Biological individuality

9.1 Microbial Biodiversity and Bacterial Modes of Living

Microbial biodiversity is essential for life and conserving it shall be a primary concern both to sustain human health and ecosystems wellbeing. As a matter of facts, microbial biodiversity decisively affects the functioning of ecosystems (Falkowski

J. Marques da Silva
BioISI – Biosystems and Integrative Sciences Institute and Department of Plant Biology, Faculdade de Ciências, Universidade de Lisboa, Lisboa, Portugal
e-mail: jmlsilva@fc.ul.pt

E. Casetta (✉)
Department of Philosophy and Education, University of Turin, Turin, Italy
e-mail: elena.casetta@unito.it

© The Author(s) 2019

195

E. Casetta et al. (eds.), *From Assessing to Conserving Biodiversity*,
History, Philosophy and Theory of the Life Sciences 24,
https://doi.org/10.1007/978-3-030-10991-2_9

et al. 2008; van der Heijden et al. 2008), being largely responsible for their metabolism and biomass. Microbes set the level of primary production, drive decomposition and convert organic matter so that it can be used by plants, and contribute to climate regulation. Moreover, microbial biodiversity is fundamental in sustainable development, in virtue of its industrial and commercial applications such as, for instance, sustainable removal of chemical pollutants from the environment, or the employment of bacteria as pesticides (Qaim and Zilberman 2003). In terms of ecosystem services, its value is enormous.

On the basis of FAO's data, the economic value of microbial functions, once converted in terms of the ecosystem services microbial diversity contributes to, amounts to 500 billion US$ per year for soil formation on agricultural land; 90 for nitrogen fixation for agriculture; 42.5 for pharmaceuticals; 1.3 for the synthesis of industrial enzymes (Table 9.1, Bell et al. 2009: 125). But the involvement of microbes in ecosystem functioning, and in every aspect of our life is so pervasive and so intimate that it is probably impossible not to underestimate the value—the monetary value as well—of microbial diversity.

In spite of its pivotal importance, the role of biodiversity at the microbial level is usually ignored in the debates concerning the so-called biodiversity crisis as well as by conservation policies and actions, mainly centred on animals and plants. Yet, the total biomass of bacteria and archaea has been estimated to equate that of terrestrial and marine plants, making microbes the largest unexplored reservoir of biodiversity on earth (Whitman et al. 1998). To get an idea, think that each millilitre of water in oligotrophic environments, such as deep oceans or deep subsurfaces, contains approximately 1×10^5 bacterial cells, most of them in a dormant starving state (Kjelleberg 1993) and that each gram of soil contains between 2000 and 8.3 million species of bacteria (Roesch et al. 2007).

Among the reasons for this neglect there is, of course, the fact that microbes are far less appealing than charismatic taxa such as the giant panda or the Siberian tiger, and policies based on the motto "save the *Bacillus thuringiensis*" must not sound very catchy to a politician looking for votes.[1] Besides political reasons, there are difficulties connected with the assessment of microbial biodiversity. Firstly, most microbial strains are difficult to be cultured in the laboratory. Secondly, microbes typically live as interacting communities, and the number of individual bacteria to identify is simply too large to permit a proper survey of the composition of the community (think that a poor environment such as drinking water has thousands of microbial individuals per millilitre). As a consequence, the diversity of microbial communities is still largely unknown (Bell et al. 2009). Thirdly, the majority of microbial communities are multispecies communities, and there is an ongoing controversy in settling how these species should be defined and identified (Doolittle and

---

[1] And yet, the usefulness of *Bacillus thuringiensis* is well recognised, since some of its strains, during sporulation, produce toxins. These toxins target insects' larvae, acting as a natural insecticide; and they do not target humans, hence being—unlike the majority of artificial pesticides—relatively innocuous. (A more controversial matter is, of course, the use of *Bacillus thuringiensis* genes in GM crops, such as Bt corn.)

Papke 2006; Achtman and Wagner 2008; Ereshefsky 2010; O'Malley 2014).[2] Setting aside the unresolved theoretical aspects of the problem, which are not of our concern here, microbial species are normally recognised amongst microbiologists, at least for practical purposes, and expressions like "multispecies community" and "multispecies biofilm" are commonly used. For instance, a convention is largely adopted that two individual microbes are of the same species when they share an arbitrary degree of DNA sequence similarity—often, 97% is taken as the threshold (Bell et al. 2009: 124).

As said, microorganisms rarely live alone; more typically they live in communities. When such communities comprise a thin layer of cells irreversibly adhering to a (biotic or abiotic) surface or interface and produce a matrix of slime-like extracellular polymeric substances (EPS) – mainly polysaccharides, proteins and nucleic acids – in which they become embedded, they are called "biofilms". They can include microorganisms of just one kind (monospecies biofilms) or they can be composed— and more often so—of different species of bacteria, or even of different types of microorganism, both prokaryotic (bacteria, archaea) and eukaryotic (fungi, protists, algae) (Besemer 2015). Biofilms are the major mode of microbial life (Nadell et al. 2016), and multispecies biofilms are the dominant form in nature (Elias and Banin 2012). They "might lack the grandeur of a tropical rain-forest, but not their complexity, or their significance in terms of ecosystem function" (Hansen et al. 2007a).

In several cases biofilms are enemies hard to destroy, being more resistant to antibiotic treatments than bacteria at the planktonic state (as, for instance, is the case for oral biofilms or *Pseudomonas aeruginosa* biofilms which infect lungs of cystic-fibrosis-affected patients). In other cases, they play a crucial role for ecosystem functioning, contributing substantially to biogeochemical processes and to the maintenance and promotion of biodiversity. To show their role in general ecosystem maintenance and in promoting biodiversity—not only at the microbial level but also at the macrobial level, such as plants or animals—the example of plants' biofilms is illuminating.

Plant surfaces and the rhizosphere are rich of multispecies biofilms, which grow attached to plants' transport vessels, stems, leaves, and roots, helping maintain biodiversity both at the microbial and at the plants' level. One way they help conserve biodiversity at the microbial level is by ensuring the survival of bacteria they are made of in harsh environments (Angus and Hirsch 2013; Bogino et al. 2013). In fact, bacteria that compound the biofilm are provided by the EPS with a protection that they do not have in their planktonic state. EPS protects them against environmental stress factors, allowing them to better face fluctuating environmental conditions such as desiccation, changes in pH or in temperature, or scarce availability of nutrients, which may derive both from casualties and from human activities, for instance

---

[2] This is part and parcel of the so-called "species problem" (see for instance Richards 2010), i.e. the problem of defining, delimiting, and identifying species. These tasks are particularly controversial when asexually reproducing organisms—like a large part of microorganisms—are at issue; partisans of the biological species concept such as Dobzhansky (1937) and Mayr (1970), for instance, simply claim that asexually reproducing organisms do not form species at all.

as a consequence of climate change or deforestation. Biofilms have also effects on plants' health and productivity (of course, depending on the kind of biofilms, these effects can be beneficial or detrimental to the host plant). For instance, certain bacteria in the EPS matrix seem not only to improve biocontrol—i.e. the protection of plants from phytopathogens—but also stimulate plant growth by secreting specific hormones and by means of a variety of other mechanisms, hence representing sustainable alternatives to agrochemicals (Bogino et al. 2013; Lugtenberg et al. 2013). In short, "biofilms have been shown to enhance not only the fitness of individual bacteria but also plant health and productivity as a result of the cumulative selective advantage of the individual bacteria" (Bogino et al. 2013, p. 15849).

To generalise, at the ecosystem's level, a fundamental functional role is played by two kinds of multispecies autotrophic biofilms (i.e. multispecies biofilms which are able to synthesise their own organic food from inorganic substances – opposite to heterotrophic biofilms, which depend on other organisms or on dead organic matter for food, since they do not synthesise their own food), namely aquatic photosynthetic multispecies biofilms and land biological soil crusts (BSCs), which we shall discuss in more detail in Sect. 9.3. Photosynthetic biofilms form the basis of trophic chains in many aquatic environments, as is the case with diatom (a major group of microalgae)-dominated intertidal microphytobenthos (MPB) on estuaries (Macintyre et al. 1996). The EPS produced by MPB diatoms promote the aggregation of sediment particles and therefore crucially impact the stabilisation of estuarine intertidal mudflats (Paterson and Black 1999). A similar ecological role in terrestrial ecosystems is played by the EPS produced by the cyanobacteria of BSCs (Eldridge and Leys 2003; Adessi et al. 2018).

MPB are diatom-dominated communities that cover estuarine mudflats (epipelic MPB) and sandbanks (epipsamic MPB). Epipelic MPB is particularly relevant for ecosystem functioning, as it is largely responsible for the high primary productivity of intertidal mudflats, which in turn are important components of estuarine productivity (Underwood and Kromkamp 1999 – in Stal and de Brouwer 2003). This intertidal microphytobenthos living on soft-sediment habitats is composed mostly of motile diatoms exhibiting rhythmic vertical migrations across the photic zone (i.e. the uppermost layer) of the sediment in close synchronisation with tidal and day/night cycles (Round and Palmer 1966; Paterson 1986). It should also be mentioned that the hidden beauty of MPB communities has been praised by biologists (Marques da Silva 2015), who referred to them as "the secret gardens" (Macintyre et al. 1996) since under the microscope lens they reveal unexpected beauty. In addition to their indisputable ecological role then, a further contribution to biodiversity can be highlighted for MPB, namely their aesthetic value—an item included amongst ecosystem services (for instance by Costanza et al. 1997, p. 254, who refer to "aesthetic, artistic, educational, spiritual, and/or scientific values").

Diatoms, just like bacteria, excrete EPS which are believed to contribute to the cohesion of intertidal mudflats (Paterson and Black 1999). MPB are then biofilms in their own right, as cells occupy a thin layer over the surface of a substrate, and they are immersed in an EPS matrix produced by them—the two requirements to classify a microbial community as a biofilm (as seen above). They are multispecies, eukaryotic-dominated, photosynthetic biofilms.

A similar ecological role is played, in terrestrial ecosystems, by the EPS produced by the cyanobacteria of biological soil crusts (Eldridge and Leys 2003). Biological soil crusts are complex assemblages of cyanobacteria and other bacteria, green algae, mosses, microfungi, lichens. Cyanobacterial filaments and fungal hyphae (and sometimes also lichen and moss rhizoids), present on the few top millimetres of soil, stick together loose particles, making a matrix that stabilises and protects the soil surface from the erosive forces of wind and rainfall. They also function as living mulch by retaining soil moisture (Adessi et al. 2018) and discouraging annual weed growth. Although compositionally diverse, these crusts occur in all arid and semi-arid regions, where they occupy the nutrient-poor zones between vegetation clumps in many types of arid-land vegetation. In some plant communities, they may constitute up to 70% of the living cover, fixing atmospheric nitrogen and significantly contributing to soil's organic matter. Yet, biological soil crusts have only recently been recognised as having a key influence on terrestrial ecosystems. Their role on biodiversity is also significant because, in arid and semi-arid communities, there are often many more species associated with the biological soil crusts at a given site than there are vascular plants (Belnap et al. 2001 and references therein). Bare soil is initially colonised by cyanobacteria. Later, if the environmental conditions are favourable, ecological succession may progress to more structurally complex biological crusts, forming a rough, uneven carpet or skin of low stature (1–10 cm in height) dominated by bryophytes (i.e. non-vascular land plants). Here we will focus mainly on cyanobacteria-dominated BSCs, common on early successional stages of soil colonisation. Although rarely defined as biofilms, these crusts meet the criteria for being biofilms: they exist as a thin layer at the interface between soil and air and are embedded in a matrix formed by the cyanobacteria-produced EPS.

9.2 How Multispecies Biofilms Increase Phenotypic and Genetic Diversity

Besides contributing to the maintenance of biodiversity at macrobial levels of biological organisation, there are several ways in which multispecies biofilms increase biodiversity at the microbial level and help maintain it. In this section, we shall see how they increase phenotypic and genetic diversity. Then, in Sect. 9.3, we are going to argue that, in virtue of being putative evolutionary individuals they, first, prompt a general reflection on the processes originating biodiversity and, secondly, highlight some limits of the species-based approach to biodiversity conservation.

Phenotypic diversity is the diversity of phenotypes in a population. Biofilms promote the emergence of new microbial phenotypes, increasing, as we are going to show, biodiversity both at the level of the monospecies colony and at the biofilm level. Being related to different fitness values in a population, phenotypic diversity is crucial for the survival and possible evolution of a species. According to some, "phenotypic diversity is the most important aspect of biodiversity, at least in terms of evolution by natural selection of organisms" (Woods 2000, p. 44ff). To understand how and why biofilms promote phenotypic diversity, we need to briefly illustrate how they form.

Biofilm formation is a complex process typically including 5 phases. The first phase is primary adhesion, namely the initial—and reversible—attachment of cells to a surface or an interface. It is worth noting that because of the small size of bacterial cells, gravity plays a negligible role, and the approach of cells to substrates is mostly due to their motility. When cells come close to the substrate, the balance between attraction and repulsion forces will dictate the occurrence of primary adhesion. In primary adhesion, binding is still reversible. The second phase is irreversible adhesion. If cells come closer to the substrate (1.5 nm), ionic, covalent and hydrogen bonds come into play and cells enter secondary adhesion; the binding becomes permanent. It is at this stage that the production of extracellular polymeric substances increases. EPS, strengthening the link between cells and substrate, provides the biofilm with a structural support (in a way that may remind us of animals' and plants' extracellular matrix). In the third and fourth phase, the architecture of the biofilm develops and the biofilm reaches maturity. Biofilm maturation changes the conditions within the environment that surrounds the microorganisms in terms of cell population density, pH, oxygen and the presence of micronutrients. In turn, microenvironment diversity causes local environmental adaptation, producing physiological heterogeneity in the biofilm. It has been shown that in *Pseudomonas aeruginosa* biofilms, the cells at the top had high mRNA abundance for genes involved in general metabolic functions, whereas the cells at the bottom had low mRNA abundance for housekeeping genes. Accordingly, cells at the top were actively dividing, whereas cells at the bottom showed a low division rate, possibly due to a long-term exposition to a totally depleted level of oxygen (Williamson et al. 2012). Vlamakis et al. (2008) showed that in *Bacillus subtilis* biofilms, motile, matrix-producing, and sporulating cells are located at distinct regions, and that the localisation and frequency of the different cell types are dynamic throughout the development of the biofilm. It has been suggested that sporulation is a culminating feature of biofilm formation, and that spore formation is coupled with the formation of an architecturally complex community of cells. Therefore, as a consequence of the heterogeneity of the microenvironment, a difference between reproductive and metabolic cells may arise (Cabarkapa et al. 2013; Kaplan 2010). The mature biofilm is a multicellular entity endowed with a complex internal structure, metabolism, and division of labour among its part. The structural adaptations and relations it shows "are made possible by the expressions of sets of genes that result in phenotypes that differ profoundly from those of planktonically grown cells of the same species" (Stoodley et al. 2002, p. 187). Gene regulation during biofilm development controls the switch from the planktonic state of cells to their biofilm state, making possible the growth of the biofilm itself (changes in phenotypic expression in response to growth strategies have been documented, see Stoodley et al. 2002). The fifth phase is detachment, e.g. the release of individual cells or clumps of cells from the biofilm.

As said, living in a biofilm promotes the emergence of new microbial phenotypes. For instance, in a study on *Pseudomonas aeruginosa*, Sauer et al. (2002) concluded that during biofilm development, *P. aeruginosa* cells display at least three different phenotypes: planktonic, mature biofilm, and dispersion (and the dispersion-phenotype looks more similar to the planktonic one than to the mature biofilm one), with different regulation of motility, production of alginate (a principal component of the EPS), and quorum sensing (i.e. a chemical communication system that "measures" population density and regulates gene expression accordingly).

Let us substantiate our claim by considering another concrete example that shows the emergence of a new phenotype in the monospecies colony in the biofilm, as well as in the whole multispecies biofilm. Hansen et al. (2007a, b) performed an experiment on *Acinetobacter* (widely-distributed in nature and commonly occurring in soil and water) and *Pseudomonas putida* (a soil bacterium that feeds on dead organic matter). When the two bacteria live in an environment where the only source of carbon is benzyl alcohol, they stay in a sort of host-commensal relation, since *P. putida* depends on *Acinetobacter*'s partitioning benzyl alcohol in benzoate, that *P. putida* can metabolise. The experiment compared the behaviour of the bacteria first in a spatially unstructured chemostat environment (an apparatus to grow bacterial cultures at a constant rate), then on a glass surface within a biofilm flow chamber (a structure that allows direct microscopic investigation of biofilm formation; biofilms in flow chambers grow under hydrodynamic conditions, and the environment can be carefully controlled and easily changed). In the chemostat unstructured environment (i.e. at the planktonic state), the two species coexisted, but only at benzyl alcohol concentrations above 430 mM. On the glass surface, they coexisted at benzyl alcohol concentrations as low as about 130 mM. In other words, the spatial structure of the formed biofilm extended the range of resource concentrations over which the two species may survive. Looking at the biofilm's formation, Hansen et al. (2007a) observed that the structure of the biofilm after 24 h was characterised by discrete colonies of *Acinetobacter* surrounded by loose assemblages of *P. putida* and that the two species were spatially separated. But after 5 days the association changed: at separate focal points, *P. putida* grew in intimate contact with *Acinetobacter*, and after 5 further days, *Acinetobacter* colonies were covered by a layer of *P. putida*.

During the course of the biofilm's development, a rough phenotypic variant of *P. putida* colony (VP) emerged and was selected over the ancestral (wild-type) *P. putida*. While the new phenotype is a property of *P. putida*'s cells, the ensuing variation determines the emergence of a new biofilm phenotype. To show this, we shall consider three differences between the two biofilm phenotypes: persistency, biomass, fitness. When compared to the phenotype of the biofilm with the *P. putida* wild-type (WTP), the phenotype of the biofilm with the *P. putida* variant (VP) displays (i) more persistency; (ii) more biomass (ecological productivity); (iii) a different fitness.

(i) *More persistency*

Because of the extensive consumption of oxygen to metabolise benzyl alcohol, the presence of oxygen in the biofilm environment decreases, and the "mantle" of the *P. putida* variant acts as a shield, preventing *Acinetobacter*—which is covered by it—from reaching oxygen. This stressful environment induces the *P. putida* new variant to produce a more resistant, cellulose-like polymer (EPS) that prevents the biofilm from dispersing. Hansen et al. (2007b) tested the response to oxygen starvation by inducing oxygen downshift (near to zero), with the result that, while the VP showed no dispersal, in the WTP's case the first signs of detachment appeared within minutes and almost the whole biofilm detached after 25 min. The proposed explanation for the difference is that thanks, at least in part, to the more resistant EPS, the rough variant sticks robustly to *Acinetobacter* microcolonies (which, in turn, serve as an anchor point) ensuring them close access to the secreted benzoate. The production of EPS, progressively thickening over the days and securing the robustness of the biofilm, prevents the biofilm from dispersing in spite of the increased competition for oxygen.

(ii) *More biomass*

Hansen et al. (2007a, p. 534) noticed that "The productivity of the derived community was significantly greater than that of both the ancestral community and *Acinetobacter* alone .... This effect was attributable to enhanced productivity of *P. putida* and occurred despite a deleterious effect of *P. putida* on *Acinetobacter*—the species on which *P. putida* is reliant." Accordingly, even though the variation occurred in *P. putida*, not only *P. putida* biomass, but also the biofilm's biomass was increased.

(iii) *Different fitness*

There is no unique way to understand—and calculate—biofilms' fitness. In the literature, at least two different ways can be found. One way is to understand fitness as *evolutionary fitness* (i.e. differential reproductive success). In biofilms, evolutionary fitness may, in turn, be understood in two different ways: as a function of the fitness of the separate clonal lineages comprising the biofilm (O'Malley 2014, p. 112); or as the ability of a biofilm to seed new patches, which is expected to increase with its rate of dispersal. Another way is to understand biofilms' fitness as *persistence*, i.e. as the increase or decrease in the propensity to persist in response to pressures of the selective environment (Bouchard 2014). In the case under consideration, no matter how biofilms' fitness is understood, WTP and VP show differential fitness (in particular, WTP's fitness is higher than VP's fitness if understood as evolutionary fitness, while it is lower when understood as persistence).

Moving from phenotypic diversity to genetic diversity, it should be highlighted that the *P. putida* variant phenotype has been proved to be heritable and, when inoculated together with *Acinetobacter* into fresh biofilm flow chambers, the structure of the resulting community was similar to the newly formed one, VP, (and not to the ancestral one). Two independent mutations, responsible for the mantle-like phenotype, were identified in wapH (PP4943)—a gene involved in lipopolysaccharide biosynthesis. Accordingly, it may be argued that living in a biofilm also increased genetic biodiversity, both at the colony and the biofilm level, providing a new double-mutant genotype. It is normally assumed that this genetic variability occurred by random mutation processes, supplemented through natural selection in the local environment of the biofilm (Gross et al. 2012).

9.3 Multispecies Biofilms as Drivers of Evolution

"Evolutionary biologists often study the origin of biodiversity through the identification of the units at which evolution operates" (Bapteste et al. 2012, p. 18266). Throughout the history of evolutionary biology, bacteria have been often thought to follow different evolutionary rules not in line with those emphasised by the Modern Synthesis, which was modelled on the evolution of plants and animals, and according to which species or populations are considered to be the units of evolution and speciation the main process producing diversity. The peculiarity of prokaryotes was thought to be due mainly to their asexuality, that would prevent genuine speciation, to their hereditary mechanism not relying on specialised germ cells, and to the location of their genome, which is not in a cell nucleus. However, since prokaryotes have genes, display mutations, and show a process of vertical transmission, it is now commonly accepted that they fit the Modern Synthesis view of evolution (O'Malley 2014, Chap. 4). Yet, the genetic and phenotypic diversity we examined in the previous section cannot be accounted for in terms of species and speciation. Some recent works (see, in particular, Bapteste et al. 2012) have shown that many other processes rather than vertical descent contribute to generating diversity, namely processes that use genetic material from multiple sources, such as recombination, lateral gene transfer, and symbiosis. These processes produce evolutionary outcomes at different hierarchical levels, think for instance to the multiple symbiotic evolutionary origin of chloroplasts (Whatley and Whatley 1981). In the remaining we shall focus on the possibility of multispecies entities being units of selection along with "standard" organisms. If so, species, at least at the microbial level, might – quite unsurprisingly – prove not to be the most relevant units of biodiversity, and therefore speciation might not be the only process to be taken into account. Accordingly, on the one hand, a satisfactory account of biodiversity generation would ideally require to consider a much more varied set of evolutionary processes than just speciation; on the other hand, as we are going to suggest in the conclusion, species might sometimes not be the best targets of conservation actions. To put it in other words: if—as Huxley thought (1942: 126)—"… bacteria have their own evolutionary rules", we may outline two main theoretical possibilities: either these rules set them apart from the Modern Synthesis' framework, requiring the existence of two different "modes" of evolution; or these rules enforce a better understanding, as well as an inclusion, in the modern synthesis' framework. We shall take this second stance, suggesting, moreover, that this would have positive consequences not only in our understanding of the processes producing biodiversity but also in shaping conservation actions.

9.3.1 The Origin of Biodiversity

Those entities which are able to produce diversity, i.e. those entities capable of random heritable variations that eventually might increase fitness in the context of changing environments, are units of selection or evolutionary individuals, i.e. the objects that undergo the conditions identified by Darwin and formalised by Lewontin in his classic 1970 article (Lewontin 1970), namely phenotypic variation, differential fitness, heritability of fitness.

Lewontin was speaking of variation among *individual organisms* of a population or a species. Different individual organisms have different phenotypes, which have different rates of survival and reproduction. This difference in fitness must be somehow (notwithstanding the nature of the mechanism) heritable. When these three principles hold, a population or a species will undergo evolutionary change. Lewontin also pointed out that the logical skeleton of the Darwinian argument "turns out to be a powerful predictive system for changes at all levels of biological organization … any entities in nature that have variation, reproduction, and heritability may evolve", provided that they satisfy the three principles. In compliance with Lewontin's suggestion, several entities have been proposed as possible "evolutionary individuals" in addition to individual organisms, both at lower (e.g. alleles) and higher (e.g. groups) levels of the hierarchy.[3] Might some *multispecies* entities, and multispecies biofilms in particular, be evolutionary individuals?

Recently, this question caught the attention of philosophers of biology working on biological individuality (see Bouchard and Huneman 2013). This is because multispecies biofilms may be thought to show several characteristics that could make them similar to multicellular organisms such as us or other animals, i.e. the paradigmatic units of selection. Accordingly, it may be asked whether, by analogy, multispecies biofilms are amongst the units of selection, i.e. whether they are evolutionary individuals. According to Ereshefsly and Pedroso, the answer is positive, at least for some biofilms. They concede that evolutionary individuality can be a matter of degree and that not all biofilms are promising candidates to be evolutionary individuals: "the fact that some biofilms and other multispecies consortia exhibit more internal competition than others implies that some biofilms are more individual-like than other biofilms" (Ereshefsky and Pedroso 2015, p. 10131). According to these authors, there are 7 conditions a biological entity must meet to be said an evolutionary individual: (1) Having a high internal integration and being delineated from the environment; (2) Presenting division of labour and, relatedly, (3) coordination between the parts, and cooperation; (4) Bearing adaptations; (5) Transmitting traits between generations; (6) Having a reproductive bottleneck; and, relatedly, (7) Genetic homogeneity. In the next section, we will analyse these criteria in more detail with the aim of evaluating whether in comparison with heterotrophic bacterial

---

[3] Notice that while entities at a lower level (e.g. alleles) are more likely to satisfy the three conditions, whether there are evolutionary individuals at a higher level (groups, species, clades…) remains a matter of debate. The controversy concerns, in particular, reproduction—and, relatedly—heritability of fitness.

biofilms, two kinds of multispecies autotrophic biofilms (i.e. microphytobenthos and biological soil crusts) meet the above criteria.

9.3.2 Are MPB and BSCs Evolutionary Individuals?

In this section, we go through the seven criteria for biofilms' evolutionary individuality (as they have been put forward by Ereshefsky and Pedroso 2015) to see whether they are fulfilled by MPB and BSCs, in comparison to paradigmatic bacterial biofilms, namely heterotrophic biofilms.

(1) High internal integration and delineation from the environment. This criterion is met by heterotrophic bacterial biofilms, since bacteria within a biofilm "act" as a whole, they communicate (for instance through quorum sensing) and exchange genes (more than planktonic bacteria). This is so mainly in virtue of the EPS, which packs them and keeps their parts together, contributing both to internal integration and delineation from the outside.

Also in MPB, the basic element of internal integrity is the EPS matrix that keeps cells together. The EPS matrix has been demonstrated to help congregate sediment particles, decrease erosion (Paterson and Black 1999) and probably facilitate the flux of carbon between other constituents of the biofilm, including embedded meiofauna (Middelburg et al. 2000). Likewise, in BSCs the EPS matrix plays a pivotal role in favouring the internal integration of the biofilm and its delineation from the environment. The integration is possibly more complex than in MPB and heterotrophic bacterial biofilms, due to the greater diversity which characterises BSCs. BSCs are usually constituted of cells coming from different higher-level taxa too, even in the early successional stages dominated by cyanobacteria, which may include green algae, heterotrophic bacteria, microfungi (Belnap et al. 2001 and references therein), and archaea (Soule et al. 2009). In later successional stages (e.g. bryophyte or lichen-dominated BSCs), integration will be looser as the intrusion of the macroscopic rooting structures of these organisms into the cyanobacterial biofilm is expected.

(2) Division of labour, and, relatedly, (3) Coordination and cooperation amongst parts. Functional specialisation, or division of labour, is common in most multicellular, colonial and social organisms, but it is far from ubiquitous (Simpson 2012). Quorum sensing is probably the main way bacteria coordinate; for instance, it regulates the amount of extracellular DNA that bacteria produce for their EPS, regulates gene expression, and triggers dispersal of the cells of biofilm. Eukaryotes are also known to utilise quorum sensing, as is the case with many fungal species and even social insects (Amin et al. 2012 and references therein). However, quorum sensing systems *sensu stricto* are not described for diatoms, albeit some mechanisms underlying diatom cell signalling have been discovered (Falciatore and Bowler 2002), and have been described in epilithic – i.e., colonizing bedrocks – cyanobacterial communities (Sharif et al. 2008), even though not in BSCs. However, it should be

noticed that the studies of quorum sensing-like mechanisms in non-bacteria lag significantly behind those concerning bacteria.

Thanks to quorum sensing and analogous mechanisms, a division of labour can be frequently observed in heterotrophic bacterial biofilms, and it is often dependent on biofilms' internal architecture. For instance, there exist biofilms in which surface bacteria consume oxygen, intermediary bacteria convert sulphide into hydrogen sulfate, and bacteria living at greater depth cycles convert sulphate into sulphide. A further mark of individuality is cooperation among the parts (accompanied by low internal competition). It has been shown that bacteria within biofilms often excrete public goods (Czárán and Hoekstra 2009), i.e. exoproducts that are costly to produce but enhance the fitness of other cells, such as EPS or anti-biotic degradation compounds. This production is regulated through quorum sensing, and cheating is a phenomenon that frequently occurs (there are mutants which do not participate in the production of common goods while exploiting producers; or "liars", who produce the quorum sensing signal but not the exoproduct). Yet, cheaters are kept in check by "honest cooperators" through a variety of strategies (for instance, lateral gene transfer of genetic elements that infect non-cooperative mutants inducing them to produce the public good).

May division of labour and coordination and cooperation amongst parts be found in MPB and BSCs as well? In MPB, photosynthesis is mostly performed by diatoms, albeit cyanobacteria might contribute in a variable proportion (Vieira et al. 2013a). Transmission of organic carbon from diatoms to heterotrophic bacteria occurs in a short time (Middelburg et al. 2000). On the other hand, diatoms may benefit from the respiratory inorganic carbon released from sub-surface bacteria (Marques da Silva et al. 2017; Vieira et al. 2016). The relationship between these organisms, however, has been poorly characterised. Bacteria benefit from this relationship by gaining a source of readily available organic metabolites and a constant source of materials. Diatoms may gain specific organic compounds such as vitamins or growth factors from bacteria (Trick et al. 1984), but almost no evidence exists for the transfer of organic compounds from bacteria to phytoplankton.[4] Recent research has underlined that obliged bacterial mutualism is a widespread phenomenon in nature (Morris et al. 2013). Therefore, it might be expected to be present also in the bacterial/diatom populations of microphytobenthos.

Putative division of labour on the diatom fraction of microphytobenthos may be related to the different cells' sizes of diatom species (Vieira et al. 2013b). Small cells present higher photosynthetic rates and therefore are better nutrient scavengers (Geider et al. 1986). On the other hand, large cell diatoms may be better adapted to variable environments (as is the case with intertidal mudflats) because of the increased capacity to store nutrients (Turpin and Harrison 1980).

As MPB, also BSCs comprise autotrophic (e.g. cyanobacteria and green algae) and heterotrophic (e.g. microfungi and archaea) organisms. Therefore, at this broad

---

[4] The recent controversy on the dependency of marine diatoms on symbiotic marine bacteria for the supply of vitamin B12 (a metabolite that diatoms do not synthesise, but which is essential for their metabolism, Croft et al. 2005; Droop 2007) highlights the complexity of this issue.

level, a metabolic division of labour is present. In cyanobacterial mats (a different type of cyanobacteria-dominated communities), the supply of organic nitrogen is produced by a type of filamentous cyanobacteria. Filamentous cyanobacteria are also common in BSCs and therefore a similar role may be envisaged. To our knowledge, no further empirical evidence concerning division of labour in BSCs may be found in the literature. However, recent work (Kim et al. 2016) is likely to show that division of labour is a widespread phenomenon that rapidly emerges in bacterial communities.

(4) Bearing adaptations. Heterotrophic bacterial biofilms seem to meet also this criterion. Working with microphytobenthos, Jesus et al. (2009) showed that diatom-dominated biofilms adapt to sediment conditions and tidal height. Sandy sediments exhibit photosynthetic pigments characteristic of a mixed cyanobacteria/diatom assemblage, showing an alternate seasonality, with cyanobacteria increasing in the summer and diatoms dominating in the spring. Furthermore, epipsammic (inhabiting sand) diatoms were smaller than epipelic (inhabiting mud) diatoms, suggesting an adaptation of the biofilm to the substrate, since these epipsammic non-motile diatoms live in close attachment to individual sand grains (Barnett et al. 2015). Also, epipelic biofilms showed evidence of being low light-acclimated and photo-regulating by vertical migration movements, whereas epipsammic biofilms showed no vertical migration rhythms. Thus, the two types of diatom biofilms have distinct strategies to adapt to light intensity: epipelic diatoms use vertical migration to position themselves at the sediment depth of optimum light conditions, and epipsammic diatoms use specific photosynthetic and photoprotective pigments to photo-regulate. Cartaxana et al. (2015) found that elevated temperature led to an increase of cyanobacteria and a change in the relative abundance of major benthic diatom species present in the MPB community. On the contrary, no significant effect of elevated CO_2 was detected on the relative abundance of cyanobacteria and major groups of benthic diatom species.

To our knowledge, direct experimental evidence for adaptation is absent for BSCs, but the wide variety of biological crusts covering ecologically different dry lands suggest considerable adaptability.

Thus, there is little doubt that criteria (1-2-3-4) are met by heterotrophic bacterial biofilms, even by multispecies biofilms (where internal competition is expected to be higher, even though it has been shown that quorum sensing works also interspecifically – Federle and Bassler 2003). Although less information is available, MPB biofilms, and even BSCs, are likely to meet these criteria. It is with heritability—and hence reproduction—that things become more controversial. In a nutshell, the problem is that, even though biofilms have what may be called, metaphorically, a life-cycle by analogy with "standard" multicellular organisms like us, it is not clear to what extent the analogy can be stretched. Clarke (2016), for instance, objected that biofilms do not exhibit genuine life cycles, since the claims about life cycle stages can be conceptualised equally well in terms of ecological succession or changes in community structure. The colonisation of a new territory by founder species would correspond to phase 1, their modification of the environment and the consequent secondary colonisation to phase 3 and 4, seeding to phase 5, etc.

However, we do not find the objection entirely compelling, since it underestimates the role of the EPS (phase 2) for the cohesion and internal integrity of the entities under scrutiny. Yet we agree with Clarke that the issue is not an (entirely) empirical matter. Even though more data need to be collected, certain theoretical issues need to be settled in order to give a proper answer: can dispersion be seen as a form of reproduction? Do biofilms form lineages? Do they transmit faithfully acquired mutations to their "heirs"? The three further criteria help us tackle this sort of questions. Establishing whether biofilms, or more generally "multilineages clubs",[5] can be counted amongst evolutionary individuals is, in part, a theoretical matter (which notion of reproduction—and which notion of inheritance—should we endorse?), and in part a matter of collecting empirical evidence—concerning reproduction, maintenance mechanisms, and fitness—which is still lacking.

(5) Transmitting traits between generations. This criterion has to do with Lewontin's heritability of fitness, i.e. the transmission of those biofilm-level traits that confer differential fitness to biofilms. We have discussed biofilms' fitness in the previous section, arguing that, at least in the case study that we considered, biofilms may have differential fitness and that the related trait can be transmitted. More generally, Ereshefsky and Pedroso (2015) report that specific genes within bacteria have been identified that transmit traits such as quorum sensing, the capacity to engage in specific metabolic interactions, aggregation patterns, cooperative behaviours, and the mechanisms underlying lateral gene transfer, as well as those for the production of EPS. No empirical data are available for microphytobenthos and biological soil crusts. Notice that the fulfillment of criterion 5 depends on availability of relevant empirical evidence, but also on which notion of reproduction is adopted. If a strict notion of reproduction—based on how we and other paradigmatic evolutionary individuals reproduce—is advocated, this and the two further criteria are unlikely to be met by biofilms.

(6) Reproductive bottleneck and, relatedly, (7) Genetic homogeneity. An individual has a reproductive bottleneck when its life cycle begins as a single cell (or a few cells) and that cell is replicated to form the cells of an individual. A reproductive bottleneck reduces competition among the cells of an individual (being the cells generated from the germline cells genetically homogeneous), and it favours the evolution of new traits allowing small mutational changes in the germline to have major effects on the descendants of the organism. Biofilms "reproducing" by dispersion/aggregation and exchanging gene laterally lack a reproductive bottleneck *sensu stricto*, even though it might be argued that the same results of reproductive bottleneck may be obtained in other ways, for instance through an ecological bottleneck in which the size of the population decreases following environmental factors such as antimicrobial treatments. In this way, an ecological bottleneck reduces competition amongst the cells, increasing genetic relatedness (Ereshefsky and Pedroso 2015). As for genetic homogeneity, multispecies biofilms clearly lack it, being made

---

[5] Multilineages clubs are defined as "coalitions of entities that replicate in separate events and exploit some common genetic material that does not trace back to a single locus in a single last common ancestor of all the members" (Bapteste et al. 2012: 18268).

of distinct lineages and being lateral gene transfer extremely common. Microphytobenthos and biological soil crusts, being *multi-kingdom* biofilms, are even less genetically homogeneous than biofilms made of bacteria only.

It is clear from this analysis, summarised in Table 9.1, that the possibility that not only multispecies heterotrophic bacterial biofilms but also MPB and BSCs are evolutionary individuals cannot be ruled out in principle. On the one hand, in order to provide a definite answer, further studies and more empirical data are needed (see for instance Van Colen et al. 2014). On the other hand, a theoretical reflection on the traditional (Modern-synthetic) way of conceiving the notions of reproduction and inheritance might be required. What matters for the purpose of the present contribu-

Table 9.1 Criteria for biofilms' evolutionary individuality applied to two autotrophic biofilms and to a paradigmatic (heterotrophic) bacterial biofilm

	Heterotrophic bacterial biofilms	MPB	BSCs
Internal integrity	EPS matrix	EPS matrix	EPS matrix
Division of labour	Depth-dependent metabolic specialisation / production of public goods	Producers/consumers/specific metabolite synthesis/ carbohydrate *Synthesis vs.* storage	Producers/ consumers Putative cyanobacterial N_2 fixing
Coordination amongst parts	Quorum sensing; lateral gene transfer	Putative quorum sensing analogs (pheromones, nitric oxide)	Putative quorum sensing
Bearer of adaptations	Antibiotic resistance / metabolic interactions / sequential aggregation	Adaptation to sediment granulometry, tidal height and temperature and dissolved inorganic carbon (changes in population structure – in diatom species composition/size classes and in diatom/cyanobacteria ratio)	No direct empirical data available
Heritable adaptive traits	*Sensu lato* possibly met, but not *sensu stricto*; putative specific genes transmitting traits such as quorum sensing, metabolic integration, aggregation patterns, cooperative behaviors, lateral gene transfer, and production of EPS	No empirical data available	No empirical data available
Reproductive bottleneck	Missing *sensu stricto*; but possible "ecological bottleneck"	No empirical data available	No empirical data available
Genetic homogeneity	No	No	No

tion is that *if* they were discovered to be evolutionary individuals, they would be discovered to contribute to biodiversity in an even more fundamental way than by providing ecosystem services and genetic and phenotypic diversity, namely by being amongst the units at which evolution by natural selection operates.

9.4   Conclusions

In 1996, 10 years after the emerging of the term "biodiversity", a book entitled *Species. The Units of Biodiversity* appeared which has now become a reference for whoever aims at approaching the species problem (Claridge et al. 1997). The first line of the abstract of the first chapter, written by the editors of the book, i.e. M.F. Claridge, H.A. Dawah and M.R. Wilson, states that "From a practical viewpoint species are generally the units of biodiversity". The truth of the claim is taken for granted along all the chapter, where it is also stated that "species are normally the units of biodiversity and conservation", and it is added that the growing recognition of the importance of biodiversity, connected to its crisis, confers to the wrangle over the nature of species a wider significance. We believe that, also in the light of contemporary biological knowledge, our evolutionary understanding has partially changed, and that the claim that species are the units of biodiversity and conservation[6] is not so plain as it was thought and has partially changed since then. In particular, we think that entities such as multispecies biofilms may teach us some lessons, both from an evolutionary and a conservationist point of view.

Before looking at the lessons, let us "unpack" the claim that species are the units of biodiversity. This can be interpreted in at least three ways. The first is that species are important units of classification, i.e. that they are *taxonomic units*. The second is that species are important *evolutionary units*, i.e. they undergo evolutionary change. The third is that species are targets of conservation, i.e. they are *conservation units*.

First lesson, concerning species as taxonomic units. If understood in this way, the claim that species are the units of biodiversity is questionable at the microbial level because it is not clear whether species indeed exist when their constituent organisms reproduce asexually, as many microbes do (as already discussed in Claridge et al. 1997). As shown in Sect. 9.1, for instance, Ernst Mayr simply claimed that asexual organisms do not form species at all. Under the biological species concept, species are considered as quasi-discrete gene packages: organisms belonging to a species exchange genes amongst them but not with organisms belonging to other species. This is not the case for microbes, which exchange genes laterally across small and large phylogenetic distance. We have also seen that, at least operationally, the microbial world can be partitioned into species as well. However, this does not by any means guarantee that such operational taxonomic units correspond to "natural joints" as they might just be conventional categories made up by taxonomists, and it could be that, at least at the microbial level, other joints are more natu-

---

[6] See also Reydon, Chap. 8 in this volume.

ral than species. As said, it is a fact that multispecies biofilms are the main mode of organisation of microbial life, and it is another fact that multispecies biofilms do exhibit new diversity, both at the genetic and at the phenotypic level.

Second lesson, concerning species as units of evolution (and organisms as evolutionary individuals, or units of selection). Traditionally species have been considered the units of biodiversity in evolutionary biology. The reason is simple: variation and selection range over various levels (macromolecules, cells, organisms...), resulting in evolution at higher levels, typically populations or species. In other words, (and taking a realistic stance towards species), species speciate, originating new species, i.e. increasing diversity and conferring the tree of life its shape. In the Modern-synthetic view, which is modelled on animals and plants, species are made of organisms, each of which belongs to its species and not to another, in an Aristotelian fashion. Organisms, the interactors, are the bearers of variations that— when inheritable—are transmitted through vertical descent. This picture has been questioned: there is growing evidence that vertical descent is not the only pattern of inheritance. Bapteste and his colleagues (2012), for instance, focus on what they call "introgressive descent", i.e. the process through which the genetic material of an evolutionary individual "propagates into different host structures and is replicated within these host structures" (Bapteste et al. 2012: 18266). On the basis of what we have discussed in this contribution, phenotypic variation and molecular change can arise in response to selection in multispecies entities, i.e. in entities which, by definition, do not belong to any species or belong to more than one species. Moreover, we have argued that the possibility that multispecies biofilms, both autotrophic and heterotrophic, are themselves units of selection cannot be ruled out. A question then arises, which requires further reflection: if multispecies biofilms, or other multilineages consortia, may be units of selection, which are the "corresponding" units of evolution?

Third lesson, concerning species as units of conservation. The centrality of species characterises a quite traditional approach to biodiversity conservation, which involves the maintenance of viable populations of certain species, such as, for instance, indicator species—those species that are thought to indicate the state of biodiversity of a certain area—or endangered species—such as the ones listed in the IUCN red list. Several criticisms have been moved to this approach. In particular, a discrepancy between theory and practice has been highlighted: although "most conservationists claim to protect 'species', the conservation unit actually and practically managed is the individual population" (Casacci et al. 2013). Ideally, all individual populations of all species should be conserved; but conservation actions have to take into account our epistemic limits and economic constraints, and criteria for prioritisation are needed.

A way to proceed is to let evolutionary considerations indicate the direction in selecting targets for conservation actions. Notice that putting together biodiversity conservation and evolution strangely generates a sort of paradox: evolution implies change, and conservation implies keeping things as they are, or even bringing them back to their previous, allegedly pristine, state. However, conservation should also be understood in terms of preserving those ecological and evolutionary mechanisms necessary to promote natural dynamics (see Smith et al. 1993, and Vecchi and Mills, Chap. 12 in this volume), and it has been recognised that practical conservation decisions should be based on evolutionary considerations (Höglund 2009, Chap. 8.3; Eizaguirre and Baltazar-Soares 2014). The paradox dissolves through the distinction between two kinds of biodiversity conservation (Sarkar 2002). One is the analogue of ameliorative medicine and is practiced when we allow a species to decline and then try to recover it (the extreme case is the "emergency room", when we intervene only when a species is on the brink of extinction). The second kind is the analogue of preventive medicine, and it mainly consists in putting forward management procedures for the survival of the units of interest. What are such units? Sarkar (2002) suggested that conservationists should prioritise *places*. Places, in Sarkar's view, are *individual* places, specific regions on Earth "filled with the particular results of [their] individual story". More precisely, "preventive" conservation would consist in prioritisation of places for biodiversity value, whereas measuring biodiversity value would require the choice of surrogates. A different proposal is to focus on the Evolutionarily Significant Units (ESUs). The key idea of the ESUs-based approach is that, rather than preserving all phenotypic variants, it would be worthwhile to preserve those populations which "shows evidence of being genetically separate from other populations, and contributes substantially to the ecological or genetic diversity found within the species taxon as a whole" (Hey et al. 2003: 600). The premise is that as long as evolutionary processes are able to operate, their products, in particular specific adaptive phenotypes, can be replaced or recreated (Casetta and Marques da Silva 2015a).

Notwithstanding its merits (Casacci et al. 2013), the ESU-based approach is still conceptually species-centred (populations are population *of* species), i.e. it does not take into account multispecies entities. Yet, at the microbial level (and we have seen the importance of microbial biodiversity not only *per se* but also for general ecosystem functioning, in particular as far as biological soil crusts and microphytobentos are concerned), species might not be the most important units of biodiversity. Accordingly, an ESU-based approach would have the same limits as a species-based one.

However, we think that the idea of letting evolutionary consideration indicate the direction is worth following. In particular, we argued elsewhere (Casetta and Marques da Silva 2015a, b) that the species-based approach should be integrated by considering as (more) worthy of protection those entities that have evolutionary potential[7] (an indicator of the population's capacity to respond to environmental change, linking in this way environmental change and evolutionary dynamics). That suggestion makes even more sense in the case of the microbial world. In fact, we argued throughout our chapter that multispecies collectives such as biofilms might possess a certain propensity to evolve in response to environmental changes. We suggest that they should be recognised in biodiversity conservation accordingly and that, in conserving places, and in the perspective of preventive conservation actions, a priority should be given to those entities endowed with evolutionary potential.

---

[7] See also Minelli, Chap. 11, and Vecchi and Mills, Chap. 12, in this volume.

Such entities are traditionally thought to be species and populations, but, on the basis of what we have discussed, evolutionary potential might be a property of multispecies communities as well.

In sum, evolutionary potential is a crucial feature to be taken into account in conservation actions once that conservation is understood not only as ameliorative medicine but also as preventive medicine. Taking into account evolutionary potential would enlarge the focus of conservation actions from species to a larger number of entities (i.e. certain species and populations, certain multispecies communities, and possibly certain ecosystems as well), and it might meet the need of prioritisation that economic constraints dictate to biodiversity conservation practice while going beyond a mere species-based or population-based approach.

Acknowledgements Though this contribution has been conceived and discussed together, Elena Casetta is the author of Sects. 9.1, 9.2 and 9.4, while Jorge Maques da Silva is the author of Sect. 9.3. We are thankful to Philippe Huneman, Sandro Minelli, and Davide Vecchi for their useful comments on a previous version of this chapter. This work was funded by the Fundação para a Ciência e a Tecnologia (R&D project Biodecon, ref. PTDC/IVC-HFC/1817/2014) and by the European Union's Fp7 Research and Innovation Programme under the Marie Sklodowska-Curie grant agreement No 609402 – 2020 (Train to Move).

References

Achtman, M., & Wagner, M. (2008). Microbial diversity and the genetic nature of microbial species. *Nature Reviews Microbiology, 6*, 431–440.

Adessi, A., Cruz de Carvalho, R., De Philippis, R., Branquinho, C., & Marques da Silva, J. (2018). Microbial extracellular polymeric substances improve water retention in dryland biological soil crusts. *Soil Biology and Biochemistry, 116*, 67–69.

Amin, S. A., Parker, M. S., & Armbrust, E. V. (2012). Interactions between diatoms and bacteria. *Microbiology and Molecular Biology Reviews, 76*(3), 667–684.

Angus, A. A., & Hirsch, A. M. (2013). Biofilm formation in the rhizosphere: Multispecies interactions and implications for plant growth. In F. J. de Bruijn (Ed.), *Molecular microbial ecology of the rhizosphere* (Vol. 2, pp. 703–712). Hoboken: Wiley. https://doi.org/10.1002/9781118297674.

Bapteste, E., Lopez, P., Bouchard, F., Baquero, F., McInerney, J. O., & Burian, R. M. (2012). Evolutionary analyses of non-genealogical bonds produced by introgressive descent. *PNAS, 109*(45), 18266–18272.

Barnett, A., Méléder, V., Blommaert, L., Lepetit, B., Gaudin, P., Vyverman, W., Sabbe, K., Dupuy, C., & Lavaud, J. (2015). Growth form defines physiological photoprotective capacity in intertidal benthic diatoms. *The ISME Journal, 9*, 32–45.

Bell, T., Gessner, M. O., Griffiths, R. I., McLaren, J., Morin, P. J., van der Heijden, M., & van der Putten, W. (2009). Microbial biodiversity and ecosystem functioning under controlled conditions and in the wild. In S. Naeem, D. E. Bunker, A. Hector, M. Loreau, & C. Perrings (Eds.), *Biodiversity, ecosystem functioning, and human wellbeing: An ecological and economic perspective* (pp. 121–133). Oxford: Oxford University Press.

Belnap, J., Kaltenecker, J. H., Rosentreter, R., Williams, J., Leonard, S., & Eldridge, D. (2001). Biological soil crusts: Ecology and management. In P. Paterson (Ed.), *Technical reference* (pp. 1730–1732). Denver: United States Department of Interior.

Besemer, K. (2015). Biodiversity, community structure and function of biofilms in stream ecosystems. *Research in Microbiology, 166*, 774–781.

Bogino, P. C., Oliva Mde, L., Sorroche, F. G., & Giordano, W. (2013). The role of bacterial biofilms and surface components in plant-bacterial associations. *International Journal of Molecular Sciences, 14*, 15838–15859.

Bouchard, F. (2014). Ecosystem evolution is about variation and persistence, not populations and reproduction. *Biological Theory, 9*(4), 382–391.

Bouchard, F., & Huneman, P. (Eds.). (2013). *From groups to individuals. Evolution and emerging individuality*. Cambridge, MA: MIT Press.

Cabarkapa, I., Levic, J., & Djuragic, O. (2013). Biofilm. In A. Méndez-Vilas (Ed.), *Microbial pathogens and strategies for combating them: Science, technology and education* (Vol. I, pp. 42–52). Badajoz: Formatex Research Center.

Cartaxana, P., Vieira, S., Ribeiro, L., Rocha, R. J. M., Cruz, S., Calado, R., & Marques da Silva, J. (2015). Effects of elevated temperature and CO2 on intertidal microphytobenthos. *BMC Ecology, 15*, 10.

Casacci, L. P., Barbero, F., & Balletto, E. (2013). The "Evolutionarily Significant Unit" concept and its applicability in biological conservation. *Italian Journal of Zoology*, 1–12. https://doi.org/10.1080/11250003.2013.870240.

Casetta, E., & Marques da Silva, J. (2015a). Facing the big sixth. From prioritizing species to conserving biodiversity. In E. Serrelli & N. Gontier (Eds.), *Macroevolution – Explanation, interpretation, evidence* (pp. 377–403). Berlin: Springer.

Casetta, E., & Marques da Silva, J. (2015b). Biodiversity surgery. Some epistemological challenges in facing extinction. *Axiomathes, 25*(3), 239–251.

Claridge, M. F., Dawah, H. A., & Wilson, M. R. (Eds.). (1997). *Species: The units of biodiversity*. London: Chapman & Hall.

Clarke, E. (2016). Levels of selection in biofilms: Multispecies biofilms are not evolutionary individuals. *Biology and Philosophy, 31*, 191–212.

Costanza, R., d'Arge, R., de Groot, R., Farberk, S., Grasso, M., Hannon, B., Limburg, K., Naeem, S., O'Neill, R. V., Paruelo, J., Raskin, R. G., Suttonkk, P., & van den Belt, M. (1997). The value of the world's ecosystem services and natural capital. *Science, 387*, 253–260.

Croft, M. T., Lawrence, A. D., Raux-Deery, E., Warren, M. J., & Smith, A. G. (2005). Algae acquire vitamin B12 through a symbiotic relationship with bacteria. *Nature, 438*, 90–93.

Czárán, T., & Hoekstra, R. F. (2009). Microbial communication, cooperation and cheating: Quorum sensing drives the evolution of cooperation in bacteria. *PLoS One, 4*(8), e6655. https://doi.org/10.1371/journal.pone.0006655.

Dobzhansky, T. (1937). *Genetics and the origin of species*. New York: Columbia University Press.

Doolittle, F., & Papke, T. (2006). Genomics and the bacterial species problem. *Genome Biology, 7*, 116.1–116.7.

Droop, M. R. (2007). Vitamins, phytoplankton and bacteria: Symbiosis or scavenging? *Journal of Plankton Research, 29*(2), 107–113.

Eizaguirre, C., & Baltazar-Soares, M. (2014). Evolutionary conservation-evaluating the adaptive potential of species. *Evolutionary Applications, 7*(9), 963–967.

Eldridge, D. J., & Leys, J. F. (2003). Exploring some relationships between biological soil crusts, soil aggregation and wind erosion. *Journal of Arid Environments, 53*, 457–466.

Elias, S., & Banin, E. (2012). Multi-species biofilms: Living with friendly neighbors. *FEMS Microbiology Reviews, 36*(5), 990–1004.

Ereshefsky, M. (2010). Microbiology and the species problem. *Biology and Philosophy, 25*(4), 553–568.

Ereshefsky, M., & Pedroso, M. (2015). Rethinking evolutionary individuality. *PNAS, 112*(33), 10126–10132.

Falciatore, A., & Bowler, C. (2002). Revealing the molecular secrets of marine diatoms. *Annual Review of Plant Biology, 53*, 109–130.

Falkowski, P. G., Fenchel, T., & Delong, E. F. (2008). The microbial engines that drive Earth's biogeochemical cycles. *Science, 320*, 1034–1039.

Federle, M. J., & Bassler, B. L. (2003). Interspecies communication in bacteria. *Journal of Clinical Investigation, 112*(9), 1291–1299. https://doi.org/10.1172/JCI200320195.

Geider, R. J., Osborne, B. A., & Raven, J. A. (1986). Growth, photosynthesis and maintenance metabolic cost in the diatom *Phaeodactylum tricornutum* at very low light levels. *Journal of Phycology, 22*, 9–48.

Gross, R., Schmid, A., & Buehler, K. (2012). Catalytic biofilms: A powerful concept for future bioprocesses. In G. Lear & G. D. Lewis (Eds.), *Microbial biofilms: Current research and applications* (pp. 193–222). Cardiff: Caister Academic Press.

Hansen, S. K., Rainey, P. B., Haagensen, J. A. J., & Molin, S. (2007a). Evolution of species interactions in a biofilm community. *Nature Letters, 445*(1), 533–536.

Hansen, S. K., Haagensen, J. A. J., Gjermansen, M., Jorgensen, T. M., Tolker-Nielsen, T., & Molin, S. (2007b). Characterization of a *Pseudomonas putida* rough variant evolved in a mixed-species biofilm with Acinetobacter sp strain C6. *Journal of Bacteriology, 189*(13), 4932–4943.

Hey, J., Waples, R. S., Arnold, M. L., Butlin, R. K., & Harrison, R. G. (2003). Understanding and confronting species uncertainty in biology and conservation. *Trends in Ecology & Evolution, 18*(11), 597–603.

Höglund, J. (2009). *Evolutionary conservation genetics*. Oxford: Oxford University Press.

Huxley, J. (1942). *Evolution: The modern synthesis*. London: George Allen & Unwin.

Jesus, B., Brotas, V., Ribeiro, L., Mendes, C. R., Cartaxana, P., & Paterson, D. M. (2009). Adaptations of microphytobenthos assemblages to sediment type and tidal position. *Continental Shelf Research, 29*, 1624–1634.

Kaplan, J. B. (2010). Biofilm dispersal: Mechanisms, clinical implications, and potential therapeutic uses. *Journal of Dental Research, 89*(3), 205–218.

Kim, W., Levy, S. B., & Foster, K. (2016). Rapid radiation in bacteria leads to a division of labour. *Nature Communications, 7*, 10508. https://doi.org/10.1038/ncomms10508.

Kjelleberg, S. (1993). *Starvation in bacteria*. New York: Plenum.

Lewontin, R. (1970). The units of selection. *Annual Review of Ecology and Systematics, 1*, 1–18.

Lugtenberg, B. J. J., Malfanova, N., Kamilova, F., & Berg, G. (2013). Plant growth promotion by microbes. In F. J. de Bruijn (Ed.), *Molecular microbial ecology of the rhizosphere* (Vol. 2, pp. 561–574). Hoboken: Wiley. https://doi.org/10.1002/9781118297674.

Macintyre, H. L., Geider, R. J., & Miller, D. C. (1996). Microphytobenthos: The ecological role of the "secret garden" of unvegetated, shallow-water marine habitats. I. Distribution, abundance and primary production. *Estuaries, 19*, 186–201.

Marques da Silva, J. (2015). Reconciling science and nature by means of the aesthetical contemplation of natural diversity. *Rivista di Estetica, 59*(2), 93–113.

Marques da Silva, J., Cruz, S., & Cartaxana, P. (2017). Inorganic carbon availability in benthic diatom communities: Photosynthesis and migration. *Philosophical Transactions of the Royal Society London B, 372*, 20160398.

Mayr, E. (1970). *Populations, species, and evolution*. Cambridge, MA: Harvard University Press.

Middelburg, J. J., Barranguet, C., Boschker, H. T. S., Herman, P. M. J., Moens, T., & Heip, C. H. R. (2000). The fate of intertidal microphytobenthos carbon: An in situ ^{13}C-labeling study. *Limnology and Oceanography, 45*(6), 1224–1234.

Morris, B. E. L., Henneberger, R., Huber, H., & Moissl-Eichinger, C. (2013). Microbial syntrophy: Interaction for the common good. *FEMS Microbiology Reviews, 37*, 384–406.

Nadell, C. D., Drescher, K., & Foster, K. R. (2016). Spatial structure, cooperation and competition in biofilms. *Nature Reviews Microbiology, 14*, 589–600.

O'Malley, M. (2014). *Philosophy of microbiology*. Cambridge: Cambridge University Press.

Paterson, D. M. (1986). The migratory behaviour of diatom assemblages in a laboratory tidal microecosystem examined by low temperature scanning electron microscopy. *Diatom Research: The Journal of the International Society for Diatom Research, 1*, 227–239.

Paterson, D. M., & Black, K. M. (1999). Water flow, sediment dynamics and benthic biology. *Advances in Ecological Research, 29*, 155–193.

Qaim, M., & Zilberman, D. (2003). Yield effects of genetically modified crops in developing countries. *Science, 299*(5608), 900–902.

Richards, R. (2010). *The species problem*. New York: Cambridge University Press.

Roesch, L. F., Fulthorpe, R. R., Riva, A., Casella, G., Hadwin, A. K., Kent, A. D., Daroub, S. H., Camargo, F. A., Farmerie, W. G., & Triplett, E. W. (2007). Pyrosequencing enumerates and contrasts soil microbial diversity. *The ISME Journal, 1*(4), 283–290.

Round, F. E., & Palmer, J. D. (1966). Persistent, vertical-migration rhythms in benthic microflora. II. Field and laboratory studies on diatoms from the banks of the river Avon. *Journal of the Marine Biological Association of the UK, 46*, 191–214.

Sarkar, S. (2002). Defining 'biodiversity'; Assessing biodiversity. *The Monist, 85*(1), 131–155.

Sauer, K., Camper, A. K., Ehrlich, G. D., Costerton, J. W., & Davies, D. G. (2002). *Pseudomonas aeruginosa* displays multiple phenotypes during development as a biofilm. *Journal of Bacteriology, 184*, 1140–1154.

Sharif, D., Gallon, J., Smith, C. J., & Dudley, E. (2008). Quorum sensing in Cyanobacteria: N-octanoyl-homoserine lactone release and response, by the epilithic colonial cyanobacterium Gloeothece PCC6909. *The ISME Journal, 2*, 1171–1182.

Simpson, C. (2012). The evolutionary history of division of labour. *Proceedings of the Royal Society B, 279*, 116–121.

Smith, T. B., Bruford, M. W., & Wayne, R. K. (1993). The preservation of process: The missing element of conservation programs. *Biodiversity Letters, 1*(6), 164–167.

Soule, T., Anderson, I. J., Johnson, S. L., Bates, S. T., & Garcia-Pichel, F. (2009). Archaeal populations in biological soil crusts from arid lands in North America. *Soil Biology & Biochemistry, 41*, 2069–2074.

Stal, L. J., & de Brouwer, J. F. C. (2003). Biofilm formation by benthic diatoms and their influence on the stabilization of intertidal mudflat. In J. Rullkötter (Ed.), *BioGeoChemistry of tidal flats. Proceedings of a workshop at the Hanse Institute of Advanced Study, May 14–17, Delmenhorst, Germany* (pp. 109–111). Wilhelmshaven: Forschungszentrum Terramare.

Stoodley, P., Sauer, K., Davies, D. G., & Costerton, J. W. (2002). Biofilms as complex differentiates communities. *Annual Review of Microbiology, 56*, 187–209.

Trick, C. G., Anderson, R. J., & Harrison, P. J. (1984). Environmental factors influencing the production of an antibacteria Jmetabolite from a marine dinoflagellate, Prorocentrum minimum. *Canadian Journal of Fisheries and Aquatic Sciences, 41*, 423–432.

Turpin, D. H., & Harrison, P. J. (1980). Cell size manipulation in natural marine, planktonic, diatom communities. *Canadian Journal of Fisheries and Aquatic Sciences, 37*, 1193–1195.

Underwood, G. J. C., & Kromkamp, J. (1999). Primary production by phytoplankton and microphytobenthos in estuaries. *Advances in Ecological Research, 29*, 93–153.

Van Colen, C., Underwood, G. J. C., Serôdio, J., & Paterson, D. M. (2014). Ecology of intertidal microbial biofilms: Mechanisms, patterns and future research needs. *Journal of Sea Research, 92*, 2–5.

Van der Heijden, M. G. A., Bardgett, R. D., & van Straalen, N. M. (2008). The unseen majority: Soil microbes as drivers of plant diversity and productivity in terrestrial ecosystems. *Ecology Letters, 11*, 296–310.

Vieira, S., Ribeiro, L., Jesus, B., Cartaxana, P., & Marques da Silva, J. (2013a). Photosynthesis assessment in microphytobenthos with conventional and imaging pulse amplitude modulation fluorometry. *Photochemistry and Photobiology, 89*, 97–102.

Vieira, S., Ribeiro, L., Marques da Silva, J., & Cartaxana, P. (2013b). Effects of short-term changes in sediment temperature on the photosynthesis of two intertidal microphytobenthos communities. *Estuarine, Coastal and Shelf Science, 119*, 112–118. https://doi.org/10.1016/j.ecss.2013.01.001.

Vieira, S., Cartaxana, P., Máguas, C., & Marques da Silva, J. (2016). Photosynthesis in estuarine intertidal microphytobenthos is limited by inorganic carbon availability. *Photosynthesis Research, 128*, 85–92. https://doi.org/10.1007/s11120-015-0203-0.

Vlamakis, H., Aguilar, C., Losick, R., & Kolter, R. (2008). Control of cell fate by the formation of an architecturally complex bacterial community. *Genes & Development, 22*, 945–953.

Whatley, J. M., & Whatley, F. R. (1981). Chloroplast evolution. *New Phytologist, 87*, 233–247.

Whitman, W. B., Coleman, D. C., & Wiebe, W. J. (1998). Prokaryotes: The unseen majority. *Proceedings of the National Academy of Sciences of the United States of America, 95*, 6578–6583.

Williamson, K. S., Richards, L. A., Perez-Osorio, A. C., Pitts, B., McInnerney, K., Stewart, P. S., & Franklin, M. J. (2012). Heterogeneity in *Pseudomonas aeruginosa* biofilms includes expression of ribosome hibernation factors in the antibiotic-tolerant subpopulation and hypoxia-induced stress response in the metabolically active population. *Journal of Bacteriology, 194*(8), 2062–2073.

Woods, P. M. (2000). *Biodiversity and democracy: Rethinking society and nature*. Vancouver: University of British Columbia Press.

Chapter 10
Considering Intra-individual Genetic Heterogeneity to Understand Biodiversity

Eva Boon

Abstract In this chapter, I am concerned with the concept of Intra-individual Genetic Hetereogeneity (IGH) and its potential influence on biodiversity estimates. Definitions of biological individuality are often indirectly dependent on genetic sampling -and vice versa. Genetic sampling typically focuses on a particular locus or set of loci, found in the the mitochondrial, chloroplast or nuclear genome. If ecological function or evolutionary individuality can be defined on the level of multiple divergent genomes, as I shall argue is the case in IGH, our current genetic sampling strategies and analytic approaches may miss out on relevant biodiversity. Now that more and more examples of IGH are available, it is becoming possible to investigate the positive and negative effects of IGH on the functioning and evolution of multicellular individuals more systematically. I consider some examples and argue that studying diversity through the lens of IGH facilitates thinking not in terms of units, but in terms of interactions between biological entities. This, in turn, enables a fresh take on the ecological and evolutionary significance of biological diversity.

10.1 Introduction to Intra-individual Genetic Heterogeneity

> These days we have become beguiled with diversity: how animals, such as insects, and plants, such as angiosperms, have produced so incredibly many species. In the origins of multicellularity we see a most primitive example of diversification. In some ways, it is almost an ideal case because we can make an argument for its basis. (Bonner 1998)

Intra-individual genetic heterogeneity (IGH for short) is a characterisation that applies to multicellular biological entities. Simply put, it describes a state in which the cells of the biological entity under consideration contain divergent genomes. Some have argued that (similarity in) genome structure and content can give indications about the expected balance between cooperation and conflict between and

E. Boon (✉)
Eindhoven Technical University, Eindhoven, The Netherlands
e-mail: E.Boon@tue.nl

© The Author(s) 2019   219
E. Casetta et al. (eds.), *From Assessing to Conserving Biodiversity*,
History, Philosophy and Theory of the Life Sciences 24,
https://doi.org/10.1007/978-3-030-10991-2_10

even within the cells (Queller and Strassmann 2009, 2016; Strassmann and Queller 2010) – in other words, about ecological interactions between genomes. This idea is a major rationale behind investigating biodiversity in the light of IGH.

Another rationale is that IGH highlights the fundamental elusiveness of some key concepts included in definitions of biodiversity. For example, the UN Convention on Biological Diversity defines biodiversity as "the variability among living organisms from all sources including, inter alia, terrestrial, marine and other aquatic ecosystems and the ecological complexes of which they are part; this includes diversity within species, between species and of ecosystems".[1] The introduction to this volume already mentioned that the ambiguity surrounding terms such as "organisms", "species" and "ecosystems" has already been discussed for a longer time. Technological advances now enable philosophers of biology and biologists to consider in detail how actual patterns of genetic diversity match or clash with these concepts. Studying IGH is part of this effort, in the sense that patterns of genetic diversity within and between biological entities do not always coincide with our perception of these entities as biological, ecological or evolutionary units. To stay with the analogy of diagnosing and treating a patient, as was elaborated in the introduction, understanding IGH is important to better diagnose a patient. The consequences for treatment will be discussed at the end of this chapter.

Here, I am concerned with how the concept of IGH can influence our perception of biodiversity, considering that in biodiversity studies, our definition of biological individuality is often dependent on genetic sampling -and vice versa. Genetic sampling is often concentrated on a particular genome, such as the mitochondrial, chloroplast of nucelar genomes. However, if ecological function or evolutionary individuality can play out on the level of multiple divergent genomes, as I shall argue is the case in IGH, our current genetic sampling strategies may miss out on relevant biodiversity.

I will proceed as follows. As more and more examples of IGH become available, it is becoming possible to investigate the positive and negative effects of IGH on the functioning of a multicellular individual (sect. 10.2). I argue that considering diversity through IGH facilitates thinking not in terms of units, but in terms of interactions between biological entities (sect. 10.3). This, in turn, may enable a fresh take on the ecological and evolutionary significance of diversity. From the examples proposed in this chapter, we can consider how IGH as an unexplored dimension of biodiversity may help us understand what diversity is *relevant* to our research goals.

10.2   Examples of IGH

To make the concept of IGH more concrete, I will start off with a number of familiar and possibly less familiar examples of IGH. Common resolutions to genetic conflict within the organism, such as the separation of germline and soma and genetic

---

[1] https://www.cbd.int/doc/legal/cbd-en.pdf

bottlenecks, are only applicable to a narrow range of multicellular organisms -mostly metazoans. Organisms that consist of easily regenerating parts (plants, algae) or hyphal networks (fungi) seem often more genetically heterogeneous. In this context, it is interesting to note that metazoans that regenerate from body fragments (medusa, sponges, urochordates) are also more often reported to be genetically heterogeneous (Rinkevich 2004, 2005).

Cases of intra-individual genetic heterogeneity can be divided in chimeras and mosaicisms (Pineda-Krch and Lehtila 2004a; Santelices 2004a), of which examples will be discussed separately below. In the case of mosaic entities, genomes are divergent but homologous in the sense that they share a recent common ancestor. In the case of chimeras, the genetic heterogeneity is nonhomologous: cells may have originated from evolutionary highly divergent lineages. From this definition, it becomes apparent that the distinction between a mosaic and a chimeric biological entitity is ultimately a judgment on evolutionary divergence, which in itself is often based on genome similarity.

10.2.1   Mosaic Individuals

What kinds of individuals are mosaic? As mentioned above, genetic differentiation between somatic cells is often found within plants, animals and fungi that propagate by cloning of body parts. The prevalence of mosaicism in clonally reproducing plants and animals is easy to explain if mosaicism correlates with mutation rate and longevity (Gill et al. 1995). Of course, the mosaic state will be cut short when a single cell bottleneck occurs in the reproductive cycle. Still, this does not seem to stop mosaicism from occurring in multicellular entites that pass through a single cell bottleneck, i.e. metazoans such as fish (Matos et al. 2011) and corals (Schweinsberg et al. 2015).

Mosaicism in humans is a burgeoning field, since much of present-day cancer research relies on assessing genetic heterogeneity between tumor cells, which in turn determines (in some cancers) much of the treatment and prognosis. This approach relies on the argument that the genetic heterogeneity in the tumor is governed by different evolutionary dynamics as the rest of the body (see for example Jacobs et al. 2012; Laurie et al. 2012; Vijg 2014 and references therein). Mosaicism can also have much less dramatic influences in humans, and increasing reports on mosaicism in humans (Youssoufian and Pyeritz 2002; Erickson 2014; Spinner and Conlin 2014) and even the germline (Samuels and Friedman 2015; Jónsson et al. 2017) seem to underline the varying evolutionary outcomes for IGH in a multicellular individual: positive, negative or neutral.

Another example demonstrates how mosaicism can be deeply integrated in the life history of a biological entity. Arbuscular mycorrhizal fungi (AMF) are an ancient phylum of heterotrophs that form symbioses with the majority of land plants (Wang and Qiu 2006). AMF were reported to contain hundreds or thousands of genetically differentiated nuclei within the same cytoplasm (Kuhn et al. 2001; Hijri and Sanders 2005; Boon et al. 2010, 2013b, 2015). The exact number of genetically differentiated nuclei is not clear, since genetic variation has never been exhaustively sampled for any locus in these fungi. No sexual stage has been observed, although possible recombination has been reported in AMF (see Riley et al. 2016 and references therein).[2]

It is possible that a single nucleus is not a viable entity in AMF –only populations of nuclei are (see Boon et al. 2015; Wyss et al. 2016 and references therein). For example, spores of *Rhizophagus irregularis* do not germinate under a certain volume, which is positively correlated with the number of nuclei within the spore (Marleau et al. 2011). Some authors have proposed that genetic differentiation between nuclei within the AMF cytoplasm is maintained through the fusion of related hyphae, or anastomosis (Giovannetti et al. 2015 and references therein), and is lost at sporulation (Boon et al. 2013b). This means that AMF are both mosaic individuals in the sense that their nuclei share the same genealogy, and chimeras in the sense that at least some of this variation is the result of hyphal fusion between related hyphal systems.

The positive correlation between anastomosis rates and level of relatedness between hyphae supports the view that AMF do form genetically delineable entities –although maybe not on a genome level, but on that of the genome population or pangenome. This has also been suggested by Boon et al. (2015). The propensity to fuse within cultures of the same strain seems diminished by drift (Cárdenas-Flores et al. 2010). This may indicate that the nuclei within an AMF hyphal system show self-nonself recognition *as a population*. Finally the composition of the genome population has an influence on the genotype: a change in nuclear population through anastomosis changes the (symbiotic) properties of a strain (Sanders and Rodriguez 2016 and references therein).

To summarise this example, the AMF phenotype seems determined by not a single, but multiple coexisting genomes. Anastomosis, or a lack thereof, can change the AMF phenotype, which is in turn selected upon by its environment (Roger et al. 2013b, a; Angelard et al. 2014; Wyss et al. 2016; Sanders and Rodriguez 2016). AMF form symbioses with almost all land plants (Wang and Qiu 2006), and the age of their evolutionary lineage (an estimated 500 million years, coinciding with the rise of land plants (Corradi and Bonfante 2012)) testifies to the potential ecological impact and longevity of IGH as an evolutionary strategy. The presence of mycorrhiza in the soil confers a inestimable fitness advantage to plant communities: most plant taxa form symbioses with AMF in which posphorus is exchanged for plant-produced sugars, thus stimulating plant growth and overall community biodiversity (van der Heijden et al. 2016)

In this example on AMF, IGH seems to be an essential to understand AMF life history, ecology and evolution. At the same time, the precise effects of IGH on AMF life history are hard to estimate since this IGH is often not taken into account in experimental setups and field studies due to technical and conceptual challenges (Sanders and Rodriguez 2016). Still, we can add yet another layer of complexity. If

---

[2] Note that some recombination estimates may not be reliable if Glomus is indeed multigenomic.

we consider that AMF are obligate plant symbionts and themselves are associated with particular microbial communities, a picture emerges of another set of interactions. AMF are functionally dependent on their plant hosts, and probably gain major fitness benefits from their associated microbial communities (Bonfante and Anca 2009; Herman et al. 2011). Thus, since the ecological function and evolutionary longevity of the mosaic AMF is dependent on nonhomologous lineages (i.e. plants and microbes), we may also consider AMF and their associated plants and microbes as chimeras. This is not only theoretically relevant: the growth benefits that AMF confer to their plant hosts can potentially confer huge benefits to sustainable agriculture.

10.2.2 Chimeric Individuals

As in the AMF example above, most mutualisms can also be considered chimeras. This depends on the criteria for individuality that are being used, and to some extent on the degree of genetic divergence one is willing to accept between the component genomes of the chimera. For example, in the case of lichen and corals the mutualism is so tight that the historical and most intuitive view is to see the chimera as a single entity. Only with the advent of molecular techniques have scientists started to distinguish different genomes and consider the partners as separate 'individuals'. With the following two examples, I would like to illustrate how broad the definition of 'chimera' can be, and highlight how considering IGH has consequences for ecological and evolutionary inference in these examples.

My first case is chimerism in macroalgae, which is a nonmonophyletic group that encompasses brown, green and red algae (Santelices 2004b). Here, I will focus on red algae or Rhodophyta, since IGH in this taxon has been most extensively documented. The algae germinate from a disk, which can originate from multiple spores. These spores may fuse, or form individual cell walls that are subsequently surrounded by a thickened common wall. This process, called coalescence, occurs often in red algae (Santelices et al. 1999) but not between different species (Santelices et al. 2003). Coalescing disks, or crusts, may or may not fuse with each other or with new algal sporelings. Various fitness advantages were demonstrated for coalescencing disks compared to unitary disks (which have originated from a single spore). Fusion decreases the probability of mortality in early growth (Santelices and Alvarado 2008), improves erect axis formation and growth (Santelices et al. 2010) and confers an advantage later in the life cycle through differences in branching and fertility (Santelices et al. 2011). Coalescence and a fitness advantage for coalescing disks have also been reported, although less extensively, in brown (Wernberg 2005) and green (Gonzalez and Santelices 2008) algae.

Thus, like in AMF, we encounter a population of genotypes (from multiple haploid spores) which together creates a polyploid phenotype that is selected upon as a single entity (Monro and Poore 2004). In red algae, these polyploids break up again at sporulation. Possibly, selection (for cooperation) between haploid genomes occurs at the formation of the disk. How this selection takes place is unclear. It

seems reasonable to suppose that selection occurs for compatibility between particular loci or even entire genomes. It is important to note that, like in the previous
AMF example, it is not possible to associate the phenotype of a red alga with a
particular unchanging set of spore haplotypes. The phenotype of interest, i..e the
disk and the thalli that grow from it, is based on a *varying* population of multiple
haploid nuclear genomes. Therefore, the phenotype of the macroalga cannot be
reduced to its component genomes. It is in the *interaction* between these varying
genomes that a unique phenotype is established.

The ecological and evolutionary consequences of the chimeric state in macroalgae are not easy to disentangle, even more so because seaweeds can reproduce clonally as well (Fagerström and Poore 2000; Collado-Vides 2001). Nevertheless, the
above description makes clear that the life history and evolutionary constraints of
coalescing red algae cannot be described accurately without taking chimerism into
account. Again, in the light of the ecological and agricultural importance red algae
this is not a merely academic preoccupation. For example, multisporic coalescing
recruits have higher survival rates (Santelices and Aedo 2006). Thus, IGH as a state
can be manipulated by farmers to increase higher yields.

My second example of chimerism is the case of microbial multi-species consortia
or communities. With the advance of molecular techniques it has become possible to
study microbial communities in more detail than ever before, even though passing
from the fase of amassing vast quantities of data to that of interpreting them has proven
to be a challenge. Here, I would like to highlight a few patterns that have emerged with
respect to microbial diversity and function and that are relevant to the concept of
IGH. The following three points are discussed in more detail in Boon et al. (2013a).

First, microbial taxa are hard to circumscribe precisely, for a number of well-
described ontological and epistemological reasons.[3] For the purposes of this chapter, it
suffices to state that microbial taxonomy is heavily dependent on the molecular biology toolbox. This toolbox, although indispensible, has a number of limitations. The
most relevant limitation for my point is that especially early conceptions of microbial
taxa heavily relied on the assumption of genome stability. And there's the rub: in many
microorganisms, genomes can change rapidly through gene loss, gene duplication,
and the acquisition of genes from distant lineages via lateral gene transfer (LGT).

A second pattern is that taxonomic or phylogenetic thresholds (e.g. 3% genetic
differentiation) for taxon delineation fail to adequately delineate ecologically cohesive units. Even though a unifying species concept is not strictly needed for ecological analysis, also a pluralist stance needs a sound rationale and consistent approach
(or set of approaches) to define 'units'. Unfortunately, microbial diversity and community function do not always correlate. Microbes rarely act alone and are often
interdependent. It is possible that less than 1% of all known microbes can be
successfully cultured on their own (Staley and Konopka 1985), an observation also
known as 'Great Plate Count Anomaly'. It is now clear that many microbes depend
on the activity of other microbes to successfully grow and reproduce via mecha-

---

[3] Doolittle (2013) has written an extensive review on the history and challenges of microbial ontology and of course O'Malley (2014) is an invaluable resource here as well.

nisms including acquisition and exchange of metabolites (references in Boon et al. 2013a).

A final tendency is that microbial function may be a property of communities as well as of cells. Particular metabolic capacities might not be encoded within a single microbial cell, or even within a single type of microbial cell. Instead, there is increasing evidence that many microbial functions are encoded by gene networks in which genes may be easily replaced by functionally equivalent but phylogentically distant alternatives. These gene networks may be found in varying sets of microbial taxa, without a single taxa being characterised by a particular set of genes or functions. We face the same situation as in the previous examples, in which no single community genotype codes for a single community phenotype.

If microbial communities can be understood as 'chimeras', it might not be possible to lead a community function (for example, a particular metabolic product or process) back to a single taxonomic group (but see Inkpen et al. 2017). The diversity of microbes is now being explored using surveys that draw on hundreds or thousands of samples and controlled experiments, with rapid genetic assessment techniques providing much of the evidence for taxonomic and functional diversity. Since microbial interactions span all taxonomic ranks, from strain to superkingdom, understanding microbial diversity then seems to necessitate a community-centric approach (Zarraonaindia et al. 2013).

Mechanisms for the evolution of interdependence within microbial communities have been proposed in the form of a Public Goods Hypothesis ((McInerney et al. 2011) and the Black Queen Hypothesis (Morris et al. 2012). The authors of the ITSNTS model ('It's the song, not the singer') even propose 'casting metabolic and developmental interaction patterns, rather than the taxa responsible for them, as units of selection' (Doolittle and Booth 2016): in other words, microbial interaction patterns are stabler units of selection than the microbial cells that produce these patterns. For a more in-depth discussion of the evolutionary and ecological implications of seeing microbial communities such as biofilms as evolutionary individuals, see Boon (in preparation).

10.2.3 Mosaic vs. Chimeric Individuals

The reader might wonder by now whether the distinction between mosaic and chimeric individuals is at all relevant. Is the difference between the two not just a matter of degree of relatedness between individuals, rather than a difference in kind of individual? The answer to this ontological question is not at all straightforward. However, from an epistemic point of view, the differences between these two types of intra-individual genetic heterogeneity are relevant to the practise of evolutionary inference.

For example, fitness calculations between mosaic entities and chimeras are performed differently. If genetically differentiated but related cell lines work together, as in mosaic individuals, a case could be made for a special sort of kin selection. After all, there is a considerable chance that gene variants between related lineages are shared. However, if unrelated cell lines become integrated in a single entity, the balance of costs and benefits that ultimately decides between competition and cooperation cannot be explained by a more than average chance to transfer one's own genes as present in the other.

Some might disagree that there is even an epistemic difference between chimeras and mosaic biological entites. Multilevel selection theory (see Okasha 2006 and references therein) stresses that kin selection is really just a special case of group selection. Interestingly, this discussion is also highly relevant within microbial community ecology and evolution, in which the question whether microbial communities can evolve is a hot topic of debate. Boiled down to its essence, this question is really about whether entities that are composed of nonhomologous lineages can evolve as a single unit–and whether is this is a useful question to ask (Boon in preparation).

10.3   The Importance of IGH in Ecology and Evolution

Above examples lead me to two main themes for the relevance of IGH in biodiversity research. The first is that multiple *varying* genotypes can lead to a single phenotype. Since the phenotype is the actual set of traits that is selected upon, or that is ecologically relevant, extreme caution should then be exercised when a single genome (or genotype, or even a simple barcode) is taken as a proxy for the phenotype. If the more complicated genotype-phenotype relationship that is implied by IGH is ignored and genotypes that are associated with a particular phenotype are inadequately sampled, it will be difficult if not impossible to find reproducible patterns and predict community composition or ecosystem function. Second, if one of the aims of measuring biodiversity is to predict or at least understand ecosystem function, it is vital to note that while community ecology considers interactions among entities, the inference of these interactions depends critically on the level at which entities are defined.

These two themes may be made more concrete with an example: while it may be possible to describe a microbial community as performing a single ecosystem function, it may not be possible to find a specific genotype or even set of genotypes stable enough (i.e. reoccurring consistently) to characterise this functional unit. Instead, one may want to consider whether instead particular interactions between units, such as a particular exchange of metabolites or another shared fitness benefit, may be the most stable component of the interaction.[4]

This situation is a radical departure from a more traditional view, in which a one-to-one relation is assumed between genotype and phenotype. In other words, once we look away from our metazoan bias it may no longer be possible to explain phenotype and its ecological role by measuring the genotype, since this genotype, even as an amalgation of multiple component genotypes, is simply not stable enough.

---

[4] See also the ITSNTS argument of Doolittle and Booth (2016).

10.3.1 The Metazoan Bias

The examples in the previous section might seem "atypical" in the context of biodiversity. In fact, when speaking of biodiversity there is often a bias towards species that are relatively easy to identify and delineate, such as animal species. Yet vast diversity, however measured, can be found in groups such as algae, fungi, and the many phylae of microbes and virusses, which are often at the basis of ecosystem function (e.g. Wagg et al. 2014).

Still, one should wonder whether it makes sense to describe above examples as instances of IGH. In other words, how permissive can a definition of biological or evolutionary individuality be without losing its use? The term 'Intra-individual genetic heterogeneity' ultimately pivots on the definition of the 'individual'. To determine an ecological function or identify an evolutionary process, one needs to distinguish the entities that perform these functions or processes.

The discussion on biological delineation and individuation has been conducted in different contexts already and has taken a central place in recent philosophy of biology discussions (see for example Queller and Strassmann 2016; Clarke 2016; Pradeu 2016 and references therein). It becomes clear from these recent considerations that there are valid reasons to consider biological organisation from many different viewpoints. In other words, different research goals justify the use of divergent concepts of biological or evolutionary individuality and thus warrant a pluralist approach.

In this context, it does make sense to describe different kinds of biological entities as instances of IGH. For example, when we consider a system with AMF, we could choose to look at a single AMF nucleus, at a population of nuclei, or at an entire hyphal system. Enlarging our scope even more, we could choose to include the plant partners as well as the surrounding microbial communities. I propose that it is in this choice that the real point of discussion lies: how to decide on the relevant unit of diversity?

10.3.2 Biological Organization, Hierarchy and Relevance

A genotype, or even the entire genome, is often for practical purposes employed as a unique identifier for 'the biological individual'. Of course, this biological individual cannot be fully described by only its genetic code. If this were so, we would consider human identical twins to be one and the same biological individual. However, although the individual is not defined by a unique genome, a unique genome seems to havebeen a convincing criterion for assigning individuality. Why?

One possible reason is that the organization of biological diversity is considered to be hierarchical. In this view, DNA is organized in cells, cells in bodies, bodies in populations and populations in species. It is implied that without cells competing or cooperating, the body would not exist, and without bodies competing or cooperating, a population would not exist, and so forth. Leo Buss, for example, stated that "[An] explicitly hierarchical perspective on evolution predicts that the myriad complexities of ontogeny, cell biology and molecular genetics are ultimately penetrable in the context of an interplay of synergisms and conflicts between different units of selection" (Buss 1987).

This idea of hierarchy is also prominent in the literature on major transitions in evolution. Maynard Smith and Szathmáry proposed that complexity in evolution increases with time, which is achieved through a series of major transitions. They also described this complexity as mostly hierarchical (Maynard Smith and Szathmáry 1995). Others have continued or varied on this view of evolution of life on earth, yet all agree that cooperation and competition takes place at definable 'levels' (Clarke 2016 and references therein). A formalisation of this view can be found in multilevel selection theory. Proponents of this theory aim to develop and formalize the tools we need to describe and quantify the relative importance of different levels of selection (e.g. Wilson and Sober 1994). Ultimately, the interactions between these levels are proposed to lead to the diversity we observe among biological entities.

Much discussion in the major transitions literature is then about finding out how conflict at a particular level of organization is resolved, in order to explain the evolution and diversity of another level of biological organization. In this manner, a hierarchical view on biological diversity can offer a perspective with the scope to explain a large number of observations. However, it can also lead to misleading assumptions or obscure similarities. For example, the link of genome homogeneity with the delineation of the biological individual is based on the assumption that IGH leads to conflict within that individual (Michod and Roze 2001; Strassmann and Queller 2004). Yet it is not clear whether genome heterogeneity always leads to conflict. It is possible that there are cases where IGH can actually confer an advantage to the multicellular community it is part of. Some of these examples were already discussed above. The question then becomes more nuanced: when is IGH relevant?

The simple answer may be: when there is a significant effect of IGH on the possible evolutionary trajectories (sometimes referred to as 'evolvability') and ecological range that a biological unit can follow or occupy as a result of its IGH. These latter two concepts are exactly what is at stake in many biodiversity investigations. Moreover, some of the ways in which biodiversity is understood are based on taxonomic or ecologic hierarchies (e.g. Sarkar 2002). Red algae and arbuscular mycorrhizal fungi are two fairly well-documented organisms in which intra-individual heterogeneity plays an important role in understanding of both evolvability and ecological range. Furthermore, even though IGH is sparsely documented, reviews are available with more examples (Santelices 1999; Pineda-Krch and Lehtila 2004b, a), as well as a range of suggestions on how IGH could affect life history (Pineda-Krch and Lehtila 2004b; Folse 2011; Folse and Roughgarden 2012). Finally, the importance and relevance of IGH should be decided on a case-by-case basis –without assuming or dismissing its potential role off-hand.

10.4   Conclusions

I argued that the shortcut one genome-one individual has closed our eyes to the possible importance of IGH in evolution and evolution –and thus to its role in for biodiversity estimates. Arguing from the examples in this chapter, I propose that IGH can help us understand what diversity is *relevant* to our research goals. To maintain the analogy from the introduction: which characteristics of the patient and her symptoms are relevant to a diagnosis and treatment?

By expanding our practical and conceptual tools to facilitate the study of genetic heterogeneity at many different levels of biological organisation, we can start to understand diversity by focusing on interactions between entities –however defined.

References

Angelard, C., Tanner, C. J., Fontanillas, P., et al. (2014). Rapid genotypic change and plasticity in arbuscular mycorrhizal fungi is caused by a host shift and enhanced by segregation. *The ISME Journal, 8*, 284–294.

Bonfante, P., & Anca, I.-A. (2009). Plants, mycorrhizal fungi, and bacteria: A network of interactions. *Annual Review of Microbiology, 63*, 363–383. https://doi.org/10.1146/annurev.micro.091208.073504.

Bonner, J. T. (1998). The origins of multicellularity. *Integrative Biology Issues News and Reviews, 1*, 27–36. https://doi.org/10.1002/(SICI)1520-6602(1998)1:1<27::AID-INBI4>3.0.CO;2-6.

Boon, E. (in preparation). *Biofilms as evolutionary individuals: An empirical question?*

Boon, E., Zimmerman, E., Lang, B. F., & Hijri, M. (2010). Intra-isolate genome variation in arbuscular mycorrhizal fungi persists in the transcriptome. *Journal of Evolutionary Biology, 23*, 1519–1527. https://doi.org/10.1111/j.1420-9101.2010.02019.x.

Boon, E., Meehan, C. J., Whidden, C., et al. (2013a). Interactions in the microbiome: Communities of organisms and communities of genes. *FEMS Microbiology Reviews, 38*, 90–118. https://doi.org/10.1111/1574-6976.12035.

Boon, E., Zimmerman, E., St-Arnaud, M., & Hijri, M. (2013b). Allelic differences within and among sister spores of the arbuscular mycorrhizal fungus Glomus etunicatum suggest segregation at sporulation. *PLoS One, 8*, e83301.

Boon, E., Halary, S., Bapteste, E., & Hijri, M. (2015). Studying genome heterogeneity within the arbuscular mycorrhizal fungal cytoplasm. *Genome Biology and Evolution, 7*, 505–521.

Buss, L. W. (1987). *The evolution of individuality.* Princeton: Princeton University Press.

Cárdenas-Flores, A., Draye, X., Bivort, C., et al. (2010). Impact of multispores in vitro subcultivation of Glomus sp. MUCL 43194 (DAOM 197198) on vegetative compatibility and genetic diversity detected by AFLP. *Mycorrhiza, 20*, 415–425.

Clarke, E. (2016). A levels-of-selection approach to evolutionary individuality. *Biology and Philosophy, 31*, 893–911. https://doi.org/10.1007/s10539-016-9540-4.

Collado-Vides, L. (2001). Clonal architecture in marine macroalgae: Ecological and evolutionary perspectives. *Evolutionary Ecology, 15*, 531–545. https://doi.org/10.1023/a:1016009620560.

Corradi, N., & Bonfante, P. (2012). The arbuscular mycorrhizal symbiosis: Origin and evolution of a beneficial plant infection. *PLoS Pathogens, 8*, e1002600. https://doi.org/10.1371/journal.ppat.1002600.

Doolittle, W. F. (2013). Microbial neopleomorphism. *Biology and Philosophy, 28*, 351–378. https://doi.org/10.1007/s10539-012-9358-7.

Doolittle, W. F., & Booth, A. (2016). It's the song, not the singer: An exploration of holobiosis and evolutionary theory. *Biology and Philosophy, 32,* 5–24. https://doi.org/10.1007/s10539-016-9542-2.

Erickson, R. P. (2014). Recent advances in the study of somatic mosaicism and diseases other than cancer. *Current Opinion in Genetics & Development, 26,* 73–78. https://doi.org/10.1016/j.gde.2014.06.001.

Fagerström, T., & Poore, A. G. B. (2000). Intraclonal variation in macroalgae: Causes and evolutionary consequences. *Sel, 1*(1), 123–133.

Folse, H. J. (2011). *Evolution and individuality: Beyond the genetically homogenous organism.* Stanford: Stanford University.

Folse, H. J., & Roughgarden, J. (2012). Direct benefits of genetic mosaicism and intraorganismal selection: Modeling coevolution between a long-lived tree and a short-lived herbivore. *Evolution, 66,* 1091–1113. https://doi.org/10.1111/j.1558-5646.2011.01500.x.

Gill, D. E., Chao, L., Perkins, S. L., & Wolf, J. B. (1995). Genetic mosaicism in plants and clonal animals. *Annual Review of Ecology and Systematics, 26,* 423–444.

Giovannetti, M., Avio, L., & Sbrana, C. (2015). Functional significance of anastomosis in arbuscular mycorrhizal networks. In T. R. Horton (Ed.), *Mycorrhizal networks* (pp. 41–67). Dordrecht: Springer.

Gonzalez, A. V., & Santelices, B. (2008). Coalescence and chimerism in Codium (Chlorophyta) from central Chile. *Phycologia, 47,* 468–476. https://doi.org/10.2216/07-86.1.

Herman, D. J., Firestone, M. K., Nuccio, E., & Hodge, A. (2011). Interactions between an arbuscular mycorrhizal fungus and a soil microbial community mediating litter decomposition. *FEMS Microbiology Ecology, 80,* 236–247. https://doi.org/10.1111/j.1574-6941.2011.01292.x.

Hijri, M., & Sanders, I. R. (2005). Low gene copy number shows that arbuscular mycorrhizal fungi inherit genetically different nuclei. *Nature, 433,* 160–163.

Inkpen, S. A., Douglas, G. M., Brunet, T. D. P., et al. (2017). The coupling of taxonomy and function in microbiomes. *Biology & Philosophy.* Published online 1 November 2017. https://doi.org/10.1007/s10539-017-9602-2

Jacobs, K. B., Yeager, M., Zhou, W., et al. (2012). Detectable clonal mosaicism and its relationship to aging and cancer. *Nature Genetics, 44,*651–658. http://www.nature.com/ng/journal/v44/n6/abs/ng.2270.html#supplementary-information

Jónsson, H., Sulem, P., Kehr, B., et al. (2017). *Parental influence on human germline de novo mutations in 1,548 trios from Iceland.* Nature advance online publication

Kuhn, G., Hijri, M., & Sanders, I. R. (2001). Evidence for the evolution of multiple genomes in arbuscular mycorrhizal fungi. *Nature, 414,* 745–748.

Laurie, C. C., Laurie, C. A., Rice, K., et al. (2012). Detectable clonal mosaicism from birth to old age and its relationship to cancer. *Nature Genetics, 44,* 642–650. http://www.nature.com/ng/journal/v44/n6/abs/ng.2271.html#supplementary-information

Marleau, J., Dalpe, Y., St-Arnaud, M., & Hijri, M. (2011). Spore development and nuclear inheritance in arbuscular mycorrhizal fungi. *BMC Evolutionary Biology, 11,* 51.

Matos, I., Sucena, E., Machado, M., et al. (2011). Ploidy mosaicism and allele-specific gene expression differences in the allopolyploid Squalius alburnoides. *BMC Genetics, 12,* 101.

Maynard Smith, J., & Szathmáry, E. (1995). *The major transitions in evolution.* Oxford: W. H. Freeman.

McInerney, J. O., Pisani, D., Bapteste, E., & O'Connell, M. J. (2011). The public goods hypothesis for the evolution of life on Earth. *Biology Direct, 6,* 41.

Michod, R. E., & Roze, D. (2001). Cooperation and conflict in the evolution of multicellularity. *Heredity, 86,* 1–7.

Monro, K., & Poore, A. G. B. (2004). Selection in modular organisms: Is intraclonal variation in macroalgae evolutionarily important? *The American Naturalist, 163,* 564–578. https://doi.org/10.1086/382551.

Morris, J. J., Lenski, R. E., & Zinser, E. R. (2012). The Black Queen Hypothesis: Evolution of dependencies through adaptive gene loss. *MBio, 3,* e00036–e00012.

O'Malley, M. (2014). *Philosophy of microbiology.* Cambridge: Cambridge University Press.

Okasha, S. (2006). *Evolution and the levels of selection*. Oxford: Oxford University Press.

Pineda-Krch, M., & Lehtila, K. (2004a). Challenging the genetically homogeneous individual. *Journal of Evolutionary Biology, 17*, 1192–1194.

Pineda-Krch, M., & Lehtila, K. (2004b). Costs and benefits of genetic heterogeneity within organisms. *Journal of Evolutionary Biology, 17*, 1167–1177.

Pradeu, T. (2016). The many faces of biological individuality. *Biology and Philosophy, 31*, 761–773. https://doi.org/10.1007/s10539-016-9553-z.

Queller, D. C., & Strassmann, J. E. (2009). Beyond society: The evolution of organismality. *Philosophical Transactions of the Royal Society B: Biological Sciences, 364*, 3143–3155. https://doi.org/10.1098/rstb.2009.0095.

Queller, D. C., & Strassmann, J. E. (2016). Problems of multi-species organisms: Endosymbionts to holobionts. *Biology and Philosophy, 31*, 855–873. https://doi.org/10.1007/s10539-016-9547-x.

Riley, R., Charron, P., Marton, T., & Corradi, N. (2016). Evolutionary genomics of arbuscular mycorrhizal fungi. In *Molecular mycorrhizal symbiosis* (p. 421). Hoboken: Wiley.

Rinkevich, B. (2004). Will two walk together, except they have agreed? Amos 3:3. *Journal of Evolutionary Biology, 17*, 1178–1179.

Rinkevich, B. (2005). Natural chimerism in colonial urochordates. *Journal of Experimental Marine Biology and Ecology, 322*, 93–109. https://doi.org/10.1016/j.jembe.2005.02.020.

Roger, A., Colard, A., Angelard, C., & Sanders, I. R. (2013a). Relatedness among arbuscular mycorrhizal fungi drives plant growth and intraspecific fungal coexistence. *The ISME Journal, 7*, 2137–2146.

Roger, A., Gétaz, M., Rasmann, S., & Sanders, I. R. (2013b). Identity and combinations of arbuscular mycorrhizal fungal isolates influence plant resistance and insect preference. *Ecological Entomology, 38*, 330–338. https://doi.org/10.1111/een.12022.

Samuels, M. E., & Friedman, J. M. (2015). Genetic mosaics and the germ line lineage. *Genes, 6*, 216–237.

Sanders, I. R., & Rodriguez, A. (2016). Aligning molecular studies of mycorrhizal fungal diversity with ecologically important levels of diversity in ecosystems. *The ISME Journal, 10*, 2780–2786.

Santelices, B. (1999). How many kinds of individual are there? *Trends in Ecology & Evolution, 14*, 152–155.

Santelices, B. (2004a). Mosaicism and chimerism as components of intraorganismal genetic heterogeneity. *Journal of Evolutionary Biology, 17*, 1187–1188. https://doi.org/10.1111/j.1420.9101.2004.00813.x.

Santelices, B. (2004b). A comparison of ecological responses among aclonal (unitary), clonal and coalescing macroalgae. *Journal of Experimental Marine Biology and Ecology, 300*, 31–64. https://doi.org/10.1016/j.jembe.2003.12.017.

Santelices, B., & Aedo, D. (2006). Group recruitment and early survival of Mazzaella Laminarioides. *Journal of Applied Phycology, 18*, 583–589. https://doi.org/10.1007/s10811-006-9067-1.

Santelices, B., & Alvarado, J. L. (2008). Demographic consequences of coalescence in sporeling populations of Mazzaella Laminarioides (Gigartinales, Rhodophyta). *Journal of Phycology, 44*, 624–636. https://doi.org/10.1111/j.1529-8817.2008.00528.x.

Santelices, B., Correa, J. A., Aedo, D., et al. (1999). Convergent biological processes in coalescing Rhodophyta. *Journal of Phycology, 35*, 1127–1149. https://doi.org/10.1046/j.1529-8817.1999.3561127.x.

Santelices, B., Aedo, D., Hormazabal, M., & Flores, V. (2003). Field testing of inter- and intraspecific coalescence among mid-intertidal red algae. *Marine Ecology Progress Series, 250*, 91–103. https://doi.org/10.3354/meps250091.

Santelices, B., Alvarado, J. L., & Flores, V. (2010). Size increments due to interindividual fusions: How much and for how long? *Journal of Phycology, 46*, 685–692. https://doi.org/10.1111/j.1529-8817.2010.00864.x.

Santelices, B., Alvarado, J. L., Chianale, C., & Flores, V. (2011). The effects of coalescence on survival and development of Mazzaella laminarioides (Rhodophyta, Gigartinales). *Journal of Applied Phycology, 23*, 395–400. https://doi.org/10.1007/s10811-010-9566-y.

Sarkar, S. (2002). Defining "Biodiversity"; Assessing biodiversity. *The Monist, 85*, 131–155.

Schweinsberg, M., Weiss, L. C., Striewski, S., et al. (2015). More than one genotype: How common is intracolonial genetic variability in scleractinian corals? *Molecular Ecology, 24*, 2673–2685. https://doi.org/10.1111/mec.13200.

Spinner, N. B., & Conlin, L. K. (2014). Mosaicism and clinical genetics. *American Journal of Medical Genetics. Part C, Seminars in Medical Genetics, 166*, 397–405. https://doi.org/10.1002/ajmg.c.31421.

Staley, J. T., & Konopka, A. (1985). Measurement of in situ activities of nonphotosynthetic microorganisms in aquatic and terrestrial habitats. *Annual Review of Microbiology, 39*, 321–346.

Strassmann, J. E., & Queller, D. C. (2004). Genetic conflicts and intercellular heterogeneity. *Journal of Evolutionary Biology, 17*, 1189–1191.

Strassmann, J. E., & Queller, D. C. (2010). The social organism: Congresses, parties, and committees. *Evolution, 64*, 605–616. https://doi.org/10.1111/j.1558-5646.2009.00929.x.

van der Heijden, M. G., de, B. S., Luckerhoff, L., et al. (2016). A widespread plant-fungal-bacterial symbiosis promotes plant biodiversity, plant nutrition and seedling recruitment. *The ISME Journal, 10*, 389–399.

Vijg, J. (2014). Somatic mutations, genome mosaicism, cancer and aging. *Molecular Genetics Bases of Disease, 26*, 141–149. https://doi.org/10.1016/j.gde.2014.04.002.

Wagg, C., Bender, S. F., Widmer, F., & van der Heijden, M. G. A. (2014). Soil biodiversity and soil community composition determine ecosystem multifunctionality. *Proceedings of the National Academy of Sciences, 111*, 5266–5270.

Wang, B., & Qiu, Y. L. (2006). Phylogenetic distribution and evolution of mycorrhizas in land plants. *Mycorrhiza, 16*, 299–363.

Wernberg, T. (2005). Holdfast aggregation in relation to morphology, age, attachment and drag for the kelp Ecklonia radiata. *Aquatic Botany, 82*, 168–180. https://doi.org/10.1016/j.aquabot.2005.04.003.

Wilson, D. S., & Sober, E. (1994). Reintroducing group selection to the human behavioral sciences. *The Behavioral and Brain Sciences, 17*, 585–608.

Wyss, T., Masclaux, F. G., Rosikiewicz, P., et al. (2016). Population genomics reveals that within-fungus polymorphism is common and maintained in populations of the mycorrhizal fungus Rhizophagus irregularis. *The ISME Journal, 10*, 2514–2526.

Youssoufian, H., & Pyeritz, R. E. (2002). Mechanisms and consequences of somatic mosaicism in humans. *Nature Reviews Genetics, 3*, 748–758.

Zarraonaindia, I., Smith, D., & Gilbert, J. (2013). Beyond the genome: Community-level analysis of the microbial world. *Biology and Philosophy, 28*, 261–282. https://doi.org/10.1007/s10539-012-9357-8.

Chapter 11
Biodiversity, Disparity and Evolvability

Alessandro Minelli

Abstract A key problem in conservation biology is how to measure biological diversity. Taxic diversity (the number of species in a community or in a local biota) is not necessarily the most important aspect, if what most matters is to evaluate how the loss of the different species may impact on the future of the surviving species and communities. Alternative approaches focus on functional diversity (a measure of the distribution of the species among the different 'jobs' in the ecosystem), others on morphological disparity, still others on phylogenetic diversity. There are three major reasons to prioritize the survival of species which provide the largest contributions to the overall phylogenetic diversity. First, evolutionarily isolated lineages are frequently characterized by unique traits. Second, conserving phylogenetically diverse sets of taxa is valuable because it conserves some sort of trait diversity, itself important in so far as it helps maintain ecosystem functioning, although a strict relationships between phylogenetic diversity and functional diversity cannot be taken for granted. Third, in this way we maximize the "evolutionary potential" depending on the *evolvability* of the survivors. This suggests an approach to conservation problems focussed on evolvability, robustness and phenotypic plasticity of developmental systems in the face of natural selection: in other terms, an approach based on evolutionary developmental biology.

Keywords Evolvability · Functional diversity · Morphological disparity · Phenotypic plasticity · Phylogenetic diversity

11.1 A Concern for Biodiversity: Evolution's Products at Risk

A key problem in conservation biology is how to measure the biological diversity at risk of loss, or already lost at the global scale or in a given area or habitat.

A. Minelli (✉)
Department of Biology, University of Padova, Padova, Italy
e-mail: alessandro.minelli@unipd.it

E. Casetta et al. (eds.), *From Assessing to Conserving Biodiversity*,
History, Philosophy and Theory of the Life Sciences 24,
https://doi.org/10.1007/978-3-030-10991-2_11

The origin of the concept of biodiversity from within ecology (Wilson 1988) explains why biological diversity is primarily described and measured in terms of the number of species in a community, or in a local biota. However, when describing the ongoing extinction of the Anthropocene, the total number of species involved is not necessarily the most important issue. What matters in the end is what has been lost or may be lost with them, and how this loss may impact on the future of both species and communities.

What we eventually prioritize is often heavily biased by our emotional preference for a few kinds of organisms and also for selected areas or habitats. Vertebrates are given much more attention than nematodes, big cats quite more than rodents. Whole biotas of particular sites are cause of special concern, for example those of remote oceanic islands like the Hawaii or the Galápagos. In other instances, individual species become the target of dedicated conservation efforts because of the peculiar role they play in the ecosystem, for example (in the case of bees and other insects) as pollinators or (in the case of corals) as builders of reefs on which the existence of many other marine species depends. Other reasons for identifying a species as worth of special conservation effort are less obvious and perhaps, *prima facie*, just academic or antiquarian – as for example, when we decide that a given species is worth of special conservation effort only because it is the only member or the last survivor of a peculiar evolutionary lineage.

11.1.1 Beyond Species Number

The latter example deserves closer scrutiny. Low species number is not necessarily a sign of scarce success of the whole lineage, or of impending risk of extinction: in several instances, it is a consequence of ecological marginality, that is, of adaptation to infrequent habitats, or to habitats confined to very small corners of the planet's surface. Ricklefs (2005) demonstrated that this is indeed the case for a number of tribes and even families of passerine birds. In this huge group (about 6000 extant species), one to five species belonging to each of these small subgroups are morphologically quite unusual and this correlates with their adaptation to marginal habitats. For example, these birds have unusually long legs and elongated bills facilitating feeding when the birds are perching or forage on hard substrates such as bark or rock. Within each of these small groups, genetic distances among the few extant species are generally large, indicating old divergence and, by inference, low speciation rate – opposite to the trend prevailing in species-rich, successful groups inhabiting more widely distributed habitats. The fact that these groups are still around in spite of a low speciation rate suggests that in these groups also the rate of extinction is lower than the average.

Clearly, estimates of biological diversity limited to counting species number in a community or in the fauna or flora of a given area fail to capture all the information we need to obtain a satisfactory assessment of possible criticalities and, as a consequence, to formulate sensible conservation measures that might be eventually adopted.

Indeed, a number of metrics of biological diversity have been proposed (summarized in Erwin 2008) other than those that measure just *taxic diversity*, i.e. the number of species or of lower (e.g., subspecies) or higher (e.g., genera or families) classificatory units. Some alternative approaches focus instead on *functional diversity* (a measure of the distribution of the species among the different 'jobs' in the ecosystem), others on *morphological disparity*, still others on *phylogenetic diversity*. All these approaches (which should be intended as complementary rather than alternative, although their usefulness is likely to be uneven in different instances) result in estimates of biodiversity to which all species in the community or biota contribute, although not necessarily to the same extent. In the next paragraphs I will briefly comment on these metrics, before moving to a less conventional approach to conservation problems, largely based on intrinsic properties of the individual species.

11.1.2 Disparity vs. Diversity

In terms of species number, birds are more diverse than mammals (some 10,000 vs. ca. 5600 extant species worldwide), but are instead quite more uniform in terms of morphology, reproductive biology and developmental schedules. Even including less conventional kinds such as the flightless penguins and ratites (ostrich and relatives), the range of bird structural types is much narrower than the range of structural types of mammals, which include humans and whales, bats and giraffes, moles and armadillos. All birds are oviparous, whereas in mammals there are a handful of oviparous species alongside a vast majority of viviparous species. Among the latter, some, like kangaroos, are borne at a developmental stage that deserves be called a larva, whereas others develop in their mother's wombs up to a much more advanced stage and are often capable to feed for themselves in the very day in which they are born. In technical terms, the *disparity* of mammals is much higher than the disparity of birds.

The choice of characters we can consider to evaluate a group's disparity is arbitrary, but morphological traits are usually given priority, often exclusive (Foote 1997; McGhee 1999; Wills 2001; Erwin 2007), because these aspects are the most readily accessible to quantification. As noted by Gerber et al. (2008), the concept of morphological disparity (Gould 1989, 1991; Foote 1997) has proved to be an invaluable source of information, both in palaeontology (e.g., Foote 1993, 1995, 1997; Wills et al. 1994; Dommergues et al. 1996; Roy and Foote 1997) and in the study of extant organisms (e.g., Ricklefs and Miles 1994; Ricklefs 2005).

In some lineages, the level of disparity goes together with the success as measured in terms of species diversity. Examples are some huge animal and plant genera among whose representatives disparate body plans or life styles have evolved in a relatively short time. Examples include *Megaselia* (a genus of phorid flies of which 1559 species have been described, but these are probably a minor subset of those existing on Earth, and their morphological and ecological disparity are enormous);

among the flowering plants, genera combining high diversity and high disparity include *Euphorbia* (2150 species, ranging from tiny herbs to quite large trees, and also including a number of succulents) and *Lobelia* (417 species, among which are small herbaceous plants alongside woody giants). However, there are also many large animal and plant genera within which the morphological differences are minor (low disparity), and vice versa (Minelli 2016). An example of low diversity combined with high disparity is the phylum Ctenophora, with 165 species described thus far, classified in 27 families, ten of which include only one species each.

11.1.3   Functional Diversity

By measuring disparity rather than simply counting species, we move a step in the direction of acknowledging the different functional roles the individual species play in respect to their biotic and abiotic environment. This aspects has been addressed in a more direct way by a number of approaches to biodiversity which try to capture the so-called *functional diversity*, the component of diversity that influences ecosystem dynamics, stability, productivity, nutrient balance, and other aspect of ecosystem functioning (Tilman 2001) through targeted descriptors and the calculation of corresponding indices (e.g., Mason et al. 2005; Bremner 2008; Villéger et al. 2008; Laliberté and Legendre 2010; Schleuter et al. 2010; Mouillot et al. 2013; Gagic et al. 2015; Gusmao-Junior and Lana 2015). Estimates of functional diversity are based, for example, on the number and kinds of trophic groups (e.g., primary producers, primary consumers, predators, parasites) and the number and relative abundance of species belonging to each group.

11.1.4   Phylogeny vs. Function

Phylogeny, and evolution at large, feature prominently in assessments of biological diversity and disparity, but it is not always obvious why. Of course, evolution is responsible both for the origin of the species whose number is the most popular measure of biodiversity, and for their structural and functional disparity, the two aspects mirroring the two main facets of evolutionary process – the splitting of lineages (cladogenesis) and the steady changes accumulating along each lineage (anagenesis), respectively. However, the frequent focus on phylogenetic diversity as an estimate of biodiversity and a criterion for ranking species to establish conservation priorities (Buckley 2016), deserves some explanation.

Several algorithms have been proposed to calculate phylogenetic diversity, based on the cladistic relations among the taxa (more often species, but also infraspecific units) (e.g., Vanewright et al. 1991; Faith 1992; Crozier 1997; Moritz 2002; Tucker et al. 2017).

Following the work of Vane-Wright et al. (1991), Faith (1992) and later authors (summarized in Mazel et al. 2017), there are three major reasons to prioritize the survival of species representing species-poor lineages only distantly related to the others in the sample and thus providing the largest contributions to the overall phylogenetic diversity.

First, in this way we conserve the greatest amount of evolutionary history (Vane-Wright et al. 1991), an ill-defined concept at the core of which, however, there is a sensible notion: evolutionarily isolated lineages, often represented by only one or very few extant species, are frequently characterized by unique traits, such as the two continuously growing leaves of *Welwitschia* or the egg-laying habit of the monotremes, strongly contrasting with the viviparity of all other mammals (Rosauer and Mooers 2013).

Second, conserving phylogenetically diverse sets of taxa is valuable because it conserves some sort of trait diversity (e.g., Mazel et al. 2017) This is the most popular among the arguments advocated in favour of using phylogenetic diversity as a basis on which to determine priorities for conservation. Trait diversity is considered important in so far as it helps maintain ecosystem functioning (e.g., Cadotte et al. 2008; Best et al. 2013; Winter et al. 2013; Gross et al. 2017).

Unfortunately, a strict relationships between phylogenetic diversity and functional diversity cannot be taken for granted. Through an elegant set of mathematical simulations, Mazel et al. (2017) were able to demonstrate that basing on estimates of phylogenetic diversity a strategy for conserving functional diversity is not necessarily a good strategy: the relationships between these two aspects of diversity depend on the shape of the tree depicting the phylogenetic relationships among the species involved and also on the model according to which their traits evolve across the generations. Therefore, generalizations are unwarranted. Still worse, Mazel et al. (2017) found that under plausible scenarios of evolution and ecology, prioritizing species conservation based on phylogenetic diversity can actually lead to levels of functional diversity lower than those obtained by conservation priorities determined by a random listing of the same species.

The third reason often advocated to explain why conservation priorities should be based on phylogenetic diversity is that in this way we maximize the "evolutionary potential" of the surviving biota (Faith 1992; Forest et al. 2007). As explained in a later section of this article, this vague term acquires a precise meaning and content when approached from the point of view of evolutionary developmental biology.

11.1.5 Antiquarian Sensibility

We value some human artefacts because of their current usefulness or at least because of the aesthetic pleasure we obtain by looking at them; but we also value other artefacts, even if devoid of any practical use and aesthetic qualities, simply because of their age. Museums are full of nondescript pieces of metal, bone or stone, witnesses of the human presence in particular sites at particular and often remote times, and of the cultural evolution of our ancestors. Similarly, a plant or animal lineage is often regarded as one of singular conservation value just because of the very long time since it split away from its closest living relatives. Monotremes (of which the platypus and the echidnas are the only living representatives) are an obvious example: the last common ancestor they share with the other living mammals lived between 162.9 and 191.1 million years ago (dos Reis et al. 2012). This circumstance, together with the strong unbalance in species richness (5 species only in the monotremes, compared to more than thousand times as many in the sister branch–the Theria, that is marsupials plus placentals) provides a good argument for regarding the platypus and the echidnas as a group of mammals we should not risk to bring to extinction. Another example is the tuatara, a reptile quite similar to a large lizard, but anatomically peculiar enough to deserve being placed in a distinct order, the Rhynchocephalia, of which it is the only survivor, confined to about 30 small islands off the North Island of New Zealand. This group separated from the Squamata (lizards, snakes and allies) about 228 million years ago or earlier (Hipsley et al. 2009).

Sometimes we realize too late the amount of history that is cancelled with the extinction of the last survivor(s) of a plant or animal group. This happened for example with the nesophontids, small mammals of which eight different species inhabited Cuba, Hispaniola, Puerto Rico and the Cayman Islands until their recent extinction, probably caused by the introduction of black rats by European sailors ca. 500 years ago. The nesophontids are classified with the insectivores and their closest relatives are the solenodontids, also confined in the Caribbean area. The two living species of the latter family, however, are poor substitutes for the loss of the nesophontids, not simply because they are themselves on the verge of extinction, but especially because the split between the two families (Nesophontidae and Solenodontidae) is very old, more than 50 million years (Brace et al. 2016) – longer, for example, than the age of the split between the New Worlds monkeys (the platyrrhines) and the Old World monkeys, including apes and humans (the catarrhines), and broadly the same age as the split between the ruminants and the lineage including hippos, whales and dolphins (O'Leary et al. 2013).

11.2   Conserving Evolutionary Processes

As noted by Buckley (2016), "by conserving genetic or phylogenetic diversity, we are facilitating the ability of lineages to adapt to future environmental changes." Since the early times of what eventually became conservation biology, far-seeing scientists have remarked that strategies for the long-term survival of wild species must take into account the continuing evolution of populations: as a consequence, policies should be based on adequate understanding of the population-genetics principles of conservation (Frankel 1974) about which quite little was known at the time. Twenty years later, progress in this direction was still insufficient, witness the plea of Smith et al. (1993, p. 164) who stressed that "If we are to conserve biodiversity the ecological and evolutionary mechanisms generating genetic diversity and the isolating mechanisms critical for speciation must also be preserved." Things have not changed much in the following years, and Moritz (2002) still lamented that "Less progress has been made on how to prioritize habitats, species, or populations in relation to persistence, that is, ensuring that the processes that sustain current and future diversity are protected."

Some authors (e.g., Gillson 2015) have remarked the paradox of conservation: we seek to preserve systems that are incessantly in flux, because of a number of processes running at different spatial and temporal scale, partly driven by extrinsic factors such as climate change and human disturbancy, partly expressing the organisms' evolutionary dynamics that would be innatural to contrast, if ever it would prove possible. Thus, if we can try to contrast the current biodiversity crisis by limiting the human impact on the environment, and even try to reduce, at least in some areas, the disruptive effect of rapid climate change, we may better help the survival of living species and lineages by devising conservation policies based on a sound understanding of *evolvability*.

11.3   Evo-Devo: Evolvability, Robustness, Plasticity

What is evolvability? Unfortunately, this is one of those technical terms for which too many definitions have been proposed (Pigliucci 2008; Brookfield 2009; Minelli 2017). Most of these, however, agree on regarding evolvability as the ability of populations to generate heritable phenotypic variation (Brigandt 2007; Kirschner and Gerhart 1998; G. P. Wagner and Altenberg 1996), but some are quite more specific, e.g. in focussing on the capacity to evolve new adaptations (Bedau and Packard 1992). Eventually, I prefer the definition proposed by Masel and Trotter (2010, p. 406), according to which evolvability is "the capacity of a population to produce heritable phenotypic variation of a kind that is not unconditionally deleterious. This definition includes both evolution from standing variation and the ability of the population to produce new variants."

According to Hendrikse et al. (2007), focussing on evolvability is the most characteristic feature of the research programme of *evolutionary developmental biology* (also called evo-devo). This young branch of the life sciences has much to offer to conservation (Campbell et al. 2017). Up to now, conservation efforts based on the preservation of genetic variation have followed the approach to intraspecific diversity characteristic of population genetics. But this is too limiting: as remarked long ago by Waddington (1957), the expression of genetic variation is structured by development. And this is exactly where evo-devo operates, in a systematic effort to unravel the complex relationships between genotype and phenotype (the so-called genotype→phenotype map; cf. Alberch 1991; Wagner and Altenberg 1996; Pigliucci 2001; West-Eberhard 2003; Draghi and Wagner 2008).

Indeed, to understand evolvability, we must acknowledge that the path leading from genotype to phenotype is complex and not necessarily predictable (Minelli

2017). On the one hand, due to environmental influences but also to stochastic impredictability, different phenotypes can be produced by developing organisms that share identical genotypes; reciprocally, identical phenotypes can be produced by developmental systems with different genotypes. Elaborating upon Waddington's insight, students of evo-devo have demonstrated that the expression of genetic variation is largely dependent on the structure and *robustness* of the developing system (Hansen 2006; Kirschner and Gerhart 1998; Wagner 2005; Wagner and Altenberg 1996).

Together with evolvability, robustness plays a central role in evolutionary developmental biology. The robustness of a phenotypic trait can be operationally defined as the absence of variation in the face of environmental or genetic perturbations (Félix and Barkoulas 2015). According to some authors, robustness constrains and contrasts evolvability, with negative effects on biodiversity: the rationale is that, if mutations and environmental changes have little effect, there is not much variation on which selection can act. Others (e.g., Kitano 2004; Wagner 2008; Masel and Trotter 2010; Melzer and Theißen 2016; Theißen and Melzer 2016) regard this view as simplistic and even contend that robustness may promote evolvability, i.e. the ability to produce heritable phenotypic variation (Pigliucci 2008). To explain how, we must first distinguish between two aspects of robustness, genetic vs. developmental.

Genetic robustness is "robustness to perturbations both in the form of new mutations and in the form of the creation of new combinations of existing alleles by recombination" (Masel and Trotter 2010, p. 407), without visible effects on the phenotype. In this way, in the absence of exposure to novel selective challenges, populations accumulate genetic diversity on the base of which they gain easier access to a greater range of novel genotypes, some of which may eventually prove to be advantageous (A. Wagner 2005, 2008, 2011).

A similar relationships between robustness and evolvability is found in the case of developmental robustness (also known as *canalization*, a term coined by Waddington (1942)) that is, the production of the same phenotype irrespective of genetic differences (and external perturbations). This also corresponds to the fact that populations harbour amounts of unexpressed genetic variation (*cryptic genetic variation*) that is not expressed in the phenotype unless revealed by environmental change or by modification in the overall genetic background (e.g., Badyaev 2005; Flatt 2005; Gibson and Dworkin 2004; Moczek 2007; Rieseberg et al. 2003; Schlichting 2008). This cryptic variation represents a standing potential for evolvability. Exposure to novel selective pressures can be dramatically accelerated by the human impact on the environment. In other terms, environmental change does not just alter the selective regime to which a population is exposed, but can also induce novel developmental responses even in the absence of genetic change. This property of the genotype→phenotype map is known as *phenotypic plasticity* (Fusco and Minelli 2010).

Phenotypic plasticity should not be regarded as an alternative to natural selection. On the one hand, the emergence of a novel phenotype by plasticity, following exposure to previously unexperienced environmental conditions creates a new target on which selection will operate; on the other, plasticity itself is subjected to selection, being favoured in fluctuating environments (Price et al. 2003; West-Eberhard 2003; Pfennig et al. 2010). As noted by Campbell et al. (2017), this is a situation likely experienced by a population introduced in a new area or living in habitats fragmented or otherwise damaged by man.

One might argue that phenotypic plasticity, although responsible for the emergence of new phenotypes, offers no warranty of their conservation, on the long term at least. But this would be a short-sighted perspective. An environmentally controlled phenotype can eventually fall under strict genetic control. The functional divide to be crossed is sometimes a very narrow one, as demonstrated by the control of wing development in the pea aphid, *Acyrtosiphum pisum*. In this little insect, some adults (males as well as females) are winged, while the others are wingless. In the male sex, the coexistence of these two alternative phenotypes is under genetic control, while in the females wing development is controlled by the exposure to different day-lengths in a critical phase of development. In technical terms, males exhibit genetic polymorphism for this trait, while females exhibit an environmentally controlled polyphenism, ie the outcome of phenotypic plasticity. This contrast, however, rests on minor mechanistic differences, because the developmental effect of day-length on wing development in the females is mediated by the gene product of the same gene whose alternative alleles are responsible for the wing polymorphism in the male (Braendle et al. 2005a, b). This circumstance suggests how easily a polyphenism can evolve into a genetic polymorphism, eventually allowing long-term conservation of phenotypes.

11.4   A Lesson from Past Mass Extinctions?

Irrespective of the different causes involved in these events, the mass extinctions of the past should be studied very carefully by researchers interested in conservation biology, but not so much to analyze the differential tribute paid by organisms belonging to different lineages, as to look for any possible explanation of the differential success of the survivors in the post-crisis recovery. Palaeobiologists have generally focussed on the ecological determinants of this process; that is, they have regarded the ecological space left empty by extinctions as the main determinant of the renewed occupation of morphospace. To some extent, the morphological disparity often expanded into dimensions other than those that were occupied prior to the mass extinction. However, no really new body plan emerged. This was, in a sense, a large-scale test demonstrating the developmental robustness of the main traits of body architecture of the survivors, the innovations being limited to secondary, evolutionarily plastic aspects (Erwin 2008).

Confronted with this (admittedly, only incompletely documented) evidence, it seems legitimate to rethink the evolutionary criteria in the light of which biodiversity and its ongoing loss are currently evaluated. It is hard to imagine a positive correlation between the phylogenetic relationships among the survivors and the possible outcome of their long-term evolution in a post-crisis recovery. Million years ago, by preferring to save a marsupial and a placental mammal rather than two placentals, because of the larger phylogenetic distance between the first two, we would not have been able to predict that at least two different subterranean lineages would have eventually evolved in any case: today there are indeed marsupial moles (*Notoryctes*) among the marsupials and moles (*Talpa*) and mole-rats (*Spalax*) among the placentals. More than because of the history of their lineage, survivors may be differentially important for the future of biodiversity as a function of their intrinsic qualities, particularly those expressed by the parameters on which evo-devo focuses – as said, robustness and evolvability.

Acknowledgements I am grateful to Elena Casetta and Davide Vecchi for their invitation to contribute to this volume and for their precious comments on a first draft text.

References

Alberch, P. (1991). From genes to phenotype: Dynamical systems and evolvability. *Genetica, 84*, 5–11.

Badyaev, A. V. (2005). Stress-induced variation in evolution: From behavioural plasticity to genetic assimilation. *Proceedings of the Royal Society of London. Series B, Biological Sciences, 272*, 877–886. https://doi.org/10.1098/rspb.2004.3045.

Bedau, M., & Packard, N. (1992). Measurement of evolutionary activity, teleology, and life. In C. Langton, C. Taylor, J. D. Farmer, & S. Rasmussen (Eds.), *Artificial life II: Proceedings of the workshop on artificial life* (pp. 431–462). Redwood City: Addison-Wesley.

Best, R. J., Caulk, N. C., & Stachowicz, J. J. (2013). Trait vs. phylogenetic diversity as predictors of competition and community composition in herbivorous marine amphipods. *Ecology Letters, 16*, 72–80. https://doi.org/10.1111/ele.12016.

Brace, S., Thomas, J. A., Dalén, L., Burger, J., MacPhee, R. D. E., Barnes, I., & Turvey, S. T. (2016). Evolutionary history of the Nesophontidae, the last unplaced recent mammal family. *Molecular Biology and Evolution, 33*, 3095–3103. https://doi.org/10.1093/molbev/msw186.

Braendle, C., Caillaud, M. C., & Stern, D. L. (2005a). Genetic mapping of aphicarus: A sex- linked locus controlling a wing polymorphism in the pea aphid (*Acyrthosiphon pisum*). *Heredity, 94*, 435–442. https://doi.org/10.1038/sj.hdy.6800633.

Braendle, C., Friebe, I., Caillaud, M. C., & Stern, D. L. (2005b). Genetic variation for an aphid wing polyphenism is genetically linked to a naturally occurring wing polymorphism. *Proceedings of the Royal Society B: Biological Sciences, 272*, 657–664. https://doi.org/10.1098/rspb.2004.2995.

Bremner, J. (2008). Species' traits and ecological functioning in marine conservation and management. *Journal of Experimental Marine Biology and Ecology, 366*, 37–47. https://doi.org/10.1016/j.jembe.2008.07.007.

Brigandt, I. (2007). Typology now: Homology and developmental constraints explain evolvability. *Biology and Philosophy, 22*, 709–725. https://doi.org/10.1007/s10539-007-9089-3.

Brookfield, J. F. Y. (2009). Evolution and evolvability: Celebrating Darwin 200. *Biology Letters, 5*, 44–46. https://doi.org/10.1098/rsbl.2008.0639.

Buckley, T. R. (2016). Applications of phylogenetics to solve practical problems in insect conservation. *Current Opinion in Insect Science, 18*, 35–39. https://doi.org/10.1016/j.cois.2016.09.005.

Cadotte, M. W., Cardinale, B. J., & Oakley, T. H. (2008). Evolutionary history and the effect of biodiversity on plant productivity. *Proceedings of the National Academy of Sciences of the United States of America, 105,* 17012–17017. https://doi.org/10.1073/pnas.0805962105.

Campbell, C. S., Adams, C. E., Bean, C. W., & Parsons, K. J. (2017). Conservation evo-devo: Preserving biodiversity by understanding its origins. *Trends in Ecology and Evolution, 32,* 746–759. https://doi.org/10.1016/j.tree.2017.07.002.

Crozier, R. H. (1997). Preserving the information content of species: Genetic diversity, phylogeny, and conservation worth. *Annual Review of Ecology and Systematics, 28,* 243–268. https://doi.org/10.1146/annurev.ecolsys.28.1.243.

Dommergues, J.-L., Laurin, B., & Meister, C. (1996). Evolution of ammonoid morphospace during the early Jurassic radiation. *Paleobiology, 22,* 219–240. https://doi.org/10.1017/S0094837300016183.

dos Reis, M., Inoue, J., Hasegawa, M., Asher, R. J., Donoghue, P. C. J., & Yang, Z. (2012). Phylogenomic datasets provide both precision and accuracy in estimating the timescale of placental mammal phylogeny. *Proceedings of the Royal Society of London. Series B, Biological Sciences, 279,* 3491–3500. https://doi.org/10.1098/rspb.2012.0683.

Draghi, J., & Wagner, G. P. (2008). Evolution of evolvability in a developmental model. *Evolution, 62,* 301–315. https://doi.org/10.1111/j.1558-5646.2007.00303.x.

Erwin, D. H. (2007). Disparity: Morphologic pattern and developmental context. *Palaeontology, 50,* 57–73. https://doi.org/10.1111/j.1475-4983.2006.00614.x.

Erwin, D. H. (2008). Extinction as the loss of evolutionary history. *Proceedings of the National Academy of Sciences of the United States of America, 105*(Suppl 1), 11520–11527. https://doi.org/10.1073/pnas.0801913105.

Faith, D. P. (1992). Conservation evaluation and phylogenetic diversity. *Biological Conservation, 61,* 1–10. https://doi.org/10.1016/0006-3207(92)91201-3.

Félix, M.-A., & Barkoulas, M. (2015). Pervasive robustness in biological systems. *Nature Reviews Genetics, 16,* 483–496. https://doi.org/10.1038/nrg3949.

Flatt, T. (2005). The evolutionary genetics of canalization. *Quarterly Review of Biology, 80,* 287–316. https://doi.org/10.1086/432265.

Foote, M. (1993). Discordance and concordance between morphological and taxonomic diversity. *Paleobiology, 19,* 185–204. https://doi.org/10.1017/S0094837300015864.

Foote, M. (1995). Morphological diversification of Paleozoic crinoids. *Paleobiology, 21,* 273–299. https://www.jstor.org/stable/2401167.

Foote, M. (1997). The evolution of morphological diversity. *Annual Review of Ecology and Systematics, 28,* 129–152. https://doi.org/10.1146/annurev.ecolsys.28.1.129.

Forest, F., Grenyer, R., Rouget, M., Davies, T. J., Cowling, R. M., Faith, D. P., Balmford, A., Manning, J. C., Procheş, Ş., van der Bank, M., Reeves, G., Hedderson, T. A. J., & Savolainen, V. (2007). Preserving the evolutionary potential of floras in biodiversity hotspots. *Nature, 445,* 757–760. https://doi.org/10.1038/nature05587.

Frankel, O. H. (1974). Genetic conservation: Our evolutionary responsibility. *Genetics, 78,* 53–65.

Fusco, G., & Minelli, A. (2010). Phenotypic plasticity in development and evolution: Facts and concepts. *Philosophical Transactions of the Royal Society of London. Series B, Biological Sciences, 365,* 547–556. https://doi.org/10.1098/rstb.2009.0267.

Gagic, V., Bartomeus, I., Jonsson, T., Taylor, A., Winqvist, C., Fischer, C., Slade, E. M., Steffan-Dewenter, I., Emmerson, M., Potts, S. G., Tscharntke, T., Weisser, W., & Bommarco, R. (2015). Functional identity and diversity of animals predict ecosystem functioning better than species-based indices. *Proceedings of the Royal Society of London. Series B, Biological Sciences, 282,* 20142620. https://doi.org/10.1098/rspb.2014.2620.

Gerber, S., Eble, G. J., & Neige, P. (2008). Allometric space and allometric disparity: A developmental perspective in the macroevolutionary analysis of morphological disparity. *Evolution, 62,* 1450–1457. https://doi.org/10.1111/j.1558-5646.2008.00370.x.

Gibson, G., & Dworkin, I. (2004). Uncovering cryptic genetic variation. *Nature Reviews Genetics, 5,* 681–691. https://doi.org/10.1038/nrg1426.

Gillson, L. (2015). *Biodiversity conservation and environmental change: Using palaeoecology to manage dynamic landscapes in the Anthropocene*. Oxford: Oxford University Press.

Gould, S. J. (1989). *Wonderful life*. New York: Norton.

Gould, S. J. (1991). The disparity of the burgess shale arthropod fauna and the limits of cladistic analysis: Why we must strive to quantify morphospace. *Paleobiology, 17*, 411–423. https://www.jstor.org/stable/2400754.

Gross, N., Le Bagousse-Pinguet, Y., Liancourt, P., Berdugo, M., Gotell, N. J., & Maestre, F. T. (2017). Functional trait diversity maximizes ecosystem multifunctionality. *Nature Ecology & Evolution, 1*, 132. https://doi.org/10.1038/s41559-017-0132.

Gusmao-Junior, J. B. L., & Lana, P. C. (2015). Spatial variability of the infauna adjacent to intertidal rocky shores in a subtropical estuary. *Hydrobiologia, 743*, 53–64. https://doi.org/10.1007/s10750-014-2004-4.

Hansen, T. F. (2006). The evolution of genetic architecture. *Annual Review of Ecology and Systematics, 37*, 123–157. https://doi.org/10.1146/annurev.ecolsys.37.091305.110224.

Hendrikse, J. L., Parsons, T. E., & Hallgrímsson, B. (2007). Evolvability as the proper focus of evolutionary developmental biology. *Evolution & Development, 9*, 393–401. https://doi.org/10.1111/j.1525-142X.2007.00176.x.

Hipsley, C. A., Himmelmann, L., Metzler, D., & Mueller, J. (2009). Integration of Bayesian molecular clock methods and fossil-based soft bounds reveals Early Cenozoic origin of African lacertid lizards. *BMC Evolutionary Biology, 9*, 1–13. https://doi.org/10.1186/1471-2148-9-151.

Kirschner, M., & Gerhart, J. (1998). Evolvability. *Proceedings of the National Academy of Sciences of the United States of America, 95*, 8420–8427. https://doi.org/10.1073/pnas.95.15.8420.

Kitano, H. (2004). Biological robustness. *Nature Reviews Genetics, 5*, 826–837. https://doi.org/10.1038/nrg1471.

Laliberté, E., & Legendre, P. (2010). A distance-based framework for measuring functional diversity from multiple traits. *Ecology, 91*, 299–305. https://doi.org/10.1890/08-2244.1.

Masel, J., & Trotter, M. V. (2010). Robustness and evolvability. *Trends in Genetics, 26*, 406–414. https://doi.org/10.1016/j.tig.2010.06.002.

Mason, N. W. H., Mouillot, D., Lee, W. G., & Wilson, J. B. (2005). Functional richness, functional evenness and functional divergence: The primary components of functional diversity. *Oikos, 111*, 112–118. https://doi.org/10.1111/j.0030-1299.2005.13886.x.

Mazel, F., Mooers, A. O., Dalla Riva, V., & Pennell, M. V. (2017). Conserving phylogenetic diversity can be a poor strategy for conserving functional diversity. *Systematic Biology, 66*, 1019–1027. https://doi.org/10.1093/sysbio/syx054.

McGhee, G. R. (1999). *Theoretical morphology: The concept and its applications*. New York: Columbia University Press.

Melzer, R., & Theißen, G. (2016). The significance of developmental robustness for species diversity. *Annals of Botany, 117*, 725–732. https://doi.org/10.1093/aob/mcw018.

Minelli, A. (2016). Species diversity vs. morphological disparity in the light of evolutionary developmental biology. *Annals of Botany, 117*, 795–809. https://doi.org/10.1093/aob/mcv134.

Minelli, A. (2017). Evolvability and its evolvability. In P. Huneman & D. Walsh (Eds.), *Challenges to evolutionary theory: Development, inheritance and adaptation* (pp. 211–238). New York: Oxford University Press.

Moczek, A. P. (2007). Developmental capacitance, genetic accommodation, and adaptive evolution. *Evolution & Development, 9*, 299–305. https://doi.org/10.1111/j.1525-142X.2007.00162.x.

Moritz, C. (2002). Strategies to protect biological diversity and the evolutionary processes that sustain it. *Systematic Biology, 51*, 238–254. https://doi.org/10.1080/10635150252899752.

Mouillot, D., Graham, N. A. J., Villéger, S., Mason, N. W. H., & Bellwood, D. R. (2013). A functional approach reveals community responses to disturbances. *Trends in Ecology and Evolution, 28*, 167–177. https://doi.org/10.1016/j.tree.2012.10.004.

O'Leary, M. A., Bloch, J. I., Flynn, J. J., Gaudin, T. J., Giallombardo, A., Giannini, N. P., Goldberg, S. L., Kraatz, B. P., Luo, Z. X., Meng, J., Ni, X., Novacek, M. J., Perini, F. A., Randall, Z. S., Rougier, G. W., Sargis, E. J., Silcox, M. T., Simmons, N. B., Spaulding, M., Velazco,

P. M., Weksler, M., Wible, J. R., & Cirranello, A. L. (2013). The placental mammal ancestor and the post-K-Pg radiation of placentals. *Science, 339*, 662–667. https://doi.org/10.1126/science.1229237.

Pfennig, D. W., Wund, M. A., Snell-Rood, E. C., Cruickshank, T., Schlichting, C. D., & Moczek, A. P. (2010). Phenotypic plasticity's impacts on diversification and speciation. *Trends in Ecology and Evolution, 25*, 459–467. https://doi.org/10.1016/j.tree.2010.05.006.

Pigliucci, M. (2001). *Phenotypic plasticity: Beyond nature and nurture*. Baltimore: John Hopkins University Press.

Pigliucci, M. (2008). Is evolvability evolvable? *Nature Reviews Genetics, 9*, 75–82. https://doi.org/10.1038/nrg2278.

Price, T. D., Qvarnström, A., & Irwin, D. E. (2003). The role of phenotypic plasticity in driving genetic evolution. *Proceedings of the Royal Society of London. Series B, Biological Sciences, 270*, 1433–1440. https://doi.org/10.1098/rspb.2003.2372.

Ricklefs, R. E. (2005). Small clades at the periphery of passerine morphological space. *American Naturalist, 165*, 651–659. https://doi.org/10.1086/429676.

Ricklefs, R. E., & Miles, D. B. (1994). Ecological and evolutionary inferences from morphology: an ecological perspective. In P. C. Wainwright & S. M. Reilly (Eds.), *Ecological morphology* (pp. 13–41). Chicago/London: University of Chicago Press.

Rieseberg, L. H., Raymond, O., Rosenthal, D. M., Lai, Z., Livingstone, K., Nakazato, T., Durphy, J. L., Schwarzbach, A. E., Donovan, L. A., & Lexer, C. (2003). Major ecological transitions in wild sunflowers facilitated by hybridization. *Science, 301*, 1211–1216. https://doi.org/10.1126/science.1086949.

Rosauer, D. F., & Mooers, A. (2013). Nurturing the use of evolutionary diversity in nature conservation. *Trends in Ecology and Evolution, 28*, 322–323. https://doi.org/10.1016/j.tree.2013.01.014.

Roy, K., & Foote, M. (1997). Morphological approaches to measuring biodiversity. *Trends in Ecology and Evolution, 12*, 277–281.

Schleuter, D., Daufresne, M., Massol, F., & Argillier, C. (2010). A user's guide to functional diversity indices. *Ecological Monographs, 80*, 469–484. https://doi.org/10.1890/08-2225.1.

Schlichting, C. D. (2008). Hidden reaction norms, cryptic genetic variation, and evolvability. *Annals of the New York Academy of Sciences, 1133*, 187–203. https://doi.org/10.1196/annals.1438.010.

Smith, T. B., Bruford, M. W., & Wayne, R. K. (1993). The preservation of process: The missing element of conservation programs. *Biodiversity Letters, 1*, 164–167. https://www.jstor.org/stable/2999740

Theißen, G., & Melzer, R. (2016). Robust views on plasticity and biodiversity. *Annals of Botany, 117*, 693–697. https://doi.org/10.1093/aob/mcw066.

Tilman, D. (2001). Functional diversity. In S. A. Levin (Ed.), *Encyclopaedia of biodiversity* (pp. 109–120). San Diego: Academic.

Tucker, C. M., Cadotte, M. W., Carvalho, S. B., Davies, T. J., Ferrier, S., Fritz, S. A., Grenyer, R., Helmus, M. R., Jin, L. S., Mooers, A. O., Pavoine, S., Purschke, O., Redding, D. W., Rosauer, D. F., Winter, M., & Mazel, F. (2017). A guide to phylogenetic metrics for conservation, community ecology and macroecology. *Biological Reviews, 92*, 698–715. https://doi.org/10.1111/brv.12252.

Vane-Wright, R. I., Humphries, C. J., & Williams, P. H. (1991). What to protect – Systematics and the agony of choice. *Biological Conservation, 55*, 235–254. https://doi.org/10.1016/0006-3207(91)90030-D.

Villéger, S., Mason, N., & Mouillot, D. (2008). New multidimensional functional diversity indices for a multifaceted framework in functional ecology. *Ecology, 89*, 2290–2301.

Waddington, C. H. (1942). Canalization of development and the inheritance of acquired characters. *Nature, 150*, 563–565.

Waddington, C. H. (1957). *The strategy of the genes: A discussion of some aspects of theoretical biology*. London: Allen & Unwin.

Wagner, A. (2005). *Robustness and evolvability in living systems*. Princeton: Princeton University Press.

Wagner, A. (2008). Robustness and evolvability: A paradox resolved. *Proceedings of the Royal Society of London. Series B, Biological Sciences, 275*, 91–100. https://doi.org/10.1098/rspb.2007.1137.

Wagner, A. (2011). *The origins of evolutionary innovations: A theory of transformative change in living systems*. Oxford: Oxford University Press.

Wagner, G. P., & Altenberg, L. (1996). Complex adaptations and evolution of evolvability. *Evolution, 50*, 967–976. https://doi.org/10.1111/j.1558-5646.1996.tb02339.x.

West-Eberhard, M. J. (2003). *Developmental plasticity and evolution*. New York: Oxford University Press.

Wills, M. A. (2001). Morphological disparity: A primer. In J. M. Adrain, G. D. Edgecombe, & B. S. Lieberman (Eds.), *Fossils, phylogeny, and form* (pp. 55–144). New York: Kluwer/Plenum.

Wills, M. A., Briggs, D. E. G., & Fortey, R. A. (1994). Disparity as an evolutionary index: A comparison of Cambrian and recent arthropods. *Paleobiology, 20*, 93–130. https://doi.org/10.1017/S009483730001263X.

Wilson, E. O. (Ed.). (1988). *Biodiversity*. Washington, DC: National Academy Press.

Winter, M., Devictor, V., & Schweiger, O. (2013). Phylogenetic diversity and nature conservation: Where are we? *Trends in Ecology and Evolution, 28*, 199–204. https://doi.org/10.1016/j.tree.2012.10.015.

Chapter 12
Probing the Process-Based Approach to Biodiversity: Can Plasticity Lead to the Emergence of Novel Units of Biodiversity?

Davide Vecchi and Rob Mills

Abstract The history of biology has been characterised by a strong emphasis on the identification of entities (e.g., macromolecules, cells, organisms, species) as fundamental units of our classificatory system. The biological hierarchy can be divided into a series of compositional levels complementing the physical and chemical hierarchy. Given this state of affairs, it is not surprising that biodiversity studies have focused on a "holy trinity" of entities, namely genes, species and ecosystems. In this chapter, we endorse the view that a process-based approach should integrate an entity-based one. The rationale of our endorsement is that a focus on entities does not address whether biological processes have the capacity to create novel, salient units of biodiversity. This alternative focus might therefore have implications for conservation biology. In order to show the relevance of process-based approaches to biodiversity, in this chapter we shall focus on a particular process: phenotypic plasticity. Specifically, we shall describe a model of plasticity that might have implications for how we conceptualise biodiversity units. The hypothesis we want to test is whether plastic subpopulations that have enhanced evolutionary potential vis a vis non-plastic subpopulations make them amenable to evolutionarily significant units (i.e., ESU) status. An understanding of the mechanisms that influence organismic evolution, particularly when under environmental stress, may shed light on the natural "conservability" capacities of populations. We use an abstract computational model that couples plasticity and genetic mutation to investigate how plasticity processes (through the Baldwin effect) can improve the adaptability of a population

D. Vecchi (✉)
Centro de Filosofia das Ciências, Departamento de História e Filosofia das Ciências, Faculdade de Ciências, Universidade de Lisboa, Lisboa, Portugal
e-mail: dvecchi@fc.ul.pt

R. Mills
BioISI – Biosystems & Integrative Sciences Institute, Faculdade de Ciências, Universidade de Lisboa, Lisbon, Portugal
e-mail: rob.mills@fc.ul.pt

© The Author(s) 2019
E. Casetta et al. (eds.), *From Assessing to Conserving Biodiversity*, History, Philosophy and Theory of the Life Sciences 24, https://doi.org/10.1007/978-3-030-10991-2_12

when faced with novel environmental challenges. We find that there exist circumstances under which plasticity improves adaptability, where multi-locus fitness valleys exist that are uncrossable by non-plastic populations; and the differences in the capacity to adapt between plastic and non-plastic populations become drastic when the environment varies at a great enough rate. If plasticity such as learning provides not only within-lifetime environmental buffering, but also enhances a population's capacity to adapt to environmental changes, this would, on the one hand, vindicate a process-based approach to biodiversity and, on the other, it would suggest a need to take into account the processes generating plasticity when considering conservation efforts.

Keywords Biological hierarchy · Process-approach to conservation ·
Evolutionarily significant unit of conservation (ESU) · Plasticity · Baldwin effect ·
Evolutionary potential

12.1 Entity-Based and Process-Based Approaches Are Complementary

Biodiversity conservation poses a set of complex conceptual and practical challenges. The metaphor of healing and the analogy of nature as a patient serve the purpose of discriminating such challenges in three groups.[1] The state of nature as a patient must be diagnosed and the damage to biodiversity estimated in order to cure it via appropriate conservation actions. But in order to do so, we need to be able to take care of the patient in the right way. That is, we need to be able to characterise nature as a patient appropriately by identifying the relevant targets of treatment and conservation action. This contribution deals primarily with this conceptual challenge by arguing that proper treatment can be achieved not by exclusively targeting entities (i.e., units of biodiversity as conservation targets such as species or sites), but by concomitantly focusing on the processes generating such entities. In order to build our case, let us first explain why science requires endorsing a *complementary entity-and-process-based approach*.

Most people would identify biological entities such as organisms and species as paradigmatic. This intuitive knowledge (or folk biology) is often eventually refined, encompassing entities that can only be observed through microscopes such as cells and macromolecules. Biological knowledge is basically about what these entities do and how they develop or evolve. The focus on entities can be justified for at least two important reasons. First of all, nature can be thought in hierarchical terms as a series of part-whole relationships where the entities-wholes at a higher level of organisation are composed of entities-parts at a lower level. Secondly, the epistemological advantage of this compositional view is that an entity-based ontology can be

---

[1] The metaphor and analogy are developed in Casetta, Marques da Silva & Vecchi, Chap. 1, in this volume.

upheld despite ignorance of the processes leading to the generation (and, conversely, the decomposition) of wholes.[2]

The basic idea beneath entity-based classification is that nature is stratified into hierarchical levels of composition, i.e., "... *hierarchical divisions of stuff (paradigmatically, but not necessarily material stuff) organized by part-whole relations, in which wholes at one level function as parts at the next (and at all higher) levels*" (Wimsatt 2007, p. 201). Consider the physical hierarchy for example; the idea is that fundamental particles (e.g., fermions – quarks and leptons – and bosons) compose hadrons (neutrons and protons), which in turn compose atoms, which in turn compose molecules, which in turn compose all other solid, liquid and gaseous molecular aggregates that we can directly observe at the mesoscopic scale. Fundamental particles are the component parts of the hadron wholes; hadrons are the component parts of the atom wholes etc. Fundamental particles, hadrons, atoms etc. are entities belonging to different compositional levels. Biological hierarchies can be analogously divided into a series of compositional levels that complement the physical and chemical hierarchies: macromolecular wholes are composed of fundamental particles, hadrons, atoms, molecular parts; cellular wholes are composed of macromolecular parts; organismal wholes are composed of cellular parts etc. Thus, compositional hierarchies seem to identify "natural" and not purely human-dependent ontological components, even though the details of any ontology remain revisable in the light of scientific advances. The upshot of all this is that, if nature is indeed stratified into hierarchical compositional levels, then, on the one hand, all entities in the universe are ultimately composed of fundamental particles and, on the other, physical, chemical and biological hierarchies must be compositionally related. The corollary of this view is that compositionality implies some kind of physical reductionism because, on the one hand, everything is composed of basic physical stuff (e.g., quarks, leptons and bosons) and, on the other, the basic physical level of the hierarchy is primitive. This kind of compositional physicalism is unproblematic in many respects, even though this does not mean that chemistry and biology should straightforwardly be reduced to physics. One reason is that the explanation of the behaviour of chemical and biological systems might require reference to properties that are not ascribable to their physical components. Additionally, this kind of compositional physicalism rests on ontological fundamentalism, that is, the not so innocent assumption that we can make sense of the idea that a fundamental physical level exists at all (Schaffer 2003). This view implies the controversial hypothesis that quarks, leptons and bosons are not composed of anything at all, that they are "atoms" of composition in Democritus's sense. However, even though compositional physicalism is somehow obvious (because all entities are merely composed of physical stuff), it is at the same time epistemologically vacuous. Let us explain why by making reference exclusively to biology.

---

[2] Of course, there exist also processes leading to the decomposition of wholes. However, the striking feature of the history of life is that it is a history of "complexification", of generation of wholes. Thanks to Sandro Minelli for suggesting this clarification.

There are many potential reasons to argue that the compositional physicalism so far characterised provides an unsatisfactory account of biological entities. One argument is that biological entities are composed of non-physical components, a position that might be called vitalism. A somehow different argument states that biological entities possess properties that are lacked by their physical parts, a position that might be called emergentism. Here we shall focus on another kind of argument against compositional physicalism. Suppose we were to produce an inventory of all the relevant parts and their properties: could we infer the properties of wholes? No, because we would also need an understanding of the "rules of composition" governing the behaviour of parts in the production of wholes. An entity-based ontology cannot provide a satisfactory account of the properties of wholes without providing information about the nature of the processes governing the interactions between their parts. The basic point is that an entity-based ontology would provide limited knowledge of nature unless it is complemented with a process-based ontology. This is because of two reasons.

The first is that we cannot understand biological entities and their behaviour without knowledge about their structure and functional properties. Take a protein, a whole composed of a variety of amino acids with a number of physical (i.e., biophysical and biochemical) properties; the conformational properties of proteins are dependent on the properties of their component amino-acids but are not properties of these components; for instance, the specificity of proteins (i.e., their capacity to bind to a particular ligand) is given by their structure, where this structure is generated by the functional interaction of the polypeptide chains (themselves composed of amino acids) and the environment. The point is that knowledge of the physical properties of the macromolecular components of a protein is not enough in order to account for the structural and functional properties of proteins and thus to explain and predict their behaviour: composition does not account for structure and function. Thus, a static entity-based ontology consisting of an inventory of the compositional properties of parts does not exhaust the relevant biological properties of wholes.

The second reason is that the criteria for the identification of relevant biological entities must make reference to processes. For instance, consider organismal development: even if the embryo is very different from the adult capable of reproducing, they are the same biological individual. The individuality of an organism is not a property of its component parts (i.e., cells) but of the whole. Perhaps it would be better to say that it is not even a property but a process. Hennig (1966, p. 65) introduced the concept of semaphoront in order to make sense of this constrained organismal changeability. The semaphoront corresponds to the individual (e.g., a biological organism) in an infinitely small time span of its life history during which it remains unchanged. The same concept can probably be applied to any entity, physical, chemical or biological alike. But the semaphoront is a fiction, a conceptual device that should not be reified in order to vindicate a pure entity-based ontology. In this deep sense, it might be argued, biological entities should ultimately be thought in terms of dynamical processes: every supposedly static and unchangeable biological entity should be merely thought of as a portion of its life history (Dupré 2012).

At the same time, even a pure process-based ontology relinquishing any reference to entities would be epistemically useless: science also strives to classify types of entities and to uncover practically useful criteria for entity identification. This is clearly the case in biology: we want to be able to say that the embryo is the same individual as the adult, that this cell is and always will be eukaryotic during its life history, that this organism belongs to a particular species etc. Thus, in a very basic sense, an entity-based and a process-based ontology cannot but be complementary.

12.2 Entity-Based Approaches to Biodiversity Are Deficient

The intuitive allure and the epistemological advantages of entity-based compositional hierarchies are reflected in the literature on biodiversity. In fact, compositional hierarchies are prominent in biodiversity studies (Angermeier and Karr 1994, p. 691). For instance, taxonomic hierarchies stratify biological nature in terms of the part-whole relationship between species, genera, families, orders, classes, phyla, kingdoms, domains and superdomain. Ecological hierarchies stratify biological nature in terms of the part-whole relationship between populations, communities, ecosystems, landscapes, biomes and biosphere. Genetic hierarchies stratify biological nature in terms of the part-whole relationship between alleles, genes, chromosomes, genomes and pangenomes. All these hierarchies capture an aspect of biological nature (i.e., taxonomic inclusiveness, ecological nestedness, genetic organisation). To each level corresponds an entity type, i.e., a unit of biodiversity. A general problem with compositional hierarches of the above kind is that the biodiversity units are not immaculately characterisable, in the sense that sometimes it is difficult to recognise a biological entity as an entity of a certain type, as a certain biodiversity unit. For instance, whether a population of organisms constitutes a certain species might be open to debate and might depend on which species concept we take into account[3]; biofilms or host-symbiont consortia might be either considered organisms or communities depending on which characterisation of organism we take into account etc.[4] Biology does not provide clear-cut and universally-accepted criteria for the identification of a certain biological entity as an entity of a certain type because biological entities develop and evolve. But one of the relevant issues in conservation biology is whether entity-based compositional hierarchies of the above kind provide a satisfactory framework to characterise the units of conservation.[5] In this section we shall suggest that they do not. In order to do so, we shall consider three issues. The first concerns the justification for the choice of hierarchy.

---

[3] See Reydon, Chap. 8, in this volume, on the debate concerning the nature of species.

[4] See Marques da Silva and Casetta, Chap. 9, in this volume, on this issue.

[5] This is just one of the many conceptual and practical challenges posed by conserving biodiversity. See Casetta, Marques da Silva & Vecchi, Chap. 1, in this volume.

The second pertains to the justification for the choice of biodiversity units. The third concerns the rationale for the exclusive focus on entities.

12.2.1 The Limits of Conservation Fundamentalism

The taxonomic, ecological and genetic hierarchies are somehow conflicting, even though they might be related at some level. One proposal is that they are cleanly related at the species-population-genome level because, as Angermeier and Karr (1994, p. 691) argued, *"... any population has a taxonomic identity (species), which is characterized by a distinct genome."* This essentialist proposal is flawed at least in the sense that it is assumed that a species-specific genome exists, while what exists is a gene-pool (i.e., the totality of the genes of a given species existing at a given time, see Mayr 1970, p. 417). The species genome is thus a statistical artefact reconstructed with reference to this gene pool and ideally comprising all the genomic constituents of all genomes of present (but not past and future) organisms belonging to a species. If the species-population-genome relationship were clean, it would follow that by saving all present members of one species we would conserve all species-specific genomic variation, which is clearly not the case at least in the sense that some genomic variants have been surely lost in the course of evolution and others will be acquired. Of course, other ways of carving nature might exist and other compositional hierarchies, possibly linking the three hierarchies used so far, might be devised. Sarkar (2002) has for instance proposed that two compositional hierarchies should be used, one spatial (i.e., biological molecules, macromolecules, organelles, cells, organisms, populations, meta-populations, communities, ecosystems, biosphere) and one taxonomic (alleles, genes, genotypes, subspecies, species, genera, etc. until kingdoms, domain and superdomain). The advantage of this proposal is its parsimony, particularly the merging of the genetic and taxonomic hierarchies (which implies that genomic units – i.e., functional or structural genomic components – should be considered taxonomic ones).

Given a multiplicity of hierarchies, is there a possible justification for choosing one particular compositional one? For reasons that will be uncovered in this section, we strongly doubt it. However, let us suppose for the sake of argument that no hierarchy can be privileged. The following question is whether some units should be chosen as fundamental units of conservation. Clearly, it is practically impossible to conserve all diversity at all levels of a hierarchy, as it is practically impossible to focus conservation effort on the all-comprehensive top-level unit (i.e., biota and biosphere). How should we choose relevant units then? As a matter of fact, conservation practice seems to bypass this foundational question. As Sarkar (2005, p. 182) relates, the convention in conservation practice is to choose the "holy trinity" of genes, species and ecosystems as units of conservation. Note that these three types of entity belong to different hierarchies, however compositional hierarchies are characterised. Thus, conservation biologists seem to think that, for instance, conserving genes is not sufficient to conserve ecosystems and vice versa. Are they correct? Let us analyse the holy trinity in detail by starting with genes.

As we have seen in the first section, compositional hierarchies betray a reductionist bias. Unlike in physics, in the life sciences this bias is not articulated as a problem concerning ontological fundamentalism (the idea that a compositional level is primitive): obviously no biological compositional level is primitive and ontologically fundamental given that all biological entities are made of physical stuff. But an analogous problem presents itself nonetheless: is there any reason to think that a particular biological compositional level is causally privileged? Usually this question is framed in terms of reduction: suppose that biological compositional level x is adopted as privileged, would it be possible to reduce all biological phenomena to interactions between entities at that level? Generally, the answer to this question has been negative, with few interesting exceptions, for instance in developmental biology (Rosenberg 1997; Wolpert 1994). Nonetheless, a tendency to consider the molecular level as the biologically privileged level is clearly present in many branches of biology. The reason is that it is thought that the behaviour of biological wholes should, in order to be properly understood, be unpacked in terms of molecular interactions. When we move to conservation practice, the related reductionist idea seems to be that genes are the fundamental unit of conservation because, by conserving all genomic variation, we concomitantly preserve much of the phenotypic variation that characterises the populations constituting the species and higher taxonomic levels. Sarkar (2002, p. 152) notes that this position can be justified only if some form of "global genetic reductionism" (i.e., the thesis according to which "all biological features are, in some significant way, reducible to the genes") is vindicated. In a very clear sense, global genetic reductionism is wrong, fundamentally because phenogenesis at all levels (from transcription, translation and protein folding up to cellular differentiation and morphogenesis) is causally influenced by a variety of environmental inputs. Thus, saving all genes would not save all possible phenotypic outcomes unless we also conserved all possible developmental environments, which verges on the impossible.[6] Interestingly, note that developmental environments (e.g., the folding environments of proteins considered in Sect. 12.1) are fundamentally ecosystems. Also note that genes are units of the taxonomic hierarchy while developmental ecosystems are units of the spatial or ecological one. This explains why a compositional hierarchy cannot be privileged over the others and why, as a matter of fact, focusing conservation efforts on units of two different hierarchies might turn out to be a necessary rather than an incoherent conservation strategy (for a similar argument, see Sarkar 2002, p. 152).

---

[6] There remains a possible sense in which the conservation of genomic variation goes a long way to achieve conservation of all biodiversity: if it were established that speciation (as the epitome of a lineage diversification process) completely depends on genomic change, then we would have a good argument. The issue concerns the origin of biodiversity: if it turns out that genomic change is central, then some diluted form of global genetic reductionism might be rescued in the face of phenotypic plasticity (perhaps the variation produced through plasticity would be ineffectual *per se* for speciation; see West-Eberhard 2003 for an opposite argument).

As already argued above, the idea that a peculiar genome characterises a biological species remains common in biology (despite being generally rejected in philosophy of biology). This seemingly clear link between genetic properties and species partly explains why the latter are considered an element of the trinity. After all, species are the repositories of all the genomic and phenotypic variation among its constituent organisms, where this variation is the raw material on which speciation processes work. According to Mayr (1969), species are the most fundamental unit of biological organisation. Interestingly, Mayr argued that only sexually reproducing organisms form species and that the category is not applicable to many unicellular groups of organisms (e.g., bacteria), thus betraying a bias that still characterises conservation practice too. Mayr's (1969, p. 316) argument was that species serve a specific biological function because dividing the total genetic variability of nature into discrete packages prevents the production of "disharmonious incompatible gene combinations". Conservation efforts that target species could therefore be justified as we would save all the possible "harmonious genetic combinations".[7] However, even if we endorse the view that species are important units of biological organisation, this would not be enough to justify an exclusive focus on this biotic unit in conservation practice. One reason is that estimating biodiversity through species count is problematic.[8] For instance, species diversity would not account for diversity at other levels of the same hierarchy. The fact that there are more terrestrial than marine species does not translate into more diversity at the next hierarchical level; in fact, there are more marine than terrestrial phyla, i.e., diversity and disparity clash (Grosberg et al. 2012); hence, by conserving an equal number of marine and terrestrial species, we might not conserve equal marine and terrestrial biodiversity at the phylum level. Conversely, species diversity would not account for genetic diversity, that is, for diversity at another level of a different hierarchy (or even of the same hierarchy if the general taxonomic hierarchy proposed by Sarkar mixing genetic and taxonomic units is endorsed); hence, for instance, by choosing to conserve indiscriminately either species S1 and S2 of genus G because one of the two is functionally redundant (in the sense that they play an equivalent ecological role in the ecosystem), we might not be able to conserve equal biodiversity at the genetic level; the reason is that one of the species might harbour more genetic diversity (its gene pool might be larger); so, supposing chimps and bonobos play equivalent ecological roles in the ecosystems, conserving bonobos with presumably much smaller gene pools than chimps (Prado-Martinez et al. 2013) would amount to failing to conserve genomic diversity.

Similar arguments apply to exclusive focus on ecosystems as the unit of conservation. This means that it is clearly difficult to justify biodiversity fundamentalism.

---

[7] By adding the hypothesis that genes are the most important causes of phenogenesis, we end up with the strong hypothesis that by conserving the species' characteristic gene pool (i.e., an aggregate of genomes) we are also conserving the entirety of their possible phenotypic manifestations (that is, all protein and cell types as well as all supra-cellular organismal traits), i.e., all genetic and phenotypic biodiversity.

[8] See Borda-de-Água, Chap. 5 and Crupi, Chap. 6, in this volume on this issue.

A similar position has been argued for by Angermeier and Karr (1994, p. 691). They generalise the failure of biodiversity fundamentalism by also arguing that even a focus on a single hierarchy is bound to fail, as it would lead to ignore most biodiversity. Noss (1990, p. 357) has made this point quite succinctly by arguing that *"No single level of organization (e.g., gene, population, community) is fundamental ...."*. Of course, the idea of taking into account 3 units of different hierarchies instead of one unit is exactly tailored to avoid such problems. But, as Sarkar (2002, p 138) has argued, "...even this catholic proposal falls afoul of the diversity of biological phenomena ...". Thus, Sarkar argues, even avoiding biodiversity fundamentalism in some of its two forms (either focusing exclusively on a hierarchy or on a unit) would not allow accounting for "endangered biological phenomena" that are in principle amenable to conservation, such as the synchronous flowering of particular bamboo species at a distance. Sarkar argues that in order to save this peculiar phenotypic outcome, conservation efforts should neither be directed to conserve the genome of the clumps of these bamboo species, nor even conserving the species; rather, what should be conserved are the environments in which this behaviour is expressed; only by also preserving the habitats and sites where these biological phenomena occur we would be able to conserve them. Note that this argument is analogous to the one proposed above concerning the conservation of developmental environments. Developmental environments are, like habitats and sites, entities belonging to the spatial compositional hierarchy, that is, a different hierarchy than that to which genes and populations belong. We conclude that for all these reasons there is no justification for focusing exclusively on one compositional hierarchy in conservation practice. As we have showed, at least one spatial and one taxonomic unit are concomitantly needed as conservation units in order to encompass all phenotypic biodiversity (e.g., protein conformations and developmental outcomes, genetic and phenotypic variants) and all biological phenomena (e.g., synchronous flowering of bamboo). The corollary of this conclusion is that no biodiversity unit can be the fundamental unit of conservation. Rather, a variety of units are needed to encompass all biodiversity. We shall now suggest that the limits of biodiversity fundamentalism (both in its hierarchy and unit variants) and of multi-unit approaches to conservation is arguably a symptom of a more general malaise concerning entity-based approaches to conservation practice. The fundamental question is thus whether a different kind of approach should be favoured. In particular, we ask whether there exists a rationale for the exclusive focus on entities.

12.2.2 Towards an Entity and Process-Based Approach to Conservation

As we argued in the first section, one limit of compositional hierarchies pertains to their lack of structural and functional information. The problem is thus whether conservation strategies can be devised in the absence of detailed knowledge concerning the structural properties and functional interactions between the entities constituting the compositional levels of the hierarchy. Structural hierarchies aim to represent the organisation (e.g., the topology or network of interactions) between the parts of the relevant entity-whole, while functional hierarchies map the processes governing the causal interactions between the various parts of the relevant entity-wholes. In this sense, a structural characterisation of, for instance, a cell is the topology of the network of interactions between its components parts. The structural characterisation is not merely a list of cellular components (it is not purely compositional), but it is an organised list whereby their interactions are identified. A structural characterisation is more informative than a compositional one, but is less informative than a functional one. From a functional point of view, a cell is literally an ecosystem whereby energy and matter acquired from the environment is processed internally in such a fashion as to manufacture its component parts (Luisi 2003). This means that a functional representation of a cell specifies the causal nature of the interactions between its sub-cellular components. Consider secondly that, given that functional hierarchies aim to represent the causal interactions between the elements of a hierarchy, they do not provide merely entity-based ontologies. For instance, a functional characterisation of the cell makes reference to the metabolic interactions between nutrients, constituent proteins and other macromolecules, organelles, membrane receptors etc. In this sense, it does not purely provide an entity-based ontology but also a process-based one. The upshot is that the genetic, taxonomic and ecological hierarchies for characterising biodiversity in terms of genetic organisation, taxonomic inclusiveness and ecological nestedness are, given their compositional ethos, insufficient to capture the structural and functional aspects of biodiversity (Franklin 1988). It is for this reason that compositional hierarchies should be complemented with structural and functional hierarchies, as suggested by Noss (1990, p. 359). As soon as we look at functional hierarchies, we grasp that the focus is also on processes, not merely on entities.

The crucial question is whether knowledge of functional interactions and process is necessary in order to provide a satisfactory characterisation of biodiversity and especially of the units of conservation. Consider functional interactions first. Many species are involved in complex biological relationships such as predation and pollination. Compositional hierarchies provide information concerning the relata (i.e., the entities involved in a relation) of such interactions, but this information is oblivious to process. Pollination is an ecological function that can be realised in multifarious ways by a variety of species of insects, birds, bats, snails etc. on the one hand and flowering plants on the other. Perhaps some species play a fundamental ecological role in the pollination process (as keystone species, Sarkar 2005, p. 15) and our conservation efforts should be focused on these.[9] It is therefore clear that knowledge

---

[9] The concept of keystone species can be characterised in terms of ecological centrality (when a species has many functional relationships with many different species). This characterisation, however, seems to imply a lack of specialisation on the part of the species. For instance, pollination seems to be realised in large part by highly specialised species (both plants with very few pollinators and animals pollinating very few plants) which do not have, as a consequence, many functional relationships with many different species. We would argue that the ecological centrality of a keystone species depends on its specialised functional role: a keystone species would thus be one

of this ecological role might inform conservation efforts. However, knowledge of this kind is clearly not provided by compositional hierarchies. Consider processes now. Generally speaking, two types of processes governing the behaviour of the entities identified by compositional hierarchies can be identified. First of all, those leading to the differentiation of parts. Secondly, those that, given the differentiated parts, govern their combination (i.e., combinogenesis). All natural sciences are somehow concerned with understanding the nature of the processes of part differentiation and those governing their combination. Biology certainly strives to understand differentiation and combinogenetic processes: biology is both about differentiation of part-entities (e.g., production of genetic variants, new species etc.) and about the combinogenesis of whole-entities (i.e., the emergence of new biological individuals). For instance, what are the processes that govern allelic, population and species differentiation? Theories of genomic change and speciation are part and parcel of biology of course. And what are the processes that govern genome and ecosystem formation? Equally, theories concerning genome evolution and ecology are part and parcel of biology. An entity-based approach to biodiversity is thus parasitic on biological theories concerning, among others, genomic and phenotypic change as well as biological and ecological theories concerning, among others, genome evolution, phenotypic evolution, speciation and ecosystem stability, where all these theories make a reference to processes (e.g., mutation, phenotypic plasticity, predation) impinging on a variety of biological entities belonging to various levels of various compositional hierarchies. Thus, given that reference to such processes remains invisible in compositional hierarchies, they seem by their own nature epistemologically deficient. This point is particularly relevant because it influences the characterisation of the units of conservation. Does a focus on units of biodiversity, which are biological entities, make sense without a complementary focus on their maintenance and generative processes?

12.3   Does a Process-Based Approach to Biodiversity Make Sense?

We have argued so far that an entity-based approach ignores the functional relations between the elements of the hierarchy. In a nutshell, it ignores the influence of processes of differentiation and combination of parts. In conservation science, a process-based approach would shift the focus on the processes that *originate and maintain* biodiversity. As we shall relate, the shift from entities to process has been advocated by many conservation practitioners. The argument that we shall propose does not advocate a switch to exclusive focus on process. More reasonably, we suggest that a process-based approach should integrate an entity-based one (Faith

---

performing (almost) exclusively a particular function (e.g., pollination) for other species. See Sect. 12.3 for a clear example of keystone species (Morris et al. 2012). Thanks to Alessandro Minelli for drawing attention to this putative tension.

2016). After all, what could it mean to conserve a process? Not much. As we already argued in Sect. 12.1, if an exclusive entity-based approach to biology does not make sense, even an exclusive process-based approach does not. The reason is obvious: processes are important because they create and maintain new entities, new units of potential conservation. Thus, to use the pollination example again, the issue is not whether we should either choose the relata (e.g., the populations) or the relationship as units of conservation. We cannot think of any other sensible way of conserving relationships and processes than by conserving their relata and their actors (i.e., the entities involved in the process). As we shall explain below, the shift to process is most prominently a shift in the ways in which we characterise the units of conservation. Particularly important in the present context is Ryder's (1986) proposal to characterise conservation units as evolutionary significant units (i.e., ESUs), that is, as populations of organisms that, for historical and evolutionary reasons, play peculiar causal roles in the processes targets of conservation. From a process-based perspective, the ultimate focus of conservation practice is on entities such as ESUs (Moritz 1999, p. 223).[10] Relatedly, an important issue about the characterisation of a process-based approach concerns the kind of processes that should be taken into account. Noss (1990) considers as potential targets conservation processes that are partially abiotic such as energy cycles. Noss's is an interesting suggestion. However, it should be highlighted that, again, the focus is, ultimately, inevitably on the entities that play specific causal roles in processes. For instance, in marine environments some bacteria seem to play the role of keystone species as they might exclusively perform some specific function. For example, a limited number of bacteria (e.g., of the genus *Alteromonas*) seem to process hydrogen peroxide in the ocean, performing a crucial metabolic function that benefits the incredibly large communities of the cyanobacterium *Prochlorococcus* (Morris et al. 2012). Without these bacteria, the ecosystem would probably suffer. Conservation efforts could thus be directed to save this important geochemical process, but inevitably such efforts would focus on preserving the important ecological function that *Alteromonas* bacteria play. We thus suggest that the focus should be on the processes that govern what we called entity differentiation and combinogenesis in Sect. 12.2.2, that is, most prominently the ecological and evolutionary processes that cause the origin of ESUs.

Many inter-linked themes prominent in the conservation literature explain the shift towards a complementary entity-process-based approach to conservation. This conceptual shift finds its theoretical support in the deeper integration with the evolutionary sciences and with ecology. Most generally, Norton (2001) argues that conservation science has experienced a transition from a static to a dynamic view focused on evolving systems and ecosystem processes. This interpretive hypothesis is probably supported by a shift in the characterisation of the units of conservation from static entities – e.g., species characterised essentialistically in terms of species-specific genetic and phenotypic features – to historical ones with peculiar historical

---

[10] Sarkar (2002, note 15, p. 152) has argued that focus on process is aimed to conservation of biological "integrity" rather than biodiversity. However, if the focus of a process-based approach is on entities such as ESUs, it is clearly committed to biodiversity conservation.

and evolutionary capacities. This transition has been nurtured by the dissatisfaction with prominent species approaches to conservation aimed at the maximisation of number of species saved per spatial area which are, by definition, fundamentalist and entity-biased. Ultimately, the idea is that the conservation focus should be put on the evolutionary and ecological causes of biodiversity and the preservation of process rather than on their causal effects and on the preservation of pattern. Particularly important are the attempts to identify centres of evolutionary diversity with the aim of maximising evolutionary heritage on one hand (a consequence of the integration of phylogenetic analyses with conservation biology) and the focus on the evolutionary (e.g., genetic, cf. Frankel 1974) potential of populations and historical lineages. Smith et al. (1993) argue that knowledge of the ecological and evolutionary mechanisms generating genetic diversity and of the isolating mechanisms of speciation must be part and parcel of conservation practice. Conservation practices that focus on protecting species-rich sites are doomed to fail for reasons that parallel those for which the counting-species approach did. First, such focus does not necessarily provide information on the frequency of rare species, which might not occur in areas of highest species diversity (Smith et al. 1993, p. 164). Secondly, it does not necessarily provide any information on the functional role of species and on the nature of the community dynamics of the relevant ecosystems (Smith et al. 1993, p. 165). Thirdly, it neither necessarily identifies regions with peculiar evolutionary history nor identifies lineages that are phylogenetically unique (ibid.). For all these reasons, Smith et al. propose an approach to conservation that integrates ecological and molecular information. Related to the third point above, Mace et al. (2003) have suggested that, rather than directing conservation efforts to save species, these should be directed to saving independent branches of the tree of life, that is, distinctive lineages with a long and unique evolutionary history. The rationale for this conservation strategy is that phylogenetic information permits to distinguish "cradles" of diversity from "museums". A process-based approach informed by phylogenetic information (and hence by knowledge about evolutionary history) identifies as priority conservation taxa those that display a unique evolutionary history instead of focusing efforts on conserving patterns of species richness (Mace et al. 2003, p. 1709). Along the same lines of integration of molecular data, Moritz (1999) has proposed to address conservation problems by focusing on the maintenance and restoration of those ecological and evolutionary processes that can recreate adaptive phenotypes. In order to conserve such processes, we should aim to conserve their "effectors", i.e., the ESUs or populations with evolutionary potential in which they play causal roles. Moritz argues that molecular studies are particularly important to infer evolutionary history. Molecular information will give us details about the evolutionary relationships between the populations of conservation focus to the extent that, for instance, "… translocations among populations that historically exchanged genes would be considered, whereas human-mediated mixing of historically isolated gene pools would be discouraged." (Moritz 1999, p. 223). This approach aims to conserve ESUs through the restoration of connectivity between isolated populations in anthropogenically fragmented ecosystems and the destruction of "genetic ghettos" (Moritz 1999, p. 224). In synthesis, a process-based approach to conservation might be seen as proposing an integration of varieties of ecological and evolutionary information with the aim of identifying relevant ESUs. One general characterisation of ESUs that can be extrapolated from the conservation literature reviewed so far refers to populations of organisms possessing a property of conservation interest, such as a peculiar history (i.e., being a distinctive lineage) and a crucial functional role in ecosystem welfare (i.e., being a keystone species). Even though preservation seems to be, by definition, the aim of conservation biology, it is interesting to observe that Smith et al. (1993, p. 164) have argued that the aim of conservation science is "…to promote and preserve natural dynamics." What could promotion amount to? A promotion (rather than preservation) characterisation of ESUs might refer to properties of populations such as the ability to cope with environmental stress (i.e., adaptability) or an enhanced capacity to diversify into lineages with distinctive genetic and phenotypic features. In the latter two cases, it might be said that the population ESU displays "evolutionary potential" (Casetta and Marques da Silva 2015), a property that might depend either on possessing particular genomic properties or on its tendency to respond to environmental change purely phenotypically, where such properties might be important for populations' adaptability and diversification.[11] In the following section we shall focus on populations that display evolutionary potential in the latter sense. The hypothesis we would like to test is whether plastic populations of a species might be considered ESUs amenable to conservation. In particular, we would like to show that plastic subpopulations that have enhanced evolutionary potential vis a vis non-plastic subpopulations make them amenable to ESU status.

12.4 Can Phenotypic Plasticity Confer Evolutionary Potential?

In this section we shall thus focus on a particularly evolutionary process, i.e., phenotypic plasticity (Fitzpatrick 2012; Forsman 2015; Miner et al. 2005; Valladares et al. 2014; West-Eberhard 2003). By plasticity we refer to the ability of the organism to react to environmental inputs with an appropriate phenotypic change during embryogenesis (developmental plasticity) and further developmental stages (phenotypic plasticity). Two main types of plasticity exist: reaction norms and polyphenisms. In reaction norms the genome allows a continuous range of potential phenotypes. On the other hand, polyphenisms are discontinuous (either/or) phenotypes elicited by the environment. The essence of plasticity is that the genome does not wholly dictate the nature of the phenotypic outcome. It is reasonably straightforward to intuit about selective advantages to phenotypic plasticity: where there exist different or varying environmental conditions that are experienced, either a) by different individuals across a population, or b) by the same/each individual through its

---

[11] See Minelli, Chap. 11, this volume, for the relationship between evolutionary potential and evolvability.

lifetime, a unique phenotype (narrow reaction norm) would be less fit than plastic responses. This is, of course, provided that an appropriate phenotype can be expressed, either sensitive to environmental conditions or genetic (West-Eberhard 1986). An example of a genetic switch is the X-Y sex determination system in mammals. Some species of buttercups (e.g., *Ranunculus flammula*) exemplify polyphenism through environmental sensitivity: they develop one of two distinct leaf types, depending on whether underwater or on land (Cook and Johnson 1968). A particularly advanced form of environmental sensitivity – potentially producing continuous phenotypic responses – is learning: the capacity to *change* behaviour in particular situations, according to past life experiences (Staddon 1983).

While these phenotypic flexibilities are interesting in their own right, and indeed potential benefits of plasticity are easy to identify (notwithstanding discussion regarding what those benefits trade off against), could there be a deeper evolutionary issue here? Could phenotypic plasticity not only have proximate effects, but also impact the course of evolution? The understanding that traits produced through plasticity are not heritable goes as far back as the nineteenth century with August Weismann's experiments showing a soma/germ-line separation. And the hypothesis now commonly known as Lamarckian evolution, that traits acquired during lifetime would be passed on to further generations – e.g., the strong biceps of a blacksmith – is not considered compatible with genetic inheritance (discounting epigenetic inheritance). But there is an intriguing suggestion that phenotypic changes could influence selection in an evolving population, and thus indirectly lead to genetic encoding of formerly acquired traits (Baldwin 1896; Osborne 1896; West-Eberhard 2003). The basic notion is that the relatively rapid exploration of phenotype space via plastic response can introduce a selective gradient towards genetic specification of that phenotype, and thus the slower genetic variation can be "guided" by lifetime exploration (Hinton and Nowlan 1987). The selective landscape experienced by a plastic population is modulated by that plasticity, in comparison to the landscape experienced by non-plastic populations. But the modulation to fitness of specific genotypes does not require that the phenotypic traits discovered are heritable, i.e., it occurs without so-called "Lamarckian" inheritance. This process has become known as the Baldwin effect (a term coined by Simpson 1953). The effect depends on the existence of phenotypic plasticity (Bradshaw 1965) having already evolved but this in itself can only facilitate the first of two phases: selection among the various phenotypes expressed for those most appropriate to the present environmental conditions. The second phase, genetic assimilation, is not a necessary consequence of the existence of plasticity, nor does it depend on a reduction in the level of plasticity.[12] The Baldwin effect has been the subject of a plethora of computational studies (see e.g., Turney et al. 1996; Paenke et al. 2009; Sznajder et al. 2012), following the seminal work of Hinton and Nowlan (1987). Almost all of these works considered evolution in single-peaked fitness landscapes; but in Mills and Watson (2006) we showed that, via a Baldwinian process, a learning population is able to cross a fit-

---

[12] Mills and Watson 2005 further discuss how canalisation, although often implicated in studies on the Baldwin effect, is not actually a necessary mechanism for the effect.

ness valley. Here we use the same model to illustrate various scenarios, including that learning is able to repeatedly guide genetic evolution in a variable environment.

12.4.1 A Model of Plasticity

We model a population of individuals each with a string of n binary variables to represent their genotype, which specifies the phenotypes that the individual will express throughout its lifetime, through a trivial (but non-deterministic) genotype–phenotype (G-P) mapping. Specifically, for each lifetime trial i, the phenotype p_i is based on the genotype with mutation-like variation applied at a rate of μ_L, independently applied at each locus. The phenotypes from the T trials are independent from each other, and can be thought of as a cloud of points surrounding the genotypically-specified location. The individuals are bestowed with a simple capacity to learn, which is facilitated through the way that fitness is calculated: during each lifetime trial, a learning individual recalls the best solution found so far, whether it is the newest phenotypic strategy, or whether it was found long ago (see Hinton and Nowlan 1987). At the end of each generation, the individuals are selected in proportion to their fitness,[13] and reproduce asexually. During reproduction, point mutation is applied to each gene, i.e., each gene is transmitted to the offspring with a probability of 1-μ_G, otherwise with probability μ_G a new random allele is drawn (note that this model does not rule out the possibility of multiple mutations but that they are uncorrelated when they occur). The population size m is constant through time. In this model there is no way for an individual to perform less lifetime exploration, i.e., there is no mechanism for canalisation (Waddington 1953). This simplification is not meant to imply that there would never be a selective advantage to such a reduction, but rather to keep the spotlight on the consequences of plasticity.

Simulation Experiment 1 We consider the evolution of a population on a simple and abstract fitness landscape, where there are two rare phenotypes p_1 and p_2 that receive high fitness and all other phenotypes are equally bad. Here, our main question is to investigate whether the form of phenotypic learning in this model is sufficient for the population to evolve across the fitness valley between the two peaks. Accordingly, the first peak/phenotype confers high fitness ($f(p_1) = H$) and the second peak confers lower fitness ($f(p_2) = L$). The environment remains like this for s generations, after which the quality of peaks switches, such that $f(p_1) = L$ and $f(p_2) = H$.

Parameters used in this experiment: $H = 100$, $L = 10$, $f(p|p \neq p_1, p \neq p_2) = 1$, $n = 16$ genes. The separation d of the two peaks is 5 bits, and the switching interval s is 50

---

[13] Since all individuals experience the same number of learning trials before selection, it could be seen as if selection only occurring on adult organisms; however, the model confers benefits to successful learning earlier in the lifetime, even though we do not explicitly include phenomena such as probabilistic death without reproduction.

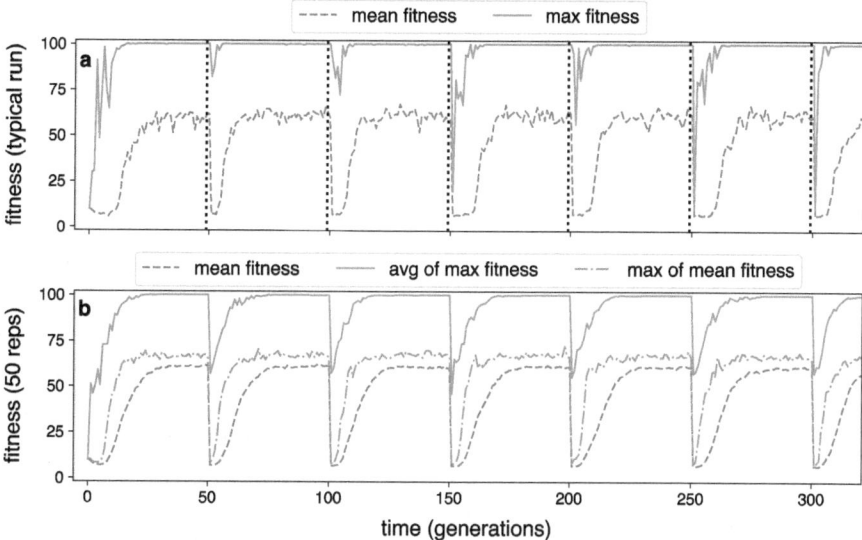

Fig. 12.1 Fitness measures of populations evolving in a switching environment. (**a**) one example run; (**b**) averages over 50 replicates. The switches in environment are marked by the dotted vertical lines

generations. We set the mutation rate μ_G at 1/20 and the lifetime variation rate μ_L at $2/n$, the number of trials per individual $T = 256$, and the population size m at 200.

To see what is happening in the population, we can observe the fitness over time (Fig. 12.1). Initially all organisms of the population possess the genotype specifying phenotype p_2, and within only a few generations some individuals in the population find a high-fitness phenotype, p_1. Any mutation that brings the genotype closer to directly specifying p_1 will be favoured since discovering the phenotype earlier in the lifetime results in higher fitness. Accordingly, such high-fitness genotypes propagate through the population, as is reflected in the rise in mean fitness. After each switch in the environment (dashed vertical lines), we see a sharp drop in fitness, reflecting the fact that the population was adapted to a previous challenge. However, phenotypic plasticity enables individuals to rapidly re-discover p_2, which is now the highest-fitness phenotype in the environment.

Simulation Experiment 2 Rather than fixing the rate of environmental switching, here we leave this parameter s open; and to ascertain the capacity of a plastic population to cope with such environmental change we run simulation experiments for various different values of T.

From the results in Fig. 12.2 we see two different trends: (1) populations experiencing a large number of trials T can achieve high fitness, provided the environment does not change too rapidly. When the interval is very short there is insufficient time for the population to find the high peak and assimilate it genetically. Note however that the high peak is found phenotypically by some fraction of the population: with-

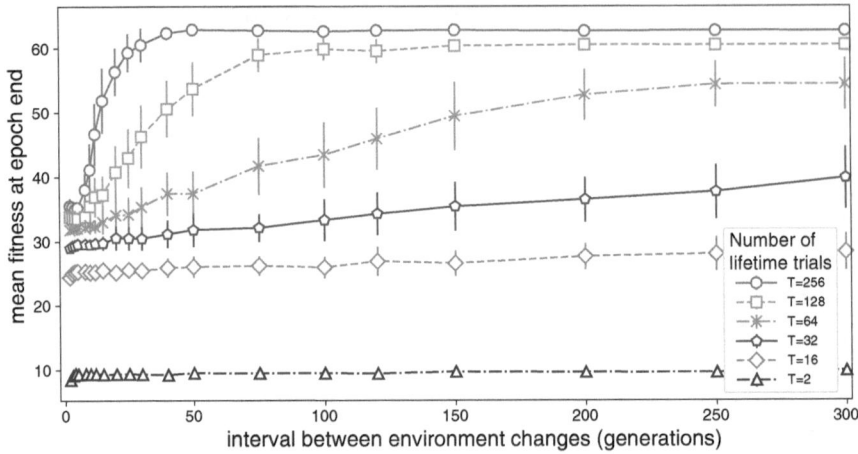

Fig. 12.2 Fitness of the population in the final generation before a switch. Data show the mean and standard deviation of mean fitness, across 40 replicates

out doing so, the mean fitness could not be greater than $f(L) = 10$. (2) The smaller the value of T, the less capacity the population has to adapt to the new challenge presented by the switched environment. At its extreme, with $T = 2$, the population wholly fails to adapt under any environmental switching rate tested.

The dynamics of a non-plastic population are qualitatively different. As the mutation model permits multiple loci to change simultaneously during reproduction, a multi-locus valley could, in principle, be crossed. However, the expected time this takes grows exponentially with the number of genes that must change at once. For the 5-bit valley and $\mu_G = 1/20$ (as used above), it takes a mean of over 75,000 generations, and even though a higher rate would reduce this, even an optimal rate of $\mu_G = 5/16$ takes a mean of 340 generations (mutation rates that are so high introduce difficulties in terms of drift and retaining high-fitness peaks even if/when discovered, besides severe penalties to average fitness). Importantly, these generation times are very high in comparison to the environmental switching frequencies that a learning population is able to thrive in.

The above model and experiments illustrate how one variety of phenotypic plasticity can enable a population to rapidly evolve across fitness valleys, a pattern of evolutionary change that cannot be experienced by non-plastic populations. On the flip-side, the benefits to the plastic population are lessened under more and more constant environments. At some point any benefits would be outweighed by the costs of learning (e.g., energetic cost of memory, risks). Although such aspects are omitted from the model here, and thus trade-offs are not directly visible in the results, the logic is straightforward: if genetic adaptation alone is sufficient in some stable environment, we should not expect to see plasticity playing any significant role.

Besides these two examples of environments, what other responses to changing environments might we expect to see in evolution? In the extreme case where many novel challenges appear within one lifetime, there may be plastic responses that do not become assimilated into the genome. In the absence of any regularity in those challenges, if the new challenges in one's lifetime are unrelated to the challenges faced by their ancestors, plasticity may be favourable but it is hard to see how any specific genetic adaptation would arise. If changes occur over a few generations, a Baldwinian-type interaction between plasticity and genetic evolution leading to genetic assimilation may result. Alternatively, if environmental changes are particularly repetitive, we may expect to see polyphenic/polymorphic genotypes and environmentally-sensitive switching (see West-Eberhard 1986, 2003) as mentioned above. If the environmental changes are strongly structured, we may additionally see modular architectures evolve in the genotypes (Parter et al. 2008) that are able to more quickly adapt to new challenges (Watson et al. 2014).

12.5   Conclusion

Our argument has been that focusing solely on entities, be they genes, species or ecosystems, is inherently problematic for conservation practice. We first argued that what we called *biodiversity fundamentalism* is untenable. It is both untenable as a thesis concerning the exclusive focus on one compositional hierarchy and as a thesis concerning the existence of a fundamental unit of conservation. Secondly, we argued that the genetic, taxonomic and ecological hierarchies for characterising biodiversity in terms of genetic organisation, taxonomic inclusiveness and ecological nestedness are, given their compositional ethos, insufficient to capture the functional dimension of biodiversity, particularly the evolutionary processes that maintain and originate new biodiversity units. Thirdly, we have proposed a *complementary entity-and-process-based approach to conservation practice*. Within this context, we distinguished between two types of important properties that evolutionarily significant units (i.e., ESUs) of conservation interest might exhibit: those amenable to conservation because they *preserve* natural dynamics (e.g., being a distinctive lineage) and those that *promote* them (e.g., being a population with a greater capacity for adaptation to change or stress). We focused on the latter because conservation strategies are aimed to identify not only "museums" but also "cradles" of biodiversity. Given this background, the hypothesis we wanted to test is whether plastic populations of a species might be considered ESUs with relevance for conservation. In particular, we wanted to show that plastic subpopulations that have enhanced evolutionary potential vis a vis non-plastic subpopulations make them amenable to ESU status. The model indeed shows that plasticity yields evolutionary potential, which is displayed in environments that switch in a few to a few tens of generations. Thus, populations with adaptation capacities available might possess an interesting property to consider when deciding on how to focus conservation efforts. Given that plastic populations might be important for species' adaptability and diversification, they might be considered ESUs potentially amenable to conservation. This vindicates, on the one hand, a process-based approach to biodiversity and, on the other, suggests the need to take into account the processes generating plasticity when considering conservation efforts.

Acknowledgements Many thanks to Sandro Minelli, Philippe Huneman and Elena Casetta for feedback. Davide Vecchi acknowledges the financial support of the Fundação para a Ciência e a Tecnologia (Grant N. SFRH/BPD/99879/2014 and BIODECON R&D Project. Grant PTDC/IVC-HFC/1817/2014). Rob Mills acknowledges support by UID/MULTI/04046/2013 centre grant from FCT, Portugal (to BioISI).

References

Angermeier, P. L., & Karr, J. R. (1994). Biological integrity versus biological diversity as policy directives: Protecting biotic resources. *Bioscience, 44*(10), 690–697.

Baldwin, J. M. (1896). A new factor in evolution. *American Naturalist, 30*, 441–451.

Bradshaw, A. D. (1965). Evolutionary significance of phenotypic plasticity in plants. *Advances in Genetics, 13*, 115–155.

Casetta, E., & Marques da Silva, J. (2015). Facing the big sixth: From prioritizing species to conserving biodiversity. In E. Serrelli & N. Gontier (Eds.), *Macroevolution: Explanation, interpretation and evidence* (pp. 377–403). Cham: Springer.

Cook, S. A., & Johnson, M. P. (1968). Adaptation to heterogeneous environments. I. variation in heterophylly in *ranunculus flammula* l. *Evolution, 22*(3), 496–516.

Dupré, J. (2012). *Processes of life: Essays in the philosophy of biology*. Oxford: Oxford University Press.

Faith, D. P. (2016). Biodiversity. In The Stanford encyclopedia of philosophy (Summer 2016 edition), ed. Edward N. Zalta. http://plato.stanford.edu/archives/sum2016/entries/biodiversity. Accessed 25 Sept 2018.

Fitzpatrick, B. M. (2012). Underappreciated consequences of phenotypic plasticity for ecological speciation. *International Journal of Ecology*. https://doi.org/10.1155/2012/256017.

Forsman, A. (2015). Rethinking phenotypic plasticity and its consequences for individuals, populations and species. *Heredity, 115*, 276–284.

Frankel, O. H. (1974). Genetic conservation: Our evolutionary responsibility. *Genetics, 78*, 53–65.

Franklin, J. F. (1988). Structural and functional diversity in temperate forests. In E. O. Wilson (Ed.), *Biodiversity* (pp. 166–175). Washington, DC: National Academy Press.

Grosberg, R. K., Vermeij, G. J., & Wainwright, P. C. (2012). Biodiversity in water and on land. *Current Biology, 22*(21), R900–R903.

Hennig, W. (1966). *Phylogenetic systematics*. Urbana: University of Illinois Press.

Hinton, G. E., & Nowlan, S. J. (1987). How learning can guide evolution. *Complex Systems, 1*(3), 495–502.

Luisi, P. (2003). Autopoiesis: A review and a reappraisal. *Naturwissenschaften, 90*, 49–59.

Mace, G. M., Gittleman, J. L., & Purvis, A. (2003). Preserving the tree of life. *Science, 300*(5626), 1707–1709.

Mayr, E. (1969). The biological meaning of species. *Biological Journal of the Linnean Society, 1*, 311–320. https://doi.org/10.1111/j.1095-8312.1969.tb00123.x.

Mayr, E. (1970). *Populations, species, and evolution*. Cambridge: Harvard University Press.

Mills, R., & Watson, R. A. (2005). Genetic assimilation and canalisation in the Baldwin effect. In M. S. Capcarrère et al. (Eds.), *Advances in artificial life* (pp. 353–362). Berlin: Springer.

Mills, R., & Watson, R. A. (2006). On crossing fitness valleys with the Baldwin effect. In L. M. Rocha (Ed.), *Proceedings of the tenth international conference on the simulation and synthesis of living systems* (pp. 493–499). Cambridge: MIT Press.

Miner, B. G., Sultan, S. E., Morgan, S. G., Padilla, D. K., & Relyea, R. A. (2005). Ecological consequences of phenotypic plasticity. *Trends in Ecology and Evolution, 20*(12), 685–692.

Moritz, C. (1999). Conservation units and translocations: Strategies for conserving evolutionary processes. *Hereditas, 130*(3), 217–228. https://doi.org/10.1111/j.1601-5223.1999.00217.x.

Morris, J. J., Lenski, R. E., & Zinser, E. R. (2012). The black queen hypothesis: Evolution of dependencies through adaptive gene loss. *MBio, 3*(2), e00036–e00012. https://doi.org/10.1128/mBio.00036-12.

Norton, B. G. (2001). Conservation biology and environmental values: Can there be a universal earth ethic? In C. Potvin et al. (Eds.), *Protecting biological diversity: Roles and responsibilities*. Montreal: McGill-Queen's University Press.

Noss, R. F. (1990). Indicators for monitoring biodiversity: A hierarchical approach. *Conservation Biology, 4*(4), 355–364.

Osborn, H. F. (1896). Oytogenic and phylogenic variation. *Science, 4*(100), 786–789. https://doi.org/10.1126/science.4.100.786

Paenke, I., Kawecki, T. J., & Sendhoff, B. (2009). The influence of learning on evolution: A mathematical framework. *Artificial Life, 15*(2), 227–245.

Parter, M., Kashtan, N., & Alon, U. (2008). Facilitated variation: How evolution learns from past environments to generalize to new environments. *PLoS Computational Biology, 4*(11), e1000206.

Prado-Martinez, J., Sudmant, P. H., Kidd, J. M., Li, H., Kelley, J. L., Lorente-Galdos, B., Veeramah, K. R., Woerner, A. E., O'connor, T. D., Santpere, G., & Cagan, A. (2013). Great ape genetic diversity and population history. *Nature, 499*(7459), 471–475.

Rosenberg, A. (1997). Reductionism redux: Computing the embryo. *Biology and Philosophy, 12*, 445–470.

Ryder, O. A. (1986). Species conservation and systematics: The Dilemma of subspecies. *Trends in Ecology and Evolution, 1*, 9–10.

Sarkar, S. (2002). Defining "biodiversity"; assessing biodiversity. *The Monist, 85*(1), 131155.

Sarkar, S. (2005). *Biodiversity and environmental philosophy: An introduction* (Cambridge studies in philosophy and biology). Cambridge: Cambridge University Press.

Schaffer, J. (2003). Is there a fundamental level? *Noûs, 37*, 498–517.

Simpson, G. G. (1953). The Baldwin effect. *Evolution, 7*(2), 110–117.

Smith, T. B., Bruford, M. W., & Wayne, R. K. (1993). The preservation of process: The missing element of conservation programs. *Biodiversity Letters, 1*(6), 164–167.

Staddon, J. E. R. (1983). *Adaptive behavior and learning*. Cambridge: Cambridge University Press. (Internet edition 2003).

Sznajder, B., Sabelis, M. W., & Egas, M. (2012). How adaptive learning affects evolution: Reviewing theory on the Baldwin effect. *Evolutionary Biology, 39*(3), 301–310.

Turney, P., Whitley, D., & Anderson, R. (1996). Evolution, learning, and instinct: 100 years of the Baldwin effect. *Evolutionary Computation, 4*(3), iv–viii. https://doi.org/10.1162/evco.1996.4.3.iv.

Valladares, F., Matesanz, S., Guilhaumon, F., Araújo, M. B., Balaguer, L., Benito-Garzón, M., Cornwell, W., et al. (2014). The effects of phenotypic plasticity and local adaptation on forecasts of species range shifts under climate change. *Ecology Letters, 17*, 1351–1364. https://doi.org/10.1111/ele.12348.

Waddington, C. H. (1953). Genetic assimilation of an acquired character. *Evolution, 4*, 118–126.

Watson, R. A., Wagner, G. P., Pavlicev, M., Weinreich, D. M., & Mills, R. (2014). The evolution of phenotypic correlations and "developmental memory". *Evolution, 68*(4), 1124–1138.

West-Eberhard, M. J. (1986). Alternative adaptations, speciation, and phylogeny A review. *Proceedings of the National Academy of Sciences, 83*(5), 1388–1392.

West-Eberhard, M. J. (2003). *Developmental plasticity and evolution.* Oxford: Oxford university press.

Wimsatt, W. C. (2007). *Re-engineering philosophy for limited beings: Piecewise approximations to reality.* Cambridge: Harvard University Press.

Wolpert, L. (1994). Do we understand development? *Science, 266,* 571–572.

Chapter 13
Between *Explanans* and *Explanandum*: Biodiversity and the Unity of Theoretical Ecology

Philippe Huneman

Abstract Biodiversity is arguably a major topic in ecology. Some of the key questions of the discipline are: why are species distributed the way they are, in a given area, or across areas? Or: why are there so many animals (as G. Evelyn Hutchinson asked in a famous paper)? It appears as what is supposed to be explained, namely an *explanandum* of ecology. Various families of theories have been proposed, which are nowadays mostly distinguished according to the role they confer to competition and the competitive exclusion principle. *Niche* theories, where the difference between "fundamental" and "realised" niches (Hutchinson GE, Am Nat 93:145–159, 1959) through competitive exclusion explains species distributions, contrast with *neutral* theories, where an assumption of fitness equivalence, species abundance distributions are explained by stochastic models, inspired by (Hubbell SP, The unified neutral theory of biodiversity and biogeography. Princeton University Press, Princeton, 2001).

Yet, while an important part of community ecology and biogeography understands biodiversity as an *explanandum*, in other areas of ecology the concept of biodiversity rather plays the role of the *explanans*. This is manifest in the long lasting stability-diversity debate, where the key question has been: how does diversity beget stability? Thus explanatory reversibility of the biodiversity concept in ecology may prevent biodiversity from being a unifying object for ecology.

In this chapter, I will describe such reversible explanatory status of biodiversity in various ecological fields (biogeography, functional ecology, community ecology). After having considered diversity as an *explanandum*, and then as an *explanans*, I will show that the concepts of biodiversity that are used in each of these symmetrical explanatory projects are not identical nor even equivalent. Using an approach to the concept of biodiversity in terms of "conceptual space", I will finally argue that the lack of unity of a biodiversity concept able to function identically as *explanans* and *explanandum* underlies the structural disunity of ecology that has been pointed out by some historians and philosophers.

P. Huneman (✉)
Institut d'Histoire et de Philosophie des Sciences et des Techniques, CNRS/Université Paris I Panthéon Sorbonne, Paris, France

© The Author(s) 2019
E. Casetta et al. (eds.), *From Assessing to Conserving Biodiversity*,
History, Philosophy and Theory of the Life Sciences 24,
https://doi.org/10.1007/978-3-030-10991-2_13

Keywords Explanation · Species richness · Functional diversity · Phylogenetic
diversity · Modern Synthesis · Neutral theory · Niche · Stability

13.1   Introduction

Amongst the questions that theoretical ecologists have been debating for decades
one finds: why are species distributed the way they are, in a given area, or across
areas? How is biodiversity related to areas? Why are there so many species in
tropical regions? In general, why are there so many animals (as Hutchinson asked
in a famous paper)? Is the amount of species currently decreasing and at what
tempo? Why are so many species getting extinct in some environments now?
Those questions have to do with what we have been calling, since Walter G. Rosen
coined the word in the 80s (Takacs 1996) and Wilson (1988, 1992) popularized it,
"biodiversity".

However, there are many ways of measuring biodiversity, tracking its progress
or, more realistically, its erosion: different measurement methods defined by different indexes, such as Shannon index, Simpson index, etc. (Gosselin 2014; Noss
1990), as well as various ways of capturing it in relation to the ecological scale, such
as beta diversity, gamma diversity,[1] etc. Moreover, there are several concepts of
biodiversity, some attributing species a privileged role and others including also
genes, or ecosystems, as is attested in the definition of biodiversity used in international conventions, such as the Convention on Biological Diversity (1992):
"'Biological diversity' means the variability among living organisms from all
sources including inter alia, terrestrial, marine and other aquatic ecosystems, and
the ecological complexes of which they are a part; this includes diversity within
species, between species and of ecosystems". And even at the level of species diversity, species richness as the mere amount of species is often considered too rough a
biodiversity concept. In order to design robust diversity indices, ecologists or conservation biologists often add species evenness, and then consider the width of
diversity, named as disparity – some wanting also to integrate the consideration of
abundances within the concept of diversity (Blandin 2014). In addition to mere species counting, however, some dimension of species similarity sometimes ought to
be included in the concept of diversity: mitigating species diversity by functional or
phylogenetic similarity results in the concepts of phylogenetic diversity or functional diversity, whose use is especially required in ecophylogenetics (Mouquet
et al. 2012) for the former and in functional ecology for the latter.

---

[1] Those terms were introduced by Whittaker to capture aspects of the local and regional distributions of diversity. *Alpha diversity* refers to species diversity on sites or habitats at a local scale as
well as to the ratio of local to regional diversity, *beta diversity* compares the species diversity
between ecosystems or across environmental gradients; *gamma diversity* is the total diversity in a
landscape and therefore the compound of the former two.

Thus, while it was tempting in the beginning to consider biodiversity as a key question and a key *explanandum* of ecology, the diversity of biodiversity prevents us from straightforwardly claiming this. It may be argued, in turn, that this diversity seems to echo a lack of unity that affects ecology itself. It has indeed often been complained in ecology that the field lacks the unity that characterises the sister field of evolutionary biology. In 1989, Hagen already saw ecology as affected by a deep cleavage between a holological perspective and a mereological perspective, the latter using a demographic approach to ecosystems and communities while the former relies on a systemic view of the ecological objects, with or without appealing to evolutionary schemes of thought and natural selection (Hagen 1989). He concluded that this cleavage is essential to the discipline, and in turn allows ecology to explore a wide variety of objects and problems. More recently, Vellend (2016) has explicitly drawn a parallel between evolutionary biology and ecology and argued that ecology never had a unified framework similar to the one that structured evolutionary biology from the 50s onwards, and that allowed this science to flourish by providing researchers with common concepts, methods, key examples, key issues, and references.

Would it make sense to consider that the diversity of biodiversity is involved in the lack of unity of theoretical ecology? Or, more precisely, which disunity would be induced by this diversity, and is it unredeemable?

This will be the main question of the chapter. I will start by considering the issue of the long sought unity of ecology (13.2). Then I will explicate what I call the "explanatory reversibility" of biodiversity in ecology, namely its capacity to be *explanandum* and *explanans* in a science, as an essential feature of its theoretical role (13.3). Section 13.4 will consider more precisely the aspects of diversity as an *explanandum* of various ecological programmes, involving distinct explanatory schemes. Section 13.5 turns to diversity as an *explanans*, focusing on the relations between various kinds of stability and distinct notions of diversity, and characterizing the differences between such diversity and the way diversity is used in the explanatory programmes formerly described. In Sect. 13.6, I propose an account of the ecological notion of diversity in terms of a "conceptual space", in which various biodiversity concepts used in the varied explanatory strategies I described are specifically constructed. I use it in order to explicate the specific profile of the explanatory reversibility of diversity in ecology, and draw conclusions about the lack of unity in ecology and the epistemic status of the notion of diversity. The major argument developed there relies on the fact that the two explanatory projects concerning diversity target different "regions" of the total conceptual space of biodiversity so described.

13.2 The Unity of Ecology

It is often heard that ecology lacks unity – be it to complain about the missing unity (Vellend 2010), or to claim that it is a richness proper to this scientific discipline (Hagen 1989). Inversely, "unifying principles" or theories have been constantly pursued for ecology (e.g. Margalef 1963; Hubbell 2001; Loreau 2010; Vellend 2016). Before considering the specific theoretical role of a biodiversity concept in ecology, and the possible unifying role it could play as an object or a pervasive concept, I will review the most general divides that seem to prevent such unity. After having listed some subfields, I will attempt at ordering this disunity by indicating the major lines of division (summarized in Fig. 13.1 below).

A quick glance at ecological subdisciplines shows the overall variety of questions and methods that characterizes the field. *Behavioural ecology* studies the traits ("behaviours") of organisms, hypothesized as adaptations to their (possibly social) environment; *community ecology* is about communities, i.e. sets of various species in the same region, considered from the viewpoint of the diversity and succession of species occurring within it. *Population ecology* mostly considers few species and focuses on the dynamics of the abundances of each of them given their major ecological interactions (predation or competition). *Biogeography* is interested in the distribution of species across higher scale dimensions, namely regions. *Functional ecology* considers the interactions between various species from the viewpoint of their net effect on the shared environment, especially by addressing networks of trophic relations and ascribing its species a role in the ecosystem (Loreau 2010). *Ecosystem ecology* as advocated by the Odum brothers (e.g. Odum 1953) develops such approach and uses schemes of thermodynamic thinking in considering ecosystems (i.e. communities plus their abiotic environment) under the perspective of semi-closed systems exchanging matter and energy with their environment (Hagen 1992). On the other hand, *Ecological genetics* initiated by E.B. Ford (Ford 1964) – a student of Fisher – considers the dynamics of population in various species from the standpoint of the changes of gene frequency within each species. Finally, *evolutionary ecology* (Roughgarden 1979) borrows tools from ecological genetics and approaches ecological patterns as results of evolutionary processes.

Fig. 13.1 The divides of ecology. Each thick line represents one of the four dimensions. The thin lines stand for the position of three historically important views in ecology: in black, Clements (1916); in red, Allee et al. (1949); in blue Nicholson and Bailey (1935) (Color figure online)

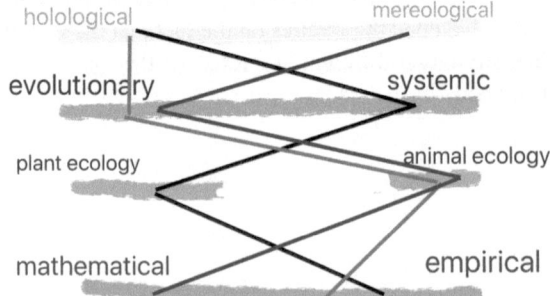

Broadly speaking, evolutionary biology and ecology have complex relationships (McIntosh 1986, pp. 256–263; Collins 1986; Harper 1967; Stearns 1982; Antonovics 1976; Huneman 2019). Besides its interest in explaining adaptation and evolution (phylogenetic patterns), evolutionary biology is interested in explaining the patterns of diversity and unity across diversity (i.e. homologies and analogies) that characterize extant and past taxa (as well as molecular patterns). And ecologists are generally interested, as I will argue more extensively in the following, in diversity within and across communities and ecosystems. Thus, both disciplines, at their own timescale, focus on the same *explanandum,* that is *diversity.*

Haeckel famously defined ecology as the "science of the struggle for existence", thus directly tying it to evolutionary biology to the extent that natural selection is seen by Darwinians as the key *explanans* and cause of evolution, adaptation and diversity. In principle, the emphasis on natural selection can be more or less strong in ecology, and this characterizes the whole field of ecological sciences: they are somehow ordered along a gradient which goes from evolutionary ecology to ecosystem ecology. At one extremity, evolutionary ecology adopts a very *evolutionary* viewpoint, considering ecosystems as the scene of competition, cooperation and mutualistic interactions, all occurring in evolutionary time and therefore being always dynamic. The other extremity of the continuum may be represented by most trends in ecosystem ecology, which adopts a very *systemic* viewpoint (sometimes akin to thermodynamics) insofar as ecosystems are open dissipative systems, more or less chaotic, dealt with in thermodynamic or statistical mechanics terms (Hagen 1992).

This divide is not the same as Hagen's distinction between *mereological* and *holological* perspectives on ecology mentioned before, since the holological view would accept an evolutionary understanding of ecology that takes communities or ecosystems as targets of selection. For instance, within a holological view echoing a Clementsian concept of community, an evolutionary parallel between communities and organisms, both being shaped by natural selection, is explicitly drawn in one of the major works on animal ecology in the mid twentieth century, namely the *Principles of animal ecology* written by Chicago ecologists Clyde Allee, Thomas Park, Orlando Park, Schmitt and Alfred Emerson, and praised in the *American Naturalist* by Dobhzansky. They say: "a community may be said to have a characteristic anatomy, an equally characteristic physiology and a characteristic heredity", therefore community is the "smallest [unit] that can be (…) selfsustained", and is precisely "*a resultant of ecological selection*" (Allee et al. 1949, 437).

Besides this gradient around the use of selection, which ranges across ecology, from the most systemic explanatory schemes (e.g. Odum's ecosystem ecology) to the most evolutionary understanding (e.g. Roughgarden's evolutionary ecology), ecology has been disciplinary cleaved between *plant ecology* and *animal ecology* since the 1900s. The two traditions were developed quite separately, starting respectively with the major advances of Warming (1909) and then Clements (1916) in plant ecology on the one hand, and the attempt at systematising animal ecology by Elton in the 1920s (e.g. Elton 1927), on the other hand.

Clements' idea of succession in communities, analogous to the development of organisms, was a key concept for much of plant ecology (Horn 1975; Lortie et al.

2004). Even though the individualistic concept of community put forth by Gleason in the 30s (Gleason 1939) won over much of ecology, one can still see a difference between plant and animal ecology to the extent that the attention paid to succession and assemblages has been more prevalent in plant than in animal ecology. And as noted by Harper (1967), it is harder to track the offspring of plants and determine their reproductive success, which partly explains why evolutionary perspectives were less favoured in plant ecology than in animal ecology.

In turn, animal ecologists have been massively worried about the question of the regulation of population size, which was probably the major controversial theme of ecology in the 50s, as can be noticed in the major gathering of evolutionary biologists and ecologists at the Cold Spring Harbour Symposium in 1957, devoted to population biology. Most of the talks – by Anderwartha, Birch, Lack, Chitty, Orians, etc. – were about population regulation in animal ecology. Much of this interest in regulation of population stemmed from a concern about pests. Charles Elton, a pioneer of invasion ecology, was the founder of the Bureau of Animal Populations in Oxford and one of its important tasks was pest control (see Chew (2011) on historical overview of invasion ecology, and Richardson (2011) on Elton's legacy in the field). Understanding the reasons of population regulation, population cycles and possible overpopulation was a crucial requisite for a successful control. One may argue that this context explains the difference between plant and animal ecology regarding the prevalence of the population regulation issue.[2]

Orthogonal to this divide between plant and animal ecology, there is an important tension between a more *empirically* oriented ecology and a mostly *mathematical* ecology (e.g. Schoener 1972). In a 1949 paper on population regulation (Solomon 1949). ME Solomon, British ecologist of the Bureau of Pest Control, noticed that ecologists are divided into two camps, one that starts from biology and generalizes, and one that builds mathematical models first and then tries to fit in the biological facts – e.g. Thomas Park's experiments on flour beetles, or perturbation experiments on populations (Smith 1952) vs. Lotka and Volterra's equations. This divide (see Kingsland 1995) still persists in various modes, as indicates the need recently felt by some theoretical ecologists to vindicate the use of mathematical theorizing (Servedio et al. 2014).

Regarding those five distinctions, each theoretical construction can be situated on each of the axes constituted by the gradient occurring between the poles of the distinction. In Fig. 13.1 I sketched the position of very influential works taken from distinct periods of the history of ecology (Allee et al.'s *Treatise* (1949), Nicholson and Bailey's model of host-parasite dynamics (1935), and Clements' plant ecology (1916)).

---

[2] Actually, Clements and Shelford (1939) intended to close the gap between plant and animal ecology, by applying a very general concept of community. They say: "the development of the science of ecology has been hindered in its organization and distorted in its growth by the separate development of plant ecology on the one hand and animal ecology on the other." (p.v) Ten years on and with a similar goal of systematizing ecological knowledge and providing basic principles, Allee et al. (1949), while acknowledging that principles of ecology should be general, still restrained to animal ecology for reasons of immaturity of the field.

However, in the face of these various divides within ecology, someone could argue that *biodiversity* defines an object of investigation that crosses frontiers between traditions, paradigms, and explanatory strategies. A major issue in ecology is indeed *coexistence* – why is it that certain various species coexist and others not? How can they do so? It is a question in both plant ecology and animal ecology, approached from mereological as well as holological perspectives, and through mathematical or more empirically oriented perspectives as well. Community ecology is openly concerned with explaining biodiversity patterns, and biogeography enquires about species-area laws, which are patterns about how biodiversity is scattered across various kinds of areas (MacArthur and Wilson 1967). However, even functional ecology gives a key role to diversity (Loreau 2010), at least under the mode of "functional diversity", namely the differences between species partitioned according to equivalence classes defined by ecological functional roles (producer, nutrient cycler, etc.) (Dussault and Bouchard 2017).

In addition, the emergence of the word 'biodiversity' in the 1980s could also indicate that there is an object proper to ecology here. In the following I shall focus on the explanatory logics of diversity in ecology, in order to assess (and eventually infirm) the hypothesis that biodiversity constitutes a shared object amongst various ecological theories and traditions, and that its concept could help define a unifying framework for ecology.

13.3   The Explanatory Reversibility of Diversity

Many ecologists' researches indeed focus on diversity. They range from very general questions about what causes diversity in general – Hutchinson asking "Why are there so many animals?" (Hutchinson 1961)–, to questions about the way diversity is distributed locally and regionally – species-area laws in biogeography, and the mathematical models explaining them in McArthur and Wilson's *Theory of Island Biogeography* (1964), species abundance distributions in community ecology, or the patterns of succession of plant species in communities (as illustrated in Clements' works), as well as questions of medium degree of generality about how it is possible that many species coexist *generaliter*. One could view all these questions as various modes of an overarching coexistence question: how is coexistence (amongst *diverse* species or organisms) possible and realised at various scales?

Besides *explaining* diversity under its various modes, ecology is concerned with biodiversity in another and very different way. A longstanding debate in ecology regards what has been labelled the "diversity-stability hypothesis" (Ives and Carpenter 2007; Pimm 1984). Simply put, it is the claim that diversity – especially species richness – begets stability (mostly in the form of the constancy of species abundances). The more species an ecosystem includes, the more stable it seems to be (namely, it contains the same species for a long time, with abundances fluctuating around a steady means). In this sense tropical forests, which are species rich ecosystems, have been providing examples of this pattern for many decades. The intuition of this fact was very robust, but its explanation has been overlooked for a long time.

However, in 1974 when Robert May started to investigate this hypothesis mathematically, by modelling networks of species and increasing the diversity value, it turned out that diversity does not beget stability but on the contrary prevents it (May 1974). Assuming that a system which is "stable only within a comparatively small domain of parameter space (…) may be called *dynamically fragile*", clearly "such a system will persist only for tightly circumscribed values of the environmental parameters" (May 1975). The result of May's models is that a "wide variety of mathematical models suggest that as a system becomes more complex, in the sense of more species and a more rich structure of interdependence, it becomes more dynamically fragile". (ib.) Researchers then tried to address this gap between these mathematical models and some data that tended to show a stability-friendly effect of diversity. The question of stability then became: what does explain the fact that some empirically attested diversity does not conform to May's mathematical models?

Ecological stability is actually a crucial issue for theoretical reasons. After Darwin's revolution, Linnaeus' explanations for stability (namely, each species fulfills a role in a well-balanced nature; see Pimm 1993) were no longer possible, and, inversely, the constancy of ecosystems constitutes a challenge if the world is an ever-changing Darwinian world led by competition. Ecological stability is also challenging for practical reasons, since understanding what makes ecosystems robust could allow us to manage and protect them. (In fact, almost since its beginnings scientific ecology has been concerned with the damages inflicted by human industry and agriculture to natural ecosystems and ultimately to the environments in which human societies live).

Diversity is therefore a two-faceted concept: it is a major *explanandum* for ecology under various guises, but when the question concerns the stability of ecosystems, diversity becomes an *explanans*. We witness here a major epistemological feature of evolutionary and ecological questions, namely the "explanatory reversibility" of key concepts. Some concepts may indeed be the *explanandum* in some contexts and the *explanans* in others, and this reversibility attests to their theoretical significance. In evolutionary biology, notions such as plasticity (Nicoglou 2015), robustness (Wagner 2005; Huneman 2018) or mutation rates display this epistemic feature, which was first recognised by Fisher in connection with some major properties of the genetic system (dominance, recessivity, etc.) that condition evolution and are at the same time a product of past evolution (Fisher 1932).

In the field of ecology, diversity constitutes one of the concepts whose epistemic profile displays such reversibility. In the following I will explore this reversibility in more detail, and examine the role it may play in the structure of ecology.

13.4   Diversity as an *Explanandum*: Conceptual and Historical Aspects of the Ecological Coexistence Issue

From the early times of ecology, diversity as an explanandum has been understood as a question of coexistence. I shall recapitulate this matter, and then consider a theoretical framework used to address it, namely the concept of ecological niche and its formulation by Hutchinson. I shall then turn to rival conceptions, mostly structured today around the idea of a "neutral theory". In each case, I will emphasize the aspects of the concept of biodiversity that are prominently addressed in each explanatory scheme.

The *coexistence* question may be arguably one of the key issues handled by ecologists since the early twentieth century. For decades, plant ecologists have embraced Clements' concept of community, which is slightly like an organism and displays a process of succession analogous to the development of an organism.[3] Clements and the animal ecologist Sheldon in *Bioecology* (1939) generalized this idea to plant and animal communities. Allee et al. (1949) major treatise on animal ecology took the concept of community on board – i.e. "the natural unit of organization in ecology" (437) – as well as the parallel between organisms and communities, since like Clements they consider communities to have a "metabolism". Their question here is about explaining the composition rules of an assemblage of species in a given community, and whether there are laws governing these species' procession.

However, in the 40s and 50s, the coexistence question seemed to be supplanted by a different issue, i.e. the explanation of *population regulation*: why does a population of a species generally fluctuate over a specific abundance, with regular cycles? From Elton (1927) to Hutchinson (1957) at least, the regulation issue was the other major problem for ecologists, especially animal ecologists – with, as mentioned above, a practical concern for invasions and pest control. To some extent, the regulation issue was more mathematically tractable than that of coexistence, as attested in the seminal models by Lotka and Volterra (Volterra 1926) and by Nicholson and Bailey (1935), which mostly deal with two or three species. Nicholson and Bailey explicitly acknowledged that handling many species would require very sharp mathematical skills (1935, 597).

Yet, when Hutchinson (1957) formulated his influential concept of niche as a hyperspace of environmental parameters in which a subspace of the hyperspace defines the viability conditions for a species, the coexistence question came again at the center of theoretical ecology.[4] At the time, such question was often traced back to an appeal to some form of group selection, as exemplified by Allee et al. (1949) animal ecology treatise and shared by many ecologists, as indicated above (see

---

[3] "Development is the basic process of ecology, as applicable to the habitat and community as to the individual and species." (Clements and Shelford 1939, 4)

[4] See Pocheville (2015) for a conceptual history of 'ecological niche' that relates Hutchinson to earlier views by Grinnell and Elton.

Mitman (1988) on these ideas of collectives and group selection). Coexistence in a community could be thought along the same lines as organismic integration, given that natural selection – individual in the latter case, collective in the former case – underlied both systems and their cohesiveness. David Lack's work on clutch size (Lack 1947, 1954) however progressively provided powerful arguments to think that individual natural selection, and not group selection, was the reason of population regulation, and a little bit later the idea of group selection met the devastating critique issued by Williams (1966). All this made the group selection approach to the coexistence question harder in principle. Hutchinson's idea of niche to some extent thereby set the frame for more fruitful approaches to various modes of the coexistence question.

More precisely, Hutchinson published his conception in the "Concluding remarks" to the 1957 Cold Spring Harbour Symposium, where prominent ecologists and evolutionary biologists debated population ecology and mostly the regulation issue.[5] The volume was a final landmark in the debate over competition-centered (inspired by Nicholson (1933) initial model of regulation by density-dependent factors, and mostly represented by Lack (1954)) and density-independence-centered explanations of population regulation that emphasized factors such as climate (Anderwartha and Birch 1954).[6] Hutchinson's view of the niche followed his assessment of the debate, which tried to fairly acknowledge some epistemic value in both positions – mainly Lack's view of density-dependent regulation by competition and Anderwartha and Birch (1954) view of regulation by density-independent factors.

This concept of niche was used by Hutchinson to make sense of the role of competition in the regulation process. But more importantly, it also allows a grasp on the *coexistence* issue. Here, what explains coexistence is indeed the fact that first, each species has a "fundamental niche", and second, that the portion of a fundamental niche shared by two species will be exclusively inhabited by the best competitor (Fig. 13.2). "Fundamental niches" once restricted by the process of competition – so, finally, natural selection – yield the "realised niches", which explain where a species will actually be found in the environment. In a classic study, Joseph Connell (1961) studied two species of barnacles, *Balanus balanoid* and *Chtamalus stemallus*, which have a stratified distribution along the coast of Scotland. The *Balanus* live on the border between see and rock, while the *Chtamalus* live just above it (Fig. 13.3a). *Balanus* cannot really live much higher because they cannot resist dessication during low tides. But if we take out the *Balanus*, the *Chtamalus* now appear to occupy also the space inhabited by the *Balanus*, in addition to their known territory (Fig. 13.3b). Thus, the fundamental niche of *Chtamalus* is the whole region of the rocks on which *Balanus* and *Chtamalus* live, but their realised niche is the territory where one finds them along with the *Balanus*, because the latter are a better competitor and wash *Chtamalhus* away from this portion of their fundamental

[5] On Hutchinson's work and influence on ecology see Slack (2010).

[6] Collins (1986) and Huneman (2019) argue that this episode was indeed instrumental in introducing the evolutionary viewpoint in ecology.

Fig. 13.2 Realised and
fundamental niches (the
circle dots species is a
better competitor than the
crosses species)
(Hutchinson (1957))

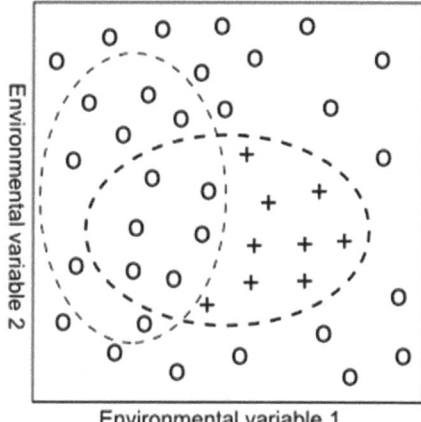

Environmental variable 1

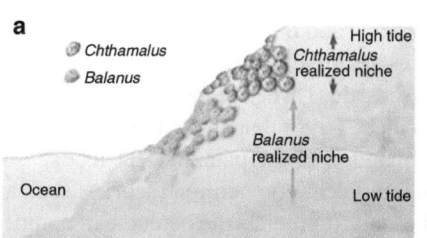

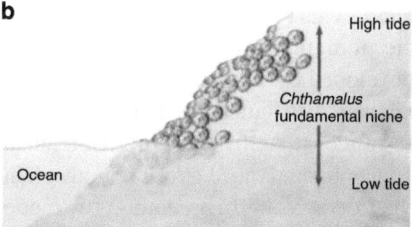

Fig. 13.3 (**a**) Coexistence of Balanus and Chtamallus. (**b**) When Balanus are taken out, Chtamalus reveal their fundamental niche by occupying it

niche. Partitioning the total environmental hyperspace into realised niches through the competitive exclusion principle eventually explains coexistence. The basic idea is that similar species cannot coexist for a long period of time, so one of them ultimately restricts the realised niche of the other: this idea yields a family of "limiting similarity" theories of coexistence that have been elaborated since the 60s. "The generalization (..) that two species with the same niche requirements cannot form mixed steady-state populations in the same region has become one of the chief foundations of modern ecology." (Hutchinson 1957)

As Hutchinson noticed, this theoretical tool is however not able to wholly explain coexistence. The "paradox of plankton", as he himself formulated (Hutchinson 1961), is the fact that while the hyperspace of environmental parameters in the ocean is very small, since there are few parameters (light, pH, temperature, etc.) distributed over a small range of values, thousands of plankton species exist instead of a few ones as predicts the theory of the niche. Hutchinson considered various accounts to explain this, especially the view that in reality the parameter values vary at the same time as the competitive exclusion process operates, which entails that the equilibrium partition of the niche hyperspace predicted by the principle of competitive exclusion cannot be reached,[7] and many more species actually exist.

The niche theory elaborated by Hutchinson has been crucial to address the coexistence issue. Hutchinson (1957) used it to explain why there are so many animals: the huge variety of plants, on which animals feed, makes for a very large hyperspace for possible animals, and hence many realised niches and species. But another crucial ecological aspect is the distribution of species abundances, which is not taken into account by Hutchinson's theory. As seen by Fisher and then Preston in the 30s, the species abundance distribution (SAD) seems to realize constant patterns, which can be expressed by either a lognormal curve (Fisher et al. 1943) or a logseries (Preston 1948). Witnessing such a regularity in SADs, across various kinds of ecosystems on the planet, raised the question of explaining such patterns. Consequently, theories of coexistence approached by limiting similarity elaborated since the 60s have been refined to understand such patterns of biodiversity as SADs. A most recent elaboration of this theory is called the R* theory. Stemming from MacArthur and Levins (1967) paper on patchy environments and the competition between species foraging a finite set of identical resources heterogeneously distributed, and later developed by Tilman (1982), this theory asserts that "when resources are heterogeneously distributed, the number of species can be larger than the number of limiting resources, thereby resolving Hutchinson's paradox of the plankton. R* theory is a conceptual advance over previous phenomenological-competition theories, such as the Lotka–Volterra predator–prey model, because it predicts the outcome of competition experiments before they are performed." (Marquet et al. 2014).

Biodiversity as an *explanandum* means more than the coexistence question and SAD patterns (see e.g. MacArthur 1972). In fact, another pattern discovered by ecologists carrying out censuses (especially on plants) was about the relation of surface area and species number. Called "species-area law", this was also a major pattern to be explained. In the seminal book by MacArthur and Wilson, *Theory of Island Biogeography* (1967), the authors recognise that the distribution of species is a major *explanandum*. However, their aim consists in switching from the "natural history" of species, which is mostly collecting patterns of those distributions, to a mechanistic explanation (more on this below). *Theory of Island Biogeography* starts by elaborating the "inland-island model". Islands are small territories separated from one another by the sea, and all of them are at some distance from the inland. The inland constitutes a reservoir for species. Individuals of those species colonise islands, but the chances of colonizing an island depend both on the distance of inland to X and on the size of X. The mathematical model therefore intends to

---

[7] "At any point the illuminated zone of the ocean or a lake the phytoplankton is normally quite diversified. There is no opportunity for niche specialization and the fundamental trophic requirements of all forms will cause them to draw on the same food supply. Such population cannot therefore represent equilibria, but since in general the plankton, though continually changing, remains in a highly diversified state, one can only suppose that the direction of competition is continually undergoing change with the progress of the seasons and concomitant thermal and chemical changes in the water and that no opportunity for the establishment of a single species equilibrium condition ever occurs." (Hutchinson 1959)

explain varieties of species-area laws on the basis of these three parameters: island size, amount of islands, distances to inland.

Theory of Island Biogeography uses a hypothetico-deductive model-based method: like Fisher or Kimura's population genetics, it starts by building models and then considers how data and patterns fit to the model. *Theory of Island Biogeography* also works at a higher scale than community ecology. "Islands" of course are theoretical entities, not physical islands; they correspond to territories that are poorly communicating genes and organisms to other territories: they can be valleys, forest patches, etc. (Island biogeography has theoretical affinities with the concept of "metapopulation" elaborated at the same time by Richard Levins (1969)).

Noticeably, most of *Theory of Island Biogeography* considers the dispersion and colonisation of species and not the relative fitness differences between individuals. Hubbell (2001, 2006, 2010) will consider this as the first "dispersal assembly" model of coexistence, and will contrast it with the "niche assembly" models deriving from the limiting similarity theory (Leigh 2007). His own theory, called the "neutral unified theory of biogeography and ecology" (Hubbell 2001), intends to elaborate a dispersal assembly model, which is therefore 'neutral' in the sense that, like Kimura's evolutionary models (Kimura 1985; He and Hubbell 2005), there is no fitness differences between elements (alleles for Kimura, species for Hubbell). His model integrates both regional and local community scales, and therefore allows to explain species abundance distribution as well as species-area laws. The key change for his theory (as compared to *Theory of Island Biogeography*) is that the neutrality assumption, called "ecological equivalence", is defined in terms of per capita birth and death rates rather than in terms of species fitness (as McArthur and Wilson were doing) (Munoz and Huneman 2016, Hubbell 2005; Purves and Turnbull 2010). It met predictive success: "It was surprising to find that spatial neutral models give rise to frequency distributions of precision that are very similar to those estimated from biological surveys, as a consequence of the spatial patterns produced by local dispersal alone" (Bell et al. 2006) – which concerned several kinds of communities (see McGill et al. 2006): plants, coral reefs or fish (Muneepeerakul et al. 2008).

Thus, a major divide in contemporary community ecology is today defined by the meaning ascribed to the neutral theory: whether ecologists follow Hubbell in considering that it is a good theory for biodiversity (Bell 2000), especially because it has far less parameters than the rival R* theory (Marquet et al. 2014), or think that the niche paradigm is still the best explanation since neutral models are not explanations (see amongst others Chave 2004; Hubbell 2005; Holt 2006; Allouche and Kadmon 2009; Leibold and McPeek 2006; Doncaster 2009; Rosindell et al. 2011). Without delving into the controversy, it has to be noted that the neutral theory appears as a unified, scale-encompassing theory of biodiversity while the limiting similarity paradigm proposes explanations that are generally different for several aspects of coexistence (species abundance distribution, species area laws, etc.) (Huneman 2017).

In all those theories, the *explanandum* is a range of patterns of coexistence that are defined mostly in terms of *species diversity*, and often the species richness is the major aspect – even though species abundances are also taken crucially into account in SADs.

But what happens if we turn to diversity as an *explanans*?

13.5   Diversity as an *Explanans*

In this section, I will consider the so-called diversity-stability hypothesis and the particular notions of diversity that have been involved in the attempts to clarify, formulate and test this hypothesis for four decades. I will first consider approaches that focus on the way diversity as species richness is organized through interactions in ecological networks; then I shall turn to the notions labelled "functional diversity" and "phylogenetic diversity".

The diversity-stability hypothesis has been for a long time an assumed but unproven hypothesis, evidenced by many observations – somewhat like famous mathematical conjectures that are not proven but seem established by the behaviours of known numbers. As Orians writes, "The belief that natural ecosystems become more diverse and, hence, more stable with time after a disturbance is widely accepted and regularly repeated in ecology textbooks (...) the correlations, not to mention causations, are still obscure." (Orians 1975, p. 139) Indeed, Orians notices that even the correlation between diversity and stability could not exclude that such a common cause as environmental constancy yields both.[8]

To this extent, the real meaning of the terms involved (which diversity? what stability?) was not really investigated.[9] Thus, as indicated above, May's mathematical findings that diversity *per se*, crudely defined as the number of species interacting in an ecosystem, does not beget stability, triggered a reflexive turn in the study of diversity as a stability-promoter.[10]

Robert May's results were in fact showing that an ecosystem with randomly interacting species is less stable – in the sense of constancy of species' abundances – while the amount of species increases. Yet some evidence of a stability-begetting effect of diversity existed in the field, as stated, for example by McNaughton (1977): "The weight of evidence resulting from explicit tests of the diversity-stability hypothesis (...) suggests, not that the hypothesis is invalid, but that it is correct".

---

[8] "Environmental constancy facilitates diversity while reducing perturbations that might affect stability" (Orians 1975, p. 139)

[9] "The concepts are normally discussed with poorly defined terms, reflecting an uncertainty about what concept(s) of stability are useful in ecology" (ib.)

[10] Notice that the regulation issue, arguably another key issue of theoretical ecology, also concerns an aspect of stability, since it is about the steadiness of one species population abundance. Therefore, the diversity as *explanans* is involved in the other major question of ecology, provided one assumes that the key questions are regulation and coexistence.

Thus, ecologists then enquired about what aspects of diversity were in those cases accounting for this effect. First, in contrast with May's models, the connections in an ecological network are often not random. It may be that many predators have only one prey, for instance, and that a few superpredators have many preys. In any case, few ecosystems are such that species have an even chance of having a given number of preys or predators. Thus, the question switched towards the identification of the properties of ecological networks that would be such that increasing diversity would increase stability. Given a fixed degree of diversity as species richness, many networks are possible: a first rough characterization of their differences is their particular value of connectance, namely, the ratio of the amount of realised connections (here, interactions) between species to the total amount of possible connections.

More generally, a perspective on the question of the role of ecological diversity in stability is the general investigation of topological properties of graphs realised by ecological networks of interactions. Diversity, as species richness *per se*, does not increase stability but some network topologies make it likely to promote stability: this hypothesis supports the general move towards an investigation of ecological networks and their role in stability (Solé et al. 2002; Dunne 2006; Dunne et al. 2002; Kéfi et al. 2016). Some of the results emphasize the key role of species networks topologies in guaranteeing some stability. Scale-free networks, in which the distribution of the degrees[11] of the nodes follows some power law,[12] are stable because this topology entails a very low probability for a random species extinction to reach one of the hubs of the network and hence alter the overall structure, and ultimately the functioning of the community (Solé et al. 2002). This probability becomes lower with the increasing size of the network, i.e. with the increase in species richness.

Small-world structures[13] of ecological networks, when they are realised, also beget stability. This is because the high clustering coefficient means that the overall pattern of interaction is mostly preserved if some cluster in the network is altered. On the other hand, the short path length means that a species which loses its privileged interacting species in its neighbouring cluster can still be related to its other interacting species via the other species in its network, to which it is highly connected (Strogatz 2001; Solé and Goodwin 1988). In this case, similarly, increasing

---

[11] In a given network, made up of nodes (or vertices) and edges that connect some nodes, the "degree" of a node is the amount of edges on this node.

[12] Intuitively, there are a few nodes with many connections (they will be called hubs), slightly more hubs with a bit less connections, and so on, and a large majority of nodes with only very few connections. Formally speaking, the number of nodes of degree n + 1 will be 1/10 the number of nodes of degree n. (Or any mathematically power law of the same kind). Wealth in human societies is known to follow power laws; and one frequent generating process for power law nodes distributions is the "preferential attachment", namely, the probability of having a new connection is proportional to the extant amount of connections. Sometimes called "rich get richer", this process is clearly instantiated by financial mechanisms (Albert and Barabasi 2002).

[13] Small-world is a kind of network characterized by the fact that it is highly clustered (a cluster being a set of nodes more significantly connected between themselves than to other nodes) and at the same time has a short path length (the path length being the average number of edges between two randomly taken nodes) (Watts and Strogatz 1998).

the amount of species, hence the size of the network, strengthens this stability-enhancing property.

May's counterintuitive findings about stability not yielded by diversity in general are therefore corrected or supplemented by those network analyses of the topology of ecological network; however, it is not clear exactly what is meant in both cases by "stability". Thus, the meaning of "stability" in all these models had to be questioned. As Tilman (1994) made clear, even if species richness does not, in theory, beget stability as constancy of species *abundances*, it has a positive effect on the constancy of *biomass* of an ecosystem. That is clearly another meaning of "stability", which relies on diversity. And early on, Holling (1973) had introduced "resilience" understood as the ability of an ecological system to restore its key parameters after a perturbation. Resilience has various modes and can be empirically measured. Moreover, "persistence" named the fact that an ecosystem does not "lose" a species, even though the abundances of all species vary a lot and do not come back to the initial state.

Yet, notions of stability are themselves even more numerous than that and it is not even clear if there is one overarching meaning. Orians (1975) distinguishes: *Constancy* – "a lack of change in some parameter of a system, such as the number of species, taxonomic composition, life form structure of a community, or feature of the physical environment"; *Persistence* – "the survival time of a system or some component of it"; *Inertia–* "the ability of a system to resist external perturbations"; *Elasticity* – "the speed with which the system returns to its former state following a perturbation" (which is similar to Holling's resilience); *Amplitude* – "the area over which a system is stable"; *Cyclical Stability* – "the property of a system to cycle or oscillate around some central point or zone"; *Trajectory Stability* – "the property of a system to move towards some final end point or zone despite differences in starting points". (Fig. 13.4) Stability, in other words, depends on the kind of perturbations one considers, and for Orians, in addition, all measures should be related to fitness: "For these relationships to be insightful, perturbations [or perturbation types] should be related to the evolutionary histories of the organisms experiencing the perturbations, and measured in terms of the total investments that must be made to increase or maintain fitness during those perturbations." (ib. p. 143) This indicates a bias in favor of evolutionary approaches to ecology, which may not be found in other theories of stability, especially when one turns to functional or ecosystems ecology.

In any case, the question of which diversity begets stability, and how it is possible that a certain diversity begets a certain stability, presupposes that one clarifies which stability is at stake. Not all diversity properties are likely to beget the same stability property.[14]

Of course, the ecological networks can be understood also from the perspective of their dynamics (Ulanowicz 1983; Szyrmer and Ulanowicz 1987), and especially

---

[14] On the various meanings of stability in ecology, and the possibility of formally making sense of some of them in the context of phase spaces, attractors and measures of Lyapounov exponents, see Justus (2008).

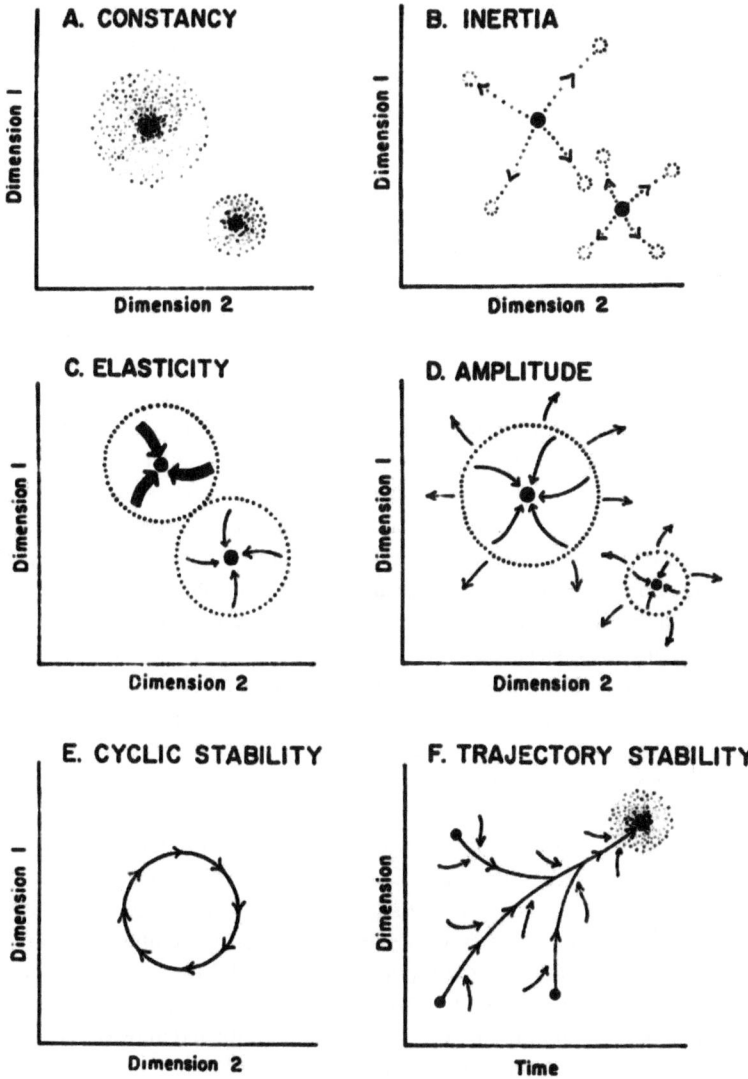

Fig. 13.4 The six kinds of stability in ecology. (After Orians 1975)

by considering not only the constraints put by the topology on the possible dynamics (Huneman 2015), but also, by capturing the major behaviours of the dynamics of fluxes within the networks and the possible evolution of the networks likely to follow (Ulanowicz 1986). This approach is perfectly compatible with a functional ecology that would consider ecosystems as open thermodynamic systems and model their inner behaviour, such as what Odum (1953) theorised. It allows researchers to understand the role that increasing diversity (as species richness) plays in the productivity of ecosystems, or ecosystem functioning, or some key features of ecosystem functioning.

The network perspective is not the only way to capture the possible contributions of diversity, mostly as species richness, to stability, or to some aspects of it. Functional ecologists started to define "functional differences" understood in terms of functional roles of a species played in an ecosystem (Blandin 2014). From this perspective, two species can be biologically different but functionally equivalent. Such functional diversity may be likely to play a role that species diversity cannot play in the emergence and maintenance of some stability. However, functional diversity and species richness are not wholly orthogonal. As Tilman (1996) argued, species diversity induces a lot of microscale environmental heterogeneities, which in turn allow for a wide variety of ecological roles. But this connection is just plausible and does not allow one to always consider species richness as a proxy for functional diversity.

Experiments have recently confirmed the stability-enhancing role of functional diversity. The bumphead parrotfish *Bolbometopon muricatum* is the largest parrotfish in the oceans and is considered a keystone species in the coral reef (Huey and Belwood 2009). It is a major target for fishermen (and hence an imperilled species) but is also heavily consuming reef substrate: "the most conspicuous and perhaps most powerful effect *B. muricatum* has on reef ecology is delivered via individuals' intense direct consumption of reef substrate." (McCauley et al. 2014)

The experimental change of this parrotfish to another parrotfish, or the reintroduction of other parrotfish species after its removal, show that the equilibrium of the coral reef is threatened. Species diversity in this case is not changed (Bellwood et al. 2003). But given that the functional role of the bumphead parrotfish is unique, it follows that functional diversity is decreased while species richness remains constant. In this case functional diversity, and not species diversity, is what contributes to ecosystem stability.

Functional diversity seems thus to positively relate to productivity and stability of ecosystems. However, as argued by Cadotte et al. (2009) "functional group richness is a problematic measure for two reasons. First, the removal or addition of "functionally redundant" species may have effects on community dynamics and processes, indicating that there are important functional differences not captured by broad groupings. (...) The second reason is that functional group richness tends to predict only a limited amount of variation in productivity and may even explain less variation than having randomly assigned groups."

Thus, more recently ecologists have started to consider *phylogenetic* diversity and its role in ecosystem functioning and conservation biology, under the name of "ecophylogenetics". Here, phylogenetic diversity is understood as "the amount of evolutionary history represented in the species of a particular community", and "commonly used measures of phylogenetic diversity are the total branch length of a phylogenetic tree that contains all species present in a community, or the sum of pairwise distances between species weighed by their relative abundances." (Mouquet et al. 2012) Ecologists found, for instance, that plant productivity is enhanced in communities with phylogenetically distantly related fungal species compared to closely related species. "This result suggests, under the hypothesis of a strong phylogenetic signal of the traits considered, that the loss of an entire lineage could have strong negative ecological consequences since distinct lineages are likely to perform different functions." Thus, to this extent one can use phylogenetic diversity "as a proxy of unmeasured functional diversity for the purpose of assessing its connection to ecosystem functioning" (Mouquet et al. 2012).

The three diversities, species richness, phylogenetic diversity and functional diversity, are in general quite decoupled. This is manifest in a study by D'Agata et al. (2014) on the human impact on biodiversity loss in coral reefs. In the reef area, human density varies on a gradient spanning from 1,7 to 1720 inhabitants/km². The researchers investigated the effect of this density upon the three biodiversities. It turned out that the impact starts to be sensible at a threshold of around 20 inhabitants/km²; however, the effect is very different regarding each kind of diversity. Considering the extreme impact, at 1705 inhabitants/km² the effects are: on species richness: 12%; on functional diversity: 46%; on phylogenetic diversity: 36%. Thus, first, species richness is a very bad predictor of human impact on biodiversity loss and should be not used as an indicator for coral management, one should prefer functional and phylogenetic biodiversity instead; second, the slope of the impact after the threshold, on each diversity, is significantly different, therefore they cannot be taken as proxies for each other (Fig. 13.5).

Fig. 13.5 Differential effects on human density in three kinds of biodiversity. (After D'Agata et al. (2014))

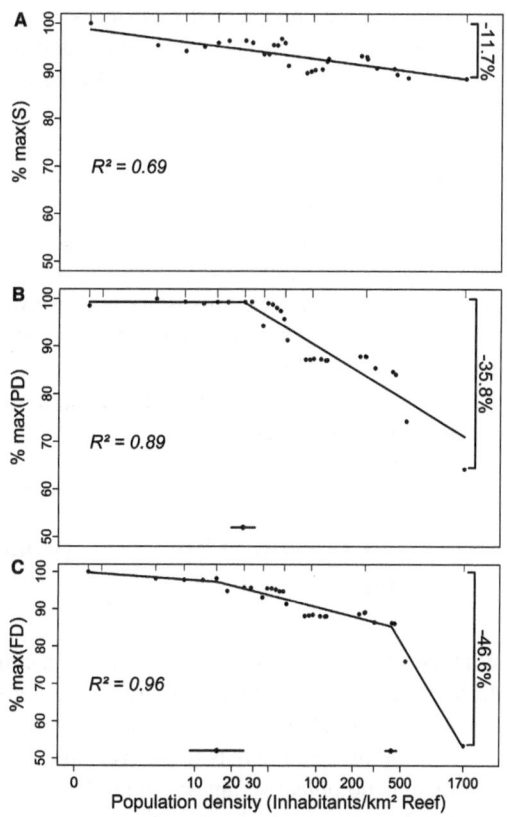

To sum up, diversity as an *explanans* is diffracted into several concepts of diversity such that each plays, within different explanatory perspectives, a specific explanatory role regarding productivity, stability and other ecosystem functioning aspects. Those diversities are not translatable and are in general weakly correlated, even though locally under some conditions they can be quite aligned.

It seems therefore that the explanatory reversibility of diversity includes a gap between the *explanandum* and the *explanans*, since the *explanandum* is mostly concentrated upon species richness, unlike the *explanans*. In turn, the *explanandum* is instantiated in various *patterns* of biodiversity that may link space and diversity, while the *explanans* generally does not include biodiversity patterns (or at least the same biodiversity patterns: SADs, species-area distributions etc.). Two general conclusions can be drawn here: as an *explanans*, ecological diversity is much more diffracted than as an *explanandum*; and the explanatory reversibility of the concept is not transparent, complete or univocal.

One can usefully compare this explanatory reversibility to the explanatory reversibility of robustness in evolutionary biology. Here, robustness, understood either as a capacity to function notwithstanding disturbances, or as an ability to maintain a set of functions in a very wide range of circumstances (Kitano 2004) also covers distinct meanings. Especially, the two key types of robustness for evolutionary biologists are "mutational robustness", as a robustness defined with regard to genetic mutations, and "environmental robustness", as a robustness defined with regard to environmental changes (de Visser et al. 2003). Biologists debate about whether one has been the effect of the other, and then, given that robustness is a very general property of living systems at all levels (Wagner 2005), they ask two kinds of questions: what made robustness evolve (robustness as an *explanandum*)? What does robustness do in evolution and how does it affect it (robustness as an *explanans*)? But such explanatory reversibility of robustness (Huneman 2018) is such that the two types of robustness are together considered, both, in the *explanans* side and in the *explanandum* side. This is not the case with the biodiversity concept in ecology. In the last section, I shall attempt to account for the structure of the concept of diversity in a way that will make sense of this specific explanatory reversibility of the concept. Ultimately, this will decide upon the role of "biodiversity" as a crucial concept for unifying ecology.

13.6   A "Conceptual Space" Approach to the Diversity Concept

What do I mean when I say that some X – a community or an ecosystem – is more diverse than Y? Does it include more species, or species more diverse, or more functionally diversified, or is X phylogenetically more extended on the tree of life than Y?

No principled way exists to answer this question. One could be tempted to say that there is no objective answer at all. However, another approach consists in saying that there are many objective facts enveloped in a judgment about X being more diverse than Y, and that the concept of biodiversity is then in each case built or constructed upon this set of objective facts. Various answers to the question are then yielded by various ways of constructing this concept of biodiversity.

Such an approach could be developed in the following terms: consider each of the properties used to construct biodiversity indices and to measure biodiversity as axes in a hyperspace. Species richness would obviously be one, as would then be species evenness, disparity, species abundance, phylogenetic distance, functional differences. Those axes describe facts about each community or ecosystem that can be objectively measured: the number of species at the local scale, their abundances, the functional redundancies or the amount of the phylogenetic trees covered by the species in a community or metacommunity are not in the eye of the beholder, they can be settled independently of epistemic preferences, explanatory strategies or methodological choices (or, at least, their objectivity is not different or less objective than generally establishing facts in science). Thus each community or ecosystem occupies a point (or a small neighbourhood, considering that the values evolve in time) in this space, defined by how much it scores on each of these axes (Fig. 13.6). Functional diversity is the projection of this point on the axis "functional diversity"; same for phylogenetic diversity; etc.

But of course each axis may not be as important as the others regarding a given diversity measure – for instance, some concepts of diversity used in conservation biology would overtone functional diversity or species abundances; and diversity in ecophylogenetics, but also in biogeography, could overemphasise the axis of "phylogenetic diversity". Many diversity indices are indeed constructed by considering the values on

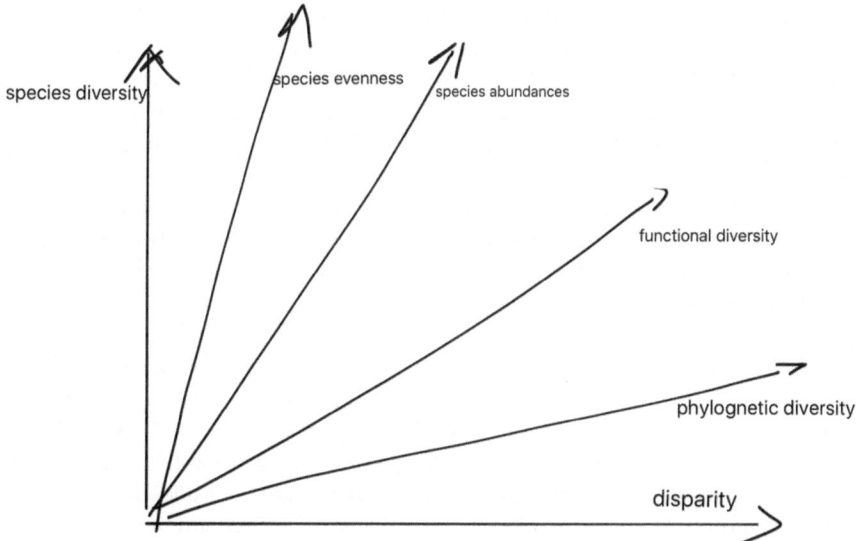

Fig. 13.6 The conceptual space of biodiversity and its axes. Notice that axes can be weighed and scaled differently in order to construct a specific biodiversity concept

several axes but not all of them, and then by possibly weighing the various axes differently; e.g. species richness or disparity could be differently weighed. If one wants to represent in our conceptual space the way a specific biodiversity concept, and then biodiversity measure, is constructed, one could assign different scales to each axis. This plurality of choices regarding the importance, weight or scales of each axis results in a plurality of possible concepts of biodiversity. And in turn, each explanatory project in ecology regarding diversity – as an *explanans*, or as an *explanandum* – will involve one (or a few) specific biodiversity concepts among this plurality.

In this approach, "biodiversity" appears as a possible construction built upon the objective values that X (community, ecosystem) scores on various axes. Each way of constructing it, by making projections on some axes, or taking only a few axes, possibly scaled or weighed in different ways, provides a different concept of biodiversity. Each of these concepts in turn is based on objective facts, but includes some epistemic and possibly *non-epistemic* values that governed the construction of this concept from those facts. For instance, the biodiversity concept used in conservation may emphasize the dimension of abundance, since the probability of extinction of a species – which is in general something conservation biologists intend to prevent – is inversely proportional to abundance. But the weighing of the axes here, and the overweighing of abundances, relies on the non-epistemic value of our interest in conserving species. Inversely, some biodiversity concepts used when one wants to design, maintain or maximise ecosystem services, may favor the functional diversity; here too, the reasons for weighing axes differently relies on non-epistemic values, namely our interest in flourishing ecosystem services.

Now, the explanatory reversibility of the concept of diversity can be approached in this context. Considering that biodiversity is defined in this conceptual space determined by the axes I mentioned, it appears that diversity as *explanans* and diversity as *explanandum* target different regions of this space (Fig. 13.7). According to analyses in Sects. 13.4 and 13.5, the *explanans* is heavily concentrated around the axes on functional and phylogenetic diversity, while the *explanandum* would be rather located around the axes of species richness, evenness and abundances. The overall conceptual space of diversity is therefore not identically involved in the two explanatory takes on diversity, and this characterizes the epistemic nature of such an explanatory reversibility, as compared to the explanatory reversibility of the concept of robustness mentioned above. The latter is "complete", while the former is not – in the sense that the conceptual space (respectively, of diversity and of robustness) is in the latter case completely and identically concerned by both explanatory projects, and in the former, partially and differently concerned by each explanatory project. But (unlike diversity) robustness cannot claim to be a shared and pervasive object in evolutionary biology, and therefore the "completeness" of its explanatory reversibility does not carry consequences for the question of the theoretical unity of evolutionary biology, unlike in the case of ecological diversity considered here.

This approach to diversity as a conceptual space was not only intended to provide a representation for the incompleteness of the explanatory reversibility of diversity. It is more generally intended to make sense of the fact that the epistemic status of diversity in ecology does not allow for a theoretical unity based on such

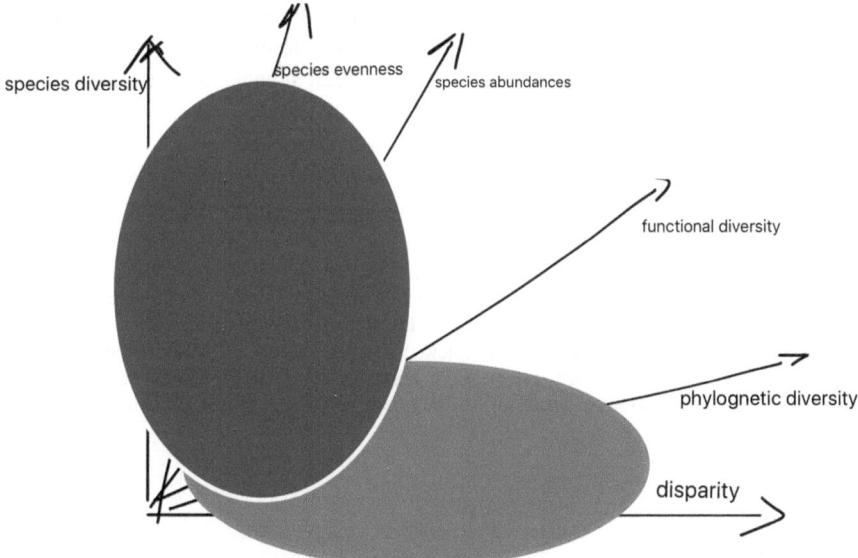

Fig. 13.7 Regions of the conceptual space of biodiversity targeted by explanatory projects: in red, when biodiversity is the *explanans*, in blue, when it is the *explanandum* (Color figure online)

concept, because various explanatory questions in ecology target different conceptual areas of this space. Thus, even though there is here some unity, due to the fact that there is one conceptual space, it is the unity conferred by a same general object. But the fact that the space is so to say differently exploited by various approaches and traditions makes it difficult to think that for this sole reason a theoretical unity can embrace all those approaches and traditions.

Through this "conceptual space approach" to diversity, one sees that diversity is not a purely subjective property, or a property that only exists is the eye of the (scientific) beholder; that many crucial explanatory projects in ecology diversely target diversity; and that at the same time, all these projects cannot be theoretically unified through this reference to diversity as something objective.

13.7 Conclusion

To wrap up the arguments made here, biodiversity is arguably a key issue in ecology, and many theories and explanatory strategies are concerned by it. Diversity is manifestly an explanatory reversible concept, and at first stake this could mean that it could play a role in unifying ecological theories and tools. However, because of the specific incompleteness of the explanatory reversibility of the concept of diversity, illustrated by the way in which its conceptual space is variously targeted by explanatory projects, it appears that the discourses, theories and explanatory strategies of ecologists could not be theoretically unified as a set of scientific approaches to diversity. Even if it is perhaps illusory to think that a single concept could unify a theoretical field, in a non-superficial manner at least, the present enquiry shows that in order to search for unifying principles for ecology, one should not start by focusing on biodiversity. Ecologists share a concern for biodiversity, but the geography of the concept of diversity is such that this shared concern cannot become a principle of unification or an essential part of a unifying strategy.

Acknowledgements I am grateful to the audience of the Urbino conference on biodiversity, organized by Elena Casetta and Andrea Borghini in 2015, where a first version of this chapter has been presented. I warmly thank Davide Vecchi and Elena Casetta, whose comments significantly improved the manuscript, as well as Matt Chew, Sébastien Dutreuil, Antoine Dussault and two anonymous reviewers. This work was possible thanks to the CNRS Laboratoire International Associé ECIEB Paris-Montréal and the GDR "Les savoirs de l'environnemental" (CNRS GDR3770 Sapienv).

References

Albert, R., & Barabasi, A.-L. (2002). Statistical mechanics of complex networks. *Reviews of Modern Physics, 74*, 47–97.

Allee, W. C., Park, O., Emerson, A. E., Park, T., & Schmidt, K. P. (1949). *Principles of animal ecology*. Philadelphia: W. B. Saunders Company.

Allouche, O., & Kadmon, R. (2009). A general framework for neutral models of community dynamics. *Ecology Letters, 12*, 1287–1297.

Andrewartha, H. G., & Birch, L. C. (1954). *The distribution and abundance of animals*. Chicago: University of Chicago Press.

Antonovics, J. (1976). Plant population biology at the crossroads. Input from population genetics. *Systematic Botany, 1*, 234–245.

Bell, G. (2000). The distribution of abundance in neutral communities. *The American Naturalist, 155*, 606–617.

Bell, G., Lechowicz, M. J., & Waterway, M. J. (2006). The comparative evidence relating to the neutral theory of community ecology. *Ecology, 87*, 1378–1386.

Bellwood, D. R., Hoey, A. S., & Choat, J. H. (2003). Limited functional redundancy in high diversity systems: Resilience and ecosystem function on coral reefs. *Ecology Letters, 6*, 281–285.

Blandin, P. (2014). La notion de biodiversité: sémantique et épistémologie. In E. Casetta & J. Delord (Eds.), *La biodiversité en questions* (pp. 31–82). Paris: Matériologiques.

Cadotte, M. W., Cavender-Bares, J., Tilman, D., & Oakley, T. H. (2009). Using phylogenetic, functional and trait diversity to understand patterns of plant community productivity. *PLoS One, 4*, e5695.

CBD. (1992). *Convention on biological diversity*, United Nations. http://www.cbd.int/doc/legal/cbd-en.pdf.

Chave, J. (2004). Neutral theory and community ecology. *Ecology Letters, 7*, 241–253.

Chew, M. K. (2011). Invasion Biology: Historical Precedents. In D. Simberloff & M. Rejmánek (Eds.), *Encyclopedia of biological invasions* (pp. 369–375). Berkeley: University of California Press.

Clements, F. E. (1916). *Plant succession* (Carnegie Institute Pubi. no. 242). Washington, DC: Carnegie Institute.

Clements, F., & Shelford, V. (1939). *Bio-ecology*. New York: J. Wiley & Sons, Inc.; London.

Collins, J. P. (1986). Evolutionary ecology and the use of natural selection in ecological theory. *Journal of the History of Biology, 19*, 257–288.

Connell, J. (1961). The influence of interspecific competition and other factors on the distribution of the barnacle chthamalus stellatus. *Ecology, 42*(4), 710–723.

D'Agata, S., Mouillot ,D., Kulbicki, M., Andrefouet, S, Bellwood, D., Cinner, J., Cowman, P., Kronen, M., Pinca, S., Vigliola, L. (2014). Human-mediated loss of phylogenetic and functional diversity in coral reef fishes. *Current Biology, 24*(5), 555–560.

de Visser, J. A. G. M., Hermisson, J., Wagner, G. P., Ancel Meyers, L., Bagheri-Chaichian, H., Blanchard, J. L., & Chao, L. (2003). Evolution and detection of genetic robustness. *Evolution, 57*, 1959–1972.

Doncaster, C. P. (2009). Ecological equivalence: A realistic assumption for niche theory as a testable alternative to neutral theory. *PLoS One, 4*, e7460.

Dunne, J. (2006). The network structure of food webs. In M. Pascual & J. Dunne (Eds.), *Ecological networks: Linking structure to dynamics in food webs*. Oxford: Oxford University Press.

Dunne, J. E., Williams, R. J., & Martinez, N. D. (2002). Food web structure and network theory: The role of connectance and size. *PNAS, 99*, 12917–12922.

Dussault, A., & Bouchard, F. (2017). A persistence enhancing propensity account of ecological function to explain ecosystem evolution. *Synthese, 194*, 1115.

Elton, C. (1927). *The ecology of animals*. New York: Wiley.

Fisher, R. (1932). The evolutionary modification of genetic phenomena. *Proceedings of the 6th International Congress of Genetics, 1*, 165–172.

Fisher, R. A., Corbet, A. S., & Williams, C. B. (1943). The relation between the number of species and the number of individuals in a random sample from an animal population. *Journal of Animal Ecology, 12*, 42–58.

Ford E.B. (1964). *Ecological genetics*. London : Chapman and Hall.

Gleason, H. A. (1939). The individualistic concept of the plant community. *The American Midland Naturalist, 21*, 92–110.

Gosselin, F. (2014). Diversité du vivant et crise d'extinction: des ambiguïtés persistantes. In E. Casetta & J. Delord (Eds.), *La biodiversité en questions* (pp. 119–138). Paris: Matériologiques.

Hagen, J. (1989). Research perspectives and the anomalous status of Modern Ecology. *Biology and Philosophy, 4*, 433–455.

Hagen, J. (1992). *The Entangled bank. The origins of ecosystem ecology*. New Brunswick: Rutgers University Press.

Harper, J. L. (1967). A Darwinian approach to plant ecology. *Journal of Ecology, 55*, 247–270.

Holling, G. (1973). Resilience and stability of ecological systems. *Annual Review of Ecology and Systematics, 4*, 1–23.

Holt, R. D. (2006). Emergent neutrality. *Trends in Ecology & Evolution, 21*(10), 531–533.

Horn, H. S. (1975). The ecology of secondary succession. *Annual Review of Ecology and Systematics, 5*, 25–37.

Hubbell, S. P. (2001). *The unified neutral theory of biodiversity and biogeography*. Princeton: Princeton University Press.

Hubbell, S. P. (2005). Neutral theory in community ecology and the hypothesis of functional equivalence. *Functional Ecology, 19*, 166–172.

Hubbell, S. P. (2006). Neutral theory and the evolution of ecological equivalence. *Ecology, 87*, 1387–1398. 31.

Hubbell, S. P. (2010). Neutral theory and the theory of island biogeography. In J. B. Losos & R. E. Ricklefs (Eds.), *The theory of island biogeography revisited* (pp. 264–292). Princeton: Princeton University Press.

Huey, J., & Bellwood, D. (2009). Limited functional redundancy in a high diversity system: Single species dominates key ecological process on coral reefs. *Ecosystems, 12*, 1316–1328.

Huneman, P. (2015). Diversifying the picture of explanations in biological sciences: Ways of combining topology with mechanisms. *Synthese, 195*, 115–146.

Huneman, P. (2019). "How the Modern Synthesis came to ecology." *Journal of the History of Biology*. Forthcoming, 52, 4.

Huneman, P. (2017). Stephen Hubbell and the paramount power of randomness in ecology. In O. Harman & M. Dietrich (Eds.), *Dreamers, visionaries and revolutionaries in the life sciences*. Chicago: University of Chicago Press, pp. 176–195.

Huneman, P. (2018). Robustness as an explanandum and explanans in evolutionary biology and ecology. In M. Bertolaso, S. Caianiello & E. Serelli (Eds.), *Biological Robustness. Emerging Perspectives from within the Life Sciences*. Dordrecht: Springer, pp. 95–121.

Huneman, P., & Walsh, D. (Eds.). (2017). *Challenging the modern synthesis: Adaptation, development and inheritance*. New York: Oxford University Press.

Hutchinson, G. E. (1957). Concluding remarks. *Cold Spring Harbor Symposia on Quantitative Biology, 22*, 415–427.

Hutchinson, G. E. (1959). Homage to Santa Rosalia or why are there so many kinds of animals. *American Naturalist, 93*, 145–159.

Hutchinson, G. E. (1961). The paradox of the plankton. *American Naturalist, 95*, 137–145.

Ives, R., & Carpenter, J. (2007). Stability and diversity of ecosystems. *Science, 317*(5834), 58–62.

Justus, J. (2008). Ecological and Lyapunov stability. *Philosophy of Science, 75*, 421.

Kéfi, S., Miele, V., Wieters, E. A., Navarrete, S. A., & Berlow, E. L. (2016). How structured is the Entangled bank? The surprisingly simple organization of multiplex ecological networks leads to increased persistence and resilience. *PLoS Biology, 14*(8), e1002527.

Kimura, M. (1985). *The neutral theory of molecular evolution*. Cambridge: Cambridge University Press.

Kingsland, S. (1995). *Modeling nature: Episodes in the history of population ecology* (2nd ed.). Chicago: University of Chicago Press.

Kitano, H. (2004). Biological robustness. *Nature Reviews Genetics, 5*, 826–837.

Lack, D. (1947). The significance of clutch size. *Ibis, 89*, 302–352.

Lack, D. (1954). *The natural regulation of animal numbers*. Oxford: Oxford University Press.

Leibold, M. A., & McPeek, M. A. (2006). Coexistence of the niche and neutral perspectives in community ecology. *Ecology, 87*, 1399–1410.

Leigh, E. G. (2007). Neutral theory: A historical perspective. *Journal of Evolutionary Biology, 20*, 2075–2091.

Levins, R. (1969). Some Demographic and Genetic Consequences of Environmental Heterogeneity for Biological Control. *Bulletin of the Entomological Society of America, 15*(3), 237–240.

Loreau, M. (2010). Linking biodiversity and ecosystems: Towards a unifying ecological theory. *Philosophical Transactions of the Royal Society B, 365*(1537), 49–60.

Lortie, C. J., Brooker, R. W., Choler, P., Kikvidze, Z., Michalet, R., Pugnaire, F. I., & Callaway, R. M. (2004). Rethinking plant community theory. *Oikos, 107*, 433–438.

MacArthur, R. H. (1972). Coexistence of species. In J. A. Behnke (Ed.), *Challenging biological problems* (pp. 253–259). New York: Oxford University Press.

MacArthur, R., & Levins, R. (1967). The limiting similarity, convergence, and divergence of coexisting species. *American Naturalist, 101*, 377–385.

MacArthur, R. E., & Wilson, E. O. (1967). *The Theory of island biogeography*. Princeton: Princeton University Press.

Margalef, R. (1963). On certain unifying principles in ecology. *American Naturalist, 97*, 357–374.

Marquet, P., Allen, A., Brown, J., Dunne, J., Enquist, B., Gilloly, J., Gowaty, P. A., Green, J., Harte, J., Hubbell, S. P., O'Dwyer, J., Okie, J., Ostling, A., Ritchie, M., Storch, D., & West, G. (2014). On theory in ecology. *Bioscience, 64*, 701–710.

May, R. M. (1974). *Stability and complexity in model ecosystems*. Princeton: Princeton University Press.

May, R. (1975). Stability in ecosystems: Some comments. In W. H. van Dobben & R. H. Lowe-McConnell (Eds.), *Unifying concepts in ecology. Report of the plenary sessions of the first international congress of ecology, The Hague, the Netherlands, 1974* (pp. 161–168). Dordrecht: Springer.

Mccauley, D. J., Young, H. S., Guevara, R., Williams, G. J., Power, E. A., Dunbar, R. B., Bird, D. W., Durham, W. H., & Micheli, F. (2014). Positive and negative effects of a threatened parrotfish on reef ecosystems. *Conservation Biology, 28*, 1312–1321.

McGill, B., Maurer, B. A., & Weiser, M. D. (2006). Empirical evaluation of neutral theory. *Ecology, 87*, 1411–1423.

McIntosh, R. P. (1986). *The background of ecology. Concept and theory*. Cambridge: Cambridge University Press.

McNaughton, S. J. (1977). Diversity and stability of ecological communities: A comment on the role of empiricism in ecology. *American Naturalist, 111*(979), 515–525.

Mitman, G. (1988). From the population to society: The cooperative metaphors of W.C. Allee and A.E. Emerson. *Journal of the History of Biology, 21*, 173–192.

Mouquet, N., Devictor, V., Meynard, C. N., Munoz, F., Bersier, L.-F., Chave, J., Couteron, P., Dalecky, A., Fontaine, C., Gravel, D., Hardy, O. J., Jabot, F., Lavergne, S., Leibold, M., Mouillot, D., Münkemüller, T., Pavoine, S., Prinzing, A., Rodrigues, A. S. L., Rohr, R. P., Thébault, E., & Thuiller, W. (2012). Ecophylogenetics: Advances and perspectives. *Biological Reviews, 87*, 769–785.

Muneepeerakul, R., Bertuzzo, E., Lynch, H. J., Fagan, W. F., Rinaldo, A., & Rodriguez-Iturbe, I. (2008). Neutral metacommunity models predict fish diversity patterns in Mississippi-Missouri basin. *Nature, 453*, 220–222.

Munoz, F., & Huneman, P. (2016). From the neutral theory to a comprehensive and multiscale theory of ecological equivalence. *Quarterly Review of Biology, 91*(3), 321–342.

Nicholson, A. J. (1933). 'The Balance of Animal Populations'. *Journal of Animal Ecology, 2*, 132–178.

Nicholson, A., & Bailey, V. (1935). "The Balance of Animal Populations-Part 1". *Proceedings of the Zoological Society*: 551–598.

Nicoglou, A. (2015). The evolution of phenotypic plasticity: Genealogy of a debate in genetics. *Studies in History and Philosophy of Biological and Biomedical Sciences C, 50*, 67.

Noss, R. F. (1990). Indicators for monitoring biodiversity: A hierarchical approach. *Conservation Biology, 4*(4), 355–364.

Odum, E. P. (1953). *Fundamentals of ecology*. Philadelphia: W. B. Saunders Co.

Orians, G. (1975). Diversity, stability and maturity in natural ecosystems. In W. H. van Dobben & R. H. Lowe-McConnell (Eds.), *Unifying concepts in ecology. Report of the plenary sessions of the first international congress of ecology, The Hague, the Netherlands, 1974* (pp. 139–150). Dordrecht: Springer.

Pimm, S. L. (1984). The complexity and stability of ecosystems. *Nature, 307*, 321–326.

Pimm, S. L. (1993). *The balance of nature? Ecological issues in the conservation of species and communities*. Chicago: University of Chicago Press.

Pocheville, A. (2015). The ecological niche: History & recent controversies. In T. Heams, P. Huneman, G. Lecointre, & M. Silberstein (Eds.), *Handbook of evolutionary thinking in the sciences*. Dordrecht: Springer.

Preston, F. W. (1948). The commonness and rarity of species. *Ecology, 29*, 254–283.

Purves, D. W., & Turnbull, L. A. (2010). Different but equal: The implausible assumption at the heart of neutral theory. *Journal of Animal Ecology, 79*, 1215–1225.

Richardson, D. (Ed.). (2011). *Fifty years of invasion ecology: The legacy of Charles Elton*. London: Wiley.

Rosindell, J., Hubbell, S. P., & Etienne, R. S. (2011). The unified neutral theory of biodiversity and biogeography at age ten. *Trends in Ecology and Evolution, 26*, 340–348.

Roughgarden, J. (1979). *Theory of population genetics and evolutionary ecology: An introduction*. London: Prentice Hall.

Schoener, T. W. (1972). Mathematical ecology and its place among the sciences. *Science, 178*, 389–391.

Servedio, M. R., Brandvain, Y., Dhole, S., Fitzpatrick, C. L., Goldberg, E. E., et al. (2014). Not just a theory–The utility of mathematical models in evolutionary biology. *PLoS Biology, 12*(12), e1002017.

Slack, N. (2010). *G. Evelyn Hutchinson and the invention of modern ecology*. New Haven: Yale University Press.

Smith, F. E. (1952). Experimental methods in population dynamics. *Ecology, 33*, 441–450.

Solé, R., & Goodwin, B. (1988). *Signs of life: How complexity pervades biology*. New York: Basic Book.

Solé, R. V., Ferrer, R., Montoya, J. M., & Valverde, S. (2002). Selection, tinkering and emergence in complex networks. *Complexity, 8*, 20–33.

Solomon, M. E. (1949). The natural control of animal population. *Journal of Animal Ecology, 18*, 1: 1–1:34.

Stearns, S. (1982). The emergence of evolutionary and community ecology as experimental science. *Perspectives in Biology and Medicine, 25*(4), 621–648.

Strogatz, S. (2001). Exploring complex networks. *Nature, 410*, 268–276.

Szyrmer, J., & Ulanowicz, R. E. (1987). Total flows in ecosystems. *Ecological Modelling, 35*, 123–136.

Takacs, D. (1996). *The Idea of Biodiversity. Philosophies of Paradise*. Baltimore and London, John Hopkins University Press.

Tilman, D. (1982). *Resource competition and community structure*. Princeton: Princeton University Press.

Tilman, D. (1994). Competition and biodiversity in spatially structured habitats. *Ecology, 75*, 2–16.

Tilman, D. (1996). Biodiversity: Population versus ecosystem stability. *Ecology, 77*, 350–363.

Ulanowicz, R. E. (1983). Identifying the structure of cycling in ecosystems. *Mathematical Biosciences, 65*, 219–237.

Ulanowicz, R. E. (1986). *Growth and development: Ecosystems phenomenology*. New York: Springer.

Vellend, M. (2010). Conceptual synthesis in community ecology. *Quarterly Review of Biology, 85*, 183–206.

Vellend, M. (2016). *The theory of ecological communities*. Princeton: Princeton University Press.

Volterra, V. (1926). Variazioni e fluttuarzioni del numero d'individui in specie animali conviventi. *Memoria della Reale Accademia Nazionale dei Lincei, 2*, 31–113.

Wagner, A. (2005). *Robustness and evolvability in living systems*. Princeton: Princeton University Press.

Warming, E. (1909). *Oecology of plants*. Oxford: Clarendon Press.

Watts, D. J., & Strogatz, S. H. (1998). Collective dynamics in "small-world" networks. *Nature, 393*, 440.

Williams, G. C. (1966). *Adaptation and natural selection*. Princeton: Princeton University Press.

Wilson, E. O. (1992). *The diversity of life*. Cambridge, MA: Belknap Press

Wilson, E. O. (Ed.). (1988). *Biodiversity*. Washington, DC: National Academy Press.

Chapter 14
Functional Biodiversity and the Concept of Ecological Function

Antoine C. Dussault

Abstract This chapter argues that the common claim that the ascription of ecological functions to organisms in functional ecology raises issues about levels of natural selection is ill-founded. This claim, I maintain, mistakenly assumes that the function concept as understood in functional ecology aligns with the selected effect theory of function advocated by many philosophers of biology (sometimes called "The Standard Line" on functions). After exploring the implications of Wilson and Sober's defence of multilevel selection for the prospects of defending a selected effect account of ecological functions, I identify three main ways in which functional ecology's understanding of the function concept diverges from the selected effect theory. Specifically, I argue (1) that functional ecology conceives ecological functions as *context-based* rather than *history-based* properties of organisms; (2) that it attributes to the ecological function concept the aim of explaining ecosystem processes rather than that of explaining the presence of organisms within ecosystems; and (3) that it conceives the ecological functions of organisms as *use* and *service* functions rather than *design* functions. I then discuss the extent to which the recently proposed causal role and organizational accounts of ecological functions better accord with the purposes for which the function concept is used in functional ecology.

Keywords Functional biodiversity · Function · Selected effect theory · Ecosystem selection · Superorganism

A. C. Dussault (✉)
Centre interuniversitaire de recherche sur la science et la technologie (CIRST), Université du Québec à Montréal (UQAM), Montréal, QC, Canada

Département de philosophie, Collège Lionel-Groulx, Sainte-Thérèse, QC, Canada
e-mail: antoine.cdussault@clg.qc.ca

14.1   Introduction

In the last decades, *functional biodiversity* has become a central focus in ecology and environmental conservation (e.g. Tilman 2001; Naeem 2002; Petchey and Gaston 2006; Nock et al. 2016). This follows from the recognition by an increasing number of ecologists of the explanatory and predictive limitations of more traditional "species richness" measures of biodiversity. This recognition has led ecologists and conservationists to consider, alongside the number of species present in a community, the particular features of organisms of those species and how those features determine their potential relationships with their environments (see Hooper et al. 2002, 195; DeLaplante and Picasso 2011, 173; Nunes-Neto et al. 2016, 296–297). Consideration of those features has fostered among ecologists an interest in the ways in which organisms can be grouped or classified on the basis of their *functional traits*, which are deemed to be of more direct ecological importance than those on which the more standard taxonomic measures of biodiversity are based.

Those functional groupings include:

Guilds: Groupings of organisms on the basis of similarities in resource use. Two organisms are members of a same guild if they tend to use a similar resource in a similar way (Simberloff and Dayan 1991; J. B. Wilson 1999; Blondel 2003).

Functional response groups: Groupings of organisms on the basis of similar expected response to environmental changes. Two organisms are members of the same functional response group if they tend to respond similarly to similar changes in environmental conditions (Catovsky 1998; J. B. Wilson 1999; Hooper et al. 2002; Lavorel and Garnier 2002)

Functional effect groups: Groupings of organisms on the basis of similar roles in ecosystem processes. Two organisms are members of the same functional effect group if they tend to contribute similarly to some important ecosystem process (e.g. nutrient cycling, primary productivity, energy flows) (Catovsky 1998; Hooper et al. 2002; Lavorel and Garnier 2002; Blondel 2003).[1]

Among those three modes of functional classification, the first two—*guilds* and *functional response groups*—are commonly used to explain the assembly of ecological communities and how their species composition changes in response to changes in their environments. The third—*functional effect groups*—is commonly used to explain

---

[1] It should be noted that functional ecologists have adopted various modes of functional classification with different emphases, and have used diverse terminologies to refer to them. For instance Wilson (1999) draws a contrast between *alpha guilds* and *beta guilds* which is essentially equivalent to the contrast made above between *guilds* and *functional response groups*. Similarly, Catovsky (1998), and Lavorel and Garnier (2002) draw a contrast between *functional response groups* and *functional effect groups* similar to the one made above, but define *functional response groups* also in reference to resource use (a basis for classification that I associated with *guilds*). And likewise, Blondel (2003) draws a contrast between *guilds* and *functional groups*, and his concept of *functional group* is essentially equivalent to the above concept of *functional effect group*. I think that my above identification of three main modes of functional classification adequately reflects the complementary epistemic aims in relation to which ecologists use functional classifications.

ecosystem processes through delineating the particular contributions of organisms of different species to those processes (see discussion in Sect. 14.3.2 below).

A particularity of the third mode of functional classification—*functional effect groups*—is that it involves the ascription of *roles* or *functions* to organisms within ecosystems (Catovsky 1998, 126; Symstad 2002, 23–24; Jax 2010, 54). As remarked by Jax (2010, sec. 4.2) and DeLaplante and Picasso (2011, sec. 3.2), such ascriptions of ecological functions to organisms within ecosystems raise important philosophical issues. One of them concerns the meaning of the *function* concept and its relationship to claims about natural selection. Given the association made by many biologists and ecologists between the concept of function and the evolutionary concept of *adaptation* (Williams 1966), the idea that organisms fulfil functions within ecosystems has been claimed to raise issues about the levels at which natural selection customarily operates (see Calow 1987, 60; DeLaplante and Picasso 2011, 184). As we shall see, a linkage of the notion of ecological function to community and ecosystem selection assumes an elucidation of this notion along the lines of the *selection effect* theory of function advocated by many philosophers of biology (e.g. Wright 1973; Millikan 1989; Neander 1991; Godfrey-Smith 1994).[2] According to this theory, which some refer to as "The Standard Line" on functions given its many adherents (Allen and Bekoff 1995, 13–14), the function of a part or trait of a biological entity is the effect for which this part or trait was preserved by natural selection operating on the ancestors of that entity. A selected effect elucidation of the concept of ecological function would therefore entail that ascribing a function to an organism within an ecosystem amounts to saying that at least some of the traits of this organism have been shaped by ecosystem-level selection. Relatedly, a selected effect elucidation of the ecological function concept, as we shall also see, would in some way revive the old idea of communities and ecosystems as tightly integrated *superorganisms* shaped by natural selection (Allee et al. 1949; D. S. Wilson and Sober 1989).

In this chapter, I will argue that the common association between function ascriptions in functional ecology and issues about levels of selection is ill-founded. As just mentioned, this association assumes an understanding of ecological functions along the lines of the *selection effect* theory of function, and I will maintain that the understanding of the function concept at play in functional ecology does not in fact align with this theory. I will do so through identifying important ways in which functional ecology's use of the ecological function concept diverges from the understanding conveyed by the selected effect theory. This will highlight that, when they ascribe functions to organisms within ecosystems, functional ecologists are not committed to views of ecosystems as units of selection. Their understanding of ecological functions and ecosystem functional organization, as I will emphasize, attributes to ecosystems a lower degree of part-whole integration than what would be entailed by the selected effect theory. The discussion of the ecological function concept presented in this chapter will therefore reinforce the near consensus that has recently emerged among philosophers of biology and ecology, according to which the ecological

---

[2] For overviews of philosophical theories of function, see McLaughlin (2001), Wouters (2005), Walsh (2008), Saborido (2014), and Garson (2016).

function concept should be elucidated along the lines of non-selectionist alternatives to the selected effect theory of function (Maclaurin and Sterelny 2008, sec. 6.2; Odenbaugh 2010; Gayon 2013; Nunes-Neto et al. 2014).

My discussion will be organized as follows. In Sect. 14.2, I will discuss the common contention that the use of the function concept in ecology raises issues about levels of selection. I will explore the implications of Wilson and Sober's defence of multilevel selection for the prospects of defending a selected effect account of ecological functions. In Sect. 14.3, I will dispute the claim that the ecological function concept raises issues about levels of natural selection. I will do so by highlighting three important ways in which functional ecology's understanding of the ecological function concept diverges from the selected effect theory. Finally, in Sect. 14.4, I will briefly discuss two non-selectionist accounts of ecological functions that have recently been proposed by philosophers of biology and ecology; namely, the *causal role* account (Maclaurin and Sterelny 2008, sec. 6.2; Odenbaugh 2010; Gayon 2013), and the *organizational* account (Nunes-Neto et al. 2014). I will maintain that neither of these two accounts fully accords with how ecological functions are understood in functional ecology.

14.2   Ecological Functions and Levels of Selection

As mentioned in the introduction, the ascription of ecological functions to organisms in functional ecology is often taken to raise issues about levels of natural selection. As DeLaplante and Picasso (2011, 184) recall:

> [A]ttitudes toward function language in ecology have been influenced by the group selection debate that took place in the 1960s (Wynne-Edwards 1962; Williams 1966). The critique of group selection was based on the affirmation that within orthodox evolutionary theory, natural selection acts primarily at the level of individual organisms (or, indeed, the level of individual genes), and rarely if ever at the level of groups. [...] Evolutionary ecologists tend to associate the language of functions with organism-environment relationships relevant to selection and adaptation (*e.g.*, "functional traits"). But if natural selection only acts at the level of individuals within species populations, then the language of functions should only apply at this level [...]. Consequently, evolutionary ecologists are inclined to be skeptical of function attributions at the community and ecosystem level.[3]

Along similar lines, in the inaugural issue of the journal *Functional Ecology*, Calow (1987, 60) maintains that a focus on the functions fulfilled by organisms within communities "implies that the way they contribute to the balanced economy of the community is an important criterion of selection".

Such a linkage of the notion of ecological function to community or ecosystem selection assumes an understanding of this notion along the lines of the *selected effect* theory of function developed in the philosophy of biology (Wright 1973;

---

[3] For are more detailed discussion of the issues raised by the group selection debate for functional approaches to ecology, see Hagen (1992, chap. 8).

Millikan 1989; Neander 1991). Some support for this assumption can be found in the fact that the selected effect theory has, to some extent, established itself as "The Standard Line" on functions in the philosophy of biology (Allen and Bekoff 1995, 13–14). Since its initial introduction, it has been adopted by many prominent philosophers of biology (e.g. Griffiths 1993; Mitchell 1993; Godfrey-Smith 1994). According to the selected effect theory, the function of a part or trait of a biological entity is the effect for which this part or trait was preserved by natural selection operating on ancestors of this entity. Thus, ascribing a function to an organism within an ecosystem would amount to saying that at least some of the traits of this organism have been shaped by ecosystem-level selection. In other words, ascribing a function to an organism within an ecosystem would amount to saying that organisms from its lineage have the traits on account of which they are classified in a particular *functional effect group* partly because their having those traits conferred a selective advantage to the ecosystem they are part of. Thus, functional ecologists' ascribing ecological functions to organisms within ecosystems would commit them to the idea that communities and ecosystems are units of natural selection. The view of ecosystem functional organization implicitly adopted in functional ecology would therefore be similar to that espoused by mid-Twentieth century ecologists who believed that communities and ecosystems were tightly integrated *superorganisms* subject to community or ecosystem-level selection (e.g. Allee et al. 1949).

Although, as remarked by DeLaplante and Picasso (see quote above), many biologists and ecologists are sceptical about the idea that natural selection customarily operates at the level of communities and ecosystems, some support for this idea can be found (as they also remark) in Wilson and Sober's defence of multilevel selection (see e.g. Wilson and Sober 1989; Sober and Wilson 1994). Wilson and Sober's main focus is population-level selection, but they also apply their multilevel selectionist approach to communities and ecosystems. Wilson and Sober's defence of multilevel selection improves upon previous defences in part by identifying an unrealistic assumption underlying classical arguments against it. This assumption is that individual organisms within populations interact randomly with each other and therefore have equal chances of mating with any other member of their population. Contrary to this assumption, Wilson and Sober emphasize, the heterogeneity of many environments entails that, in practice, populations in the ecological world tend to be structured in ways that make their individual members more likely to interact with only a small subset of their whole population. This, as Wilson and Sober explain, creates conditions favourable to the operation of natural selection on single-species groups of organisms and even communities and ecosystems (Wilson and Sober 1989, 341–4).

They illustrate the possibility of community-level selection with the example of *phoretic associations*. Phoretic associations are communities formed by a winged insect associated with many wingless organisms (e.g. mites, nematodes, fungi and microbes) that rely on the winged insect for transportation from one resource patch to another. When the winged insect reaches a new resource patch (e.g. carrion, dung, or stressed timber), it brings along a whole community of "phoretic associates" which then colonize the patch. Wilson and Sober explain how natural selection might operate on phoretic associations as a whole:

> Consider a large number of resource patches, each of which develops into a community composed of the insects, their phoretic associates, plus other species that arrive independently. The community of phoretic associates may be expected to vary from patch to patch in species composition and in the genetic composition of the component species. Some of these variant communities may have the effect of killing the carrier insect. Others may have the effect of promoting insect survival and reproduction, and these will be differentially dispersed to future resource patches. Thus, between-community selection favors phoretic communities that do not harm and perhaps even benefit the insect carrier. At the extreme, we might expect the community to become organized into an elaborate mutualistic network that protects the insect from its natural enemies, gathers its food, and so on. (Wilson and Sober 1989, pp. 348–9)

Such a scenario, they emphasize, is not only a theoretical possibility. Empirical data from studied phoretic communities show no negative effects on the carrier insect in most cases and positive effects in many cases. In a subsequent paper, Wilson (1997), 2020–22) discusses other likely cases of community selection that conform to his and Sober's approach, as well as a likely case of ecosystem selection involving micro-ecosystems forming at the surface of lakes and oceans.[4]

Wilson and Sober's defence of community and ecosystem selection thus seems to provide grounds for interpreting at least some of the functions fulfilled by organisms within communities and ecosystems along the lines of the selected effect theory of function. For instance, the selected effect theory entails that some phoretic associates in Wilson and Sober's phoretic association case have functions within the phoretic association. This is the case of phoretic associates that are part of the association partly because some of their traits conferred a selective advantage to the phoretic association as a whole. Similar function ascriptions would be implied by the selected effect theory in relation to organisms involved in the other cases of community and ecosystem selection described by Wilson (1997). In line with those observations, Wilson and Sober themselves conceive their defence of multilevel selection as legitimizing the view that some communities and ecosystems are *functionally organized* entities (Wilson and Sober 1989, 337–344; see also Wilson 1997). They even claim that communities and ecosystems that are units of selection according to their approach can genuinely be regarded as *superorganisms* (Wilson and Sober 1989, 349).[5]

However, it should be emphasized that Wilson and Sober's defence of multilevel selection lends at best very limited support to the application of the selected effect theory in ecology. Wilson and Sober are careful to emphasize that their defence of community and ecosystem selection is professedly modest. They see it as an important strength of their approach that it does not consist in an "overly grandiose" superorganism theory that attributes "functional design [...] to ecosystems in general"

---

[4] For related discussions of artificial ecosystem selection experiments, see Swenson et al. (2000a), Swenson et al. (2000b) and Blouin et al. (2015).

[5] For a discussion of Wilson and Sober's defence of multilevel selection in relation to the selected effect theory of function, see Basl (2017, sec. 4.2).

(Wilson and Sober 1989, 352). As they insist, their approach entails that "[n]ot all groups and communities are superorganisms, but only those that meet the specified (and often stringent) conditions" (Wilson and Sober 1989, 343). Functional ecologists, in contrast, envision their approach as a framework for the study of ecosystems in general. Such a broad scope is not legitimized by Wilson and Sober's approach. Therefore the support lent by Wilson and Sober's defence of multilevel selection to the application of the selected effect theory of function in ecology seems too limited for the purposes of functional ecology.

In the next section, I will argue that significant aspects of the use of the function concept in functional ecology point to an understanding of function that diverges from the selected effect theory. This will show that, contrary to what is sometimes suggested (see above), the ascription of ecological functions to organisms in functional ecology does not hinge on claims that ecosystems are units of natural selection.

14.3 Ecological Functions in Functional Ecology

14.3.1 Ecological Context vs. Selective History

Historically and conceptually, contemporary functional ecology's construal of the function concept derives from the renowned community ecologist Charles Elton's (1927, 1933) understanding of the *ecological niche*. Elton's understanding of the niche was tied to a functionalist view of ecological communities, which drew an analogy between feeding interactions within ecological communities and economic exchanges in human societies.[6] In Elton's coinage, the term "niche" referred to "what [an animal] is *doing* in its community", and emphasized an animal's "*relations to food and enemies*" in contrast to "appearance, names, affinities, and past history." (Elton 1927, 63–64, emphasis in the original) The niche concept was "used in ecology in the sense that we speak of trades or *professions* or *jobs* in a human community" (Elton 1933, 28, emphasis added). Thus, Elton's understanding of the niche was tied to a picture of ecological communities in analogy with human societies (with an economic focus), rather than with individual organisms. The niches of organisms, as he conceived them, were analogous to the economic roles fulfilled by individuals within human societies, rather than with the functions of organs within organisms. This communitarian-economic analogy attributed to ecological

---

[6] Elton's understanding of the niche contrasted with the one previously adopted by Joseph Grinnell (1917), the other originator of the niche concept, who used the niche concept to denote a species' particular *environmental requirements* (see Leibold 1995, 1372–1373). The contrast between Grinnell's and Elton's niches parallels the contrast presented in the introduction between on the one hand, *guilds* and *functional response group* and on the other hand, *functional effect groups* (see Hooper et al. 2002, 196). For discussions of the contrast between Grinnell's and Elton's niche concepts, see also Schoener (1989), Griesemer (1992), and Pocheville (2015).

communities of a lower degree of part-whole integration than the one characteristically found in individual organisms. Notably, Elton (1930) emphasized that individual organisms retain a significant degree of autonomy with respect to the communities in which they are involved, and he rejected the view (held by some later Twentieth-century ecologists) that natural selection customarily operates on ecological communities as a whole (McIntosh 1985, 167; Haak 2000, 32).

Contemporary functional ecology's understanding of ecological functions is in many respects similar to Elton's functional understanding of the niche. A first important aspect of this understanding that does not align with the selected effect theory concerns the basis on which ecological functions are ascribed to organisms in functional ecology. In functional ecology, the ecological functions of organisms within ecosystems are conceived as *context-based* properties of those organisms, which they bear on account of their actual and potential interactions with other organisms. This context-based understanding contrasts with that conveyed by the selected effect theory, according to which the functions of biological items are *history-based* properties of those items (i.e. properties borne by those items on account of their selective history). The conceptual dissociation of the ecological function concept from evolutionary considerations is made explicit by some functional ecologists. Petchey and Gaston (2006, 742), for instance, state that "[f]unctional diversity [in ecology] generally involves understanding communities and ecosystems based on what organisms do, rather than on their evolutionary history".

Functional ecology's context-based understanding of ecological functions is aptly portrayed by Jax (2010, 79):

> In contrast to parts of an organism, a particular species has no clearly defined role within an ecosystem: a bird may have the function of being prey to other animals—but only if these carnivorous animals are parts of the specific system. If there are no predators in the system, the same species or even individual will not have the role "prey". Even if we can say that the bird actually has the role of being prey, we can also find other roles, e.g. its role to distribute seeds and nutrients, to be predator for insects, etc. That is, like a person within a human society, who may be teacher, spouse, child, politician etc., either at the same time or at different times, it can have several roles. Roles can change and the same person as well as the same species can even take opposing roles in time […]. "The" one and only role of a species does not exist. Roles are strongly context-dependent.

On this context-based understanding, the ascription of ecological functions to organisms within ecosystems does not entail claims about selective history. For instance, an ecologist's depiction of a rabbit as fulfilling the role of a prey (or primary consumer) within an ecosystem does not entail the claim that rabbits and their traits were selected for serving as food for predators. Rabbits eat grass and grow muscles for their own survival and, as a by-product, acquire traits that make them nutritious and palatable for those predators. Likewise, an ecologist's reference to foxes as fulfilling the role of regulator of herbivore populations within an ecosystem does not entail the claim that foxes and their traits were selected for regulating herbivore populations. Foxes chase and eat preys to feed themselves and, as a by-product, exert a form of control over their preys' populations.

It should be noted, however, that contemporary functional ecology expands upon Elton's approach to the study of ecological communities in two important ways.

First, it expands upon Elton's approach by integrating ecosystem ecology's thermo-dynamic and biogeochemical outlook on the ecological world (see Hagen 1992, chaps. 4–5). Thus, whereas Elton used the niche primarily to study how interspecific interactions within communities explain the regulation of populations within them and the maintenance of their structural features (Hagen 1992, 52; Pocheville 2015, 549), the ascription of ecological functions to organisms in contemporary func-tional ecology is more primarily tied to the aim of studying how the traits of organ-isms determine their potential contributions to ecosystem processes (see K. W. Cummins 1974; Naeem 2002). Thus, in contemporary functional ecology, the eco-logical functions of organisms are their particular contributions to ecosystem pro-cesses (e.g. nutrient cycling, primary productivity, energy flows). Contemporary functional ecologists ascribe functions to organisms in order to delineate their par-ticular contribution to the realization and maintenance of those processes.

Second, contemporary functional ecology expands upon Elton's focus on feed-ing (or trophic) interactions between organisms, by also considering ecological functions acquired by organisms through *non-trophic* interactions with other organ-isms. Those non-trophic interactions are ones in which organisms affect each oth-er's lives through other means than the direct provision of food (in the form of living or dead tissues). Important non-trophic ecological functions include those fulfilled by *ecosystem engineers*, i.e. organisms that create, modify and maintain habitats in ways that affect the lives of other organisms (e.g. beavers build dams and in so doing create habitats and make many resources available for numerous other organ-isms) (Jones et al. 1994, 1997; Berke 2010). Non-trophic ecological functions also include those of *pollinators* and *seed dispersers* (see Blondel 2003, 227–228).

Those two significant expansions notwithstanding, it remains the case that eco-logical function ascriptions as conceived in functional ecology do not involve claims about selective history. For instance, an ecologist's saying that, by building a dam, a beaver fulfils the role of a pond provider with respect to the numerous organisms for which the pond is a favourable habitat does not entail the claim that beavers were selected for providing habitats to those organisms. Beavers build dams and create ponds for their own benefit and, as a by-product, provide habitats to numerous organisms.

An important research aim associated with functional ecology's context-based understanding of function is that of studying the *functional equivalence* between phylogenetically-divergent organisms. Elton (1927, 65), for instance, remarked that the arctic fox, which subsists on guillemot eggs and seal remains left by polar bears, occupies essentially the same niche as the spotted hyæna in tropical Africa, which feeds upon ostrich eggs and zebra remains left by lions. Although they have evolved their traits in distinct selective contexts, arctic foxes and spotted hyæna occupy simi-lar niches. Along similar lines, contemporary functional ecologists have identified functional equivalences, for instance, between ants, birds and rodents, which simi-larly contribute to seed dispersal in some desert ecosystems, and between humming-birds, bats and moths, which similarly contribute to the pollination of Lauraceae (a family of plants from the group of angiosperm that usually have the form of trees or shrubs) (see Blondel 2003, 226). The acknowledgement of functional equivalences between phylogenetically-divergent organisms conflicts with the understanding of function conveyed by the selected effect theory, in that this theory would entail that two organisms can have similar ecological functions only to the extent that their traits have evolved in similar selective contexts.

14.3.2   The Explanatory Aim of Ecological Functions

A second important aspect of functional ecology's understanding of functions that diverges from the selected effect theory concerns the *explanatory aim* attributed to the function concept. In functional ecology, as seen in the preceding section, the *explanandum* of ecological function ascriptions is ecosystem processes. The ecological functions of organisms are their particular contributions to the ability of ecosystems to realize and maintain those processes. This contrasts with the *explanandum* of function ascriptions according to the selected effect theory. According to the selected effect theory, the *explanandum* of ecological function ascriptions is the *presence* of the biological items to which functions are ascribed within a system (typically an organism). For instance, according to the selected effect theory, saying that pumping blood is the function of the heart entails not only saying that pumping blood is the way in which hearts contribute to blood circulation in animals with circulatory systems. It also entails saying that animals with circulatory systems have hearts because hearts pump blood (i.e. that hearts *are present* within those organisms because they pump blood). The selected effect functions of biological items explain the presence of those items because, by definition, those functions are the effects for which those items were preserved by natural selection.

To make plain that the *explanadum* of ecological function ascriptions in functional ecology is not the *presence* of organisms within ecosystems, we must recall functional ecology's three main modes of functional classification identified in the introduction. As seen in the introduction, functional ecologists use three main modes of functional classification: (1) *guilds* (groupings based on similar resource use), (2) *functional response groups* (groupings based on similar response to environmental factors), and (3) *functional effect groups* (grouping based on similar roles in ecosystem functioning). As also seen in the introduction, the mode of functional classification that is concerned with functions of organisms within ecosystems is the third one (i.e. *functional effect groups*). However, the modes of functional classification that are primarily involved in the theoretical frameworks used by functional ecologists to explain the presence of organisms within ecosystems are the two other ones (*guilds* and *functional response groups*). Those functional classifications are the ones primarily involved in theories developed for explaining the *assembly* of ecological communities and how communities respond to changes in environmental conditions (through changes in species composition). According to those theories (see Keddy 1992; Díaz et al. 1999), the ability of some particular organisms to establish and maintain themselves in a given community depends, first, on their ability to tolerate the local environmental conditions, and, second, on their ability to exploit the resources available in this community (which requires them to be able to successfully compete with other organisms also using those resources or to share those resources with them). The former ability depends upon the *functional response group* to which organisms belong, and the latter one depends upon their *guild*. The *functional effect groups* to which organisms belong play no significant role in explaining the assembly of ecological communities and their responses to environmental changes.

To be sure, if some regular coincidence could be found between, on the one hand, *guilds* and *functional response groups*, and on the other hand, *functional effects groups*, then one could argue that an explanatory connection nevertheless exists between the ecological functions of organisms and their presence within ecosystems. Functional ecologists, however, emphasize the frequent non-coincidence of those groupings (see e.g. Lavorel and Garnier 2002; Blondel 2003). For instance, birds can disperse some plants' seeds in three different ways: (1) through catching seeds in their plumage and then accidentally dropping them elsewhere (epizoochory), (2) through swallowing fruits and then regurgitating or defecating them elsewhere (endozoochory), or (3) through caching dry fruit seeds for future use and then "forgetting" them (synzoochory). Birds that disperse some plants' seeds in those three ways all belong to the same *functional effect group*. However, insofar as only the birds that disperse seeds in the two latter ways (endozoochory and synzoochory) use the seeds as resources, those birds and those that disperse seeds in the former way (epizoochory) do not belong to the same *guild* (see Blondel 2003, 227–228). Likewise, some varieties of dung beetles feed upon the non-digestive part of large herbivores' green food. Those dung beetles do so in three different ways: (1) through dwelling inside the dung, (2) through burying pieces of the faeces from 0.5 to 1 meter under the dung, and (3) through making a ball of dung, laying eggs within it and rolling it to a place where they can bury it. All dung beetles use the dung as a resource and therefore belong to the same *guild*. However, insofar as the different ways of using the resource lead to different decomposition processes, the three types of dung beetles do not belong to the same *functional effect group* (see Blondel 2003, 228).

It may be objected that the *functional effect groups* to which organisms belong must at least partly explain their presence within ecosystems, given that organisms depend upon the achievement of ecosystem processes for their own existence, and, for this reason, depend, at least indirectly, upon the reliable fulfilment of their own functional contributions to those processes. By fulfilling their ecological functions, in other words, organisms must indirectly contribute to the realization and maintenance of their own conditions for existence, such that they are indirect causes of their continued presence within the ecosystem (or at least of the continued presence of organisms of their *functional effect group*).

I think, however, that this kind of causal link between the fulfilment of their ecological functions by organisms and their presence within ecosystems can, at best, be very weak. Strictly speaking, what organisms contribute to realizing, by fulfilling their ecological functions, is not the conditions necessary for their own presence within an ecosystem, or even for the presence of organisms from their *functional effect group*. What they contribute to realizing is, more accurately, the conditions necessary for the presence of organisms from the *guild* or *functional response group* to which they belong. Abilities to exploit the conditions organisms contribute to realizing by fulfilling their ecological function are determined by membership in *guilds* and *functional response groups*, not by membership in *functional effect groups*. This is well illustrated by a phenomenon studied by ecologists as the "negative selection effect" (Jiang et al. 2008). The "negative selection effect" occurs when some ecological function stops being fulfilled as a result of the displacement of a species that fulfils this function (i.e. that belongs to a particular *functional effect group*) by another species that does not fulfil it (i.e. that does not belong to the same *functional effect group*). The reason why the latter species displaces the former one is that both species use the same resource (i.e. belong to the same *guild*) and the latter species is better at competing for this resource. Thus, suppose, that a species S fulfils the ecological function F within the ecosystem E, and that, by doing so, S contributes to the realization of environmental condition C and to the availability of resource R within E. S therefore belongs to the *functional effect group f* (which encompasses organisms that are able to fulfil F), and also belongs to the *guild r* and the *functional response group c* (which encompass, respectively, organisms that use resource R and that require environmental conditions C). Now, we can see more clearly that, by contributing to the realization of C and the availability of R, organisms from S only weakly promote their own presence (or the presence of other species from f) in E. What organisms from S promote by contributing to the realization of C and the availability of R is, in fact, the presence of any species from *guild r* and *functional response group c*. By doing so, therefore, organisms from S promote their own presence within E only provided that there is no other species S_l that also belongs to c and r and that is more efficient than S in exploiting R. If such a species comes around, then the fulfilment of their ecological function by organisms from S will instead promote the presence of S_l within the ecosystem, and consequently S's own displacement by S_l. And if S_l does not belong to f and S was the only species that fulfiled F within E, then F will stop being fulfiled in E. Likewise, by contributing to the realization of C and the availability of R, organisms from S may promote the presence of other species from *functional effect group f* only to the extent that those other species belonging to f also belong to r and c. There, however, is no reason to assume that, on a general basis, species that belong to f will also belong to r and c. The possibility of such a "negative selection effect," I think, makes clear that the *functional effect groups* to which organisms belong have only limited relevance to the aim of explaining why they are present within ecosystems.

14.3.3   By-Products and the Notion of "Functioning as"

As indicated in Sect. 14.3.1, in functional ecology, ecological functions may be ascribed to organisms on the basis of traits that are evolutionary by-products rather than selected effects (on this point, see also Maclaurin and Sterelny 2008, 115; and

Odenbaugh 2010, 251). This observation points to a third important aspect of functional ecology's understanding of functions that does not align with the selected effect theory. This aspect can be highlighted by drawing the connection between functional ecology's understanding of the function concept and Achinstein's (1977, 350–6) delineation of three distinct meanings of "function" in ordinary language: *design*, *use* and *service* functions. An entity's *design* function consists in what this entity was *designed* or *created* to do (e.g. the function of a mouse trap is to catch mice); whereas an entity's *use* function consists in what it is *used for* (e.g. this table is used for sitting), and an entity's *service* function consists in what it *serves as* (e.g. a watch's second hand serves as a dust sweeper). A table's functioning as a seat or the second hand of a watch's functioning as a dust sweeper do not entail that tables and second hands have been (intentionally) designed for those functions. This distinction between design functions on the one hand, and use and service functions on the other hand, is sometimes also expressed in terms of a contrast between the notion of *being the function of* (e.g. breathing is the function of the nose) and that of *functioning as* (e.g. the nose functions as an eyeglass support) (e.g. Boorse 1976, 76; Bedau 1992, 787–789).

In light of this distinction, the selected effect theory of functions can be interpreted as concerned with *design* functions, that is, as concerned with specifying *the function of* some biological item (as is reflected in selected effect theorists' typical association of function with *design*, see e.g. Wright 1973, 164–65; Millikan 1984, 17). In contrast, functional ecology's context-based functions can be conceived as concerned with *use* and *service* functions, that is, as concerned with specifying what an ecological item can *functions as* in relevant ecological contexts. For instance, rabbits that are preyed upon by foxes in an ecosystem *function as* primary consumers within that ecosystem. In turn, foxes that prey upon those rabbits and exert some control on their population *function as* regulators of the rabbit population within that ecosystem. And likewise, beavers that build dams within an ecosystem and by doing so create habitats and make many resources available for numerous other organisms *function as* pond providers within that ecosystem. Similar to the cases of a table's functioning as a seat and the watch's second hand's functioning as a dust sweeper, rabbits' functioning as primary consumers, foxes' functioning as regulators of rabbit populations and beavers' functioning as pond providers within an ecosystem do not entail claims that rabbits, foxes and beavers were (evolutionarily) designed for fulfilling those functions. Functional ecology thus seems to make use of an ordinary notion of function that is conceptually distinct from the one that the selected effect theory is meant to elucidate. It is not concerned with functions that organisms are (evolutionarily) *designed* to fulfil within ecosystems, but, with functions that they (more fortuitously) fulfil as a result of being (context-dependently) involved in use and service interactions with other organisms.

Above, I maintained that functional ecology attributes to ecological communities a lower degree of part-whole integration than the one characteristically found in individual organisms (in line with Elton's analogy between ecological communities and human societies). Interpreting ecological functions as use and service functions provides some illumination of this idea. A notable feature of individual organisms seems to be their characteristic *teleological integration* (see Queller and Strassmann 2009, 3144). The parts of organisms seem, in some biologically relevant sense, to be *designed* for fulfilling their functions within those organisms. In Achinstein's terminology, the parts of organisms have *design* functions. For instance, hearts do not merely fulfil the role of pumping blood within organisms with circulatory systems, they are (evolutionarily) *designed* for doing so.

Insofar as functional ecology conceives the functions fulfilled by organisms within ecosystems as *use* and *service* functions (in contrast to *design* functions), then functional ecology does not attribute to ecosystems the kind of teleological integration commonly attributed to individual organisms. From the theoretical perspective of functional ecology, ecosystems are functionally organized in a much weaker way than paradigm individual organisms. They are functionally organized not in virtue of being superorganisms shaped by ecosystem-level selective processes, but, more weakly, in virtue of being more or less self-maintaining networks of organisms involved in use and service interactions with each-other. Those use and service interactions collectively generate the ecosystem processes in relation to which functional ecologists ascribe functions to organisms. This view of ecosystem functional organization contrasts with that espoused by mid-Twentieth century ecologists who depicted ecosystems as tightly unified *superorganisms* shaped by community or ecosystem-level natural selection.

14.4   What Is an Ecological Function, Then?

In the previous section, I identified three aspects of functional ecology's understanding of ecological functions that do not align with the selected effect theory of function:

1. Functional ecology conceives ecological functions as *context-based* rather than *history-based* properties of organisms
2. Functional ecology attributes to the ecological function concept the aim of explaining ecosystem processes rather than that of explaining the presence of organisms within ecosystems
3. Functional ecology conceives the ecological functions of organisms as *use* and *service* functions rather than *design* functions

Those three aspects, I think, indicate that, contrary to what is often assumed (see Sect. 14.2), the ascription of ecological functions to organisms in functional ecology does not hinge on claims that natural selection customarily operates at the level of ecosystems. Functional ecology's understanding of the function concept diverges from "The Standard Line" on function according to which functions in biology must be understood as naturally selected effects.

Through highlighting the three aspects just mentioned, the above discussion reinforces the near consensus that has recently emerged among philosophers of biology and ecology, according to which the ecological function concept should be elucidated along the lines of non-selectionist alternatives to the selected effect theory of function (see Nunes-Neto et al. 2013).[7] Philosophers who share this consensus have proposed accounts of ecological functions along the lines of Cummins's (1975) *causal role* theory (Maclaurin and Sterelny 2008, 114–115; Odenbaugh 2010, 251–252; Gayon 2013, 76–77), or along those of Mossio et al. (2009) *organizational* theory of function (Nunes-Neto et al. 2014). How do these accounts stand with respect to functional ecology's use of the function concept?

In some significant respects, the *causal role* theory of function accords with functional ecology's use of the function concept as characterized above. The causal role theory ascribes functions to the parts of biological entities in a way that is entirely independent of their selective history. Function ascriptions, in the causal role theory, serve to identify the particular contributions of the parts of a system to the activities or capacities of that system. This use of the function concept concords with functional ecology's understanding of ecological functions as contributions of organisms to ecosystem processes (see Cooper et al. 2016, sec. 4). Moreover, in line with the above linkage of functional ecology's understanding of functions with Achinstein's notions of *use* and *service* functions (see Sect. 14.3.3), the causal role theory does not confer a privileged epistemic status to the notion of *being the function of* over that of *functioning as* (see Cummins 1975, 762; Craver 2001, 55). Thus, the causal role theory seems to better accord with functional ecology's use of the function concept.

However, a significant limitation of the causal role theory in relation to functional ecology, I think, is its ultimate relativization of functions to the epistemic interests of researchers. According to the causal role theory, parts of a system can be ascribed functions in relation to any capacity or activity of this system that researchers are interested in explaining, provided that the relation between this capacity or activity and the individual contributions of the system's parts is complex enough.[8] As many critics of the causal role theory point out, one problem with this liberal take on functions is that it implausibly entails that functions can be ascribed to the parts of a system on account of their contributions to capacities that amount to deteriorations of those systems (e.g. that a function can be ascribed to a tumour on account of its contribution to the capacity of an organism to die from cancer, see Neander 1991, 181). Thus, on a causal role account, ecological functions could, for instance, be ascribed to organisms from an invasive species on account of their contribution to the ecosystem's capacity to collapse (the fragilization of ecosystems and their possible collapse resulting from the establishment of invasive species is indeed something that ecologists are interested in explaining). Such a degree of inclusive-

---

[7] Dissenters from this consensus are Bouchard (2013) and Dussault and Bouchard (2017), who argue that ecological functions should be understood as contributions to ecosystem fitness (conceived as ecosystem resilience). It should nonetheless be noted that Dussault and Bouchard do not advocate a selected effect account of ecological functions, but rather a forward-looking evolutionary account derived from Bigelow and Pargetter's (1987) dispositional theory of function.

[8] For more details on how causal role theorists substantiate this complexity requirement, see Cummins (1975, 764), Davies (2001, chap. 4), and Craver (2001, sec. 3.2).

ness, I think, does not appropriately reflect the fact that functional ecologists tend to ascribe functions to organisms mainly in relation to capacities or activities of ecosystems that contribute to those ecosystems' ability to maintain themselves. Those processes include primary productivity, nutrient cycling, water uptake, storage of resources, etc. (see, enumerations of ecosystem processes in Walker 1992, 20; and Blondel 2003, 226). Thus, the common objection that the causal role theory is overly liberal also seems to apply in the case of ecological functions.

An organizational account of ecological functions would avoid this problem. The organizational theory defines the functions of the parts of a system as their contribution to the ability of the system to maintain its organization (see Mossio et al. 2009). Such a linkage between functions and the self-maintenance of systems excludes function ascriptions in relation to capacities that amount to deteriorations of systems (see Nunes-Neto et al. 2014, 137–138). In this respect, the organizational theory of function seems to restrict function ascriptions in a way that is consistent with the use of the concept in functional ecology.

However, an important limitation of the organizational theory in relation to functional ecology, I think, is that it shares with the selected effect theory the idea that function ascriptions in part explain the presence of function bearers within systems. According to the organizational theory, a biological item has a function within a system if, on the one hand, it contributes to the maintenance of the organization of this system, and if, on the other hand, it is in turn maintained by the organization of the system (Mossio et al. 2009, 16–20). Thus, according to the organizational theory, the function bearing parts of a system indirectly contribute to (and therefore explain) their own presence within this system through contributing to that system's maintenance. In this regard, the organizational theory is similar to the selected effect theory (though, in contrast to the selected effect theory, the organizational theory does not make it a requirement that natural selection be the process through which the function bearing parts of systems promote their own presence). The organizational theory therefore attributes to function ascriptions an explanatory aim that is foreign to functional ecology's understanding of the concept. As seen in Sect. 14.3.2, ecological functions as understood in functional ecology are not conceived as explanatory of the presence of organisms within ecosystems. The presence of organisms within ecosystems is explained by their belonging to some *guilds* and *functional response groups*, not by their belonging to some *functional effect groups*. Ecological function ascriptions and the grouping of organisms in *functional effect groups* serve to explain the realization and maintenance of ecosystem processes through delineating the particular contribution of organisms to those processes.

Hence, neither the *causal role* nor the *organizational* account of ecological functions fully accord with functional ecology's use of the function concept. The observations made in this section, however, suggest that functional ecology requires an account of functions that combines aspects of those two accounts while eschewing some of their other aspects. An elaboration of such an account must be deferred to future work.

14.5 Conclusion

In the preceding sections, I criticised the common supposition that the ascription of ecological functions to organisms in functional ecology hinges on claims that natural selection customarily operates at the level of ecosystems. This supposition, I maintained, rests on the incorrect assumption that the function concept as understood in functional ecology aligns with the selected effect theory of function advocated by many philosophers of biology (sometimes deemed "The Standard Line" on functions). After exploring the implications of Wilson and Sober's defence of multilevel selection for the prospects of defending a selected effect account of ecological functions, I identified three main ways in which functional ecology's understanding of the function concept diverges from the selected effect theory. Specifically, I argued (1) that functional ecology conceives ecological functions as *context-based* rather than *history-based* properties of organisms; (2) that it attributes to the ecological function concept the aim of explaining ecosystem processes rather than with that of explaining the presence of organisms within ecosystems; and (3) that it conceives the ecological functions of organisms as *use* and *service* functions rather than *design* functions. I then briefly discussed the recently proposed accounts of ecological functions along the lines of the causal role and organizational theories of function, and concluded that functional ecology requires an account of functions that selectively draws on those two accounts.

Acknowledgements The author would like to thank Léa Derome, Anne-Marie Gagné-Julien, Philippe Huneman and anonymous referees for valuable comments, as well as O'Neal Buchanan for linguistic revision of the manuscript. The work for this chapter was supported by a postdoctoral fellowship from the Social Sciences and Humanities Research Council of Canada (SSHRC, 756-2015-0748) and a research grant from the Fonds de recherche du Québec – Société et culture (FRQSC, 2018-CH-211053).

References

Achinstein, P. (1977). Function statements. *Philosophy of Science, 44*, 341–367.
Allee, W. C., Emerson, A. E., Park, O., Park, T., & Schmidt, K. P. (1949). *Principles of animal ecology*. Philadelphia: Saunders Co.
Allen, C., & Bekoff, M. (1995). Function, natural design, and animal behavior: Philosophical and ethological considerations. In N. S. Thompson (Ed.), *Perspectives in ethology* (Vol. 11, pp. 1–46). New York: Plenum Press.
Basl, J. (2017). A trilemma for teleological individualism. *Synthese, 194*, 1057–1074. https://doi.org/10.1007/s11229-017-1316-0.
Bedau, M. (1992). Where's the good in teleology? *Philosophy and Phenomenological Research, 52*, 781–806.
Berke, S. K. (2010). Functional groups of ecosystem engineers: A proposed classification with comments on current issues. *Integrative and Comparative Biology, 50*, 147–157. https://doi.org/10.1093/icb/icq077.

Bigelow, J., & Pargetter, R. (1987). Functions. *Journal of Philosophy, 84*, 181–196.

Blondel, J. (2003). Guilds or functional groups: Does it matter? *Oikos, 100*, 223–231. https://doi.org/10.1034/j.1600-0706.2003.12152.x.

Blouin, M., Karimi, B., Mathieu, J., & Lerch, T. Z. (2015). Levels and limits in artificial selection of communities. *Ecology Letters, 18*, 1040–1048. https://doi.org/10.1111/ele.12486.

Boorse, C. (1976). Wright on functions. *Philosophical Review, 85*, 70–86.

Bouchard, F. (2013). How ecosystem evolution strengthens the case for functional pluralism. In P. Huneman (Ed.), *Functions: Selection and mechanisms* (pp. 83–95). Springer: Dordrecht.

Calow, P. (1987). Towards a definition of functional ecology. *Functional Ecology, 1*, 57–61. https://doi.org/10.2307/2389358.

Catovsky, S. (1998). Functional groups: Clarifying our use of the term. *Bulletin of the Ecological Society of America, 79*, 126–127. https://doi.org/10.2307/20168223.

Cooper, G. J., El-Hani, C. N., & Nunes-Neto, N. (2016). Three approaches to the teleological and normative aspects of ecological functions. In N. Eldredge, T. Pievani, E. Serrelli, & I. Tëmkin (Eds.), *Evolutionary theory: A hierarchical perspective* (pp. 103–124). Chicago: University of Chicago Press.

Craver, C. F. (2001). Role functions, mechanisms, and hierarchy. *Philosophy of Science, 68*, 53–74. https://doi.org/10.1086/392866.

Cummins, K. W. (1974). Structure and function of stream ecosystems. *BioScience, 24*, 631–641. https://doi.org/10.2307/1296676.

Cummins, R. C. (1975). Functional analysis. *Journal of Philosophy, 72*, 741–764.

Davies, P. S. (2001). *Norms of nature: Naturalism and the nature of functions*. Cambridge, MA/London: The MIT Press.

DeLaplante, K., & Picasso, V. (2011). The biodiversity–ecosystem function debate in ecology. In K. DeLaplante, B. Brown, & K. A. Peacock (Eds.), *Philosophy of ecology* (pp. 219–250). Oxford/Amsterdam/Waltham: Elsevier.

Díaz, S., Cabido, M., & Casanoves, F. (1999). Functional implications of trait–environment linkages in plant communities. In E. Weiher & P. Keddy (Eds.), *Ecological assembly rules* (pp. 338–362). Cambridge: Cambridge University Press.

Dussault, A. C., & Bouchard, F. (2017). A persistence enhancing propensity account of ecological function to explain ecosystem evolution. *Synthese, 194*, 1115–1145.

Elton, C. S. (1927). *Animal ecology*. New York: The Macmillan Company.

Elton, C. S. (1930). *Animal ecology and evolution*. Oxford: Clarendon Press.

Elton, C. S. (1933). *The ecology of animals* (3rd ed.). London: Methuen.

Garson, J. (2016). *A critical overview of biological functions* (SpringerBriefs in philosophy). Cham: Springer.

Gayon, J. (2013). Does oxygen have a function, or where should the regress of functional ascriptions stop in biology? In P. Huneman (Ed.), *Functions: Selection and mechanisms* (pp. 67–79). Dordrecht: Springer.

Godfrey-Smith, P. (1994). A modern history theory of functions. *Noûs, 28*, 344–362.

Griesemer, J. R. (1992). Niche: Historical perspectives. In E. F. Keller & E. Lloyd (Eds.), *Keywords in evolutionary biology*. Cambridge, MA: Harvard University Press.

Griffiths, P. E. (1993). Functional analysis and proper functions. *British Journal for the Philosophy of Science, 44*, 409–422.

Grinnell, J. (1917). The niche-relationships of the California thrasher. *Auk, 34*, 427–433.

Haak, C. (2000). *The concept of equilibrium in population ecology*. Doctoral dissertation, Halifax, Nova Scotia: Dalhousie University.

Hagen, J. B. (1992). *An entangled bank: The origins of ecosystem ecology*. New Brunswick: Rutgers University Press.

Hooper, D. U., Solan, M., Symstad, A., Diaz, S., Gessner, M. O., Buchmann, N., Degrange, V., et al. (2002). Species diversity, functional diversity and ecosystem functioning. In M. Loreau, S. Naeem, & P. Inchausti (Eds.), *Biodiversity and ecosystem functioning: Synthesis and perspectives* (pp. 195–208). Oxford: Oxford University Press.

Jax, K. (2010). *Ecosystem functioning*. Cambridge/New York: Cambridge University Press.

Jiang, L., Pu, Z., & Nemergut, D. R. (2008). On the importance of the negative selection effect for the relationship between biodiversity and ecosystem functioning. *Oikos, 117*, 488–493. https://doi.org/10.1111/j.0030-1299.2008.16401.x.

Jones, C. G., Lawton, J. H., & Shachak, M. (1994). Organisms as ecosystem engineers. *Oikos, 69*, 373–386.

Jones, C. G., Lawton, J. H., & Shachak, M. (1997). Positive and negative effects of organisms as physical ecosystem engineers. *Ecology, 78*, 1946–1957. https://doi.org/10.1890/0012-9658(1997)078[1946:PANEOO]2.0.CO;2.

Keddy, P. A. (1992). Assembly and response rules: Two goals for predictive community ecology. *Journal of Vegetation Science, 3*, 157–164. https://doi.org/10.2307/3235676.

Lavorel, S., & Garnier, E. (2002). Predicting changes in community composition and ecosystem functioning from plant traits: Revisiting the holy grail. *Functional Ecology, 16*, 545–556. https://doi.org/10.1046/j.1365-2435.2002.00664.x.

Leibold, M. A. (1995). The niche concept revisited: Mechanistic models and community context. *Ecology, 76*, 1371–1382. https://doi.org/10.2307/1938141.

Maclaurin, J., & Sterelny, K. (2008). *What is biodiversity?* Chicago: University of Chicago Press.

McIntosh, R. P. (1985). *The background of ecology: Concept and theory*. Cambridge/New York: Cambridge University Press.

McLaughlin, P. 2001. *What functions explain: Functional explanation and self-reproducing systems* (Cambridge studies in philosophy of biology). Cambridge/New York/Melbourne: Cambridge University Press.

Millikan, R. G. (1984). *Language, Thought, and Other Biological Categories: New Foundations for Realism*. Cambridge, MA/London: MIT Press.

Millikan, R. G. (1989). In defense of proper functions. *Philosophy of Science, 56*, 288–302.

Mitchell, S. D. (1993). Dispositions or etiologies? A comment on Bigelow and Pargetter. *Journal of Philosophy, 60*, 249–259.

Mossio, M., Saborido, C., & Moreno, A. (2009). An organizational account of biological functions. *British Journal for the Philosophy of Science, 60*, 813–841.

Naeem, S. (2002). Functional biodiversity. In H. A. Mooney & J. G. Canadell (Eds.), *Encyclopedia of global environmental change* (pp. 20–36). Chichester/Rexdale: Wiley.

Neander, K. (1991). Functions as selected effects: The conceptual analyst's defense. *Philosophy of Science, 58*, 168–184.

Nock, C. A., Vogt, R. J., & Beisner, B. E. (2016). Functional traits. In *Encyclopedia of life sciences* (pp. 1–8). Chichester: Wiley. https://doi.org/10.1002/9780470015902.a0026282.

Nunes-Neto, N., Moreno, A., & El-Hani, C. N. (2013). The implicit consensus about function in philosophy of ecology. In N. Nunes-Neto, C. N. El-Hani, & A. Moreno (Eds.), *The functional discourse in contemporary ecology* (pp. 40–65). Salvador: Doctoral dissertation, Universidade Federal da Bahia.

Nunes-Neto, N., Moreno, A., & El-Hani, C. N. (2014). Function in ecology: An organizational approach. *Biology and Philosophy, 29*, 123–141.

Nunes-Neto, N., Do Carmo, R. S., & El-Hani, C. N. (2016). Biodiversity and ecosystem functioning: An analysis of the functional discourse in contemporary ecology. *Filosofia e História da Biologia, 11*, 289–321.

Odenbaugh, J. (2010). On the very idea of an ecosystem. In A. Hazlett (Ed.), *New waves in Metaphysics* (pp. 240–258). Basingstoke: Palgrave Macmillan.

Petchey, O. L., & Gaston, K. J. (2006). Functional diversity: Back to basics and looking forward. *Ecology Letters, 9*, 741–758. https://doi.org/10.1111/j.1461-0248.2006.00924.x.

Pocheville, A. (2015). The ecological niche: History and recent controversies. In T. Heams, P. Huneman, G. Lecointre, & M. Silberstein (Eds.), *Handbook of evolutionary thinking in the sciences* (pp. 547–586). Dordrecht: Springer. https://doi.org/10.1007/978-94-017-9014-7_26.

Queller, D. C., & Strassmann, J. E. (2009). Beyond society: The evolution of organismality. *Philosophical Transactions of the Royal Society B: Biological Sciences, 364*, 3143–3155. https://doi.org/10.1098/rstb.2009.0095.

Saborido, C. (2014). New directions in the philosophy of biology: A new taxonomy of functions. In M. C. Galavotti, D. Dieks, W. J. Gonzalez, S. Hartmann, T. Uebel, & M. Weber (Eds.), *New directions in the philosophy of science* (The philosophy of science in a European perspective) (pp. 235–251). Dordrecht: Springer. https://doi.org/10.1007/978-3-319-04382-1_16.

Schoener, T. W. (1989). The ecological niche. In J. M. Cherrett & A. D. Bradshaw (Eds.), *In Ecological concepts: The contribution of ecology to an understanding of the natural world* (pp. 79–114). Oxford: Blackwell Scientific Publications.

Simberloff, D., & Dayan, T. (1991). The guild concept and the structure of ecological communities. *Annual Review of Ecology and Systematics, 22*, 115–143. https://doi.org/10.1146/annurev.es.22.110191.000555.

Sober, E., & Wilson, D. S. (1994). A critical review of philosophical work on the units of selection problem. *Philosophy of Science, 61*, 534–555.

Swenson, W., Arendt, J., & Wilson, D. S. (2000a). Artificial selection of microbial ecosystems for 3-chloroaniline biodegradation. *Environmental Microbiology, 2*, 564–571.

Swenson, W., Wilson, D. S., & Elias, R. (2000b). Artificial ecosystem selection. *Proceedings of the National Academy of Sciences, 97*, 9110–9114. https://doi.org/10.1073/pnas.150237597.

Symstad, A. J. (2002). An overview of ecological plant classification systems. In R. S. Ambasht & N. K. Ambasht (Eds.), *Modern trends in applied terrestrial ecology* (pp. 13–50). New York: Springer.

Tilman, David. 2001. Functional diversity. In Encyclopedia of biodiversity, 3:109–120. Amsterdam Elsevier. doi:https://doi.org/10.1016/B0-12-226865-2/00132-2.

Walker, B. H. (1992). Biodiversity and ecological redundancy. *Conservation Biology, 6*, 18–23.

Walsh, D. M. (2008). Function. In P. Stathis & M. C. London (Eds.), *The Routledge companion to philosophy of science*. New York: Routledge.

Williams, G. C. (1966). *Adaptation and natural selection: A critique of some current evolutionary thought*. Princeton: Princeton University Press.

Wilson, D. S. (1997). Biological communities as functionally organized units. *Ecology, 78*, 2018–2024.

Wilson, J. B. (1999). Guilds, functional types and ecological groups. *Oikos, 86*, 507–522.

Wilson, D. S., & Sober, E. (1989). Reviving the superorganism. *Journal of Theoretical Biology, 136*, 337–356.

Wouters, A. (2005). The function debate in philosophy. *Acta Biotheoretica, 53*, 123–151.

Wright, L. (1973). Functions. *Philosophical Review, 82*, 139–168.

Wynne-Edwards, V. C. (1962). *Animal dispersion in relation to social behavior*. Edinburgh: Oliver and Boyd.

Chapter 15
Integrating Ecology and Evolutionary Theory: A Game Changer for Biodiversity Conservation?

Silvia Di Marco

Abstract Currently, one of the central arguments in favour of biodiversity conservation is that it is essential for the maintenance of ecosystem services, that is, the benefits that people receive from ecosystems. However, the relationship between ecosystem services and biodiversity is contested and needs clarification. The goal of this chapter is to spell out the interaction and reciprocal influences between conservation science, evolutionary biology, and ecology, in order to understand whether a stronger integration of evolutionary and ecological studies might help clarify the interaction between biodiversity and ecosystem functioning as well as influence biodiversity conservation practices. To this end, the eco-evolutionary feedback theory proposed by David Post and Eric Palkovacs is analysed, arguing that it helps operationalise niche construction theory and develop a more sophisticated understanding of the relationship between ecosystem functioning and biodiversity. Finally, it is proposed that by deepening the integration of ecological and evolutionary factors in our understanding of ecosystem functioning, the eco-evolutionary feedback theory is supportive of an "evolutionary-enlightened management" of biodiversity within the ecosystem services approach.

Keywords Ecosystem functions · Evolution · Niche construction · Ecosystem engineering · Conservation biology

15.1 Introduction

Currently, one of the central arguments in favour of biodiversity conservation is that it is essential for the maintenance of ecosystem services, that is, the benefits that people receive from ecosystems (MA 2003, 2005). However, as remarked by Georgina Mace and colleagues, although both biodiversity and ecosystem scientists implicitly acknowledge that biodiversity plays different roles at the different levels

S. Di Marco (✉)
Centro de Filosofia das Ciências da Universidade de Lisboa, Lisbon, Portugal
e-mail: sdmarco@fc.ul.pt

© The Author(s) 2019
E. Casetta et al. (eds.), *From Assessing to Conserving Biodiversity*,
History, Philosophy and Theory of the Life Sciences 24,
https://doi.org/10.1007/978-3-030-10991-2_15

of the ecosystem services hierarchy, their approach to biodiversity conservation remains fundamentally different. Conservation biologists typically struggle to develop an evidence base that supports the protection of biodiversity, in particular charismatic and endangered species, as a good endowed with cultural, scientific and even "intrinsic" value, while ecologists focus on the contribution provided by biodiversity, usually understood as functional diversity, to ecosystem processes and services (Mace et al. 2012). Face to the challenges posed by the ecosystem services approach to biodiversity conservation, this mismatch amongst professionals is a reason of concern. Still, the growing interest amongst ecologists for the feedbacks between organisms and ecosystems promises to shed new light on the interactions between biodiversity, ecosystem processes and ecosystem services, and has the potential to influence biodiversity conservation planning.

In this regard, various authors stress the fact that since the introduction of the concept of ecosystem service in conservation policies, community and ecosystem ecologists have paid more and more attention to biodiversity, especially species and genes diversity, as a driver of ecosystem functioning (Naeem 2002; Loreau 2010). In particular, Michel Loreau has argued that if ecologists are to understand and model the effects of biodiversity on the functioning of ecosystems, they have to develop new theories to connect the dots that link the evolution of species traits at the individual level (evolutionary biology), the dynamics of species interactions (community ecology) and the overall functioning of ecosystems (ecosystem ecology) (Loreau 2010). An endeavor whose difficulties cannot be understated, especially if one takes into account the "explanatory reversibility" of the concept of biodiversity in ecology,[1] and the philosophical issues posed by both the notion of ecosystem *function* and the idea that organisms play a *role* in an ecosystem.[2]

Bracketing these questions, as well as the problems posed by the polysemy of 'biodiversity',[3] the present chapter aims to spell out the interaction and reciprocal influence between conservation science, evolutionary biology, and ecology, in order to understand whether a stronger integration of evolutionary and ecological studies might help clarify the relationship between biodiversity and ecosystem functioning, and influence biodiversity conservation practices within the ecosystem services approach.

To this aim I will first describe the divide between what Mace et al. (2012) have called the "ecosystem services perspective" and the "conservation perspective" within the ecosystem services approach, and present Loreau's view on the possible integration of ecological and evolutionary studies. Subsequently, I will analyse the eco-evolutionary feedback theory by Post and Palkovacs (2009), as an example of such integration. In particular, I will argue that this theory helps operationalise the evolutionary concept of niche construction (Laland et al. 1999; Odling-Smee et al. 2003), and offers theoretical instruments to develop a more sophisticated understanding of the relationship between ecosystem functioning and biodiversity.

---

[1] See Huneman, Chap. 13, in this volume.

[2] See Dussault, Chap. 14, in this volume.

[3] See Toepfer, Chap. 16, and Meinard et al., Chap. 17, in this volume.

Finally,[4] I will argue that by deepening the integration of ecological and evolutionary factors in our understanding of ecosystem functioning, the eco-evolutionary feedback theory is supportive of an "evolutionary-enlightened management" (Ashley et al. 2003) of biodiversity within the ecosystem services approach.

15.2 On the Relationship Between Biodiversity and Ecosystem Services

Ecosystem services are the benefits that humans derive, directly or indirectly, from the ecosystems or, phrased differently, they are "the functions and processes of ecosystems that benefit humans" (Costanza et al. 2017). They are classified into *provisioning services*, such as food, clear water, timber, and fuel; *regulating services*, such as flood protection, pests control, and climate regulation; *supporting services*, corresponding to basic ecosystem processes such as primary production, soil formation, and nutrients cycle; and *cultural services*, corresponding to a range of cultural benefits – e.g., aesthetic, recreational, or spiritual – that people receive from ecosystems.

15.2.1 Ecosystem Services in Brief

The idea of ecosystem service is a socio-economic concept that dates back to 1977, when *Science* published the article "How much are Nature's services worth?" by Walter Westman, but gained momentum in the academia only in 1997, with the publication of the book *Nature's Services: Societal Dependence on Natural Ecosystems* (Daily 1997) and an article by Robert Costanza and colleagues on the value of the world's ecosystem services and natural capital (Costanza et al. 1997). The goal of these publications was to make explicit the contribution of ecosystems to human well-being, and put an economic value on it (between 16 and 54 trillion USD per year at the time), in order to make transparent the trade-offs involved in any decision concerning the use of land and natural resources. This monetary approach stirred a fierce debate, which is still ongoing, but eventually the concept of ecosystem service met biodiversity conservation: first, in 2001, with the launch of the Millennium Ecosystem Assessment (MA) by the United Nations Environment Programme, and later, in 2007, with The Economics of Ecosystems and Biodiversity (TEEB) initiative promoted by the German Government and the European Commission. These programmes are focused, respectively, on the ecological and economic aspects of ecosystem services, and are based on a utilitarian view of biodiversity (biodiversity must be preserved as an ecosystem service in itself, or as a

---

[4] With an argument intersecting that expounded by Alessandro Minelli, Chap. 11, in this volume.

component of the environment necessary for the maintenance of other ecosystem services), and on the implicit (and controversial) assumption that the protection of the ecosystem services leads to the protection of biodiversity (Mace et al. 2012).

15.2.2   Ecosystem Services and Biodiversity: Epistemological and Ethical Troubles

Biodiversity is considered a cultural service or an actual good (which might be marketable or not) when it provides non-material benefits to human beings. Wildlife, uncontaminated landscapes, totemic, charismatic and rare or endangered species have a particular appeal to human beings, because they respond to aesthetic, spiritual, religious, educational and recreational values. In these cases, people value the *diversity* of life as such—or some specific actualization of that diversity, as for instance charismatic species—and not some product or purported effect of biodiversity (e.g., variety of food or possibility to discover new drugs).[5] For all the other services, the relationship between biodiversity and human benefits is all but clear and needs to be examined on a case by case basis (Harrison et al. 2014). As a general rule, there is stronger evidence for the effects of biodiversity on ecosystems stability than on ecosystem services (Cardinale et al. 2012; Srivastava and Vellend 2005), and although it is generally agreed that biodiversity plays an insurance role, by potentially buffering ecosystems against environmental changes (Cottingham et al. 2001; Hooper et al. 2005; Loreau 2010a), data reviews and meta-analysis on the threefold relationship between biodiversity, ecosystem functioning, and ecosystem services are hampered by the lack of unified definitions and measures of biodiversity, and by the complexity and multi-faceted nature of each of the factors of the equation (Cardinale et al. 2012; Mace et al. 2012). Also, in many cases it is difficult to establish if the biodiversity effect is due to diversity as such (e.g., at the level of species, genes, or traits) or to other factors such as composition or biomass.

As mentioned above, within the ecosystem services approach, ecosystem services and biodiversity are often used as synonyms, thus implying that they are the same thing and that, by protecting one, we are automatically protecting the other (Costanza et al. 2017; TEEB 2010). On the contrary, within the conservationist perspective, biodiversity is an ecosystem service or a good *per se,* and as such it does not necessarily contribute to other ecosystem services and is potentially in conflict with them. Both positions have pitfalls. For what concerns the conservationist perspective, the main problem is that it is blind to the functional role of biodiversity, and often focuses on charismatic or endangered species. In so doing it loses sight of the greater variety of units, levels and scales at which biodiversity occurs, and perpetuates a static vision of life both at the species and ecosystem level. On the contrary, within the ecosystem services perspective, the functional role of biodiversity

---

[5] But for a problematisation of the relationship between biodiversity and cultural services see, for instance, Sarkar 2005, Cardinale et al. 2012.

is acknowledged, but in practice ecologists account for its contribution to the ecosystem almost exclusively in terms of simple trophic structures and the related stocks and flows of energy, nutrients and biomass. This poses epistemological problems related to the different aims, conceptual frameworks, and methodologies adopted in different scientific disciplines, where such problems call for theoretical and empirical solutions. Also, values of biodiversity other than its contribution to ecosystem functioning are not taken into account, thus posing an ethical problem (Mace et al. 2012).

The ethical criticism is the one most often leveraged against the ecosystem services approach (Reyers et al. 2012), and can be framed within a number of related debates: the controversy on the monetary nature of the concept of ecosystem service (e.g., McCauley 2006; Redford and Adams 2009); the debate about the instrumental *versus* intrinsic value of biodiversity (e.g., Norton 1986; Sarkar 2005; Maquire and Justus 2008; Justus et al. 2009); or the opposition between ecocentrism and anthropocentrism in environmental ethics (e.g., Singer 1975; Thompson and Barton 1994; Naess 1973). In this chapter, I let aside the ethical issues and focus on the epistemological problems instead, trying to understand whether a stronger integration between ecology and evolutionary theory might make a difference in conservation planning within the ecosystem services approach.

15.2.3   Ecosystem Services and Biodiversity: An Ecologist's Perspective

For those who embrace the conservation perspective, there is a potential opposition between biodiversity and ecosystem services, and some authors see the ecosystem services approach as an unwarranted thwarting of the original mission of conservation, namely, the protection of biodiversity or, more generally, nature, for its own sake (e.g., McCauley 2006; Redford and Adams 2009). From this perspective, the ecosystem services approach is detrimental to biodiversity conservation. However, if one tackles this criticism from an epistemological point of view, letting aside the controversy concerning the value of biodiversity, it becomes apparent that the endorsement of the concept of ecological service in many conservation policies has produced at least one major benefit for biodiversity science in that it has given special impulse to the study of the effects of biodiversity on ecosystem functioning in experimental and theoretical ecology (Loreau 2010). According to Loreau, this had relevant consequences for ecology both at the epistemological and disciplinary level. At the epistemological level, it has revived and reshaped the diversity-stability debate—that has run through ecosystem ecology since the 1950s (e.g., MacArthur 1955; May 1973; Pimm 1984)[6]—, and has given momentum to the study of the respective roles of individual-level and ecosystem-level selection in shaping ecosys-

---

[6] See Huneman, Chap. 13, in this volume, for a discussion of the notions of diversity used in the formulation and test of the stability hypothesis (biodiversity as an *explanans*).

tem properties—a controversial issue in both ecology and evolutionary biology (see Williams and Lenton 2007; Loreau 2010b). More importantly, it has changed the way ecosystem and community ecologists approach the study of biodiversity, giving prominence to the idea that biodiversity, especially species and genes diversity, is a driver of ecosystem functioning (Naeem 2002; Loreau 2010), and populations cannot be studied as homogeneous biomass pools in which individuals operate in identical ways to influence the nutrient and energy flows amongst the ecosystem compartments (Bassar et al. 2010).

At the disciplinary level, the need to better understand the effects of biodiversity on ecosystem functions at different spatial and temporal scales has made more evident and urgent the importance of integrating community ecology, ecosystem ecology and evolutionary biology (Loreau 2010, b).[7] Indeed, the development of the ecosystem services approach in environmental protection and biodiversity conservation has not only turned the study of the relationship between biodiversity and ecosystems into a pressing scientific matter, imposing a research agenda on ecologists (i.e., to understand the role and relevance of biodiversity for the delivery of ecosystem services). It has also implicitly indicated the scientific hypothesis to be tested, namely that biodiversity is necessary for ecosystem processes and that the loss of biodiversity hampers the functioning of ecosystems in the short and/or long term, thus affecting the provision of ecosystem services.

To answer the practical questions raised by the ecosystem services approach it is necessary to understand how ecosystems function and predict how they might change under a variety of environmental and anthropic pressures, such as climate change, habitat loss and degradation, overharvesting and diffusion of invasive exotic species. All these factors affect biodiversity as much as ecosystems as a whole. Loreau agrees with Mace and colleagues that current models of interaction between biodiversity and ecosystem functioning, based mostly on the modelling of evolutionary complex food webs, have several limitations. He stresses that important insights might come from theories such as ecosystem engineering (Jones et al. 1994, 1997; Wright and Jones 2006) and niche construction (Laland et al. 1999; Odling-Smee et al. 2003), which try to account for the ability of organisms to transform their habitat with relevant consequences both at the ecological and evolutionary level. In the last decade, there has been a surge of interest for eco-evolutionary theories (Whitham et al. 2006; Fussman et al. 2007), particularly in theoretical ecology (Kokko and Lopez-Sepulcre 2007). In what follows I present and discuss David Post and Eric Palkovacs' eco-evolutionary feedback (EEFB) theory, because it is an interesting example of ecological re-elaboration and clarification of the niche construction theory (henceforth NCT) originally formulated by Kevin Laland and John Odling-Smee, and also because Post and Palkovacs suggest that an integration of ecological and evolutionary theories would have relevant consequences not only for our understanding of ecosystem functioning, but also for biodiversity conservation.

---

[7] But see Huneman, Chap. 13, in this volume, for a criticism of this endeavour.

15.3   Eco-Evolutionary Feedback Theory

An eco-evolutionary feedback is "the cyclical interaction between ecology and evolution such that changes in ecological interactions drive evolutionary change in organismal traits that, in turn, alter the form of ecological interactions, and so forth" (Post and Palkovacs 2009). This description of the reciprocal causation between ecological and evolutionary change clarifies the ecological relevance of NCT by making a clear distinction between the process of niche construction, defined as "the effect of an organism on its environment" (Post and Palkovacs 2009), and the evolutionary feedbacks that occur in response to the environmental changes caused by organisms. Niche construction *sensu stricto* (Post and Palkovacs 2009) includes both active engineering and the effects caused by the by-products of biological process, while the evolutionary feedback can be the result of heritable traits change or phenotypic plasticity. By explicitly separating the general process of EEFB into two sub-processes (niche construction + evolutionary feedback), EEFB theory makes clear that not all the biotic processes that shape the environment can cause subsequent evolution, because many factors can prevent the evolutionary feedback. However, when the feedback occurs, it has important consequences at both the evolutionary and ecological level, because it can affect the direction of evolution and alter the role of species in the ecosystem. It also highlights that both processes, even when they do not occur together, have important ecological and evolutionary consequences, hence deserving in-depth study. Finally, unlike NCT, at least in its initial version, EEFB allows for cases in which the recipient population of the modified selective pressure can be different from the population that produced the environmental transformation in the first place (see Odling-Smee et al. 2013; Barker and Odling-Smee 2014).

For an EEFB to occur, three conditions need be satisfied: (1) organisms must have a phenotype that *strongly* impacts the environment, i.e., they must structure or construct their niche (e.g., nutrients cycling and translocation, habitat construction and modification, consumption)[8]; (2) the changes produced in the environment must cause selection on a population and that this population has sufficient genetic capacity to evolve in response to changes in the environment; (3) the time-scales of the ecological and evolutionary responses have to be congruent, i.e., the constructed niche must persist for a duration that is sufficient to select the relevant traits (this corresponds to the concept of ecological inheritance in NCT).

For what concerns (2) it should be noticed that, as in adaptive evolution more generally, the evolutionary factors that determine whether a population will evolve or go extinct are a combination of genetic factors (e.g., high levels of genetic variation are expected to favour evolutionary change); demographic factors (e.g.,

---

[8] Potentially, all organisms are niche constructors, because all organisms interact with the environment. However, as it will be explained below, a key factor for the identification of meaningful cases of niche construction in the EEFB theory is the strength (magnitude and/or extent) of the interaction between an organism and the environment (which includes other organisms), and the spatial and temporal scale of the effects of such interaction.

population size and genetic drift); and ecological factors (e.g., the rate of deforestation or the introduction of a toxic compound).

For what concerns (3), what counts as a *sufficient duration* will depend on the niche, as well as on the species and traits under consideration. In any case, there must be an overlap of ecological and evolutionary time: the constructed niche must persist long enough to produce evolutionary effects, and evolution must be fast enough to feed back on the constructed niche and further influence it. Since what matters is the congruence between ecological and evolutionary time, in principle evolution does not need to be rapid for EEFB to emerge. Slow niche construction, such as the oxygenation of earth's atmosphere by cyanobacteria, can create eco-evolutionary feedbacks as much as rapid evolution associated with rapid niche construction. However, the study of EEFB associated with rapid evolution has the advantage of being more easily amenable to empirical tests, and is more likely to be relevant in terms of biodiversity protection and ecosystem services conservation practices.

15.3.1 EEFB and Contemporary Evolution: Three Empirical Cases

The existence of rapid contemporary evolution, i.e., the evolution of heritable traits over a few generations (Stockwell et al. 2003; Jones et al. 2009),[9] is neither particularly controversial in ecology nor in evolutionary biology. What is controversial is the overall ecological and evolutionary relevance (prevalence and magnitude) of this phenomenon. As a matter of fact, in spite of the accumulation of studies that in the course of the last 40 years have shown that a strict distinction between ecological and evolutionary time is unwarranted, ecologists still tend to ignore potential effects of evolution on ecological interactions, because they assume that evolution occurs on a much slower time scale than ecological dynamics (Bassar et al. 2010). On the other hand, evolutionary biologists tend to ignore the action of organisms on their environment, because it is considered too weak and flimsy to significantly change selection pressures (Laland and Sterelny 2006). Eco-evolutionary theories challenge these entrenched views. In fact, there is growing evidence that contemporary evolution is a widespread phenomenon—which concerns many traits and many organisms from all kingdoms—and the evidence for potential cases of eco-evolutionary feedbacks is growing. Here I summarise three of the five empirical cases reviewed by Post and Palkovacs (2009): alewives' speciation caused by patterns of migration, its influence on zooplankton communities, and the subsequent evolution of foraging traits; the effect of the life histories of Trinidad guppies on

---

[9] Rapid evolution, contemporary evolution and microevolution are sometimes used as synonyms, and definitions vary (e.g., Thompson 1998, Kinnison and Hairston 2007, Ashley et al. 2003). Here I follow Post and Palkovacs 2009 and use contemporary evolution to refer to the overlap of ecological and evolutionary times, irrespectively of the actual duration of the process.

nutrient cycling and its potential feedback on male guppies' phenotype; the soil-mediated impact of *Populus* leaf tannins levels on the development of adapted roots.

15.3.1.1 Alewives and Zooplankton

Along North America East coast, the ecological isolation of lakes from the ocean has led to the phenotypic differentiation of alewife (*Alosa pseudoharengus*) land-locked populations that differ from the original anadromous population in feeding morphology and prey selectivity. Anadromous fishes migrate up rivers from the ocean to spawn and then go back to the open sea. In this case, the alewives only temporarily affect the community structure of lacustrine zooplankton (niche construction via predation, Post and Palkovacs 2009) before they go back to the ocean, thus the duration of the constructed niche is not long enough to cause an eco-evolutionary feedback. On the contrary, in the landlocked populations, intense year-round predation pressure eliminates large-bodied preys and produces a lacustrine zooplankton community of relatively low biomass of small-bodied zooplankton throughout the year (persistent constructed niche). This exerts a strong selection for traits related to foraging on small zooplankton, so that the landlocked population has developed smaller mouth gape and narrower spacing between gill rakes compared to the ancestral anadromous population (evolutionary feedback). In this case there is strong evidence for a complete EEFB.

15.3.1.2 Trinidad Guppies and Nutrients Cycling

Observations in the wild have shown that the life-histories (age and size at maturity) of Trinidad guppies (*Poecilia reticulata*) are affected by predation pressure. In high-predation environments, guppies reach maturity at an earlier age and smaller size, and they reproduce more frequently giving birth to smaller offsprings, with important effects for the population phenotype. Mesocosm experiments have shown that under conditions of equal biomass, populations characterised by a high number of small individuals (high-predation environment) drive higher nutrients flows compared to populations with fewer larger individuals (low-predation environment), thus increasing the rates of primary production, i.e., algal biomass (constructed niche). This, in turn, might influence further differentiation amongst guppies' populations, for instance, by influencing traits such as male colour patterns, which are under natural and sexual selection, and are sensitive to the levels of algae-derived carotenoids in the environment (potential eco-evolutionary feedback).

15.3.1.3 *Populus* and Soil Nutrients Levels

Poplar trees are foundation species whose chemical effects on leaf litter strongly influence community dynamics and ecosystem processes. Observational studies have shown that intraspecific variation in condensed tannin levels in poplar trees' leaves controls decomposition and nitrogen mineralisation rates, as well as the composition of the microbial community in the soil, thus creating a microhabitat (constructed niche). Since high concentrations of tannins inhibit nutrients release from leaves litter, poplar trees with high tannin levels will have to cope with low nutrients levels. According to EEFB theory, these trees should display some form of adaptation. Indeed, a strong positive correlation between leaf tannin levels and the development of finer roots has been observed, thus providing indirect evidence for eco-evolutionary feedback. However, ecological factors such as the presence of other plant species, herbivores and nutrients loading might disrupt or reduce the strength of the feedback by altering the ecology of the soil.

It is worth noticing that it is not always clear whether contemporary evolution is due to heritable traits or phenotypic plasticity. However, as remarked by Palkovacs et al. (2012), although such distinction is fundamental to our understanding of evolutionary and ecological processes, in the context of conservation biology it might be more important, and urgent, to link phenotypic change and ecosystem dynamics, regardless of the specific causes of change. Also, considering that plasticity itself is a hereditary trait that evolves and can direct future phenotypic change, it is not always useful to draw a thick line between plasticity and genetic change in terms of potential ecological causes and effects (Ghalambor et al. 2007; Palkovacs et al. 2012).[10] What is most relevant here is to highlight that the species more likely involved in EEFB are also the most relevant in terms of ecosystem functioning, because they strongly affect the community and the ecosystem where they live. They can be keystone, foundation, or dominant species, ecosystem engineers, or species that alter nutrient cycles through translocation or recycling.

15.3.2 EEFB, Niche Construction, and Ecosystem Engineering

What all these organisms have in common is that they are strong interactors.[11] To be a strong interactor, however, often depends on the ecological context: foundation species in one habitat might be rare in another, weak interactors in species-rich communities might have strong effects in species-poor communities, and species that move nutrients will have very different impacts in low- compared to high-nutrient environments (Post and Palcovaks 2009; Paine 1966; Menge et al. 1994).

---

[10] See also Minelli, Chap. 11, in this volume.

[11] For a detailed discussion of the differences between strong interactors, in particular between keystone species and ecosystem engineers see Boogert et al. 2006.

Thus, the ability of a species to construct a persistent niche often depends on the overall conditions of the ecosystem and the community, which means that it can vary in space and time. In turn, the eco-evolutionary feedback, with its potential to alter and respond to environmental selective pressure, can lead to the differentiation of a population whose ecological role is different from that of the original population, thus affecting community and ecosystem dynamics. Indeed, there might be instances in which the change of the traits of a species is at least as important as its presence/absence in terms of ecological effects. In the case of the alewives from North American coastal lakes, for example, there is evidence that the differentiation of the landlocked population has influenced the evolution of one of its preys, *Daphnia ambigua*, and this is likely to cause further effects on trophic cascades, because *Daphnia* is itself a strong interactor (a dominant grazer for zooplankton) (Palkovacs et al. 2012).

A main feature of EEFB theory is that it highlights the fact that organisms actively build their environment and that species, species traits, and species ecological impacts are dynamic and vary across space and time. A consequence of this is that within the research framework set by eco-evolutionary theories, the functional role of biodiversity in an ecosystem cannot be understood simply in terms of more or less complex trophic webs. This simplifying idealisation has been at the core of the success of ecosystem ecology in the study of terrestrial global biogeochemistry, but it has been increasingly called into question by ecologists themselves at least since the 1990s (Loreau 2010). In particular, the concept of ecosystem engineering introduced by Clive Jones and colleagues (Jones et al. 1994, 1997; Wright and Jones 2006), often considered the ecological counterpart of Laland and Odling-Smee's NCT, has shown that connectance webs that describe the processes driven by ecosystem engineers should be studied along with trophic webs, if we are to accurately model the interactions between communities and ecosystems. Importantly, these studies have shown that the laws of conservation of mass and energy, as well as the stoichiometry rules used to model trophic webs, cannot be used to predict the structure and outputs of ecosystem engineering networks, for which specific qualitative and quantitative models have been proposed (Jones et al. 1997; Boogert et al. 2006).

Ecosystem engineers are "organisms that directly or indirectly modulate the availability of resources (other than themselves) to other species, by causing physical state changes in biotic and abiotic materials. In so doing they modify, maintain and/ or create habitats" (Jones et al. 1994). Within EEFB theory, they are seen as strong candidates for eco-evolutionary feedbacks, together with keystone species (species, usually predators, whose impact on their community or ecosystem is much larger than would be expected from their abundance), dominant species (species that outnumber their competitors in abundance or total biomass), and foundation species (species that strongly influence the structure of the community, e.g., by creating habitats). Accordingly, studying ecosystems from an EEFB theory perspective implies to parse strong ecological interactors according to a range of qualitative and quantitative models, e.g., strong *per capita* interactions that produce effects in the short term *vs.* weak but continuous *per capita* interactions that produce cumulative effects in the long term. Trophic webs, then, are but one of the interaction networks that compose the overall connectivity of the ecosystem. The other crucial ecological relationships that need attention are the non-trophic interaction webs described by ecosystem engineering theory, and the environmentally-mediated gene-associations (EMGAs) theorised by Odling-Smee and colleagues (Odling-Smee et al. 2003, 2013; Barker and Odling-Smee 2014), in a development of the original NCT, prompted by the insights provided by ecosystem engineering and eco-evolutionary theories.

15.3.3   EEFB and Environmentally-Mediated Gene-Associations

EMGAs are "indirect but specific connections between distinct genotypes mediated either by biotic or abiotic environmental components in the external environment [...]. They map sources of selection stemming from one population's genes onto genotypes in another population that evolve in response to those modified sources" (Odling-Smee et al. 2013). These *indirect* evolutionary interactions mediated by the environment emerge when the niche constructed by a population—via its physiological processes as well as active engineering—influences the selective pressure acting on the same population or, more often, on a different population of different species. For example, in the case of Trinidad guppies, predators, through differential predation pressure, can influence guppy populations' life histories, leading to the differentiations of populations of larger or smaller guppies, characterised by different rates of excretion that determine differential inorganic nutrients distribution. This, in turn, affects algal growth, which has the potential to feed back on the selection of male guppy colour patterns through the concentration of carotenoids released by algae in the environment.

The idea of EMGAs helps formalise the causal chain of EEFB in genetic terms, and can be used to visualise the ramifications of evolutionary and ecological effects deriving from niche construction via biotic or abiotic mediations. In its original form, it gives epistemic priority to the genetic component within the EEFB's causal chain, but in those cases in which the niche construction is underpinned by non-heritable variation, environmentally-mediated *genotypic*-associations are replaced by environmentally-mediated *phenotypic*-associations (EMPAs), thus emphasising that the phenotype should not be thought of as the mere epiphenomenon of genetic information, but as the dynamic result of the combination of heritable variation with a number of non-heritable factors, such as plasticity, epigenetics and population structure (Odling-Smee et al. 2013).

It follows that, in order to respond to the requirements of EEFB theory, the study of ecosystem processes and functioning should be articulated along two interrelated axes, which force ecosystem ecology to revise its operational simplifying idealisations. On the one hand, the study of the sub-process of niche construction requires the development of ecosystem models that account for high degrees of connectance at the different scales of the ecosystem, integrating trophic and competitive webs with more complex interaction webs, as well as EMGAs or EMPAs; on the other hand, the study of the sub-process of evolutionary feedbacks needs to be carried out taking into account both genetic and non-heritable phenotypic variation, because both can be sources of functional evolution and adaptation. Accordingly, functional diversity must be understood as a dynamic epiphenomenon that can potentially emerge from both genetic and non-genetic factors that need to be studied on a case by case basis.

All in all, what emerges from EEFB theory is a highly dynamic picture of ecosystems, populations and communities, in which the structure of biodiversity—used here as a shorthand for diversity at the level of species, genes, traits, communities, etc.—can vary more easily than both ecologists and evolutionary biologists are prone to believe, and where the causal chain of change does not go exclusively from the environment to the organism (ecological change as a cause of trait change), but can go from the organism to the environment (trait change as a cause of ecological change). In the next section, I explore the potential consequences of this shift of perspective for conservation biology.

15.4 Eco-Evolutionary Feedback Theory: Some Consequences for Biodiversity Conservation

The study of EEFB pushes ecologists to recognise that contemporary evolution creates phenotypic differences that can alter the role of a species in a community or ecosystem at ecological time-scales. This implies that evolution can no longer be considered mere background noise in the study of ecosystem dynamics, and extant and potential novel biodiversity become a fundamental component of the study of ecosystem dynamics. For Post and Palkovacs: "the study of eco-evolutionary feedbacks focuses attention on the bidirectional interactions that unify ecology and evolution, and highlights the importance of conserving both ecological and evolutionary diversity in nature" (Post and Palkovacs 2009). But how, exactly, could EEFB theory guide biodiversity conservation? As referred to in Sect. 15.1, a criticism leveraged by conservation biologists to ecologists within the ecosystem services approach is that they account for biodiversity's contribution to ecosystem functions almost exclusively in terms of simple trophic structures (Mace et al. 2012). What kind of instruments does EEFB theory offer to tackle this issue?

15.4.1 Ecosystem Engineers First?

Considering that EEFB theory has many points in common with the ecosystem engineering theory, some important insights about the impact of eco-evolutionary theories on biodiversity conservation can be found in Crain and Bertness 2006 and in Boogert et al. 2006. For these authors, ecosystem engineers should be the primary targets of biodiversity conservation policies, because they shape habitats and ecosystems, with all their related species and functions. Since ecosystem engineers are responsible for a much higher and more complex level of inter-species connectance than the trophic webs generated by other organisms, the loss of ecosystem engineers is more likely to have far reaching negative consequences on both communities and whole ecosystems (Crain and Bertness 2006. See also Jones et al. 1997; Wright and Jones 2006). Although species that are ecosystem engineers under certain circumstances may not be so under others, it is possible to identify fundamental engineering roles in ecosystems, independently of the specific species involved. Accordingly, to grant stability to ecosystem structure and functioning, conservation policies should focus on protecting the activity of key engineers, rather than the species composition of an ecosystem (Boogert et al. 2006; see also Odling-Smee et al. 2003). This is a classical argument in favour of the preservation of functional diversity rather than species diversity, and is usually criticised for being too narrow a criterion for selecting the aspects of biodiversity worth protection (Mace et al. 2012). To preserve ecosystem functioning, in fact, we do not need to protect all the species that perform a given function and their genetic variability. For instance, we do not need to protect all the species of trees in a forest, and their intraspecific variation, to ensure biomass production, oxygen emission, and CO_2 sequestration. From this perspective, the most efficient course of action would be to select the species that better perform the function of interest, and focus our conservation efforts on them. This approach is likely to leave aside rare species, which represent a primary target for conservation biology, because their functional role on an ecosystem is often negligible. In this respect, not only do aspects of biodiversity not related to ecological functions become irrelevant, but the replacement of ecosystem engineering species using artificial solutions becomes an acceptable option (e.g., replacing of caterpillars by artificially created leaf ties, see Lill and Marquis 2003). Here, at least in principle, the choice to favour technological solutions over biodiversity conservation will be constrained by considerations of efficacy and cost-effectiveness (Boogert et al. 2006), rather than by an *a priori* obligation to avoid species extinction, or a precautionary approach whereby a species (or a genome) that has no particular functional import under the present conditions might become relevant under different conditions, because of ecological changes or because our knowledge of the benefits we obtain from that particular species/genome changes (see Maclaurin and Sterelny 2008; Sarkar 2005). Thus, although there are compelling reasons for choosing ecosystem engineers as targets of biodiversity conservation, this choice must be further qualified and refined.

15.4.2 Genetic Diversity: Better Safe than Sorry

Niche construction (of which ecosystem engineering is just one possible case) is only one half of the EEFB process. The other is evolutionary feedback. To the extent that evolutionary feedbacks have the potential to produce relevant ecological effects, they should be taken into account in conservation policies aimed at preserving ecosystem functions. Since one of the conditions for EEFB is that the population(s) affected by the constructed niche must have sufficient genetic capacity to evolve in response to new selective pressures before going extinct, it follows that it is important to preserve not only functional diversity, but also genetic diversity (that might include phenotypic plasticity), because this ensures that niche constructing species, or other species potentially affected by the constructed niche, will maintain their ability to respond to environmental modifications. In this respect, it should be noted that trait change *per se* is not a guarantee of ecosystem stability, because phenotypic variation can be both a driver of or a buffer against ecological change. In the empirical cases described in Sect. 15.3.1, in fact, the putative evolutionary feedback works clearly as a stabiliser of functions only in the case of poplar trees, while in the cases of the alewives and Trinidad guppies the evolutionary feedback potentially causes a cascade of changes in the community structures whose consequences in terms of ecosystem functions need further clarification. This only makes the need to improve our understanding of eco-evolutionary interactions more compelling, in order to be able to predict when they could buffer the ecosystem and when they would magnify potential functional disruptions. Sweeping generalisations are not warranted in this relatively recent domain of inquiry, but there is evidence that contemporary evolution is most common, although less evident, when it counteracts phenotypic changes caused by environmental pressure, thus buffering ecosystem functions (Ellner et al. 2011; Palkovacs et al. 2012). Preserving the genetic diversity that feeds contemporary evolution, then, seems a safe bet.

Without entering into the debate on what genetic diversity exactly is, how to measure it, and what to do to preserve it (see Mace and Purvis 2008 for a list of problems in this field), we can say that, by providing a clear and well-structured theoretical framework for the empirical study of the reciprocal interaction between evolutionary and ecological processes, EEFB theory offers decisive evidence for the necessity of keeping into account evolutionary dynamics in the study of ecosystem functioning. Accordingly, it provides arguments to support the importance of "evolutionary-enlightened management" in biodiversity conservation (Ashley et al. 2003). In fact, whether the eco-evolutionary feedbacks magnify ecological change or buffer against it, they must be taken into account if we are to preserve ecosystems functioning.

15.4.3 EEFB Theory and Evolutionary-Enlightened Management

For the proponents of evolutionary approaches to biodiversity protection, conservation policies are hampered by the misplaced idea that while human disturbance is very fast, adaptation is a very slow process, thus irrelevant to conservation planning, whose temporal horizon seldom exceeds a few decades (Mace and Purvis 2008). Typological thinking concerning both species and ecosystems is another hindrance to evolutionary-enlightened management, since it promotes the idea that evolutionary change has relevant consequences at an ecological and human time-scale only when it concerns organisms with short generation time (e.g., microorganisms).

Consequently, it is argued, its effects on the whole of biodiversity are negligible (Ashley et al. 2003; Santamaria and Mendez 2012). Mary Ashley and colleagues also remark that, although in conservation planning it is theoretically acknowledged that species respond to change both ecologically and evolutionarily, in practice the importance of evolutionary responses is often neglected. For instance, research models on potential impacts of rising temperature and CO_2 concentrations generally make predictions concerning possible ecological adaptations based on the present ecologies of extant species, without taking into account evolutionary factors such as climate adaptation and the potential disruption to gene flow caused by climate change. Similarly, conservation approaches based on population viability analysis are based on models that assume that the life histories and demographic character-istics of a species are fixed (Ashley et al. 2003). Still, as seen in the example of the Trinidad guppies, environmental factors can significantly affect life histories, with relevant consequences for the structure of a population. This can in turn produce changes in the environment, e.g. in the recycling of nutrients, creating the condi-tions for further evolutionary feedbacks.

Rapid contemporary evolution is the main preoccupation of evolutionary-minded conservationists, not least because anthropic drivers of rapid evolution, such as hab-itat loss and degradation, overharvesting, and the introduction of exotic species, are also the factors that have led to the current extinction crisis (Stockwell et al. 2003; Palkovacs et al. 2012). EEFB theory reinforces this preoccupation because it draws on the evidence that rapid contemporary evolution is a widespread phenomenon. At the same time, one of its theoretical tenets is that eco-evolutionary feedbacks can occur at any timescale, thus highlighting that just as evolutionary factors must be taken into account not only in the long, but also in the short term, ecological effects of evolutionary change might become salient over long timescales. This happens, for instance, when a newly evolved trait constructs a niche whose effects slowly accumulate over time, because it has little *per capita* impact or because external factors intervene to dissipate or swamp the niche. Thus, the effects of EEFB can be time-lagged (Odling-Smee et al. 2013), and this makes predictions more complex, thus more prone to error, but also more realistic.

The implementation of evolutionary-enlightened management for biodiversity conservation would imply the development of research programmes that incorporate evolution into applied ecology and resource management; the assessment of popula-tions' short-term evolutionary potential using direct measures of genetic variation rather than the proxy of neutral molecular variation; and the use of quantitative genetics to assess the genetic variability of traits that are likely to be under selective pressure in hypothetical scenarios (Ashley et al. 2003). Ecological and evolutionary interactions are extremely complex and it is very hard to create workable predicting models. EEFB theory *per se* does not provide a direct answer to this problem, but offers a theoretical framework that can favour the development of such models. Post and Palkovacs' simple move of refining the NCT by splitting the EEFB into two well defined sub-processes allows to break down intricate eco-evolutionary pathways into more tractable components, which can be analysed at different spatial and temporal scales (from long-term whole-ecosystem observation to short-term, small-scale experiments). Subsequently, the general picture can be reconstructed by retracing the network of interactions, their strength and their variation over time (see Odling-Smee et al. 2013). As pointed out in Barker and Odling-Smee (2014), in order to be able to make predictions about the evolution of whole ecosystems and of their components, we need to bring together theories that are general and realistic enough to afford a "local theoretical unification" with precise and realistic models that describe the details of particular complex systems, providing "explanatory concrete integration" (Mitchell 2002). Theories such as EEFB are good candidates for making this synthesis, because they favour the integration of ecosystem ecology, population ecology, and evolutionary biology, and their respective methodological frameworks. If EEFB theory proved successful, then, we would be able to overcome the problem of having too simplified an account of ecosystem functioning and it would be possible to clarify the role of functional diversity within ecosystem processes.

15.5 Conclusions

Since the late 1990s, the development of the concept of ecosystem service for conservation policies has given new momentum to the study of the effects of biodiversity on ecosystem functioning in experimental and theoretical ecology, revitalising the traditional diversity-stability debate and fostering the development of ecosystem evolution theories.

EEFB theory emphasises the active role of organisms in shaping their environment and supports the idea that contemporary evolution is a common and widespread phenomenon. This means that species, their traits, and their ecological impacts are dynamic and vary across space and time. As a consequence, the functional contribution of biodiversity to ecosystem processes cannot be understood simply in terms of mass and energy conservation and stoichiometry rules for trophic webs, but must include, at least, the more elaborated connectance webs proposed by ecosystem engineering theory, and models of environmentally-mediated gene or phenotype associations proposed in recent developments of the NCT. Also, since contemporary evolution can be either a source of ecological change (potential disruption of ecosystem functions) or a buffer against change (preservation of ecosystem functions), in order to make predictions on the evolution of ecosystems and their capacity to sustain ecosystem services, we need to better understand eco-evolutionary interactions from the population to the whole-ecosystem level. On the whole, EEFB theory provides a non-typological image of both species and ecosystem, and challenges static visions in both ecology and evolutionary biology. On the one hand, it defies the idea that evolution is too slow to be relevant in the modelling of ecosystem processes; on the other hand, it undermines the idea that the action of organisms on their environment is too ephemeral to direct selective pressures. All in all, this calls for an evolutionary-enlightened management of biodiversity.

Ultimately, by emphasising the fact that organisms are active agents of ecological and evolutionary change rather than passive objects of selection, EEFB theory causes a shift of perspective on the role of biodiversity in the transformation of ecosystems. In fact, if "organisms and their local environments [are] integrated systems that evolve together" (Barker and Odling-Smee 2014), then species and genetic diversity are at least as important as functional diversity for the evolution and future functioning of an ecosystem. Now, to be able to make predictions about the potential evolution of ecosystems is a fundamental feature of the ecosystem services approach. By definition, what matters the most within the ecosystem services approach is to preserve functional ecosystems, so that humans can receive benefits from them. Accordingly, biodiversity is valued for what it can deliver in terms of ecological functions (with the sole exception of cultural services, where biodiversity can be relevant for its existence value). But in a scenario of locally co-evolving organisms and ecosystems, functions can be preserved only if we can preserve the possibility of organismal change. This implies to protect species and genetic diversity together with functional diversity. While the latter grants ecosystem functioning in the present, the former influences the ability of the ecosystem to continue to function under changing conditions, which can be generated in the long as well as the short term by the internal dynamics of eco-evolutionary change or by external ecological pressures, often of anthropic origin. In ecosystem services parlance, this increases the insurance value of biodiversity. Importantly, the idea of evolving species in evolving ecosystems defies static and typological thinking in ecosystem services policies as much as in traditional biodiversity conservation, thus fostering dynamic approaches and long-term planning.

Acknowledgments This work was funded by the Fundação para a Ciência e a Tecnologia through a postdoctoral grant within the R&D project Biodecon (PTDC/IVC-HFC/1817/2014).

References

Ashley, M. V., Willson, M. F., Pergams, O. R., O'Dowd, D. J., Gende, S. M., & Brown, J. S. (2003). Evolutionarily enlightened management. *Biological Conservation, 111*(2), 115–123. https://doi.org/10.1016/S0006-3207(02)00279-3.

Barker, G., & Odling-Smee, F. J. (2014). Integrating ecology and evolution: Niche construction and ecological engineering. In G. Barker, E. Desjardins, & T. Pearce (Eds.), *Entangled life. Organism and environment in the biological and social sciences* (pp. 187–211). Dordrecht: Springer. https://doi.org/10.1007/978-94-007-7067-6.

Bassar, R. D., Marshall, M. C., López-Sepulcre, A., Zandonà, E., Auer, S. K., Travis, J., Pringle, C. M., Flecker, A. S., Thomas, S. A., Fraserg, D. F., & Reznicka, D. N. (2010). Local adaptation in Trinidadian guppies alters ecosystem processes. *Proceedings of the National Academy of Sciences, 107*(8), 3616–3621. www.pnas.org/cgi/doi/10.1073/pnas.0908023107.

Boogert, N. J., Paterson, D. M., & Laland, K. N. (2006). The implications of niche construction and ecosystem engineering for conservation biology. *AIBS Bulletin, 56*(7), 570–578. https://doi.org/10.1641/0006-3568(2006)56[570:TIONCA]2.0.CO;2.

Cardinale, B. J., Duffy, J. E., Gonzalez, A., Hooper, D. U., Perrings, C., Venail, P., Narwani, A., Mace, G. M., Tilman, D., Wardle, D. A., Kinzig, A. P., Daily, G. C., Loreau, M., Grace, J. B., Larigauderie, A., Srivastava, D. S., & Naeem, S. (2012). Biodiversity loss and its impact on humanity. *Nature, 486*, 59–67. https://doi.org/10.1038/nature11148.

Costanza, R., d'Arge, R., de Groot, R., Farber, S., Grasso, M., Hannon, B., Limburg, K., Naeem, S., O'Neill, R. V., Paruelo, J., Raskin, R. G., Sutton, P., & van den Belt, M. (1997). The value of the world's ecosystem services and natural capital. *Nature, 387*(6630), 253–260.

Costanza, R., de Groot, R., Braat, L., Kubiszewski, I., Fioramonti, L., Sutton, P., Farber, S., & Grasso, M. (2017). Twenty years of ecosystem services: How far have we come and how far do we still need to go? *Ecosystem Services, 28*, 1–16. https://doi.org/10.1016/j.ecoser.2017.09.008.

Cottingham, K. L., Brown, B. L., & Lennon, J. T. (2001). Biodiversity may regulate the temporal variability of ecological systems. *Ecology Letters, 4*(1), 72–85. https://doi. org/10.1046/j.1461-0248.2001.00189.x.

Crain, C. M., & Bertness, M. D. (2006). Ecosystem engineering across environmental gradients: Implications for conservation and management. *AIBS Bulletin, 56*(3), 211–218. https://doi. org/10.1641/0006-3568(2006)056[0211:EEAEGI]2.0.CO;2.

Daily, G. C. (1997). *Nature's services: Societal dependence on natural ecosystems*. Washington, DC: Island Press.

Ellner, S. P., Geber, M. A., & Hairston, N. G. (2011). Does rapid evolution matter? Measuring the rate of contemporary evolution and its impacts on ecological dynamics. *Ecology Letters, 14*(6), 603–614. https://doi.org/10.1111/j.1461-0248.2011.01616.x.

Fussmann, G. F., Loreau, M., & Abrams, P. A. (2007). Eco- evolutionary dynamics of communities and ecosystems. *Functional Ecology, 21*, 465–477. https://doi. org/10.1111/j.1365-2435.2007.01275.x.

Ghalambor, C. K., McKay, J. K., Carroll, S. P., & Reznick, D. N. (2007). Adaptive versus non-adaptive phenotypic plasticity and the potential for contemporary adaptation in new environments. *Functional Ecology, 21*(3), 394–407. https://doi.org/10.1111/j.1365-2435.2007.01283.x.

Harrison, P. A., Berry, P. M., Simpson, G., Haslett, J. R., Blicharska, M., Bucur, M., Dunford, R., Egoh, B., Garcia-Llorente, M., Geamănă, N., & Geertsema, W. (2014). Linkages between biodiversity attributes and ecosystem services: A systematic review. *Ecosystem Services, 9*, 191–203. https://doi.org/10.1016/j.ecoser.2014.05.006.

Hooper, D. U., Chapin, F. S., Ewel, J. J., Hector, A., Inchausti, P., Lavorel, S., Lawton, J. H., Lodge, D. M., Loreau, M., Naeem, S., & SchmidT, B. (2005). Effects of biodiversity on ecosystem functioning: A consensus of current knowledge. *Ecological Monographs, 75*(1), 3–35.

Jones, C. G., Lawton, J. H., & Shachak, M. (1994). Organisms as ecosystem engineers. *Oikos, 69*, 373–386. http://www.jstor.org/stable/3545850.

Jones, C. G., Lawton, J. H., & Shachak, M. (1997). Positive and negative effects of organisms as physical ecosystem engineers. *Ecology, 78*(7), 1946–1957. https://doi. org/10.1890/0012-9658(1997)078[1946:PANEOO]2.0.CO;2.

Jones, L. E., et al. (2009). Rapid contemporary evolution and clonal food web dynamics. *Philosophical Transactions of the Royal Society B, 364*, 1579–1591. https://doi.org/10.1098/rstb.2009.0004.

Justus, J., Colyvan, M., Regan, H., & Maguire, L. (2009). Buying into conservation: Intrinsic versus instrumental value. *Trends in Ecology and Evolution, 24*(4), 187–191. https://doi. org/10.1016/j.tree.2008.11.011.

Kinnison, M. T., & Hairston, N. G. (2007). Eco-evolutionary conservation biology: Contemporary evolution and the dynamics of persistence. *Functional Ecology, 21*(3), 444–454. https://doi. org/10.1111/j.1365-2435.2007.01278.x.

Kokko, H., & Lopez-Sepulcre, A. (2007). The ecogenetic link between demography and evolution: Can we bridge the gap between theory and data? *Ecology Letters, 10*, 773–782. https://doi. org/10.1111/j.1461-0248.2007.01086.x.

Laland, K. N., Odling-Smee, F. J., & Feldman, M. W. (1999). Evolutionary consequences of niche construction and their implications for ecology. *Proceedings of the National Academy of Sciences, 96*(18), 10242–10,247. https://doi.org/10.1073/pnas.96.18.10242.

Laland, K. N., & Sterelny, K. (2006). Perspective: Seven reasons (not) to neglect niche construction. *Evolution, 60*(9), 1751–1762. https://doi.org/10.1111/j.0014-3820.2006.tb00520.x.

Lill, J. T., & Marquis, R. J. (2003). Ecosystem engineering by caterpillars increases insect herbivore diversity on white oak. *Ecology, 84*(3), 682–690. https://www.jstor.org/stable/3107862.

Loreau, M. (2010). Linking biodiversity and ecosystems: Towards a unifying ecological theory. *Philosophical Transactions of the Royal Society, B: Biological Sciences, 365*, 49–60. https://doi.org/10.1098/rstb.2009.0155.

Loreau, M. (2010a). *The challenges of biodiversity science.* Oldendorf/Luhe: International Ecology Institute.

Loreau, M. (2010b). *From populations to ecosystems: Theoretical foundations for a new ecological synthesis (MPB-46).* Princeton/Woodstock: Princeton University Press.

MA, Millennium Ecosystem Assessment. (2003). *Ecosystems and human well-being: A framework for assessment.* Washington, DC: Island Press.

MA, Millennium Ecosystem Assessment. (2005). *Ecosystems and human well-being: Synthesis.* Washington, DC: Island Press.

Maclaurin, J., & Sterelny, K. (2008). *What is biodiversity?* Chicago: University of Chicago Press.

May, R. M. (1973). *Stability and complexity in model ecosystems. Monographs in population biology.* Princeton: Princeton University Press.

MacArthur, R. H. (1955). Fluctuations of animal populations and a measure of community stability. *Ecology, 36*, 533–535. https://doi.org/10.2307/1929601.

Mace, G. M., & Purvis, A. (2008). Evolutionary biology and practical conservation: Bridging a widening gap. *Molecular Ecology, 17*, 9–19. https://doi.org/10.1111/j.1365-294X.2007.03455.x.

Mace, G. M., Norris, K., & Fitter, A. H. (2012). Biodiversity and ecosystem services: A multilayered relationship. *Trends in Ecology & Evolution, 27*(1), 19–26. https://doi.org/10.1016/j.tree.2011.08.006.

Maquire, L. A., & Justus, J. (2008). Why intrinsic value is a poor basis for conservation decisions. *BioScience, 58*, 910–911.

Menge, B. A., Berlow, E. L., Blanchette, C. A., Navarrete, S. A., & Yamada, S. B. (1994). The keystone species concept: Variation in interaction strength in a rocky intertidal habitat. *Ecology Monographs, 64*, 249–286. https://doi.org/10.2307/2937163.

McCauley, D. J. (2006). Selling out on nature. *Nature, 443*, 27–28.

Mitchell, S. D. (2002). Integrative pluralism. *Biology and Philosophy, 17*(1), 55–70.

Naeem, S. (2002). Ecosystem consequences of biodiversity loss: The evolution of a paradigm. *Ecology, 83*, 1537–1552. https://doi.org/10.1890/0012-9658(2002)083[1537:ECOBLT]2.0.CO;2.

Naess, A. (1973). The shallow and the deep, long range ecology movement. A summary. *Inquiry, 16*, 95–100.

Norton, B. G. (1986). *The preservation of species: The value of biological diversity.* Princeton: Princeton University Press.

Odling-Smee, F. J., Laland, K. N., & Feldman, M. W. (2003). Niche construction: The neglected process in evolution. In *Monographs in population biology.* Princeton: Princeton University Press.

Odling-Smee, J. F., Erwin, D. H., Palkovacs, E. P., Feldman, M. W., & Laland, K. N. (2013). Niche construction theory: A practical guide for ecologists. *The Quarterly Review of Biology, 88*(1), 3–28.

Paine, R. T. (1966). Food web complexity and species diversity. *Am. Nat. 100*, 65–75. https://doi.org/10.1086/282400.

Palkovacs, E. P., Kinnison, M. T., Correa, C., Dalton, C. M., & Hendry, A. P. (2012). Fates beyond traits: Ecological consequences of human-induced trait change. *Evolutionary Applications, 5*(2), 183–191. https://doi.org/10.1111/j.1752-4571.2011.00212.x.

Pimm, S. L. (1984). The complexity and stability of ecosystems. *Nature, 307*, 321–326. https://doi.org/10.1038/307321a0.

Post, D. M., & Palkovacs, E. P. (2009). Eco-evolutionary feedbacks in community and ecosystem ecology: Interactions between the ecological theatre and the evolutionary play. *Philosophical Transactions of the Royal Society B, 364*, 1629–1640. https://doi.org/10.1098/rstb.2009.0012.

Redford, K. H., & Adams, W. M. (2009). Payment for ecosystem services and the challenge of saving nature. *Conservation Biology, 23*, 785–787. https://doi.org/10.1111/j.1523-1739.2009.01271.x.

Reyers, B., Polasky, S., Tallis, H., Mooney, H. A., & Larigauderie, A. (2012). Finding common ground for biodiversity and ecosystem services. *BioScience, 62*(5), 503–507. https://doi.org/10.1525/bio.2012.62.5.12.

Santamaría, L., & Mendez, P. F. (2012). Evolution in biodiversity policy–current gaps and future needs. *Evolutionary Applications, 5*(2), 202–218. https://doi.org/10.1111/j.1752-4571.2011.00229.x.

Sarkar, S. (2005). *Biodiversity and environmental philosophy: An introduction.* New York: Cambridge University Press.

Singer, P. (1975). *Animal liberation: A new ethics for our treatment of animals.* New York: Random House.

Srivastava, D. S., & Vellend, M. (2005). Biodiversity-ecosystem function research: Is it relevant to conservation? *Annual Review of Ecology, Evolution, and Systematics, 36*, 267–294. https://doi.org/10.1146/annurev.ecolsys.36.102003.152636.

Stockwell, C. A., Hendry, A. P., & Kinnison, M. T. (2003). Contemporary evolution meets conservation biology. *Trends in Ecology & Evolution, 18*, 94–101. https://doi.org/10.1016/S0169-5347(02)00044-7.

TEEB. (2010). *The economics of ecosystems and biodiversity: Ecological and economic foundations.* Ed. P. Kumar. London/Washington, DC: Earthscan.

Thompson, S. C. G., & Barton, M. A. (1994). Ecocentric and anthropocentric attitudes toward the environment. *Journal of Environmental Psychology, 14*(2), 149–157. https://doi.org/10.1016/S0272-4944(05)80168-9.

Thompson, J. N. (1998). Rapid evolution as an ecological process. *Trends in Ecology & Evolution, 13*(8), 329–332. https://doi.org/10.1016/S0169-5347(98)01378-0.

Westman, W. E. (1977). How much are nature's services worth? *Science, 197*(4307), 960–964.

Whitham, T. G., Bailey, J. K., Schweitzer, J. A., Shuster, S. M., Bangert, R. K., LeRoy, C. J., Lonsdorf, E. V., Allan, G. J., DiFazio, S. P., Potts, B. M., & Fischer, D. G. (2006). A framework for community and ecosystem genetics: From genes to ecosystems. *Nature Reviews Genetics, 7*, 510–523. https://doi.org/10.1038/nrg1877.

Williams, H. T. P., & Lenton, T. M. (2007). Artificial selection of simulated microbial ecosystems. *Proceedings of the National Academy of Sciences, 104*, 8918–8923. https://doi.org/10.1073/pnas.0610038104.

Wright, J. P., & Jones, C. J. (2006). The concept of organisms as ecosystem engineers ten years on: Progress, limitations, and challenges. *BioScience, 56*(3), 203–209. https://doi.org/10.1641/0006-3568(2006)056[0203:TCOOAE]2.0.CO;2.

Part III
Conserving Biodiversity: From Science to Policies

Chapter 16
On the Impossibility and Dispensability of Defining "Biodiversity"

Georg Toepfer

The complexity of the biodiversity concept does not only mirror the natural world it supposedly represents; it is that plus the complexity of human interactions with the natural world, the inextricable skein of our values and its value, of our inability to separate our concept of a thing from the thing itself. Don't know what biodiversity is? You can't.

David Takacs (1996, p. 341)

Abstract The impressive success of the concept 'biodiversity' in the last decades, in particular in the arena of politics, is in a large part due to its power to amalgamate facts and values: the fact that living beings show variety on every level of their existence, and the assumed values that are associated with this variety. These values are far from obvious or objective, they are rather normatively prescribed. They are already at work in the process of selecting the level of analysis, e.g. the level of genes, species, or ecosystems. The concept thus ties together many different discourses from the fields of biology and bioethics, aesthetics and economy, law and global justice. One important consequence of the concept's integrative power is the impossibility of its general definition. Just as 'life', 'time' or 'world' the word is an "absolute metaphor" or "non-concept" in the sense of Hans Blumenberg: it cannot have a fixed meaning just because it mediates between various contexts and disciplines. Any attempt to define 'biodiversity' in general terms is thus futile and does not capture the role it fulfills in contemporary discourse. Rather than trying to define the concept, reconstructing the interaction of its various contexts is a more promising approach. These include, besides the obvious reference to biology and nature conservation, ancient narratives about divine creation, paradise and Noah's ark as

G. Toepfer (✉)
Leibniz-Zentrum für Literatur- und Kulturforschung, Berlin, Germany
e-mail: toepfer@zfl-berlin.org

© The Author(s) 2019
E. Casetta et al. (eds.), *From Assessing to Conserving Biodiversity*,
History, Philosophy and Theory of the Life Sciences 24,
https://doi.org/10.1007/978-3-030-10991-2_16

well as political ideas of pluralism, egalitarianism and non-hierarchical representation of individuals, or the values of market economy. In order to understand the current success and discursive role of the concept, I will analyze some of the underlying ideas, especially with respect to the representation of biodiversity in images.

16.1   The Integrative Power of 'Biodiversity'

In order to understand why 'biodiversity' has become such a successful term in the last decades it is necessary to look beyond biology to the broader socio-cultural and political contexts in which diversity became an important issue. From a merely biological point of view or from the standpoint of conservation biology it is not obvious that diversity should be more important than other abstract properties of biological systems such as stability, resilience or wilderness; or, to be more practical, the protection of one particularly endangered species or of an ecosystem. For the purpose of understanding why we now live in the "decade of biodiversity", as declared by the United Nations in 2010, it is not enough to point at increased biological knowledge in the modern age or at the crisis of mass extinction. It is necessary to focus on cultural and political values present in scientific issues—or at least in the public understanding of scientific issues.

The parallels between cultural and biological diversity are underlined by two conventions of the United Nations (Heyd 2010): The Rio "Convention on Biological Diversity" of 1992 explicitly calls for the protection and use of "biological resources in accordance with traditional cultural practices" (UNEP 1992). And the 2001 UNESCO "Declaration on Cultural Diversity" claims that "[a]s a source of exchange, innovation and creativity, cultural diversity is as necessary for humankind as biodiversity is for nature" (UNESCO 2001).

At least on a rhetorical level, 'diversity' functions as a versatile concept bringing together diverse fields. It can be linked to *economic* considerations: to plants and animals as entities providing "ecosystem services" regarding the supply of food, fibres or pharmaceuticals, or for the regulation of climate, water balance, etc. (Millennium Ecosystem Assessment 2005). At the same time, the term maintains its strong non-instrumental *ethical* dimension: it expresses a non-anthropocentric value of plants and animals. With this in mind the Holy Father, Pope Francis, in his 2015 encyclical *Laudato si'*, referred to biodiversity and ascribed intrinsic value to non-human creatures (Pope Francis 2015, no. 118; 140). Finally, 'biodiversity' has the dignity of a scientific term, as it seems to refer to something objective and measurable.

'Biodiversity' obviously forms an efficient basis for the integration of heterogeneous discourses and their public communication. Yet, by putting many things together—ethics, religion, science, business and politics—the term has an undifferentiating effect. On a political level this effect has also been welcomed because to be politically effective a term should not only describe a natural state of affairs but declare it as an important and good thing.

The main strength of the term seems to be that it does not take sides in fundamental ideological dichotomies such as scientific/emotional, profane/sacred or utilitarian/intrinsic value. It remains neutral and thus can be used in either position. And, surprisingly, considering its integrative power and ideological neutrality, 'biodiversity' seems not to be abstract: It refers primarily to concrete individuals and species – well-liked species for the most part, the so-called "charismatic megafauna" such as polar bears, lions and elephants. These tangible references make 'biodiversity' a much more attractive concept, than, for example, 'stability', 'ecosystem services' or 'sustainability'. This suggestive concreteness is, of course, a huge advantage in the communication with the general public; it is a potent instrument for connecting nature and people (Díaz et al. 2015).

In addition to its integrative function and its concreteness, 'biodiversity' fits very well into our pluralistic present because the concept renounces an overarching, universally valid (world) order and expresses a de-hierarchization and pluralization of perspectives. It refers to the heterogeneous interests and intrinsic worth of every single individual. With respect to human and non-human living beings the concept of diversity is successful, because it conveys respect and responsibility, tolerance and pleasure of heterogeneity. Since the 1980s, 'diversity' has become a central concept in social emancipation movements. It emphasizes cultural difference and includes a critical reflection of one's own cultural-relative standpoint. But, again, 'diversity' functions by integration because it refers not only to current concerns but has also a deeply rooted historical dimension leading far back to the very first written texts of mankind (see Sect. 16.3). The story of biodiversity has been so successful because it is related to some deeply rooted ideals about the world, not least the idea of paradise: for one essential characteristic of the Garden of Eden is that it is full of plants and animals of different species coexisting in a joyful and peaceful manner.

Its ongoing scientific usage and at the same time latent connection to ancient cultural-religious ideas (such as paradise) makes 'biodiversity' a powerful concept for mediating between science and the broader public. It is a paradigmatic example of what has been called "post-normal science", where "facts are uncertain, values in dispute, stakes high and decisions urgent" (Funtowicz and Ravetz 1992, p. 138). This characterization applies particularly well to the status of 'biodiversity': First, the investigation of biodiversity has to cope with uncertainties on the factual as well as the axiological or ethical level. We simply do not know enough about the amount and function of biological diversity; we do not know, for example, whether there are currently three or 100 million species of animals on earth, and we do not know how they contribute to the stability of our ecosystems. Second, we do not know how we should evaluate biodiversity: instrumentally or intrinsically. Third, stakes are high because we are currently facing a loss of biological species probably on the level of one of the five mass extinction events in earth history. Finally, decisions are urgent because this is an irreversible loss and we do not know whether there will be a tipping point when things get worse at an increased speed and scale.

Furthermore, biodiversity studies are paradigmatic for post-normal science because they examine a field where laypersons are becoming experts. Big data provided by millions of people taking part in observation surveys particularly for birds and insects is an important basis for decision-making in conservation biology. These extended peer communities with their socially distributed expertise are especially important for the knowledge of local conditions.

These factors have transformed biodiversity studies from the old paradigm of scientific discovery ('Mode 1'), characterized by the hegemony of theoretical and experimental science, to a new paradigm of knowledge production (the 'Mode 2'), in which knowledge is "socially distributed, application-oriented, trans-disciplinary, and subject to multiple accountabilities" (Nowotny et al. 2003, p. 179). The investigation of biodiversity is a post-normal Mode 2-science because it is "issue-driven" and "mission-oriented" rather than theoretical and driven by curiosity. In situations of Mode 2-science experts are incapable of providing conclusive answers to the associated problems. They can provide their views but decisions have to be made in public forums such as the Intergovernmental Science-Policy Platform on Biodiversity and Ecosystem Services (IPBES).

Being immersed in issues of philosophy, cultural history and economy, biodiversity is not being a merely scientific problem anymore, but rather one where science and politics meet. As an object of public attention and a focal point of conflicting interests, 'biodiversity' concerns the management of an issue rather than the solution to a problem.

16.2　On Defining 'Biodiversity'

As the concept of 'biodiversity' is vague or versatile, it still has to be defined to be of any value in public or scientific argumentation. In the mid-1980s 'biodiversity' was explicitly introduced with a non-scientific, but political intent. The term was coined in preparation of the *National Forum on BioDiversity,* which took place in September 1986 in Washington, D.C. The American botanist Walter G. Rosen who was involved in preparing the conference explained how he came up with the term in a later interview: Creating the term, he said, was "easy to do: all you do is to take the 'logical' out of 'biological'" (Rosen 1992, in Takacs 1996, p. 37). Linguistically speaking it was an easy task: Rosen simply used 'biodiversity' instead of 'biological diversity' which was already a well-established technical term in biology. His aim was, as he said, to create room for "emotion" and "spirit".

And this is the situation we are in now: 'Biodiversity' is a term full of emotion and spirit, expressing an ethical concern following the mass-extinction of species due to human actions. Compared to this strong ethical pulse, the explicit definitions of the term that have been given after the conference in Washington are rather weak. Most of them given by biologists, for example, are very broad. In one of the first explicit definitions of 'biodiversity' formulated by the US Office of Technology Assessment (OTA) in 1987, it is defined as "the variety and variability among living organisms and the ecological complexes in which they occur" (OTA 1987, p. 3). A few years later, Solbrig's much cited definition explains the term as "the property of living systems of being distinct, that is different, unlike" (Solbrig 1991, p. 9). The

Convention on Biological Diversity from 1992 sees it as "the variability among living organisms from all sources" (UNEP 1992, art. 2). In one of the implicit and open definitions Sarkar explains biodiversity as "what is being conserved by the practice of conservation biology" (Sarkar 2002, p. 132), and a few years later, Norton demands that any definition must be rich enough to capture "all that we mean by, and value in, nature" (Norton 2006, p. 57).

From these definitions it is obvious that the term was designed to be open, versatile, polyvalent and adaptable to changing situations (Casetta and Delord 2014, p. 251). It has been characterized as an "umbrella concept" encompassing the entire field that has formerly been called "nature protection" or "nature conservation" (Potthast 2014, p. 132). Because of its oscillation between scientific and non-scientific contexts Gutmann sees it as a "pragmatic concept" or "metaphor" (2014, p. 66). Regular movements across discoursive borders allow the concept to touch upon and somehow integrate many diverse aspects of nature and its use by humans. This mediating quality depends on the term's "performativity" by mobilizing attitudes and reactions in diverse contexts (Casetta and Delord 2014, p 251). For this performativity to be effective, the multifarious character of 'biodiversity' is essential; it allows for the fact that "[i]n biodiversity each of us finds a mirror for our most treasured natural images, our most fervent environmental concerns." (Takacs 1996, 81).

Thus, 'biodiversity' is exactly what a politically successful concept ought to be: sufficiently open in order to be meaningful to many people and powerfully employed in political processes. It amalgamates scientific and political developments, public concern and cultural changes in society.

16.3   Representing Biodiversity

Apparently, the success of 'biodiversity' as a concept in the public discourse results, at least in part, from its reference to political and social concerns about diversity in non-biological contexts. Another reason for its power in social discourse might be that it brings into play ancient formats of representing the multitude of things. Representing biodiversity—understood as species' inventory—starts with the beginning of writing, roughly 5000 years ago. In the world's earliest examples, which are lists from Mesopotamian cultures, a huge variety of trees, domestic animals, fish and birds appear alongside lists of goods which have been traded, as well as other inventory-like lists of things in the world: metals, vessels, official functions, and geographic places (Veldhuis 2014). The early list of bird species in proto cuneiform from around 3000 BC featured more than 100 entries, including ravens, which presumably were of no direct use or benefit to humans. The lists seem to be inventories of species or of kinds of things in the world regardless of their utilitarian value.

Since Mesopotamian times, such lists have been used by natural historians to log, check and order species of living beings. Linnaeus' *Systema naturae* is basically still a list; in its tenth edition of 1758 it contains about 4200 species on 800 pages. Today's records of diversity are also organized as lists, for example the *Encyclopedia of Life,* the largest online database of systematic biology.

The discovery of biodiversity could be seen as a science of lists—*Listenwissenschaft* as Assyriologists have called the presentation of knowledge in this form in Mesopotomian cultures (Schneider 1907, p. 368). The list's essential logic is paratactical: lists do not primarily *explain*, as hypotactical theory-centered science does, but they first of all describe and arrange things on the same level. Lists are apt devices for the egalitarian, non-hierarchical presentation of things.

Biodiversity images put this logic into the visual sphere by showing an egalitarian representation of diverse living beings. Paradigmatic images of this type appear in the Flemish still life of the late sixteenth and seventeenth century, for example in the work of Joris Hoefnagel or Jan van Kessel the Elder. The scattered arrangement of decontextualized animal bodies, a "strewn pattern image" (Schütz 2002, p. 66), shows what biodiversity essentially is: a state of difference. Biodiversity is the sum of different individuals with different lifeforms and lifestyles. They are presented according to the principle of addition, a colorful diversity that does not manifest a closed totality or a system of interaction. Central to this depiction is that there is no top or bottom, no hierarchy and no evaluation within one set. The principle of representation is a paratactic egalitarianism, the line-up of individuals with an equal standing, a juxtaposition of forms.

In the popular visual culture of our days you can find many examples of images that follow this paratactic, egalitarian logic. One example is a photo project by nature photographer David Liittschwager: In *A World in One Cubic Foot* he took a bright green metal cube—measuring precisely one cubic foot—and set it in various ecosystems around the world, from Costa Rica to New York Central Park (Liittschwager 2012). He documented what moved through that small space in a period of 24 h and photographed the plant and animal life he encountered in that period of time. An image of local biodiversity was then created by compiling all these photographs according to paratactic logic. In another example, Christopher Marley arranged his photographs of beetles in a kind of biodiversity mosaics (Marley 2008). This resulted in holistic figures such as squares or circles. No mosaic stone here resembles the other, and no element may be missing for the whole to be complete.

There is a striking parallel between the iconic logic of contemporary biodiversity pictures and some seventeenth-century Flemish still lives. The still lives, however, do not visualize ethical concerns about extinction of species, but refer to their creation. Their reference point is natural theology: the animals are considered immediate manifestations of God's will and thus provide access to God's plan equal to the Holy Scriptures. Although the modern concern with extinction and the late seventeenth-century focus on creation are rather distinct, they have one essential thing in common: the emphasis on and evaluation of individuals and species. In biodiversity images it is individuals and species that are appreciated in the first place, whereas their interactions with their environment and each other are pushed into the background. This decontextualizing, egalitarian logic, "specimen logic" as Jenice Neri has called it (Neri 2011, p. xiii), is essential for understanding our conceptualization and appreciation of biodiversity.

In the last decades, this specimen logic, the paratactic order of individuals, has become a dominant mode of presenting animals. It is manifest in installations at natural history museums such as the *Hall of Biodiversity* in the American Museum of Natural History in New York or the *Wall of Biodiversity* in the Museum für Naturkunde in Berlin. Without further explanation, these installations group together a great variety of stuffed animals that do not naturally occur next to each other in the same location. The installations are not about explaining and understanding, but about creating astonishment and amazement by this variety. The focus is on the aesthetic quality of the individual objects and on the feeling that each species is threatened by humans. Thus, questions of nature conservation and ethics are addressed. In parallel with older biodiversity images, the museum installations are characterised by the fact that (1) they stress the individual character of different species by presenting them in an extremely naturalistic way, (2) they decontextualize each species by displaying only one individual devoid of its ecological setting, and (3) they arrange the specimens in a non-hierarchical, egalitarian order. In contrast to earlier forms of presentation in natural history museums, biodiversity installations abandon showing causal, functional or systemic relations. The rationale of this form of presentation is that the mere sequence of different animal bodies is intended to be free of ideological or cultural preconceptions.

In these museum installations, the individuals play the central role: Each specimen not only represents the living organism its parts once belonged to, but, as there is only one specimen for each species, each specimen also represents its entire species. In these representations biodiversity is an ethic not for individuals in the first place but for species; it is about the moral dignity of species. The installation exemplifies yet another form of representation, a political representation: If the showcase of the installations is seen as a kind of parliament then each species has one vote in it; there is an equal representation for each species.

Interpreted in this way, the representations of biodiversity in natural history museums correspond, of course, to the normative discourse of egalitarian pluralistic democracies. Accordingly, it may be seen in terms of political iconography: as an expression of pluralistic social and political ideas. In short, it displays political ideals in the guise of nature.

However, the non-hierarchical, paratactic logic of presenting animals not only corresponds to liberal ideas of an egalitarian society, it also corresponds to the store-aesthetics of the market economy where the consumer can choose among the many products of equal value offered to him as being different. The increased attention to diversity can thus also be interpreted as having been influenced or enforced by capitalist economy. Moreover, it has also been argued that the origin of the dominating "taste for colorful diversity" lies in "the market": "It is the taste formed by the contemporary market, and it is the taste for the market" (Groys 2008, p. 151).

The aesthetic of diversity thus has many different roots. It can be found, amongst others, in the very old history going back to the Mesopotamian *Listenwissenschaft*, the general pleasure in the manifold of Western culture (*poikilia* in Greek, *varietas* in Latin; Grand-Clément 2015; Fitzgerald 2016), the social emancipatory movements of the second half of the twentieth century, capitalist market economy, or just postmodern taste. Just as 'biodiversity' refers to a multitude of perspectives that cannot be reduced to one coherent system there is no master-narrative for the explanation of its current success.

16.4   The Hybridization of Facts and Values in 'Biodiversity'

The various new and old traditions of diversity have in common that they are not merely descriptive but place value on variety and variability. The intersection of fact and value in the representation of diversity is particularly evident with respect to biodiversity. In the Judeo-Christian context four fundamental scenes are connected to biodiversity: scenes of Creation, of Paradise, the naming of animals by Adam and the animals boarding Noah's ark. As these are well-known religious scenes in a particular ethical context, their effect is placing values on diversity, charging it positively. This evaluative stance is also evident in Christopher Columbus's first letter from the New World (addressed to the finance minister at the Spanish court, Luis de Santángel, who supported Columbus) in which reference to the biodiversity he encountered—trees and birds of "a thousand different kinds" (*de mil maneras*)—forms an essential element of his praise of the promising land he discovered.

A more explicit appreciation of diversity can be found in Christian authors such as Augustine of Hippo who used reference to the huge diversity of animal species ("tantas diversitates animalium") for the praise of God: "how great all these things are, how magnificent, how beautiful, how amazing! And he who made them all is your God" (*Enarrationes in Psalmos*, 145, 12; transl. by Boulding 2004, p. 411). In a similar vein, Thomas Aquinas wrote: "Although an angel, considered absolutely, is better than a stone, nevertheless two natures are better than one only; and therefore a universe containing angels and other things is better than one containing angels only; since the perfection of the universe is attained essentially in proportion to the diversity of natures in it, whereby the diverse grades of goodness are filled, and not in proportion to the multiplication of individuals of a single nature." (*Scriptum super Sententiis*, lib. I, dist. 44, quaest.1, art 2; transl. by Lovejoy 1936, p. 77).

Similar views are expressed in the writings of Leibniz, who weighs one man against the whole species of lions, and writes, by the way, that he is not sure whether God would actually prefer the individual human (*Essais de théodicée*, 1710; Lovejoy 1936, p. 225). In this understanding of diversity, human beings are co-ordinated with the other species of living beings; they do not inhabit an absolute and excellent position, but compete eye-to-eye with other species. Value is placed not (only) on the intrinsic qualities of each single being but on the state of being different from others. This evaluative stance towards diversity as such can be seen as a prefiguration of the modern concept of biodiversity as an "epistemic-moral hybrid" (Potthast 2014, p. 138); it is a kind of biodiversity *avant la lettre*.

On the basis of these prefigurations it was an easy task for concerned biologists in the late twentieth century to propagate 'biodiversity' as an important issue by taking out the 'logical' from 'biological diversity'—which was taken to be a biological term—, and putting in "emotion" and "spirit" instead. This strategic reenchantment of a (supposedly) biological concept made the term very useful for the political sphere. Being full of concern and sufficiently vague, open to many ideas, even contradictory ones, the term became an effective instrument for politics.

However, in recent years the intrinsic value and hybrid character of 'biodiversity' came under attack. It has been doubted that biodiversity is a useful basis for decisions in nature conservation (Morar et al. 2015; Santana in this book). For in many cases we are not interested in the diversity of things as such. Nature conservation is often concerned not with protecting as many species as possible, but only very specific, typical or rare ones. In some cases, we are trying to limit genetic diversity, for example in cases where it leads to functional disorders, or we are trying to eradicate pathogens. Diversity in itself is not always a good thing, but only the right measure of it, so it has been argued.

Accordingly, one problem of the concept is the unconditionally positive evaluation of diversity. Another problem is that the evaluative charge of the term 'biodiversity' makes it impossible to distinguish between scientific knowledge as such and the process of evaluating this knowledge. Morar et al. (2015) argue that we should decide in an open democratic discourse which diversity we want where. The amalgamation of the two steps of gaining and evaluating knowledge into one, as the hybrid concept 'biodiversity' does, obscures the need for separating scientific facts and public review of its results: "the role of [political] judgments is obscured when decisions are presented as following automatically from empirical evidence" (Morar et al. 2015, p. 25).

This criticism problematizes exactly that aspect of the concept, which was responsible for its success: the hybridization of descriptive and normative dimensions. Obviously, the comprehensive success of the rhetoric of biodiversity was bought at no small price. Its power to hybridize makes biodiversity a useful political concept but it also stands in the way of any precise argument. Good intentions and positive effects connected to the concept cannot replace differentiated ethical reasoning. The important integrative function of the concept needs to be complemented with arguments in which the hybridization of facts and norms, of science and values, of knowledge and wonder is carefully separated again. Not scientists and their concepts but the democratic society as a whole has to decide which diversity it desires, one that includes *genetic disorders*, the *polio virus* or the *Anopheles mosquito* – or not.

16.5 Conclusion: Biodiversity as an Absolute Metaphor

Because of its involvement in various ancient traditions, and correspondingly hybrid and multifarious character, 'biodiversity' can be understood as an "absolute metaphor" or "non-concept" (*Unbegriff*) in the sense of Hans Blumenberg. Just as 'life', 'time' or 'world' the word cannot have a fixed meaning because it mediates between various contexts and disciplines. Aspects of the term can be defined within each separate context. Biology, for example, has provided many mathematically precise definitions of biological diversity as an index for measuring species richness and evenness in their distribution (but, already at this level diversity has been called a "non-concept" because it can be measured in very different ways; see Hurlbert 1971).

However, this technical understanding of biological diversity is distinct from 'biodiversity' as it functions in public debates. In these debates the concept functions as an absolute metaphor the meaning of which cannot be exhausted in any context of its use and proves "resistant to terminological claims and cannot be dissolved into conceptuality" (Blumenberg 2010, p. 5). It therefore seems that any attempt to define 'biodiversity' in precise terms is futile and does not capture the role that the word fulfills in contemporary discourses. Its interdiscursive function depends on not having a clear-cut definition but being an open concept. 'Biodiversity' not only refers to the variety and variability of the natural world but also to our conceptualization and valuation of it. This complexity is the reason for the vagueness and at the same time the power of the concept: "le plus vague est le plus puissant" (Bachelard 1947, p. 184).

References

Augustine of Hippo. (2004). *Enarrationes in Psalmos* (Vol. 6, Trans. and notes by M. Boulding). Hyde Park: New City Press.

Bachelard, G. (1947). *La formation de l'esprit scientifique*. Paris: Vrin.

Blumenberg, H. (2010). *Paradigms for a metaphorology (1960)*, (R. Savage, Trans.). Ithaca: Cornell University Press.

Casetta, E., & Delord, J. (2014). Versatile biodiversité. In E. Casetta & J. Delord (Eds.), *La biodiversité en question. Enjeux philosophiques et scientifiques* (pp. 247–253). Paris: Éditions Matériologiques.

Díaz, S., et al. (2015). The IPBES conceptual framework – Connecting nature and people. *Current Opinions in Evironmental Sustainability, 14*, 1–16.

Fitzgerald, W. (2016). *Variety. The life of a Roman concept*. Chicago: University of Chicago Press.

Funtowicz, S. O., & Ravetz, J. R. (1992). A new scientific methodology for global environmental issues. In R. Costanza (Ed.), *Ecological economics. The science and Management of Sustainability* (pp. 137–152). New York: Columbia University Press.

Grand-Clément, A. (2015). Poikilia. In P. Destrée & P. Murray (Eds.), *A companion to ancient aesthetics* (pp. 406–422). Chichester: Wiley Blackwell.

Groys, B. (2008). *Art power*. Cambridge, MA: MIT Press.

Gutmann, M. (2014). Biodiversity: A methodological reconstruction of some fundamental misperception. In D. Lanzerath & M. Friele (Eds.), *Concepts and values in biodiversity* (pp. 55–72). London: Routledge.

Heyd, D. (2010). Cultural diversity and biodiversity: A tempting analogy. *Critical Review of International Social and Political Philosophy, 13*, 159–179.

Hurlbert, S. H. (1971). The nonconcept of species diversity: A critique and alternative parameters. *Ecology, 52*, 577–586.

Liittschwager, D. (2012). *A world in one cubic foot. Portraits of biodiversity*. Chicago: University of Chicago Press.

Lovejoy, A. O. (1936). *The great chain of being*. Cambridge, MA: Harvard University Press.

Marley, C. (2008). *Pheromone*. Portland: Pomegranate Communications.

Millennium Ecosystem Assessment. (2005). *Ecosystems and human well-being: Synthesis*. Washington, DC: Island Press.

Morar, N., Toadvine, T., & Bohannan, B. J. M. (2015). Biodiversity at twenty-five years: Revolution or red herring? *Ethics, Policy & Environment, 18*, 16–29.

Neri, J. (2011). *The insect and the image. Visualizing nature in early modern Europe, 1500–1700*. Minneapolis: University of Minnesota Press.

Norton, B. (2006). Toward a policy-relevant definition of biodiversity. In J. M. Scott, D. D. Goble, & F. W. Davis (Eds.), *The endangered species act at thirty* (Vol. 2, pp. 49–58). Washington, DC: Island Press.

Nowotny, H., et al. (2003). Mode 2 revisited: The new production of knowledge. *Minerva, 41*, 179–194.

Office of Technology Assessment (OTA). (1987). *Technologies to maintain biological diversity*. Congress of the United States, OTA-F-330.

Pope Francis. (2015). *Laudato si'. Encyclical letter of the holy father Francis on care for our common home*. http://w2.vatican.va/content/francesco/en/encyclicals/documents/papa-francesco_20150524_enciclica-laudato-si.html

Potthast, T. (2014). The values of biodiversity. In D. Lanzerath & M. Friele (Eds.), *Concepts and values in biodiversity* (pp. 132–146). London: Routledge.

Sarkar, S. (2002). Defining "biodiversity", assessing biodiversity. *The Monist, 85*, 131–155.

Schneider, H. (1907). *Kultur und Denken der alten Ägypter* (Vol. 1). Leipzig: Voigtländer.

Schütz, K. (2002). Naturstudien und Kunstkammerstücke. In C. Nitze-Ertz (Ed.), *Sinn und Sinnlichkeit. Das flämische Stillleben (1550–1680). Eine Ausstellung der Kulturstiftung Ruhr Essen und des Kunsthistorischen Museums Wien* (pp. 60–109). Lingen: Luca-Verlag.

Solbrig, O. T. (1991). *Biodiversity. Scientific issues and collaborative research proposals*. Paris: UNESCO.

Takacs, D. (1996). *The idea of biodiversity. Philosophies of paradise*. Baltimore: Johns Hopkins University Press.

UNEP. (1992). *Convention on biological diversity*. https://www.cbd.int/convention/

UNESCO. (2001). *Declaration on cultural diversity*. http://portal.unesco.org/en/ev.php-URL_ID=13179&URL_DO=DO_TOPIC&URL_SECTION=201.html

Veldhuis, N. (2014). *History of the cuneiform lexical tradition*. Münster: Ugarit.

Chapter 17
The Vagueness of "Biodiversity" and Its Implications in Conservation Practice

Yves Meinard, Sylvain Coq, and Bernhard Schmid

Abstract The vagueness of the notion of biodiversity is discussed in the philosophical literature but most ecologists admit that it is unproblematic in practice. We analyze a series of case studies to argue that this denial of the importance of clarifying the definition of biodiversity has worrying implications in practice, at three levels: it can impair the coordination of conservation actions, hide the need to improve management knowledge and cover up incompatibilities between disciplinary assumptions. This is because the formal agreement on the term "biodiversity" can hide profound disagreements on the nature of conservation issues. We then explore avenues to unlock this situation, using the literature in decision analysis. Decision analysts claim that decision-makers requesting decision-support often do not precisely know for what problem they request support. Clarifying a better formulation, eliminating vagueness, is therefore a critical step for decision analysis. We explain how this logic can be implemented in our case studies and similar situations, where various interacting actors face complex, multifaceted problems that they have to solve collectively. To sum up, although "biodiversity" has long been considered a flagship to galvanize conservation action, the vagueness of the term actually complicates this perennial task of conservation practitioners. As conservation scientists, we have a duty to stop promoting a term whose vagueness impairs conservation practice. This approach allows introducing a dynamic definition of "biodiversity practices", designed to play the integrating role that the term "biodiversity" cannot achieve, due to the ambiguity of its general definition.

Y. Meinard (✉)
Université Paris-Dauphine, PSL Research University, CNRS, UMR [7243], LAMSADE,
Paris, France
e-mail: yves.meinard@lamsade.dauphine.fr

S. Coq
Centre d'Écologie Fonctionnelle et Évolutive (CEFE), CNRS, Montpellier, France
e-mail: Sylvain.COQ@cefe.cnrs.fr

B. Schmid
Institute of Evolutionary Biology and Environmental Studies, University of Zurich,
Zurich, Switzerland
e-mail: bernhard.schmid@ieu.uzh.ch

© The Author(s) 2019

353

E. Casetta et al. (eds.), *From Assessing to Conserving Biodiversity*,
History, Philosophy and Theory of the Life Sciences 24,
https://doi.org/10.1007/978-3-030-10991-2_17

Keywords Biodiversity · Definition · Conservation practice · Problem solving ·
Decision analysis

17.1   Introduction

The Convention on Biological Diversity defines biological diversity or biodiversity
as "the variability among living organisms from all sources including, inter alia,
terrestrial, marine and other aquatic ecosystems and the ecological complexes of
which they are part: this includes diversity within species, between species and of
ecosystems (United Nations 2013)." This now classical definition is largely discussed in the philosophical literature for being exceedingly vague and in need of
clarification (Sarkar 2005; MacLaurin and Sterelny 2008; Meinard et al. 2014;
Santana 2014). By contrast, most ecologists consider that this vagueness is unproblematic in practice (Mace et al. 2012).

In this chapter, we argue that this vagueness does have worrying implications in
practice, at three levels: it can impair the coordination of conservation actions, hide
the need to improve management knowledge and cover up incompatibilities between
disciplinary assumptions. Our purview in this chapter is accordingly mainly practical: we aim to address ecologists, conservation biologists and practitioners, with the
objective of convincing them that debates on the definition of biodiversity may have
concrete implications. The problems that we thereby highlight all stem from the
lack of a clear and shared definition of biodiversity. Biodiversity is certainly not the
only concept that suffers from being vaguely defined, and in many cases this vagueness does not create much problems. Accordingly, our aim here is not to claim that
vagueness is a problem in itself. Our more modest aim is to argue that, in the very
specific case of biodiversity, it does have worrying consequences.

Indeed, in this specific case, formal agreements among various actors on the term
"biodiversity" can hide profound disagreements on the nature of conservation and
ecological issues. This is reminiscent of a classical problem in decision modelling
(Bouyssou et al. 2000), for which the proven solution is for interacting actors to
articulate a commonly accepted formulation of the key questions structuring their
interaction. In line with this view, we propose that, although the notion of biodiversity does not unify biodiversity sciences in a transparent, rigorous way, such a unification may be achieved by clarifying a concept of *biodiversity practices*, understood
as coherent collaborative interdisciplinary efforts to tackle commonly identified
environmental and conservation problems. We take advantage of insights from the
philosophical literature to champion this approach and to argue that, although a
definitive definition of these biodiversity practices might be unreachable, the task to
constantly improve definitions, taking seriously conservation biologists' and conservation practitioners' value-laden stances, is crucial to the enrichment and
improvement of conservation theories and practices. If we may paraphrase Burch-
Brown and Archer (2017), although we emphasize that one cannot hope to reach a
definitive answer to the question "what is biodiversity?", our approach hence proposes a "defense of biodiversity" that consists in championing a collective effort to

constantly improve our understanding of the value-laden practices gathered under banner of "conserving biodiversity".

The remainder of this chapter is divided into three parts. In Sect. 17.2, we first show that, despite its being seemingly simple and unequivocal, the definition of "biodiversity" is exceedingly vague. Vagueness in itself is not necessarily a problem. But Sect. 17.3 uses cases studies to show that, in the case of "biodiversity", this vagueness creates problems in practice. In Sect. 17.4 we then explain our proposed solution. Section 17.5 briefly concludes.

17.2 The False Transparency of the Definition of Biodiversity

The vagueness of definitions of "biodiversity" has been extensively studied in the philosophical literature (Sarkar 2005; MacLaurin and Sterelny 2008; Meinard et al. 2014), but for lack of a concrete understanding of its implications for conservation science and practices, this debate has been largely confined to philosophical discussions without affecting real-life conservation and ecological practices (for a noticeable exception, see Delong 1996). Let us first explain why we claim that definitions of "biodiversity" are vague.

17.2.1 Diverging Definitions of "Biodiversity" Coexist

A first example will illustrate how deceptive is the idea that the definition of "biodiversity" is clear and unequivocal. Let us look at two prominent approaches to biodiversity, articulated by a leading author in conservation biology and a leading author in ecosystem ecology: Sarkar (2005) and Loreau (2010).

Loreau (2010) does not delve into definitional debates. He uses a definition very similar to the one of the CBD, stating that "biodiversity [...] includes all aspects of the diversity of life—including molecules, genes, behaviors, functions, species, interactions, and ecosystems" (p. 56). The fact that he uses such a sketchy definition suggests that he takes the definition to be unproblematic and consensual. By contrast, Sarkar (2005) explicitly tackles the definitional issue. Following Maclaurin and Sterelny (2008, p. 8), one can summarize his approach by stating that, according to his definition, "'biodiversity' [means] whatever we think is valuable about a biological system" (Maclaurin and Sterelny's interpretation of Sarkar's theory can be criticized, but for the purpose of the present chapter, we will not delve into this exegetic debate).

A striking difference between Loreau's (2010) and Sarkar's (2005) definitions is that, whereas Sarkar's definition explicitly mentions values, Loreau's definition exclusively mentions purely biological concepts and objects. Despite this major difference, Sarkar explicitly claims that he use the concept of biodiversity in an uncontroversial and widely shared sense: he even writes that his approach captures the "consensus view" (p. 145). And Loreau makes the same claim, though implicitly, since he admits that there is no need to delve into definitional issues. Despite the major difference between their respective definitions, both authors hence claim that their approach captures the general understanding of the concept.

Hence, although Loreau (2010) and Sarkar (2005) use the same term and take for granted that they understand it in the same way as everyone else, they actually understand it markedly differently. Can this kind of misunderstanding have practical implications? In the sections to follow, we argue that, in the case of biodiversity, it can.

17.2.2 The Various Disciplinary Studies "of Biodiversity" Do Not Study the Same Things

The literature presenting the numerous measures and indexes of biodiversity is extensive (Muguran and McGill 2011). It is commonplace to notice that the different disciplines (encompassing what will thereafter be termed various "biodiversity studies") respectively favor different indexes because they capture concepts that are better adapted to their subject-matter. The term "biodiversity" is used in articles from these various disciplines mostly in introductions and conclusions, whereas discipline-specific concepts such as species richness (Fleishman et al. 2006), phylogenetic distances (Faith 1992) or functional traits or attributes (Petchey and Gaston 2002; Mason et al. 2003) replace it in the methods and results sections (Meinard 2011). Similarly, environmental economists often use the term "biodiversity" to introduce and justify their research, but rapidly switch to disciplinary concepts, such as "naturalness" (Eichner and Tschirhart 2007) or "perceived diversity" (Moran 1994). The same is true of the other disciplines concerned with biodiversity. Accordingly, although they all claim to study biodiversity, the various biodiversity studies actually produce results that account for different objects, properties and processes (Maclaurin and Sterelny 2008).

The concept of biodiversity itself is never used in articulating results, in any of these disciplines. It is mostly confined to introductions and conclusions, where it plays the role of a catchword.

17.2.3 The Various Disciplinary Studies "of Biodiversity" Presuppose that they Study Various Aspects of a Common Entity

By using the notion of biodiversity in their introductions and conclusions, all these heterogeneous studies presuppose, at least implicitly, that the various objects, properties and processes that they study are aspects of a common entity: biodiversity (here we use the term "entity" in a purportedly very large sense, encompassing all sorts of ontological units, such as objects, properties, natural kinds, and so on). They do not claim that their concepts or measures represent all of biodiversity, but that there is a common entity, biodiversity, which is partially captured by their favorite measures and concepts.

In the current literature on biodiversity, the various studies simply state that their subject-matter is an aspect of the putative entity biodiversity, without explaining what this entity is supposed to be. What is this putative common entity supposed to be?

17.2.4 Defining "Biodiversity" Thanks to the Notions of Diversity or Variety Is Insufficient to Identify such a Common Entity

The literature on indexes and measures of biodiversity is notably vague on the issue of a proper identification of this common putative entity—biodiversity. The usual explanation identifies it as a specification of a more general entity: the property diversity (Maris 2010). Biodiversity would be the diversity of living things (Gaston and Spicer 2004), along genetic, phylogenetic and functional dimensions (Purvis and Hector 2000).

This approach bears some seeming credibility because "diversity", and synonyms in ordinary language such as "variety", belong to the everyday language and thus seem clear and self-evident. Intuitively, diversity is a property characterizing groups of individuals, depending on the number of individuals and on their similarities and dissimilarities. But the precise roles of numbers, similarities and dissimilarities, and the metrics used to measure them, are not elucidated at this intuitive level.

To determine whether "diversity" truly captures a coherent notion, axiomatic studies have tried to formalize the properties associated with it (Weitzman 1992; Nehring and Puppe 2002). They thereby showed that these properties are highly variable and that the notion of diversity is accordingly deeply ambiguous (Gravel 2008) (in other words, what these studies show is that, whereas it seems self-evident at first sight that diversity is a property, in fact the term "diversity" captures different sets of properties in different contexts, which makes it questionable to claim that "diversity" refers to a property properly speaking). The terms "diversity" or "variety" thus function like a term such as "adaptation". "Adaptation" has different meanings in various subfields of evolutionary biology, it has a markedly different meaning in medical physiology, and yet other meanings in ordinary language. The same holds true for "diversity" and "variety". Within disciplines or, more precisely, within subfields, these terms are relatively unambiguous and generally well-defined, but their meaning varies between disciplines or subfields. As a consequence, these terms cannot be unambiguously used in both ways at the same time. Either one relies on subfield-specific, technical and well-clarified definitions of the terms "diversity", "variety", etc.—but in that case one can no longer draw upon the self-evidence of these terms in everyday language. Or one relies on everyday language—but in that case, one has to face the fact that ordinary language does not delineate coherent notions of diversity or variety. In both cases, using the terms "diversity" or "variety" in a general definition of biodiversity is problematic, because one cannot take for granted that others will understand the notion in the intended way. Therefore, if buttressed on general terms like "diversity" or "variety", a general definition of biodiversity does not single out a unique entity, and is therefore useless to support the idea that "biodiversity" refers to a common entity.

Here again, the comparison with "adaptation" is illustrative. A rigorous evolutionary biologist would never use the term "adaptation" when talking to lay people or to physiologists without specifying that his technical understanding of the term is very specific. The evolutionary biologist knows that his interlocutors think that they understand the term "adaptation", and he knows also that, in a sense, they are right to think that they understand the term. But he also knows that they understand the term in another sense, rather than the one he has in mind. Therefore, it is natural for him to clarify the meaning of the term. This crucial step is the one that is missing in the case of "biodiversity".

The theoretical considerations developed in this Sect. 17.2 may appear purely formal, without implications for concrete conservation science and action. The goal of the following section is to demonstrate that the reverse is true.

17.3   How False Transparency Creates Concrete Problems for Conservation Science and Action

In order to explain the concrete problems created by the seemingly purely theoretical reasoning spelled out in Sect. 17.2, let us now take three concrete case studies, each illustrating a specific kind of problem.

17.3.1   The False Transparency of "Biodiversity" Can Impair the Coordination of Interacting Conservation Actions

Misunderstandings created by the false transparency of "biodiversity" can have detrimental implications at the level of practical conservation management, as can be illustrated by the story of the management of the Bel-Air valley in South-west France (Gereco, unpublished report 2014). This is a small valley (Fig. 17.1) containing a rich mosaic of aquatic and humid habitats in a karstic system close to semi-arid grasslands and upstream water meadows (surrounding the Charente River).

This valley shelters a population of otters (*Lutra lutra*) and a massive population of Louisiana crayfish (*Procambarus clarkia*). The latter is an invasive species having major detrimental impacts on the functioning of aquatic ecosystems (Angeler

Fig. 17.1 The Bel-Air valley

et al. 2001; Rodriguez et al. 2005). However, its impact on Mammals populations is modestly positive (Correira 2001), and from the point of view of otter-watchers it has the advantage to turn otters' spraints into red, greatly facilitating the observation and monitoring of otter populations. The above report also unveiled the presence of Japanese knotweed (*Reynoutria japonica* Houtt., 1777), an invasive plant with deeply damaging impacts on wetland ecosystems.

The valley is managed by an environmental association, Perennis. The downstream water meadows are protected under the Habitat Directive (HD, a cornerstone of the European Union policy to maintain biodiversity: European Commission 1992) and are accordingly managed by another environmental association, the Birds Protection League ("LPO"). Both actors act according to management schemes explicitly aimed at conserving "biodiversity".

But on closer examination, it appears that Perennis understands "biodiversity" in a Sarkar-like manner. Indeed, as amateur naturalists, they value first and foremost the emblematic otters: for them promoting biodiversity mainly means managing the otter population. Because crayfish makes it easier to observe otter, and because they have not witnessed any impact of knotweed on otters yet, they do not see invasive species as a prominent topic in their agenda to conserve "biodiversity". By contrast, directed as it is by European guidelines applicable to the entire Natura 2000 network, the LPO has to conceive of its objective to preserve biodiversity in a way that puts more emphasis on ecological functioning. In particular, following the guidelines spelled out in Evans and Arvella (2011), its management actions have to actively tackle the problems created by invasive species populations. Accordingly, for the LPO, conserving biodiversity in this area implies managing the crayfish and knotweed populations (or at least it implies a need to carve out a strategy assessing the kind of invasive mitigation actions that can be performed, and the cogency of implementing them in the light of their cost and likelihood of success).

Perennis' management strategy aims at "conserving biodiversity", but this means protecting the otter population, and does not mean tackling the invasive species issue; similarly, the LPO's strategy aims at "conserving biodiversity", but this time it means tackling the invasive species issue. Both actors could agree when comparing their objectives: they both strive to "conserve biodiversity." But if it dismisses the invasive issue when managing the valley, Perennis actually jeopardizes any attempt to tackle this very issue downstream. The formal agreement on "biodiversity" hence hides a deep disagreement on what has concretely to be done.

At this stage, one might retort that misunderstandings like the one sketched above can easily be solved if the actors talk to each other about the concrete actions they want to implement. This is certainly true, and this example is indeed somewhat schematic. Our personal experience however suggests that, in real-life management situations, such seemingly trivial disagreements can persist. This is because the term "biodiversity" provides a common vocabulary that various actors can use to express very different objectives, which can all too easily lead them to fail to see the underlying divergences. In the present work, we obviously do not claim to have quantitatively demonstrated that such problems often arise in concrete conservation situation. Our more modest claim is that it can arise.

17.3.2   The False Transparency of "Biodiversity" Hides the Need to Improve Scientific Knowledge to Solve Complex Management Problems

The case of the Bel-Air valley provided a first illustration of how a concrete management problem can remain unseen because various actors fail to see the need to compare their respective understandings of "biodiversity". In this case, the problem arises at the level of the interactions between actors implementing conservation actions. But a deeper problem can arise when innovative solutions and new management knowledge are needed to solve more complex conservation issues. In such cases, the false transparency of "biodiversity" can hide the need to improve scientific knowledge.

An example illustrating this idea is given by the management of so-called "habitats of community interest", when biological invasion mitigation conflicts with habitat conservation (see Jeanmougin et al. 2016 for a deeper investigation of this conflict). "Habitats of community interest" (HCI) are natural or semi-natural habitats constituting the Natura 2000 network, as application of HD (European Commission 1992). HCI are typically defined in European guidelines (European Commission 2013) and more detailed regional scale manuals (e.g. Bensettiti 2001–2005) by lists of floristic species. For some HCI, these lists contain numerous invasive species (see Jeanmougin et al. 2016, SI-Table 3). For example, this is the case of the HCI "Constantly flowing Mediterranean rivers with *Paspalo-Agrostidion* species and hanging curtains of *Salix* and *Populus alba* (Habitat 3280)", whose presence has been recently reported in the lower Taravo River area (Corsica, France) (Fig. 17.2) (Gereco, unpublished report 2015).

Eight of the 34 index species of this habitat (*Paspalum distichum* L., 1759, *Paspalum dilatatum* Poir., 1804, *Xanthium strumarium* L., 1753, *Symphyotrichum subulatum var. squamatum* (Spreng.) S.D.Sundb., 2004, *Dysphania ambrosioides* (L.) Mosyakin & Clemants, 2002, *Amaranthus retroflexus* L., 1753, *Cyperus eragrostis* Lam., 1791and *Erigeron canadensis* L., 1753) are considered invasive species according to various European, national or local databases. HD, as a politi-

Fig. 17.2 Paspalo-Agrostidion and curtain of Salix purpurea along the Taravo river

cal tool to maintain biodiversity, promotes the maintenance of HCI. On the other hand, the control and eradication of invasive species is also a central objective of many European initiatives to maintain biodiversity, such as the DAISIE (Delivering Alien Invasive Species Inventories in Europe) program and the recent European Directive on Invasive Species (Beninde et al. 2015).

In the case of habitats like HCI 3280, there is an antagonism between the invasive approach and the habitat approach. Indeed, if management actions achieve to mitigate populations of the above-cited invasive species, this will unavoidable imply that the area identifiable as HCI 3280 will decrease. Conversely, if management actions achieve an increase of the area occupied by HCI 3280, this will be accompanied by an expansion of populations of the above-cited invasive species. Consequently, elaborating a management scheme in areas like the lower Taravo is problematic, because two actions that are typically considered keystones of any biodiversity conservation strategy (invasive species mitigation and habitat conservation) are antagonist in such cases.

However, there is no scientific study or publication tackling this question (see Jeanmougin et al. 2016 for a bibliographic exploration quantitatively corroborating this idea). According to the database (ETC-BD 2015) constructed as part of the European-wide evaluation of the conservation status of HCI (European Union 2015), this habitat is present in no less than five countries in Europe (France, Greece, Italy, Spain and Portugal). Management schemes are hence devised and implemented all year long in the whole European Mediterranean region to manage this HCI, but there is no scientific guideline to decide how to solve the contradiction between the objectives to mitigate biological invasion and to promote the conservation status of habitat 3280.

Like most complex problems at the science-policy interface, this specific problem certainly has multifarious origins, having to do with the complex challenges in (1) translating ecological theory into practice (Knight et al. 2008), (2) defining the relevant expertise (Burgman et al. 2011), (3) choosing the relevant scientific paradigms to ensure operationality (Jeanmougin et al. 2016), (4) drawing the line between scientific information and advocacy (Brussard and Tull 2007), (5) assessing the proper place of scientific knowledge in the process of policy making (Josanoff 2012) and (6) entrenching the importance of an open diffusion of information on conservation practices (Meinard 2017a). We do not claim here to do justice to all these aspects, their interrelations and their relative importance in the genesis of problems such as the one of the above introduced lack of knowledge on HCI 3280 management and invasive species. Our more modest purpose is the following. We want to show that, by granting a key-role in the coordination between disciplinary approaches to a vague term like "biodiversity", one tends in all likelihood to render invisible the kind of knowledge gap at issue in our example. We accordingly do not claim to unfold a scientific demonstration here, but rather to hypothesize a possible mechanism that occurred to us thanks to our own field experience.

We propose that this mechanism is simply that specialists of invasive species stress the importance of controlling invasive species and present such a control as a prominent means to maintain biodiversity. But as non-specialists of habitats, they simply accept what specialists say about the self-evidence that maintaining HCI is also unquestionably good for biodiversity. Specialists of habitats behave in a symmetric way. Everyone thus agrees with the overarching objective to maintain biodiversity, everyone is careful not to question the expertise of one's interlocutor, and no one sees the need to improve knowledge and to carve out innovative management solutions in complex cases such as the one of habitat 3280.

As a consequence in the field, at the end of the story the resulting management scheme is most of the time decided more or less arbitrarily by political decision-makers or consultants on the basis of political, economic or circumstantial considerations. In the case of the Taravo River, the management scheme produced in 2014 (Lindenia, unpublished report 2015) does not mention this problem.

17.3.3 The False Transparency of "Biodiversity" Hides that Various Approaches Are Based on Incompatible Postulates

A more subtle, but no less important problem arises when interactions with non-ecological disciplines are involved. Let us start by illustrating the problem with an example: Eichner and Tschirhart (2007)'s ecological-economic study of a fishery ecosystem. Their aim is to establish how to organize fisheries given that the exploitation of a given species can have complex repercussions on the broader ecosystem. In their study system, human consumers buy items of one species (Pollock, *Theragra chalcogramma*) on markets and thereby indirectly impact other species due to between-species interactions in the ecosystem. This indirect impact then alters the provision of various ecological services. For example, the sheltering function of kelp (*Laminaria* spp.) can be altered, which has an impact on the populations of Pacific halibut (*Hippoglossus stenolepis*) and Pink salmon (*Oncorhynchus gorbusha*), which in turn alters the so-called "consumption services" (MEA 2005) for consumers of the latter species.

Eichner and Tschirhart (2007)'s solution is that if taxes on harvesting activities or caps on harvest, calibrated thanks to a precise knowledge of the functioning of the ecosystem, are implemented, demand will drop, overfishing will cease, kelp will recover, etc., and consumers will end-up being more satisfied.

The economists who authored this study claim that they provide insights that are complementary to those provided by ecologists to resolve a commonly identified problem—the problem of how to manage a complex ecological-economic system. Unfortunately, the way they see this problem is strikingly different from the way many ecologists see it. The compatibility of their prescriptions with prescriptions stemming from biological studies can accordingly become problematic.

Indeed, Eichner and Tschirhart (2007)'s understanding of the problem is based on a moral assumption—that is, an assumption about what is morally legitimate for scientists to do. They assume that consumers' preferences are given, and that the results of their study should not lead to a modification of consumers' preferences. Consequently, they do not integrate in their models the fact that being aware of the ecological impact of their act may change the behavior of the consumers. In technical economic terms, this assumption is encapsulated by the fact that human behavior is modeled using a predetermined utility function (Orléan 2011), whose parameters are not fed-back by the results of the study. Consumers are assumed to behave has if they were maximizing a function whose arguments are prices and quantities of goods they buy on markets. The knowledge of the system does not appear in the function: when a given consumer learns to know that his buying Pollock has impacts on populations of Salmon, this does not make any difference in his behavior on the Pollock market.

Such predetermined utility functions are often presented, like many other economic modelling tools, as morally neutral tools providing empirical complements to moral discussion (e.g. Scharks and Masuda 2016). When presented like this, it seems that predetermined utility function, as well as other modelling tools widely used in ecological-economic studies, can be used in conservation initiatives without interfering with the ethical motivations underlying the latter. Following the same logic, when an ecosystem ecologist works on a specific ecosystem process and an economist computes the economic value of an ecosystem service based on this process, it might look as though the two can work together and the end-result of their conjoint work is no less ethically neutral than the original ecological study of the ecological process. This repeatedly rehearsed logic is, however, largely acknowledged to be flawed: predetermined utility functions are not morally neutral modeling tools (Sen 2002; Hausman and McPherson 2006; Meinard et al. 2016). Using these tools means assuming that the results of the study should not lead to a modification of consumers' preferences: if consumers prefer x to y, the study should never aim at modifying this fact. This is not a technical constraint: implementing a feedback between the results of the model and preferences is not technically difficult (Lesourne et al. 2006). It is a moral stance: using predetermined utility functions is a means to promote the anti-paternalistic attitude to leave preferences as they are (Kolm 2005; Sagoff 2008) and to advocate that the satisfaction of preferences as they are is an acceptable, or even desirable, objective (Sagoff 2008).

This moral stance might seem reasonable enough—why should the economist think that he knows better than the consumer what the latter should prefer? But this moral stance can be problematic from the point of view of conservation sciences, because convincing people that their preferences are ill-conceived is a prominent means to achieve conservation targets. Many applied ecological studies are even openly based on moral assumptions that are diametrically opposed to the above one. Take for example the adaptive management approach (Norton 2005), according to which management practices should be seen as experiments from which managers can learn and thereby both improve their knowledge of the managed system and adjust the criteria that they use to evaluate alternative courses of actions (Lee 1993). Contrary to Eichner and Tschirhart (2007)'s model, this approach assumes that people's objectives and preferences are responsive to improvements of their knowledge of the ecological constraints (Maris and Bechet 2010), and that enabling such improvements is precisely one of the motivations to study these systems.

Eichner and Tschirhart (2007)'s solutions are solutions to the problem as they see it, constrained by moral assumptions that are not generally accepted by biologists. This does not mean that their approach is irredeemably irrelevant for biologists, but that using it to identify solutions to the problem as biologists see it requires important reinterpretations and adaptations. Eichner and Tschirhart (2007) however eschew the clarification of this point, and their argument is accepted in the ecological literature without a discussion (as illustrated by its extensive mention in Naeem et al. 2009). It is difficult to see why biologists do not assess the relevance of this model more critically. Our interpretation here is that the false transparency of the notion of biodiversity plays a role in the explanation of the existence of this blind spot. Indeed, this false transparency makes it look as though Eichner and Tschirhart (2007)'s is self-evidently relevant, since it claims to be about biodiversity. We cannot overemphasize that, obviously enough, we do not claim that the term "biodiversity" is the unique, or even the main, culprit in failures of ecological-economic studies of fisheries. The precautions articulated above when analyzing the former case study apply here as well. Our point is more modestly that, by granting a key-role in the coordination between disciplinary approaches to a vague term like "biodiversity", one tends in all likelihood to render invisible the fact that different disciplinary approaches are anchored in different moral assumptions. Like in our former case study once again, we do not claim to unfold a scientific demonstration here, but rather to hypothesize a possible mechanism, accounting for one possible cause among others behind the shortcomings of the models that we analyze.

Eichner and Tschirhart (2007)'s model is just one example, but it is a paradigmatic one, for two reasons.

First, the moral assumption mentioned above is so entrenched in the economic literature that some authors (e.g. Orléan 2011) use it to characterize the vast majority of the current economic literature. This does not mean that economic studies necessarily make this assumption, since heterodox approaches reject it (Lesourne et al. 2006), but rather that this assumption is bound to create recurrent problems in economics/ecology interactions.

Second, the problem witnessed in our example between ecology and economics exists between ecology and other disciplines. For example, numerous anthropological studies presenting themselves as studies of biodiversity acknowledge that they are based on moral postulates (e.g. Mougenot 2003). But if ecologists and anthropologists do not investigate whether their respective moral assumptions are compatible, the possibility for them to provide coherent prescriptions for action is unwarranted.

To sum up the lesson from this third case study: when various disciplines present themselves as studies of different aspects of a common object—biodiversity—they tend to ignore that, if they are based on incompatible moral assumptions—as they often are—the very meaning and usefulness of their interactions are questionable.

17.4   The Way Forward

In the former section, we explored various concrete examples that allowed us to illustrate different kinds of problems created by the false transparency of "biodiversity". The order of presentation was one of increasing complexity and increasing explanatory content: the first example was a simple case of diverging conservation objectives, the second one involved a more interpretative analysis, and the last one eventually allowed us to articulate the crux of the problem created by the false transparency of "biodiversity": collaborative works or disciplinarily studies that conceive of themselves as tackling different aspects of a common object fail to acknowledge the need to ensure that they tackle different parts of a similarly identified problem. They are caged in an illusory shared ontology of the entity biodiversity.

One might argue that the simple solution to all the problems mentioned above would be to get rid of the term "biodiversity" and stop pretending that the various "biodiversity sciences" have anything in common. Such a radical solution (championed, for example, by Santana 2014) could be counterproductive though, by discouraging interdisciplinary collaboration. This would be at odds with the widely accepted idea that interdisciplinary approaches are needed to tackle the globally pressing environmental challenges (Norton 2002; Loreau 2010b).

The aim of the present section is to delineate possible solutions based on the idea that the need to arrive at commonly identified problems should be taken seriously. We first sketch what such a requirement would concretely mean in the case of our three concrete examples, and then we take a broader view.

17.4.1   Facing the Issue of Problem Identification

A leitmotif for contemporary decision analysts, especially those working in multi-actor settings, is that decision-makers requesting decision-support often do not precisely know for what problem they really need support (Bouyssou et al. 2000; Tsoukias et al. 2013). For example, private firms can be aware that they have a problem in their production process because the output is lower than expected. But they don't know if the problem is that they are inefficient or that they were unrealistic when setting their objectives or that their overall conception of what they aim to do was flawed, etc. They know that they have a problem, but can only articulate a rough, ambiguous formulation of it. More interestingly, various stakeholders, for example involved in the management of a complex system such as a watershed, may have only a very partial understanding of the problem that that are nonetheless in charge of tackling. Clarifying a better formulation of the problem, eliminating ambiguities and vagueness often associated with the terms spontaneously used to request decision support, is accordingly a first, critical step for decision analysis (Belton et al. 1997; Rosenhead 2001). Given the nature of the problems identified in this chapter, a similar clarification, on a case-by-case basis, of the precise nature of the problem for the resolution of which (often interdisciplinary) interactions are put to use, may substantially improve the situation.

In the case of the Bel-Air valley, instead of resting content with the fact that they both manage their respective areas according to a scheme that mentions the preservation of biodiversity as an overarching aim, the two managers should answer the following questions. "What is the precise nature of the functional links between the Charente water meadows and the habitats of its tributary, the Bel-Air?", "What are the functional consequences of the absence of a control of the crayfish and Japanese knotweed populations in the Bel-Air on the ecological functioning of the nearby water meadows?", "Would it be justified that the manager of the water meadows contribute (through money or workforce) to help the manager of the Bel-Air to implement specific conservation measures?" These are difficult questions, but sweeping them under the carpet by framing the discussions with the vague consensual terms of "biodiversity" does not make them any less urgent.

In the Taravo case, the question "How to manage the river area in such a way as to promote biodiversity?" is meaningless, because, in this case, the two kinds of stakeholders in charge of the site management have distinct concrete objectives and would implement very different and potentially contradictory action. These differences and discrepancies, however, are hidden by the use of the common word "biodiversity". In this case, a clear management policy and scientific knowledge are simply lacking. The very notion that HCI 3280 is protected under European legislation is nonsense so long as there are no scientific answers to the questions: "Is it possible to define this habitat on the basis of other criteria than species lists?" and "Is it possible to preserve this habitat while controlling invasive species at the same time?" These scientific issues are currently not investigated because, mainly due to the fact that problems are formulated in the vague terms of "biodiversity", these genuine, underlying problems do not surface. Similarly, the national and local strategies regularly produced and updated by environmental institutions are of little use if they do not clarify how the various aspects of biodiversity should be ordered when they conflict in a practical management situation. If such a ranking were available, even if scientific studies turned out to demonstrate that it is impossible to define habitat 3280 without referring to invasive species, a management program could be defined for the lower Taravo on an informed, legitimate basis.

Lastly, in the case of economics/ecology interactions, they would gain much transparency if, instead of rehearsing the vulgate of the supposed biodiversity/well-being link, ecology/economics interactions were systematically anchored in a common identification of the answers to the following questions: "For the purpose of a given decision-making on a conservation issue, what kind of economic information is useful?", "Should we take an anti-paternalistic stance like most economists, or should we rather take a more pedagogic stance and admit that ecological knowledge can rightfully be used to improve everyone's decisions?" and "More generally, what kind of prescription for action is legitimate for biodiversity scientists to formulate on the basis of scientific models?"

17.4.2   A Broader View

The above paragraphs might leave the reader somewhat unsatisfied, since we simply spelled out the questions that the actors should ask themselves *in the specific cases we considered*. Isn't it possible to elaborate a more general approach, liable to help solve problems created by the false transparency of "biodiversity" in a more general setting? We believe that it is possible, and here we sketch our proposal.

17.4.2.1   What Do We Need to Define?

A common view, although often implicit, in the literature, is that the definition of the term "biodiversity" has to be an *objectivist definition*. An objectivist definition is a description of the independent, preexisting entity to which the term to be defined refers. For example, an objectivist definition of the term "Mars" is a description enabling to identify the planet to which the term refers—an object independent from and preexisting our specifying that the term "Mars" refers to it. When one claims to define "biodiversity" by specifying preexisting independent objects, properties or processes, one attempts to provide an objectivist definition.

We have argued above that the ambiguity of the current general definition of biodiversity can create damaging problems, but it is unlikely that objectivist definitions can prevent such problems from arising. Our analysis of the case of ecological-economic models rather suggests that, unless it makes an explicit reference to the value-laden aspirations that make sense of the various biodiversity sciences, a definition can hardly be useful to prevent such problems.

We therefore have to carefully examine the reason why we need a definition. We want to make sure that the various approaches gathered under the umbrella of the term "biodiversity" can provide relevant insights to coherently resolve common problems. The term "biodiversity" provides a form of unification between different approaches and disciplines. But this form of unification is defective when it comes to doing justice to this reason, because it covers up misunderstandings between approaches. What we need is another form of unification, liable to prevent such misunderstandings. Our suggestion is that this unification should rather be buttressed on a general definition of *biodiversity practices*, understood as a coherent collaborative effort from various disciplines to tackle commonly identified environmental or conservation problems.

Our suggestion is therefore to shift the focus from the definition of biodiversity conceived as a putative entity to the definition of biodiversity practices, emphasizing the value-laden aspirations underlying them. This suggestion might seem odd at first sight, because it looks as though one needs to have a prior concept of biodiversity in order to talk about "biodiversity practices". Underlying our suggestion is the idea that such a criticism stems from a linguistic illusion. Our language treats "biodiversity" as a substantive, which makes it look as though "biodiversity practices" necessarily are practices towards the entity to which the substantive refers. But we have argued that there is no such entity biodiversity to which "biodiversity" refers.

Whereas our language gives the false impression that one cannot understand what biodiversity practices are without antecedently understanding what biodiversity is, our suggestion is that the reverse is the case: in order to understand what biodiversity is, what we have to do is to start by thinking through what biodiversity practices are (Sarkar's (2005) approach is very similar to ours in this respect; for an analysis of the differences, see Meinard 2017b).

The search for such a definition of biodiversity practices is bound to be unavoidably largely tentative and interpretative, but it would be misleading to presuppose that this creates a serious problem. The reason is that, when one defines a practice, the suitable definitions cannot be definitive objectivist definitions, because the very act of defining can modify the practice, and this modification can in turn modify the definition. In such a case, the definition does not identify an independent, preexisting practice: it rather participates in the construction of the practice.

The vast literature on the definition of the terms "art" and "artistic practices" perfectly illustrates this idea. Although everybody intuitively knows what these terms mean, there is a vast literature striving to capture definitions of these terms. Unlike biodiversity scientists, art theoreticians have never accepted to rest content with the apparent self-evidence of the central notions of their field: they have endlessly kept trying to find better definitions. It turns out that, in so doing, they have greatly contributed to the enrichment of artistic practices. Indeed, the various definitions of art by prominent art theoreticians have aroused creative responses by artists, who have (more or less consciously) modified their artistic practices to highlight the restrictiveness of these definitions or to explore the avenues they had opened up (Pignocchi 2012).

We argue that, as biodiversity scientists, we should follow this example of art theoreticians. We should always include an explicit definition of the global value-laden biodiversity practices into which we see our studies as being embedded, in such a way as to dissipate misunderstandings like the one highlighted in Sect. 17.3. The point of such references is not just to harbor values, but to prevent misunderstandings. In particular, the value-laden features mentioned in such definitions must be the ones that are crucial to the identification of the general problems that biodiversity sciences should be devoted to solve. If such definitions were systematically formulated, this would launch a creative process by which other biodiversity scientists would modify their practices to criticize the shortcomings or exploit the strengths of each definition, and in turn suggest new definitions, etc.

17.4.2.2   A Tentative Definition

Let us exemplify our approach by articulating our own tentative definition of biodiversity practices:

> Biodiversity practices are studies, actions, strategies based on the aspiration that the development and diffusion of ecological knowledge can lead people to improve their course of action by developing responsible, informed and long-term decision strategies and preferences, mindful of the environmental constraints.

This definition is not the result of a grand deductive, philosophical or scientific, reasoning. It is a tentative interpretation of the studies, actions, strategies that our own experience as conservation scientists and practitioners allowed us to experience—and that our case studies above exemplified. This definition is obviously neither definitive nor uncontroversial. In some contexts, it might appear to be too vague and in need of qualifications or discussions, and the emergence of misunderstandings in the future may require reformulations. But as it stands, it is the kind of definition needed to clarify misunderstandings like the one unveiled above.

For example, if Naeem et al. (2009) had started by articulating such a definition, based on the identification of the problems tackled by biologists, they would most probably have faced difficulties to encompass Eichner and Tschirhart (2007)'s model in it, because these authors do not understand the problem of biodiversity management in the same way as most biologists do. Naeem et al. (2009) would accordingly have admitted the necessity to critically scrutinize the relevance of this model for conservation and ecological purposes. A fruitful critical discussion could have ensued and damaging misunderstanding could have been possibly dissipated.

One seeming problem with this approach is that it is likely that the problems tackled by biodiversity sciences will change over time. Encapsulating them in a single definition of biodiversity practices might accordingly risk encouraging immobility. The tentative definition of biodiversity practices just introduced is, however, liable to play a clarifying role without encouraging immobility because it harbors two crucial features. These two features characterize what we will term a "dynamic definition".

First, since it is granted the status of a tentative definition, it is open to discussion and accordingly flexible enough to continuously adapt to new insights and developments.

Second, since it is meant to be used to critically assess the relevance of various studies for one another, the very act, by a given scientist, to formulate such a definition and test it on a given study can lead him to modify his own practice instead of rejecting the study he assesses. Defining a practice can thereby lead to a modification of this very practice, and this modification can in turn modify the definition.

In this dynamic approach, the best thing that can happen now to the tentative definition of biodiversity practices introduced above is that it be taken seriously enough by biodiversity scientists for them to criticize it, thereby launching the co-evolution of biodiversity practices and their definition.

17.5   Conclusion

The term "biodiversity" is diversely understood by various users, and its general definition is vague. Here we have taken advantage of several case studies to show that this vagueness, which is usually taken by biologists to be innocuous at a theoretical level, can create problems at the concrete level of practical interactions between various approaches to biodiversity issues. The problems studied here share a common structure. In these various settings, the term "biodiversity" is used by various actors to link their respective approaches. The resulting impermeable division of labor, based on the formal but illusory agreement on the objective to study or conserve "biodiversity", hides the fact that the various approaches can promote mutually incompatible goals, eventually leading in conservation practice to self-defeating actions. To end such deadlocks, we have claimed that a clarification, on a case-by-case basis, of the precise nature of the problems for the resolution of which interdisciplinary interactions are put to use is a critical step. It can make the various misunderstandings and contradictions currently impeding management practices due to the false transparency of "biodiversity" visible and subsequently help to resolve them. This case-by-case approach then allowed us to develop a more general proposal, delineating a path towards the resolution of problems created by the false transparency of "biodiversity". The logic of this path can be summed up in four steps:

1. There is need to clarify a general definition of *biodiversity practices*, understood as a coherent collaborative effort from various disciplines to tackle environmental or conservation problems commonly identified on the basis of coherent value-laden aspirations.
2. General definitions of biodiversity practices are always tentative, because the very act of defining them can lead to a modification of our theoretical and practical approaches to biodiversity theorizing and management.
3. In our contributions, we should all make it a rule to always define the global value-laden biodiversity practices into which we see our studies as being embedded, in such a way as to prevent, as far as possible, misunderstandings with other biodiversity studies or actions.
4. We should seize every opportunity to discuss and criticize the definitions put forward by the other biodiversity scientists who have followed the steps above.

Although they have never formally articulated it, art theoreticians and artists have historically followed a similar path, which proved to be very fruitful. Our hope is that biodiversity scientists can learn from this example.

Acknowledgements This work was supported by the Fondation pour la Recherche sur la Biodiversité. We wish to thank E. Casetta, D. Flynn, A. Krajewski, L. Lhoutellier, J. Marques da Silva, X. Morin and D. Vecchi for their comments and corrections.

References

Angeler, D. G., Sanchez-Carrillo, S., Garcia, G., & Alvarez-Cobelas, M. (2001). Influence of Procambarus clarkii (Cambaridae, Decapoda) on water quality and sediment characteristics in a Spanish floodplain wetland. *Hydrobiologica, 464*, 89–98.

Belton, V., Ackermann, F., & Shepherd, I. (1997). Integrated support from problem structuring through alternative evaluation using COPE and V-I-S-A. *Journal of Multi-Criteria Decision Analysis, 6*, 115–130.

Beninde, J., Fischer, M. L., Hochkirch, A., & Zink, A. (2015). Ambitious advances of the European Union in the legislation of invasive alien species. *Conservation Letters, 8*, 199–205.

Bensettiti, F. (Ed.). (2001–2005). *Cahiers d'habitats Natura 2000. Connaissance et gestion des habitats et des espèces d'intérêt communautaire (7 volumes)*. Paris: La Documentation française.

Bouyssou, D., Marchant, T., Pirlot, M., Perny, P., Tsoukias, A., & Vincke, P. (2000). *Evaluation and decision models: A critical perspective*. Dordrecht: Kluwer Academic Publishers.

Brussard, P. F., & Tull, J. C. (2007). Conservation biology and four types of advocacy. *Conservation Biology, 21*, 21–24. https://doi.org/10.1111/j.1523-1739.2006.00640.x.

Burch-Brown, J., & Archer, A. (2017). In defense of biodiversity. *Biology and Philosophy, 32*(6), 969–997. https://doi.org/10.1007/s10539-017-9587-x.

Burgman, M., Carr, A., Godden, L., Gregory, R., McBride, M., Flander, L., & Maguire, L. (2011). Redefining expertise and improving ecological judgment. *Conservation Letters, 4*, 81–87.

Correira, A. M. (2001). Seasonal and interspecific evaluation of predation by mammals and birds on the introduced red swamp crayfish Procambarus clarkia in a freshwater marsh (Portugal). *Journal of Zoology, 255*, 533–541.

DeLong, D. C. (1996). Defining biodiversity. *Wildlife Society Bulletin, 24*(4), 738–749.

Eichner, T., & Tschirhart, J. (2007). Efficient ecosystem services and naturalness in an ecological/economic model. *Environmental and Resource Economics, 37*, 733–755.

ETC-BD. (2015). *Habitat Directive European article 17 database*. European Topic Center on Biological Diversity. www.eea.europa.eu/data-and-maps/data/article-17-database-habitats-directive-92-43-eec-1. Accessed 8 Sept 2018.

European Commission. (1992). Council Directive 92/43/EEC of 21 May 1992 on the conservation of natural habitats and of wild fauna and flora, O.J. L206, 22.7.1992, pp. 7–50. eur-lex.europa.eu/legal-content/EN/TXT/?uri=CELEX:31992L0043

European Commission. (2013). Interpretation manual of European Union habitats. EUR 28. ec.europa.eu/environment/nature/legislation/habitatsdirective/docs/Int_Manual_EU28.pdf

European Union. (2015). *The state of nature in the European Union*. ec.europa.eu/environment/nature/pdf/state_of_nature_en.pdf. Accessed 8 Sept 2018.

Evans, D., & Arvela, M. (2011). *Assessment and reporting under Article 17 of the habitats Directive – Explanatory note and guidelines for the period 2007–2012*. Final Draft. CTE/BD. circabc.europa.eu/sd/d/2c12cea2-f827-4bdb-bb56-3731c9fd8b40/Art17%20-%20Guidelines-final.pdf. Accessed 2 April 2016.

Faith, D. P. (1992). Conservation evaluation and phylogenetic diversity. *Biological Conservation, 61*, 1–10.

Fleishman, E., Noss, R. F., & Noon, B. R. (2006). Utility and limitations of species richness metrics for conservation planning. *Ecological Indicators, 6*, 543–553.

Gaston, K. J., & Spicer, J. I. (2004). *Biodiversity: An introduction* (2nd ed.). Malden: Blackwell.

Gereco. (2014). *Etude floristique de propriétés en espace naturel sensible de la Charente-Maritime*. Unpublished report.

Gereco. (2015). *Elaboration de cartographies de sites Natura 2000 en Corse-du-Sud. Site Nature 2000 Embouchure du Taravo et alentours*. Unpublished report.

Gravel, N. (2008). What is diversity? In T. A. Boylan & R. Gekker (Eds.), *Economics, rational choice and normative philosophy* (pp. 15–55). Abingdon: Routledge.

Hausman, D. M., & McPherson, M. S. (2006). *Economic analysis, moral philosophy, and public policy* (2nd ed.). Cambridge, MA: Cambridge University Press.

Jeanmougin, M., Dehais, C., & Meinard, Y. (2016). Mismatch between habitat science and habitat directive: Lessons from the French (counter)example. *Conservation Letters, 10*(5), 635–644.

Josanoff, S. (2012). *Science and public reason*. Abingdon: Routledge.

Knight, A. T., Cowling, R. M., Rouget, M., Balmford, A., Lombard, A. T., & Campbell, B. M. (2008). Knowing but not doing: Selecting priority conservation areas and the research–implementation gap. *Conservation Biology, 22*, 610–617.

Kolm, S.-C. (2005). *Macrojustice*. Cambridge: Cambridge University Press.

Lee, K. N. (1993). *Compass and gyroscope—Integrating science and politics for the environment.* Washington, DC: Island Press.

Lesourne, J., Orlean, A., & Walliser, B. (2006). *Evolutionary microeconomics.* Berlin: Springer.

Lindenia. (2015). *Etude pré-opérationnelle à la restauration, l'entretien, la gestion et la mise en valeur du Taravo. Phase 3. Programme pluriannuel de gestion.* Unpublished report.

Loreau, M. (2010). *From populations to ecosystems.* Princeton: Princeton University Press.

Loreau, M. (2010b). *The challenges of biodiversity sciences.* Oldendorf/Luhe: International Ecology Institute.

Mace, G. M., Norris, K., & Fitter, A. H. (2012). Biodiversity and ecosystem services: A multilayered relationship. *Trends in Ecology & Evolution, 27,* 19–26.

Maclaurin, J., & Sterelny, K. (2008). *What is biodiversity?* Chicago: The University of Chicago Press.

Maguran, A. E., & McGill, B. J. (Eds.). (2011). *Biological diversity. Frontiers in measurement and assessment.* Oxford: Oxford University Press.

Maris, V. (2010). *Philosophie de la biodiversité.* Paris: Buchet Chastel.

Maris, V., & Béchet, A. (2010). From adaptive management to adjustive management. *Conservation Biology, 24*(4), 966–973.

Mason, N. W. H., et al. (2003). An index of functional diversity. *Journal of Vegetation Science, 14,* 571–578.

MEA. (2005). *Ecosystems and human well-being: Biodiversity synthesis.* World Resources Institute. https://www.millenniumassessment.org/documents/document.354.aspx.pdf. Accessed 8 Sept 2018.

Meinard, Y. (2011). *L'expérience de la biodiversité.* Paris: Hermann.

Meinard, Y. (2017a). La biodiversité comme thème de philosophie économique. In G. Campagnolo & J.-S. Gharbi (Eds.), *Philosophie Economique* (pp. 319–346). Paris: Matériologiques.

Meinard, Y. (2017b). What is a legitimate conservation policy. *Biological Conservation, 213,* 115–123.

Meinard, Y., Coq, S., & Schmid, B. (2014). A constructivist approach toward a general definition of biodiversity. *Ethics, Policy & Environment, 17,* 88–104.

Meinard, Y., Dereniowska, M., & Gharbi, J.-S. (2016). The ethical stakes in monetary valuation for conservation purposes. *Biological Conservation, 199,* 67–74.

Moran, D. (1994). Contingent valuation and biodiversity: Measuring the user surplus of Kenyan protected areas. *Biodiversity and Conservation, 3,* 663–684.

Mougenot, C. (2003). *Prendre soin de la nature ordinaire.* Paris: Édition de la Maison des Sciences de l'Homme.

Naeem, S., Bunker, D. E., Hector, A., Loreau, M., & Perrings, C. (Eds.). (2009). *Biodiversity, ecosystem functioning, and human wellbeing.* Oxford: Oxford University Press.

Nehring, K., & Puppe, C. (2002). A theory of diversity. *Econometrica, 70,* 1155–1198.

Norton, B. G. (2002). *Searching for sustainability.* Cambridge: Cambridge University Press.

Norton, B. G. (2005). *Sustainability.* Chicago: The University of Chicago Press.

Orléan, A. (2011). *L'empire de la valeur.* Paris: Seuil.

Petchey, O. L., & Gaston, K. J. (2002). Functional diversity (fd), species richness, and community composition. *Ecology Letters, 5,* 402–411.

Pignocchi, A. (2012). *L'œuvre d'art et ses intentions.* Paris: Odile Jacob.

Purvis, A., & Hector, A. (2000). Getting the measure of biodiversity. *Nature, 405,* 207–219.

Rodriguez, C. F., Becares, E., & Fernandez-Alaez, C. (2005). Loss of biodiversity and degradation of wetlands as result of introducing exotic crayfish. *Biological Invasions, 7,* 75–82.

Rosenhead, J. (2001). *Rational analysis of a problematic world* (2nd rev. ed.). Wiley: New York.

Sagoff, M. (2008). *The economy of earth* (2nd ed.). Cambridge: Cambridge University Press.

Santana, C. (2014). Save the planet: Eliminate biodiversity. *Biology and Philosophy, 29,* 761–780.

Sarkar, S. (2005). *Biodiversity and environmental philosophy.* Cambridge: Cambridge University Press.

Scharks, T., & Masuda, Y. J. (2016). Don't discount economic valuation for conservation. *Conservation Letters, 9*(1), 3–4.

Sen, A. K. (2002). *Rationality and freedom*. Cambridge, MA: Harvard University Press.

Tsoukias, A., Montibeller, G., Lucertini, G., & Belton, V. (2013). Policy analytics: An agenda for research and practice. *EURO Journal on Decision Processes, 1*, 115–134.

United Nations. (2013) *Convention on biological diversity*. Rio De Janeiro.

Weitzman, M. L. (1992). On diversity. *The Quaterly Journal of Economics, 107*, 363–405.

Chapter 18
What Should "Biodiversity" Be?

Sahotra Sarkar

Abstract This paper argues that biodiversity should be understood as a normative concept constrained by a set of adequacy conditions that reflect scientific explications of diversity. That there is a normative aspect to biodiversity has long been recognized by environmental philosophers though there is no consensus on the question of what, precisely, biodiversity is supposed to be. There is also disagreement amongst these philosophers as well as amongst conservationists about whether the operative norms should view biodiversity as a global heritage or as embodying local values. After critically analyzing and rejecting the first alternative, this paper gives precedence to local values in defining biodiversity but then notes many problems associated with this move. The adequacy conditions to constrain all natural features from being dubbed as biodiversity include a restriction to biotic elements, attention to variability, and to taxonomic spread, as well as measurability. The biotic elements could be taxa, community types, or even non-standard land cover units such as sacred groves. This approach to biodiversity is intended to explicate its use within the conservation sciences which is the context in which the concept (and term) was first introduced in the late 1980s. It differs from approaches that also attempt to capture the co-option of the term in other fields such as systematics.

18.1 Introduction

Many commentators have noted that the term "biodiversity" is of very recent vintage even though biodiversity conservation has become one of the best-known components of both popular and technical discussions of environmental goals today (Takacs 1996; Sarkar 2005, 2017a). The term and associated concept(s) were only

S. Sarkar (✉)
University of Texas, Austin, TX, USA

Presidency University, Kolkata, India
e-mail: sarkar@austin.utexas.edu

© The Author(s) 2019   375
E. Casetta et al. (eds.), *From Assessing to Conserving Biodiversity*,
History, Philosophy and Theory of the Life Sciences 24,
https://doi.org/10.1007/978-3-030-10991-2_18

introduced in the context of the institutional establishment of conservation biology as an academic discipline in the late 1980s. The introduction of the term is usually attributed to Walter G. Rosen at some point during the organization of a 1986 National Forum on BioDiversity held under the auspices of the United States National Academy of Sciences and the Smithsonian Institution (Takacs 1996; Sarkar 2002).

Originally "biodiversity" was only intended as a shorthand for "biological diversity"; by the time the proceedings of the forum were published as an edited book (Wilson 1988), the new term had been promoted to become its title. The BioDiversity forum was held shortly after the founding of the U.S. Society for Conservation Biology in 1985 (Sarkar 2002). Soulé's (1985) manifesto for the new discipline of conservation biology and Janzen's (1986) exhortation to tropical ecologists to undertake the political activism necessary for conservation had appeared in the previous two years. A sociologically synergistic interaction between the use of "biodiversity" and the growth of conservation biology as a discipline then occurred and it led to a reconfiguration of environmental studies so that the conservation of biodiversity became a central concern. Conservation biology, starting in the 1990s, was conceptualized as the goal-oriented discipline devoted to the protection of biodiversity. Soulé (1985) drew a powerful analogy between conservation biology and medicine; biodiversity was the analog of health.

The existence of a goal engenders a corresponding norm for evaluating whether an action contributes to that goal and, in many contexts, of assaying the extent to which it does so. All the major programs for biodiversity conservation, *viz.*, conservation biology (Soulé 1985), conservation science (Kareiva and Marvier 2012), and systematic conservation planning (Margules and Sarkar 2007), acknowledge the normative component of biodiversity conservation. Not surprisingly, many environmental philosophers have followed suit in treating biodiversity as at least partly a normative concept (Callicott et al. 1999; Norton 2008; Sarkar 2008, 2012b).

But not all. Some philosophers (e.g., Maclaurin and Sterelny 2008), following the lead of many biologists (see Gaston 1996b and Takacs 1996), have treated biodiversity as if it were a purely scientific concept bereft of normative content. That perspective has led to a wide variety of scientific (more accurately, *scientistic*) definitions of biodiversity, each disputed, and with no prospect of resolution of these disputes. The persistence of these disputes has led to many deflationary accounts of "biodiversity" (e.g., Sarkar 2002) as well as proposals to eliminate the term completely (e.g., Morar et al. 2015; Santana 2017). These varied approaches have recently been reviewed by Sarkar (2017a) and that discussion is very briefly summarized in Sect. 18.2 of this paper.

Section 18.3 turns to the core purpose of this paper: a defense of normativism in defining biodiversity. Any such defense must address the question: whose norms? Global norms invoked by Northern conservationists must be pitted against the local norms of communities whose livelihoods are often threatened by biodiversity conservationists' interventions. Section 18.3 traces the ideological underpinnings of global normativism, then rejects it, and critically endorses the use of local values to define biodiversity. But endorsing local values is hardly unproblematic. Section 18.4 examines the problems that beset local normativism.

Accepting normativism does not mean rejecting the use of science any more in biodiversity conservation than it does in healthcare practices. For biodiversity, a partial synthesis is possible. Section 18.5 argues that a rich tradition of discussions within biology of what constitutes biodiversity can be used to lay down adequacy conditions that constrain the latitude available to normative definitions of biodiversity. It also lays out how this synthetic proposal, integrating values and (ostensibly value-free) technical science, can be used in the practice of conservation. Section 18.6 consists of some final remarks.

18.2 Approaches

Sarkar (2017a) has recently distinguished four approaches to defining biodiversity:

1. *Scientism*: Definitions falling under this rubric claim to use non-normative criteria to define and quantify biodiversity. Three such criteria have most often been deployed: richness, difference, and rarity. Each criterion has been used not only singly but also in conjunction with the others. Richness, measured by the number of units, is probably what most users of "biodiversity" have in mind when the term is not explicitly defined. It has also been partly or wholly explicitly defended by Gaston (1996a) and Maclaurin and Sterelny (2008). Difference, interpreted as complementarity, or how many new biodiversity units are introduced to those already present in an entity (such as an area or a community), has been contrasted to richness and promoted by proponents of systematic conservation planning (Sarkar 2002; Sarkar and Margules 2002; Margules and Sarkar 2007). Rarity, interpreted as endemism, along with richness has formed the basis for identifying biodiversity hotspots (Myers 1988; Myers et al. 2000).

   The main problem with these attempts, pointed out by critics such as Santana (2017), is that there seems to be no possible potential resolution of the disagreements between proponents of the different scientific definitions of biodiversity. Difficulties abound: for instance, even within ecology it has long been recognized that richness alone cannot be an adequate characterization of diversity because it does not take equitability into account (Sarkar 2007).[1]

   Efforts to decide between scientific definitions of biodiversity inevitably end up requiring the use of extra-scientific criteria. For instance, proponents of complementarity argue that its use is preferable to richness as a characterization of biodiversity because of the following argument: Consider three potential conservation areas, *A*, *B*, and *C* of which only two can be prioritized. Let *A* have

---

[1] Consider two ecological communities, *A* and *B*. Let *A* consist of 90 % species μ and 10 % species ν. Let *B* consist of 50 % species μ and 50 % species ν. Both *A* and *B* have richness 2 (assuming species are the relevant unit). Yet, there is a clear sense in which *B* is more diverse than *A*. Richness does not capture that sense.

richness 100, *B* have richness 90, and *C* have richness 50. If diversity is to be characterized as richness, the diversity ranking of the three areas would be *A* > *B* > *C* and choosing the best two would mean choosing *A* and *B*. However, suppose that *A* and *B* have 80 units in common. Then *A* and *B* together would contain 110 units. Now suppose that *A* and *C* have 30 units in common and *B* and *C* have 5 units in common. Then *A* and *C* would contain 120 units and *B* and *C* would contain 135 units. Thus the richness-based choice of *A* and *B* is the worst choice for biodiversity representation even if we use total richness as the relevant criterion for the biodiversity content of the prioritized set of conservation areas! This leads to the principle of complementarity (Vane-Wright et al. 1991; Sarkar 2012b): a new conservation area should be prioritized from the available ones on the basis of how many new units it adds to what is already present in those that have been chosen earlier.[2] The relevant point here is that the argument assumes that only two of the three potential conservation areas can be prioritized. Science does not supply this assumption. Its provenance is the existence of some resource constraint that must be respected.

Consider another choice: should richness or endemism or both be a component of biodiversity? Richness appears natural but, as seen earlier, its use is fraught with problems. How about endemism? We may opt for it out of concern for the rare and unusual. The point, though, is that these are no longer scientific claims. We have moved on to talk about values, what aspect of natural variety we deem most worthy of conservation, that is, there has been a transition to an analysis of norms. These cases are typical: extra-scientific considerations are necessary to adjudicate between conflicting scientific definitions of biodiversity.

2. *Eliminativism*: The failure of scientism in the definitional enterprise has led to one extreme response: proposals to eliminate the use of the term "biodiversity" altogether. Such a position has been forcefully argued by Morar et al. (2015) and Santana (2017). However, such a response would only become plausible if there is no other alternative to scientism. The rest of this paper argues that there is a plenitude of other available options. Suffice it here to note that banning "biodiversity" in current environmental discourse would be a daunting task and require efforts that, presumably, even eliminativists would accept as being better used to ensuring conservation in practice.[3]

3. *Deflationism*: Eliminativism as a response to the failure of scientism was preceded by a weaker strategy of deflationism. A strong form of deflationism was an assumption that, not only was there no fact of the matter about what biodiversity is, but that how it should be defined depends on local contexts, and can be gleaned by studying the practices of conservation biologists, for instance,

---

[2] Note that this choice does not guarantee that the total richness (that is, the number of unique species) would be maximized. In the example earlier, it would lead to the choice of *A* and *C* rather than *B* and *C*.

[3] For more details of these arguments, see Sarkar (2017a). Meinard, Coq, and Schmid, (Chap. 17, in this volume) give a different perspective on why eliminating "biodiversity" or even allowing it to remain irreducibly vague would lead to problems for the practice of conservation.

what is being optimized when areas are prioritized for conservation (Sarkar 2002; Sarkar and Margules 2002).

Strong deflationism was problematic for a variety of reasons, most notably perhaps because it seemed to leave no role for explicit discussion of how biodiversity should be defined, even in a given context. It was replaced by a weaker form in which normative discussion of what merits conservation determines what constitutes biodiversity (Sarkar 2008). But this takes us to normativism.

4. *Normativism*: Normativism will be developed in some detail in Sect. 18.3. What motivates this set of definitions is the recognition that the preservation of natural variety is a desirable social goal. For more than a generation, environmental ethicists have argued about the proper warrant for the admissibility of such a goal without reaching consensus (Norton 1987; Sarkar 2005) but, as environmental pragmatists have argued (e.g., Minteer and Manning 1999), these intractable foundational disputes are almost always beside the point in the practical contexts that determine how a conservation policy is formulated and whether it succeeds or fails. For environmental pragmatists, what is of paramount importance is achieving agreement on practical courses of action, shelving foundational disputes in favor of policy achievement. What matters in such contexts is to map, evaluate, and critically engage the values of legitimate decision-makers. These values are not determined by scientific inferences drawn from biological data though those data may—and should—inform the values of the decision-makers. What is critical is a community's vision of the future it desires including but not limited to its perception of its proper role in the natural world. Natural variety is one of those values and the one that is reflected in *biodiversity*; but biodiversity need not be the only natural value. Given this motivation, it remains to develop normativism more systematically. That discussion begins by moving beyond these assertions to arguments designed to establish that biodiversity must be a normative concept. In line with environmental pragmatism, there will be no further attention to foundationalist concerns in this paper.

18.3   Normativism

There are three loosely related arguments that aim to show why biodiversity must be a normative concept. To motivate these arguments consider what is perhaps the most general scientistic definition: biodiversity is the variety of life at all levels of structural, taxonomic, and functional organization. As Gaston (1996b) has documented, many biologists have defended similar definitions (e.g., McNeely et al. 1990; Wilson 1992; Johnson 1993). Is this what *biodiversity* means? If so, it does not seem plausible that biodiversity is the goal of conservation for at least two reasons: (1) There is the venerable ethical principle, *ought implies can*. Can all of biodiversity as defined above be conserved? Ecological communities left undisturbed lose species diversity through competitive exclusion. Evolving populations lose genetic (that is, allelic) diversity through natural selection. Conserving all such diversity is in practice impossible; (2) Is all such diversity in principle a desirable target of conservation? The human skin hosts thousands of microbial species though interpersonal variability is not as high as in the gut which hosts millions (Grice et al. 2009). Should we feel an imperative to conserve all the microbial diversity on the human skin or gut? Bacterial pathogens are rapidly evolving diversity to generate resistance in response to innovation in antibiotics designed to contain them. Other pathogens have shown similar, if less spectacular, responses to drugs. Should such diversity also merit active conservation?

The first argument for normativism begins with the assumption that concepts should be understood against the historical context of their introduction and use.[4] For biodiversity that context is the establishment and institutionalization of conservation biology as an academic discipline. As noted earlier, programs for conservation have always accepted the goal-orientation of the project, and the existence of that goal endows biodiversity with an irreducibly normative aspect. Proponents of conservation biology from the 1980s fundamentally disagree about goals with proponents of systematic conservation planning from the 2000s and, especially, the new conservation science from the 2010s (Kloor 2015) but they all agree with the goal-orientation of conservation. In most cultural contexts, pathogen variability is seen as removed from "biodiversity" with its attendant positive connotation.

The second argument builds on the first. As a result of the goal-orientation of conservation, biodiversity has always been used with a positive connotation. It consists of those aspects of biotic variety that should be conserved. That does not necessarily include all of natural variety. Though the rhetoric of contemporary political discourse often suggests otherwise, not all diversity is positive (Sarkar 2010). A society with extreme economic disparities is more diverse than one that is more egalitarian; but it certainly is not better. A population with both healthy and sick individuals is more diverse but less desirable than one that has only healthy individuals.

The third argument notes that, by the time "biodiversity" was introduced in the 1980s, there had been a generation-long tradition of defining and studying diversity within ecology (Sarkar 2007). Much of this work was spurred by a central theoretical hypothesis of ecology dating back to the 1950s, that diversity begets stability of ecosystems. While both the empirical and theoretical status of this claim continues to be debated today, by the mid-1980s its exploration had led to the formulation of a large variety of diversity (as well as stability) measures. These measures and the

---

[4]This claim is open to philosophical dispute: for instance, adherents of a hard distinction between the context of discovery and context of justification, etc., may deny this assumption (perhaps most famously developed by Mach in his study of physical concepts in the late nineteenth century). Those who view science through the lens of analytic metaphysics and study concepts through intuition and abstraction may also deny it. These issues will be left for another occasion. Suffice it here to note that core analytic methodologies of concept formation (for example, Carnapian explication) accept the relevance of the pragmatic context of conceptual innovation (Carnap 1950).

associated concepts they were supposed to quantify, in contrast to biodiversity, did not display normativity in their use. It is telling that this body of work was entirely ignored by conservation biologists attempting to define biodiversity in the 1990s and since. The most plausible interpretation of this lack of interest in the existing work on ecological diversity is that they viewed their own normative enterprise of designating aspects of natural variety for protection as distinct from these earlier ecological efforts. Thus, scientism was irrelevant to that enterprise. But, then, what requires explanation is why the explicit statements of definitions of biodiversity from biologists, as recorded by Gaston (1996b) and Takacs (1996), are almost always scientistic. Perhaps the explanation lies in the discomfort scientists often feel about explicit normative discussions—but this suggestive explanation is no more than sociological speculation at this point (but see Wolpe 2017).

18.3.1 Global Heritage

For biodiversity, who should set the relevant norms? In the present context this questions amount to asking who determines what aspects of natural variety should be protected. Here conservation efforts have been marred by serious ethical problems reflecting the structural inequities between the global North and the South. Conservation biology was first academically institutionalized in the United States and its agenda reflected the agenda of what has forcefully been criticized from the South as "radical American environmentalism" (Guha 1989). Soulé and his immediate followers had no hesitation in importing their values to the South, at one point arguing that the U.S. federal legal restrictions be circumvented to allow purchase of land for conservation in the South (Soulé and Kohm 1989): "Land acquisition is a very specific need … The National Science Foundation should view land purchase and maintenance in exactly the same way that it views the purchase of a piece of fancy machinery … *If there are legal barriers to direct acquisition of land in other countries by U.S. government agencies, then alternatives such as grants to such countries for the establishment and management of research reserves should be explored*" (p. 89; emphasis added). Available aid money would be better spent satisfying the desires of conservation biologists than, for instance, improving livelihoods of local people: "A potential funding source would be Public Law 480 programs which are currently operating in many developing countries" (p. 89).

If Soulé's strictures were imperialist proclamations, Janzen (1986) endorsed the missionary position when he urged: "If biologists want a tropics in which to biologize, they are going to have to buy it with care, energy, effort, strategy, tactics, time, and cash. Within the next 10–30 years (depending on where you are), whatever tropical nature has not become embedded in the cultural consciousness of local and distant societies will be obliterated.… We are the generation [that must] devote [its] life to activities that will bring the world to understand that tropical nature is an integral part of human life" (p. 306). Wilson (1992 and elsewhere) joined many others in declaring biodiversity to be a global heritage. The efforts of Northern conservationists were codified in various documents emerging from global agencies, most notably, the 1992 Rio Convention on Biodiversity.

But claims of global heritage require careful analysis and, when required, systematic deconstruction. Beyond bland assertion, what makes some natural feature or cultural artefact a *world* heritage? As we shall see there is no pat comfortable answer. Global heritage claims typically promote intervention by politically powerful external agents on decisions affecting the habitats of local residents who may have no interest in these global concerns. Moreover, these claims may not even be backed by any legitimate tangible material interest of these external agents—think of protecting a historical ruin just because of its age or a tropical rainforest because of its species richness.

The salience of these issues is borne out by looking at some particular cases: Was it wrong for the Taliban to destroy the Buddhas of Bamiyan? If so, why? And who decides? What gives the so-called international community—which is hardly a community of equals—a legitimate basis for questioning what a community in Afghanistan decides to do with some cultural artefacts present in its domain through no choice on its part? There is no reason to doubt that the strong feelings generated by the destruction of these statues probably reflects some defensible trans-cultural values. But what are they? How can they be spelt out and legitimized? How do these values serve the interest of the international community? Why do these interests override those of the local community? These questions have not received the attention they deserve. To return to the concern of this paper, turn to a biodiversity-related analog (Bevis 1995): Was it wrong for the Malaysians to log the lowland rainforests of Borneo? Why? And who decides? And so on. In this case there is an additional level of complexity. By and large, the local communities in Borneo were resistant to logging (Bevis 1995). The Malaysians opting for development were mainly economic and political elites from the mainland with the required power. The so-called international conservation community, largely activists from Europe and Australia, adopted and possibly manipulated the communities' concerns. But no one bothered to spell out whose heritage the great forests of Borneo were. And why. No matter how strongly we feel about these cases, the answers are not obvious.

Scholars have argued that concepts of heritage emerged in Europe in synchrony with the emergence of nation-states. Meskell (2014) puts it: "Intimately connected with the Enlightenment project, the formation of national identity relied on a coherent national heritage that might be marshaled to fend off the counter claims of other groups and nations" (p. 218). By the nineteenth century, in the late colonial context, the concept of heritage had begun to be applied across national boundaries, especially into the colonies. However, a concept of supranational cultural heritage only began to be formulated after World War I with tentative attempts at its legal codification originally under the auspices of the League of Nations (Boes 2013; Gfeller and Eisenberg 2016).

Full-fledged self-conscious efforts for global heritage designation and protection began with the post- World War II onset of the decolonization era and the formation of the United Nations Educational, Scientific and Cultural Organization (UNESCO) in 1945 (Gfeller and Eisenberg 2016). Claims and designations of global heritage emerged in tandem for both cultural artefacts and natural features. Arguably, especially through the Northern domination of UNESCO and other global agencies, they served to maintain Northern control of these entities in the post-colonial South even after decolonization had brought direct control to an end, for instance, when UNESCO's director Julian Huxley proposed setting aside large areas of central and east Africa as reserves (Huxley 1961; see Adams and McShane 1992 for a critique). (There will be other African examples below.) What is striking is that, beyond implicit appeals to claims of importance for some supranational group of individuals, no argument was advanced to codify why some feature is a global rather than, say, a national heritage; this is a problem that has only recently begun to receive attention (Di Giovine 2015). Instead of argument, attributions of global heritage status have systematically relied on bold assertions by proponents and demands for acquiescence on the part of those who may otherwise have resisted the globalization of their resources.

The first campaign to draw transnational attention to an ostensibly global heritage feature focused on Egypt, starting in the late 1950s, after President Nasser's modernization plan for the country included construction of the Aswan Dam. The project envisioned the submersion of a large number of historic sites and monuments of the Nile Valley, perhaps most notably the Great Temple of Ramses II at Abu Simbel. The plan generated vocal opposition from archaeologists and historians, mainly from Europe; their rhetoric suggested that Egyptians were not legitimate stakeholders in decisions about the fate of these sites (Boes 2013). Though the nationalization of the Suez Canal and his neutrality in the Cold War hardly made Nasser a popular figure in the West, conservationists were able to co-opt him to their campaign in the late 1950s. In 1960 UNESCO undertook an ambitious rescue project of relocating the monuments at risk to higher elevations. Nasser was applauded for recognizing a "right to heritage."

Parallel to the developments around Aswan, two German environmentalists, the father and son team of Bernhardt and Michael Grzimek initiated a global campaign for designating the Serengeti Plain of Tanganyika as a global heritage and "saving" it through formal protection and exclusion of local human use. The core component of their campaign was the creation of the documentary, *Serengeti Shall Not Die*, in which they explicitly and controversially drew an analogy between African wildlife and European historical monuments.[5] Immensely successful, the documentary transformed discussions of the global status of the natural heritage of the South. To continue with the Aswan parallel, shortly afterwards, and this time in India, conservationists from the North, supported by a local elite consisting largely of hunters, co-opted Prime Minister Indira Gandhi to launch Project Tiger in 1973 (Mountfort 1983) in spite of local problems due to tiger-human conflicts. There will be more on Project Tiger below.

---

[5] The German Filmbewertungsstelle Wiesbaden (FMW) dubbed this an "impermissible equation" (*unerlaubte Gleichsetzung*) and its request for the caption's removal captivated op-ed pages in the Federal Republic of Germany with discussions of censorship—see Boes (2013) for more detail.

The normative claims of conservation biology fall into this tradition and are based on the assumption that biodiversity is a global heritage. That is what makes it possible for Soulé to demand the acquisition of land in the South for the benefit of Northern conservationists. Janzen is gentler: he only wants to proselytize and convert the perceived heathens in the name of the global deity that is biodiversity. Indeed, it is commonplace for Northern conservationists to propose policies for distant lands in the South and to demand action (Dowie 2009).

For instance, in the 1980s the British parliament debated sending British troops to Kenya, Tanzania, and Mozambique to protect elephants (Neumann 2004). In the Central African Republic, in the 1990s, Bruce Hayes (a co-founder of the radical environmental organization, Earth First! in the United States), hired mercenaries to shoot at alleged poachers with no semblance of a trial, let alone a fair trial (Neumann 2004). Even when military threats are not used—unlike these African examples—economic power is often deployed against people living near or below the subsistence level if they do not conform to the demands of Northern conservationists (Dowie 2009).

To drive home the point being made, consider a hypothetical example originally constructed by Sarkar and Montoya (2011). Central Texas is home to a suite of endangered and endemic species including birds, salamanders, and arthropods (Beatley 1994; Beatley et al. 1995). In central Texas, attempts to list species under the U. S. Endangered Species Act (ESA), and then to delineate critical habitat and develop habitat conservation plans (as required by law) have long been controversial and have often led to ugly confrontations between landowners and conservationists (Mann and Plummer 1995). Now, imagine that an environmentalist from Mongolia decides to come to Texas, claim expertise on desert landscapes and cave ecology (perhaps justifiably), and demand that prime real estate around the capital city of Austin be converted into a national park. It is intriguing to speculate on the reactions from gun-toting Texans.

But, is there a salient ethical difference between this hypothetical situation and the one in which Oates (1999) (among others) demands more and better-policed national parks in west Africa? Or is it simply a question of power relations? From an ethical perspective, in both situations either we are denying the legitimacy of local sovereignty over resources or we are not. We are either accepting the legitimacy of local residents on the use of habitat or not. If we are forced to conclude that all that differentiates the two situations are power relations, Northern conservationism, as argued earlier, are continuing colonial attitudes and policies in the South (see, also, Guha 1997).

The critical normative issue here is that of parity. What one community—whether it be Northern conservationists or Mongolian desert experts—values should not be transferred without consent to the habitats of other communities. When we couple this normative claim with the realization that a definition of biodiversity is context-dependent in the sense that the valuation of biological resources varies over space (Escobar 1996), then we must turn to local values.

18.3.2   Local Values

Recall that normativism views biodiversity as consisting of entities that merit protection. What is most relevant to the present discussion is that, in practice, different groups have made different choices (Margules and Sarkar 2007). Let us begin with governmental agencies and the big non-governmental organizations (derisively dubbed "BINGOs" by Dowie (2009)) that dominate large-scale biodiversity conservation efforts.[6] In the United States, most governmental agencies adopt endangered and threatened species as biodiversity units but that is because much of conservation policy is set in the context of the legal requirements of the Endangered Species Act (ESA) of 1973. The ESA envisions protection of both animals and plants, includes subspecies under its purview, but excludes "pest" insects. In contrast, The Nature Conservancy (TNC), one of the best-known BINGOs, uses habitat types defined by characteristic ecological communities. Conservation International (CI), another BINGO, uses both globally threatened and geographically concentrated species.

Some such choice is necessary in order to provide the minimal precision required to devise conservation policy. Each of these choices reflects cultural values. For instance, US governmental agencies and CI implicitly presume that species are the bearers of value. Moreover, they implicitly presume that the extinction of every species that is admissible (excluding insect pest species) is equally (normatively) undesirable. TNC implicitly presumes that ecological communities are the bearers of value. The point is that these definitions embody cultural norms even though they are often presented as if they are universal and purely scientific definitions (Sarkar 2008).

Moreover, there are many other equally defensible choices. Sacred groves are widespread in South Asia, especially in the Western Ghats with evergreen wet forests and northeastern India, in the Eastern Himalayas. Forest communities of the Eastern Himalayas have maintained intact patches of cloud forest amidst an almost completely denuded landscape and have done so in spite of loss of most cultural associations with their sacred groves due to massive conversion of local populations to various Christian denominations starting in the mid-nineteenth century. In the state of Meghalaya, in many of these sacred groves not even deadwood can be removed.[7] The extant 29 sacred groves occupy over 25,000 ha. These are evergreen forests on a landscape dominated by limestone. Much of the ecology of the region continues to be devastated through coal mining and quarrying for limestone besides swidden farming that has an increasingly shorter cycle (five years now compared to 30 years in 1900). Traditionally each village had at least one sacred grove but local traditions were largely destroyed by the Christian missionaries. Not one of the sacred groves has been systematically inventoried except for major tree species; but they are known to be particularly rich in amphibian species that have a high degree

---

[6] For more detail and documentation of these examples, see Sarkar (2012b).

[7] Details are from Malhotra et al. (2007) and personal fieldwork.

of microendemism. At least 18 IUCN Red List amphibian species occur in this region. Cave invertebrates in the many caves and fissures under the ground have not been inventoried at all.

Some of the best-known sacred groves of Meghalaya are in the Khasi Hills near the town of Sohra (formerly known as Cherrapunji) which, with an average annual rainfall of 11,430 mm, is one of the wettest places on Earth. (The honor of being the wettest place in the world now belongs to nearby Mawsynram.) Most groves are small and occur on the top of hills but the larger ones also include valleys and the streams that run through them. The most impressive grove here is at Mawphlang which is protected because it is supposed to be inhabited by the spirit "U Basa." Its 80 ha contains at least 400 tree species; the fauna have never been inventoried. The protection regime (known as "Kw'Law Lyngdoh") is severe: Mawphlang is one of the sacred groves from which even deadwood removal is not permitted. The land around the grove is severely degraded.

The complete protection of entire ecological communities may be uncommon even though sacred groves occur throughout the South, especially in sub-Saharan African countries, most notably Ghana and Kenya. In most African countries, sacred groves target a single species or small set of species. Many cultures around the world value individual species in other ways (e.g., as totemic species) that may be of symbolic value or associated with religious practices. Some communities value entire forests. Vermuelen and Koziell (2002) report the case of the Irula hunter-gatherers, a semi-nomadic tribe from Tamil Nadu state of southern India. The tribe is well-known for its association with snakes, both in catching them and in treating snakebites. What this community values is reflected in how they choose a site for settlement. First, they assay a forest for medicinal plants, then snakes, then animals hunted for food or money (rats, rabbits, mongoose, wild cats, etc.). The assessment is complex. The size of animal populations matters and is assessed using the density of footprints. Ecological associations between vegetation type and animals are taken into account (for example, rabbits with *arugampul* (*Cynodon dactylon*), that is, Bermuda grass which, despite its name, originated in West Asia). Typically, in a twist opposite to conventional ecology, animals are taken as indicators for plants. The persistence of forests is critical to the survival of the Irula way of life.

However, this divergence of values need not lead to a vapid cultural relativism in which anything can count as biodiversity. We leave ample room for disagreement which may potentially be resolved: for instance, within a culture we may debate what we value most, whether we value every endangered species as much as we value selected endemic or charismatic ones (species of symbolic and other cultural value). Moreover, cultural values evolve and there can be crosscultural dialectics of engagement, disagreement, and change. Moreover, as we shall see in Sect. 18.5, we may adopt adequacy conditions that delimit which forms of valuing natural entities may count as valuing biodiversity. As an example, if we impose a condition that an adequate definition must value entities that cover a large portion of the taxonomic spectrum, valuing totemic species would not count as valuing biodiversity (Sarkar 2012b). These adequacy conditions will allow a partial synthesis of scientific insight and local values. But science will play a subsidiary role: even these adequacy conditions have to be culturally debated.

18.4 Problems with Local Normativism

Since a form of local normativism is being endorsed here, it behooves us to recognize and pay particular attention to potential problems. There are at least four of these.

18.4.1 Problems of Scale

The last section contrasted local values with global heritage claims about biodiversity. The designation "global" is clear enough in most contexts, referring to Earth as a whole. But "local" is far from clear: it could vary from a community defined by a municipality (or perhaps an even smaller spatial unit) to a nation-state. (Nation-states, in turn, can vary in size from the Vatican with a population of a few hundred or Lichtenstein with a few ten thousand, to China or India each with over a billion.) A few nation-states are ethnically almost homogeneous; while some cities alone embrace scores of culturally distinctive ethnic groups. Is there a natural scale at which biodiversity should be defined or at which conservation measures enacted? The former seems implausible and the latter, as we shall see below, is problematic.

As if to mimic this problem, biotic features that are typically held to merit protection also vary in spatial scale (or extent). In central Texas, microendemic salamanders sometimes have their range restricted to a single neighborhood of a city. The Barton Springs Salamander (*Eurycea sosorum*) and the Austin Blind Salamander (*Eurycea waterlooensis*) both have habitat limited to Barton Springs in the middle of the city of Austin. The Devil's Hole Pupfish (*Cyprinodon diabolis*) is endemic to a single cavern-like habitat in Nevada, United States, and has the smallest known habitat of a vertebrate species, just 0.008 ha (or 80 sq. m.) at the surface (Reed and Stockwell 2014). At the other spatial extreme, the endangered tiger (*Panthera tigris*) ranges from South Asia through Southeast Asia to Siberia (with a large gap at present, though not historically, in China) even after it has lost more than 90% of its habitat during the twentieth century. Earlier it was also present in parts of West Asia.

Different cultural concerns and values may be dominant at different spatial scales. In the case of the two salamander species just mentioned, the International Union for the Conservation of Nature and Natural Resources (IUCN) Red List, the global standard for risk designation, identifies them as "Vulnerable" but this designation is largely irrelevant to their future since the IUCN has negligible influence on conservation efforts in the United States. More pertinently, the United States Fish and Wildlife Service (USFWS) designates them as "Endangered" which affords them protection under the ESA. So does the state of Texas in its own assessment of risk for its native species. Most importantly, the protection of both salamander species has strong support within the city government of Austin and this support gets translated into actions by city agencies to maintain their habitat. The Barton Springs Salamander, in particular, is woven into the fabric of the city's cultural life. For those who view such endangered species as important components of biodiversity, this is a happy situation.

In contrast, the situation with the tiger is much more complicated. Globally, few species have dominated the consciousness of individuals for centuries as the tigers. About 70% of the world's tigers live in India (Gibbens 2017). At the national level, since the 1970s, tiger conservation has been a priority as exemplified by the 1973 launch of Project Tiger. Since 1972, the tiger has been India's National Animal (replacing the Asiatic Lion, *Panthera leo persica*, a subspecies of which the only extant population is also found in one state, Gujarat, in India). At the local level, conservation is not so simple. Tigers, as predators, often target cattle and other economically important domestic animals. They sometimes prey on humans, especially when habitat degradation and conversion, accompanied by a decrease in their non-human diet options, brings them into close contact with humans. In some tiger habitats, such as the mangrove swamps of the Sunderbans in eastern India and Bangladesh, tigers have long been positively embedded into local culture (Montgomery 1995). In many other tiger habitats in South Asia, human-tiger conflicts have led to local hostility (Gadgil and Guha 1995; Gibbens 2017).

For instance, between 2007 and 2014, in an area near the Chitwan National Park in south-central Nepal, local inhabitants intentionally killed four tigers (Dhungana et al. 2016). In India, local attitudes have been further confounded by the forced dislocation of tens of thousands of resident humans (though accurate numbers are hard to come by) during the process of the creation of Tiger Reserves under the auspices of Project Tiger (Sarkar 1999, 2005). It would come as no surprise that tiger conservation may not be welcome for communities living adjacent to tiger habitats. In fact, local resentment in India sometimes allows tiger poachers to hire local villagers to help them successfully evade anti-poaching efforts using local knowledge; there have even been acts of arson against parks and reserves by villagers adversely affected by their establishment under the aegis of Project Tiger (Gadgil and Guha 1995). Local values in many of these villages will likely not enshrine the protection of as hallowed a conservationist icon as the tiger in India. Returning to our definitional project, tigers would not necessarily be enshrined as a component of biodiversity. What is required are negotiations and tradeoffs between conservationists and victims of tiger depredation.

18.4.2 Conflicts Between Hierarchical Levels

The ambiguity of "local" shows the potential for conflicts between entities at different levels of the political (or cultural) hierarchy from communities through cities, districts, provinces, and the nation-state. These conflicts bear upon choices of a place embedded in different levels of this hierarchy. So, a locality is not only accountable to its community or city values, but also to those of the various regions of which it is a part including the nation-state that may well centralize the most relevant power for nature protection. Returning to the problem of tiger conservation in India, local communities suspicious of tiger conservation are typically pitted against conservationists at every other level of government.

The tiger case is hardly unique. In the late 1980s and early 1990s conservation efforts in central Texas were dominated by programs to protect multiple species besides the salamanders mentioned earlier. These included two migratory bird species, the Golden-cheeked Warbler (*Dendroica chrysoparia*) and the Black-capped Vireo (*Vireo atricapilla*) both of which were eventually declared as endangered by the United States Fish and Wildlife Service (Mann and Plummer 1995). Typically, such a declaration must be accompanied by the designation of "Critical Habitat" for the persistence of the species which imposes some limits on habitat use and transformation. Especially in the case of the Warbler, potential designation of Critical Habitat would have affected a wide swath of central and southern Texas. Opposition from ranchers was such that it is believed to have played a role in the defeat of incumbent Democrat Ann Richards to Republican George W. Bush in the gubernatorial election of 1994 (in spite of a promise by USFWS not to designate any Critical Habitat in a forlorn attempt to save the election for Richards). At the height of the conflict, ranchers explicitly promoted the decimation of endangered species. For these ranchers and much of rural Texas from where they came, these species would not form part of natural values that they would have chosen to protect. Yet, many of the same areas have a long history of private conservation of land and wild areas for a variety of reasons including game management for hunting.

18.4.3 Conflicts Between Localities

Conflicts occur not only across levels of a hierarchy in which a place may be embedded but between places across space. Returning to our well-worne case of tiger conservation, efforts at the national level throughout South and Southeast Asia were for a long time in conflict with China (where, perhaps, a few wild tigers persist) because of a demand for tiger body parts in a set of practices dubbed traditional Chinese medicine. In Southeast Asia many local communities (for instance, in

Borneo) value their forests which are viewed as cheap sources of timber in neighboring societies such as Japan (Bevis 1995). There are several species that are protected in their home range because they are perceived to be at risk but categorized as undesirable aliens elsewhere (Marchetti and Engstrom 2016).

It is not being suggested here that these conflicts—across geographical scales, within a hierarchy, or across localities—cannot be resolved. Resolution requires tradeoffs between different groups. Because the use of formal techniques for group decision leads to serious paradoxes (such as the Arrow's theorem—see Sarkar (2012b)), the preferred method for resolution requires deliberation, a process that has many other virtues in the resolution of environmental disputes (Norton 1994). However, there remains another problem, very similar to the conflict between places, but not quite identical; it requires cooperation, rather than tradeoffs, between communities across large geographical scales.

18.4.4   Conservation of Processes

When conservation efforts are directed towards landscapes and seascapes, their focus is typically on individual places (conservation areas), that is, culturally embedded areas with significant biodiversity content, though (as noted earlier) care must be taken to accommodate interactions between such localities. However, protecting places in isolation is rarely enough to ensure persistence of biodiversity. Persistence requires the maintenance of biophysical processes and these occur at multiple scales, from local wind-borne pollination and seed dispersal to ocean currents.

Processes themselves that can become the goal of conservation efforts include long-distance animal migrations. The spectacular 10,000-km migration of loggerhead turtles (*Caretta caretta*) between Baja California (Mexico) and Japan is well known (Shanker 2015). The annual migration of Monarch butterflies (*Danaus plexippus*) in North America is perhaps even more impressive. It is the longest insect migration known to science and the problems faced for its conservation exemplify the difficulties of conserving processes.

There are two North American migratory populations, one with habitat largely restricted to the west of the Rockies, mainly in California and adjoining states, and the other migrating from central Mexico to the north of the United States and southern Canada east of the Rockies. There are also several non-migratory populations in Florida, the Caribbean, Latin America, and elsewhere. (This means that an end to the migration phenomenon does not constitute the extinction of the species.)

The western population mainly winters in California but some insects do move further south through Arizona to Sonora in Mexico. During the Spring most individuals move to the north and east of California. The eastern population, once over a billion individuals, overwinters in a dozen or so high altitude oyamel fir forests in the Transvolcanic Belt of central Mexico, covering the trees like carpets.

All these winter roosts occur within a 100 × 100 km square (Brower and Aridjis 2013). In the spring, after a frenzy of mating, the insects fly north to Texas. Most females lay their eggs on Texas milkweeds, typically attached to the underside of a leaf with only one egg per plant. Most of the wintering generation dies.

The eggs hatch into caterpillars that feed exclusively on milkweeds, pupate, and emerge as adult butterflies. (In contrast, adults feed on the nectar of flowers of a wide range of plant species; milkweeds are no longer particularly important.) The new generation hatched in Texas continues the northward journey. The population fans out, covering much of the United States north of Texas and east of the Rockies. Some butterflies probably change course to turn south to Florida to add numbers to a local non-migratory monarch population found in that state. Most continue going north over a third and, sometimes, even a fourth generation. The northern limit of the migration spans the upper Midwest of the United States onwards to Ontario and the southern edge of Canada. Over these three or four generations the butterflies may travel up to almost 5000 km.

The return journey is even more impressive. The last generation produced in the north travels back to the tiny overwintering area in Mexico. The insects sip nectar for fuel along the way, and flying only by day while typically roosting in small groups for the night. How the insects manage to find their oyamel islands is still poorly understood. Each insect must have both a "map" and a "compass" (Agrawal 2017). Here a "map" means that the insect must know where it is: how the monarch does this remains an unsolved problem. Direction is set by a "time-compensated sun compass" by which each insect uses its internal circadian clock to sense the time of day and the position of the sun to orient itself in the correct direction. When the fall migration starts, the preferred direction is south. The compass is reset during the winter; in spring, the preferred direction becomes north.

For the last few decades, biologists have been warning that this process is endangered. (The species itself is not at risk because of the existence of many non-migratory populations.) Because the overwintering population in Mexico is the entire source of the entire northward migratory population in the spring (and, therefore, of the migratory phenomenon itself), trends in its size are directly relevant to the question whether the migration will persist in the future. These overwintering populations numbered 400 million individuals in the early 1990s but only 100 million yearly since 2010, with a historical low of about 35 million in 2013–2014 (Sarkar 2017b). What has caused this decline remains a matter of controversy.

There is some consensus the degradation and disappearance of the wintering habitat in Mexico has contributed to the migratory population decline. For the wintering habitat, Mexican authorities began systematic conservation efforts in 2000, and these now appear promising in spite of past problems (Víctor Sánchez-Cordero, personal communication). Beyond that, two conflicting hypotheses have been suggested though both could be operative. The *milkweed limitation hypothesis* predicts that spring monarch breeding populations before migration are in decline in the midwestern and northeastern United States and southern Canada. The alternative *migration survival hypothesis* proposes that the southward migratory population is suffering excessive mortality on its way south in Texas and northern Mexico (Sarkar 2017b).

If the milkweed limitation hypothesis is correct, conservation measures should be directed to milkweed restoration at the northern end of the migratory range, and many such efforts have been under way for more than a decade though, arguably, little to show in way of results. If the migration survival hypothesis is correct, efforts should be directed to providing food and shelter to the migrating population towards the southern end of the migration. If both are correct, both measures become important.

The salient point here is that maintaining the monarch migration will require collaboration across a continent-sized landscape. It is dissimilar to the case of conflicts between localities discussed earlier only because there is no potential for a solution through tradeoffs. Those who value monarch migration conservation as an important goal have a difficult task: what they are demanding is the value be attached to a process, not an entity, because the monarch as a species is not at risk of pending extinction.

18.5   A Synthetic Proposal

Where does all this leave us? Recall that, at the end of Sect. 18.3, it was noted that adequacy conditions can be adopted to constrain potential definitions of biodiversity based on local norms. It will be taken for granted that what is being targeted for protection is some aspect of nature (operationally distinguished from what are considered cultural features though, this distinction is not always trivial to maintain). The proposed constraints are intended to prevent all such natural targets of protection to be characterized as components of biodiversity, that is, what, elsewhere, I have called biodiversity constituents (Sarkar 2008, 2012b).[8] These adequacy conditions are necessary to distinguish biodiversity as a value from cases such as: what is valued is some magnificent geological formation, the desire to preserve pristine wildernesses[9], the protection of totemic species alone, the targeted protection of charismatic species such as large mammals in eastern and southern Africa, and so on. This is not to suggest that these are not important and culturally salient goals of conservation; biodiversity is not the only feature that deserves protection.[10]

More importantly, these adequacy conditions can be used to incorporate many, though not all, of the intuitions behind the many scientistic attempts to define biodiversity mentioned in Sect. 18.2. This claim will be elaborated below as the four conditions proposed here are discussed in detail. Suffice it here to know that such a

---

[8] Earlier in the literature, these were called "true surrogates" for biodiversity—see Sarkar (2002), Margules and Sarkar (2002), and Margules and Sarkar (2007).

[9] There are serious ethical problems with wilderness preservation (Woods 2001) but what is most important here is that the
     goal of wilderness preservation is not only distinct but also divergent from biodiversity conservation (Sarkar 1999).

[10] See Santana's contribution to this volume to find a similar claim in different terminology.

strategy allows a partial synthesis between the scientistic and normativist approaches, though only partial because only the intuitions behind the scientistic definitions rather than their specifics get incorporated into this strategy.

What requirements should we impose on potential biodiversity constituents? Here, four adequacy conditions will be proposed[11]:

1. *Constituent entities be biotic*: We are proposing a definition of *bio*diversity. This conditions dates back to Sarkar and Margules (2002). It allows biodiversity constituents to be habitat types, taxa, communities, genes, traits, and so on; but it excludes, for instance, physical environmental features such as rock formations or sand dunes. It also excludes human cultural diversity whether or not cultural diversity contributes to the presence or persistence of biodiversity in a given context.

    Nonbiotic features may well be good surrogates for the constituents in conservation planning. For instance, Sarkar et al. (2005) showed that sets of abiotic environmental classes are often adequate surrogates for varied classes of biota (the putative biodiversity constituents), while many authors have argued that sets of taxa are very rarely good surrogates for each other (Margules and Sarkar 2007) even though they continue to be used (Caro 2010). The success of environmental surrogates does not provide any argument that such abiotic features should be considered as components of biodiversity; rather, it shows that they are good *surrogates for biodiversity*.

2. *Emphasis must be on variability of the constituent set*: That is why it is bio*diversity*. The motivation for this criterion is best explained using an example. Neotropical rain forests have played an iconic role in conservationist campaigns since the mid-1980s, their public appeal perhaps best exemplified by Caufield's (1984) haunting account of their disappearance around the world. Yet, neotropical dry forests are far rarer and more threatened (due to ongoing land cover conversion) than rain forests. When neotropical rain forests, which are arguably over-protected in some regions such as Ecuador, are taken to be emblematic of biodiversity at the expense of neotropical dry forests to the extent of being the basis for a characterization of biodiversity, this condition is not met.

    For habitat types this means that attention should not be restricted to some subset and exclude all others entirely when biodiversity is defined. When dealing with taxonomic groups, this condition also suggests that differences at higher taxonomic levels than that of species are more salient than inter-specific differences. To put it another way, a species that is the sole member of a phylum (e.g., the aquatic species, *Trichoplax adhaerens*, the sole member of Placozoa[12])

---

[11] In my own work, these adequacy conditions have evolved over the years due to continued discussion in many forums—see, for example, Sarkar (2008, 2012b). Condition 4 is being proposed here for the first time.

[12] Note that there is some controversy over this uniqueness claim because some taxonomists feel that there is sufficient genetic diversity within this putative species to distinguish it into several morphogenetically very similar species (Voigt et al. 2004).

is more important for conservation than a species which belongs to a genus with thousands of species (e.g., any jewel beetle species of the genus Agrilus).

3. *Embrace taxonomic spread*: It is particularly important that the definition does not by fiat place arbitrary limitations on the taxa permitted to fall under the scope of "biodiversity." This requirement is probably not controversial. Part of the rhetoric of early conservation biology was that there was a need to move beyond charismatic species that had been the traditional foci of conservation campaigns and embrace the full spectrum of life as worthy of preservation. This rhetoric was often matched by the more concrete proposals that emerged from the field. Its sincerity is being accepted here.

   An important function of "biodiversity" was to codify this broadening of conservationist intent. It is arguable that not imposing some requirement that is functionally equivalent to the one being proposed here would miss the entire point of why the new term was enthusiastically adopted in the historical context of its introduction.

4. *Biodiversity constituents must be precise enough for their presence and abundance to be measured:* Within conservation biology in the 1990s, one of the motivations for defining biodiversity was to enable its measurement and quantification. For instance, Williams and Humphreys (1994) begin their discussion with two problems that have to be solved: (1) a relatively theoretical one—what is to be measured? and (2) a practical one—can the data "realistically" be collected? So, it seems reasonable to impose a measurability adequacy condition.

   Margules and Sarkar (2002) modified Williams and Humphreys' distinction to distinguish between a quantification problem and an estimation problem which together form what they called a biodiversity assessment problem. Solving the former requires the ability to measure biodiversity constituents *in principle*. That is what this condition requires. Solving the latter problem requires the operationalization of biodiversity for various purposes. For instance, in conservation planning, the detailed spatial distributions of thousands of biodiversity units are required as data. For many biodiversity constituents, obtaining such data, even though in principle possible, is not *in practice* reasonable given time and other resource constraints. What must then be found are adequate surrogates (such as the environmental classes discussed earlier) but these are not part of the definition of biodiversity.

It is instructive to analyze which sets of features survive this adequacy test and which ones do not. One standard approach, that biodiversity is all diversity at the level of genes, species, and ecosystems does not—it calls afoul of Condition 4. Sets of all at-risk species survive community; as do sets of habitat types (so long as they are defined, at least in part, using the ecological communities in them) though it is arguable that the first of these satisfies Conditions 2 and 3 only accidentally rather than as a matter of emphasis. (There is no deep reason why at-risk species—or other taxa—should be varied in their content or span much of the taxonomic hierarchy.)

These cases will probably come as no surprise to conservation biologists who embrace a scientistic attitude to biodiversity. In fact, they show how these adequacy criteria help bridge the gap between normativism and scientism. However, the adequacy conditions also admit non-standard collectives as potential constituent sets for biodiversity, for instance, the sacred groves of Meghalaya (India) discussed earlier. Conditions 1 is obviously satisfied. Condition 2 is satisfied because different kinds of forests present in the region can constitute sacred groves and, internally, they exhibit the variability of tropical cloud forests. Condition 3 is satisfied because each sacred grove is viewed as consisting of all biotic features within them. Condition 4 is satisfied because the number and type of sacred groves in any given collection is relatively easily assayed. If the earlier cases show that the adequacy conditions enable the relevance of scientific intuitions, this one shows how local normativism does not lose out. These conditions permit wide cultural divergences about what type of natural variety merits protection.

18.6   Concluding Remarks

The discussion of biodiversity in this paper has presumed the categoricity of its use in conservation biology and, more generally, biodiversity conservation. However, other areas of biology, in particular taxonomy, have also laid claim to the term over the years. How would the definitional strategy proposed here fare in these areas? Not very well, at least in the case of taxonomy. Taxonomy, by its own explicit goals, is fundamentally a descriptive enterprise; though its theoretical structure does embrace some normative issues, these are epistemological rather than axiological as seen, for instance, in the debates over cladistics (Platnick 1978). Normativism, as outlined here, is simply irrelevant to taxonomy though most taxonomists no doubt embrace many of the normative goals of biodiversity conservation.

How should we address the potential dissonance between the strategy for defining biodiversity presented here and the concerns of taxonomy? The answer given here will be cynical and based on sociological speculation that must be tested against data before the answer is deemed plausible. The speculations: Classical taxonomy had been underfunded since the dominance of molecular biology over the life sciences was established in the 1960s. By the time that conservation biology and "biodiversity" came along in the late 1980s, classical taxonomy based on macroscopic organismic rather than molecular traits, was a dying discipline. Taxonomists jumped on the biodiversity bandwagon when it became apparent that conservation was becoming a powerful current within and beyond the environmental movement. There was money for biodiversity inventory and conservation and, by endorsing that locution, taxonomists could lay claim to some of those resources.

To continue with the cynicism: Conservation biology was supposed to be a "crisis discipline" (Soulé 1985) because species were becoming extinct before biologists could even describe, let alone study, them. With respect to description, the problem was presented as a shortage of trained taxonomists available for that task.[13] The solution? More money for taxonomy. In Costa Rica, there were even moves to generate an army of sparsely-trained "parataxonomists," akin to China's barefoot doctors of the Cultural Revolution, with the task of inventory, producing lists of species at individual locations.

Taxonomy obviously does not place any taxonomic limit on what should be described: the more obscure or difficult a group of organisms, the more technical acuity could be deployed in their classification. From this perspective, the operative measure of biodiversity is species richness (or, possibly, richness at some higher taxonomic level). Success in taxonomy is determined in part by the sheer number of taxa that are successfully described. It is perhaps because they take the claims of taxonomists to be as pertinent as those of conservationists that philosophers such as Maclaurin and Sterelny (2008) embrace richness in their account of biodiversity. A major advantage of this approach is simplicity: richness is conceptually easy to grasp and relatively easy to measure in the field. But the earlier discussions in this paper should also underline the problems.

Why reject the salience of taxonomy? It is time to move beyond cynicism and speculation. The point is that the concept of richness was available to taxonomists long before the advent of "biodiversity." Not only did taxonomy not need the new concept, the neologism made no difference to the practice of taxonomy as a discipline. For taxonomists, "biodiversity" was a slogan, a source of resources for their field.

In contrast, conservation biologists required an operationalized concept of biodiversity to assess the extent to which any measure succeeds or fails because the conservation of biodiversity was the explicit goal of the field (Sarkar and Margules 2002). This is the *argument from necessity*. If we also accept that concepts are best understood in the context of their introduction and use, biodiversity must be understood in the context of conservation biology. But even if we do not endorse this argument from genesis, the argument from necessity makes conservation central to the meaning of biodiversity which, given this context, in turn requires a focus on norms and values for its definition.

Acknowledgements For comments on an earlier version, thanks are due to Elena Casetta, Davide Vecchi, and Jorge Marques da Silva.

References

Adams, J. S., & McShane, T. O. (1992). *The myth of wild Africa: Conservation without illusion.* Berkeley: University of California Press.

Agrawal, A. (2017). *Monarchs and milkweed: A migrating butterfly, a poisonous plant, and their remarkable story of coevolution.* Princeton: Princeton University Press.

Beatley, T. (1994). *Habitat conservation planning: Endangered species and urban growth.* Austin: University of Texas Press.

---

[13] Even as late as 2000, Wilson (2000) was making this claim.

Beatley, T., Fries, T. J., & Braun, D. (1995). The Balcones Canyonlands Conservation Plan: A regional, multi-species approach. In Porter, D. R. and Salvesen, D. A. Eds. *Collaborative planning for wetlands and wildlife: Issues and examples* (pp. 7592). Washington, DC: Island Press.

Bevis, W. W. (1995). *Borneo log: The struggle for Sarawak's forests*. Seattle: University of Washington Press.

Boes, T. (2013). Political animals: *Serengeti Shall Not Die* and the cultural heritage of mankind. *German Studies Review, 36*, 41–59.

Brower, L. P., & Aridjis, H. (2013). The winter of the monarch. *New York Time*, 15 March. http://www.nytimes.com/2013/03/16/opinion/the-dying-of-the-monarch-butterflies.html.   Last accessed 31 May 2017.

Callicott, J. B., Crowder, L. B., & Mumford, K. (1999). Current normative concepts in conservation. *Conservation Biology, 13*, 22–35.

Carnap, R. (1950). *Logical foundations of probability*. Chicago: University of Chicago Press.

Caro, T. (2010). *Conservation by proxy: Indicator, umbrella, keystone, flagship, and other surrogate species*. Washington, DC: Island Press.

Caufield, C. (1984). *In the Rainforest: Report from a Strange, Beautiful, Imperiled World*. Chicago: University of Chicago Press.

Dhungana, R., Savini, T., Karki, J. B., & Bumrungsri, S. (2016). Mitigating human-tiger conflict: An assessment of compensation payments and tiger removals in Chitwan National Park, Nepal. *Tropical Conservation Science, 9*, 776–787.

Di Giovine, M. A. (2015). Patrimonial ethics and the field of heritage production. In C. Gnecco & D. Lippert (Eds.), *Ethics and archaeological praxis* (pp. 201–227). New York: Springer.

Dowie, M. (2009). *Conservation refugees: The hundred-year conflict between global conservation and native peoples*. Cambridge, MA: MIT Press.

Escobar, A. (1996). Constructing nature: Elements for a poststructuralist political ecology. In R. Peet & M. Watts (Eds.), *Liberation ecologies: Environment, development, social movements* (pp. 46–68). London: Routledge.

Gadgil, M., & Guha, R. (1995). *Ecology and equity: The use and abuse of nature in contemporary India*. New Delhi: Penguin Books India.

Gaston, K. J. (1996a). Species richness: Measure and measurement. In K. J. Gaston (Ed.), *Biodiversity: A biology of numbers and difference* (pp. 77–113). Oxford: Blackwell.

Gaston, K. J. (1996b). What is biodiversity? In K. J. Gaston (Ed.), *Biodiversity: A biology of numbers and difference* (pp. 1–9). Oxford: Blackwell.

Gfeller, A. E., & Eisenberg, J. (2016). UNESCO and the shaping of global heritage. In P. Duedahl (Ed.), *A History of UNESCO* (pp. 279–299). London: Palgrave Macmillan UK.

Gibbens, S. (2017). Tiger crushed by excavator in horrific end to human-wildlife conflict. *National Geographic*.   http://news.nationalgeographic.com/2017/03/tiger-india-wildlife-human-conflict/. Last accessed 07 Oct 2017.

Grice, E. A., Kong, H. H., Conlan, S., Deming, C. B., Davis, J., Young, A. C., Bouffard, G. G., Blakesley, R. W., Murray, P. R., Green, E. D., & Turner, M. L. (2009). Topographical and temporal diversity of the human skin microbiome. *Science, 324*, 1190–1192.

Guha, R. (1989). Radical American environmentalism and wilderness preservation: A third world critique. *Environmental Ethics, 11*, 71–83.

Guha, R. (1997). The authoritarian biologist and the arrogance of anti-humanism: Wildlife conservation in the third world. *Ecologist, 27*, 14–20.

Huxley, J. S. (1961). *The conservation of wild life and natural habitats in central and east Africa: Report on a mission accomplished for UNESCO July-September 1960*. Paris: UNESCO.

Janzen, D. H. (1986). The future of tropical ecology. *Annual Review of Ecology and Systematics, 17*, 305–324.

Johnson, S. P. (1993). *The earth summit: The United Nations Conference on Environment and Development*. London: Graham and Trotman.

Kareiva, P., & Marvier, M. (2012). What is conservation science? *BioScience, 62*, 962–969.

Kloor, K. (2015). The battle for the soul of conservation science. *Issues in Science and Technology, 31*(2), 74–79.

Maclaurin, J., & Sterelny, K. (2008). *What is biodiversity?* Chicago: University of Chicago Press.

Malhotra, K. C., Gokhale, Y., Chatterjee, S., & Srivastava, S. (2007). *Sacred groves in India: An overview.* New Delhi: Aryan Books International.

Mann, C. C., & Plummer, M. L. (1995). *Noah's choice: The future of endangered species.* New York: Knopf.

Marchetti, M. P., & Engstrom, T. (2016). The conservation paradox of endangered and invasive species. *Conservation Biology, 30,* 434–437.

Margules, C. R., & Sarkar, S. (2007). *Systematic Conservation Planning.* Cambridge, UK: Cambridge University Press.

McNeely, J. A., Miller, K. R., Reid, W. V., Mittermeier, R. A., & Werner, T. B. (1990). *Conserving the world's biodiversity.* Washington, DC: International Union for Conservation of Nature and Natural Resources, World Resources Institute, Conservation International, World Wildlife Fund, and World Bank.

Minteer, B. A., & Manning, R. E. (1999). Pragmatism in environmental ethics: Democracy, pluralism, and the management of nature. *Environmental Ethics, 21,* 191–207.

Meskell, L. (2014). States of conservation: Protection, politics, and pacting within UNESCO's world heritage committee. *Anthropological Quarterly, 87,* 217–243.

Montgomery, S. (1995). *Spell of the tiger: The man-eaters of the Sunderbans.* Boston: Houghton Mifflin.

Morar, N., Toadvine, T., & Bohannan, B. J. M. (2015). Biodiversity at twenty-five years: Revolution or red herring? *Ethics, Policy & Environment, 18,* 16–29.

Mountfort, G. (1983). Project tiger: A review. *Oryx, 17,* 32–33.

Myers, N. (1988). Threatened biotas:"Hot spots" in tropical forests. *Environmentalist, 8,* 187–208.

Myers, N., Mittermeier, R. A., Mittermeier, C. G., Da Fonseca, G. A., & Kent, J. (2000). Biodiversity hotspots for conservation priorities. *Nature, 403,* 853–858.

Neumann, R. P. (2004). Moral and discursive geographies in the war for biodiversity in Africa. *Political Geography, 23,* 813–837.

Norton, B. G. (1987). *Why preserve natural variety?* Princeton: Princeton University Press.

Norton, B. G. (1994). *Toward unity among environmentalists.* New York: Oxford University Press.

Norton, B. G. (2008). Toward a policy-relevant definition of biodiversity. In G. D. Dreyer, G. R. Visgilio, & D. Whitelaw (Eds.), *Saving biological diversity* (pp. 11–20). Berlin: Springer.

Oates, J. F. (1999). *Myth and reality in the rain forest: How conservation strategies are failing in West Africa.* Berkeley, CA: University of California Press.

Platnick, N. I. (1978). Phylogenetic and cladistic hypotheses: A debate. *Systematic Zoology, 27,* 354–362.

Reed, J. M. & Stockwell, C. A. (2014). Evaluating an icon of population persistence: The Devils Hole pupfish. *Proceedings of the Royal Society of London B. 281*(1794), 20141648.

Santana, C. (2017). Biodiversity eliminativism. In J. Garson, A. Plutynski, & S. Sarkar (Eds.), *Routledge handbook of philosophy of biodiversity* (pp. 86–95). New York: Routledge.

Sarkar, S. (1999). Wilderness preservation and biodiversity conservation—keeping divergent goals distinct. *BioScience, 49,* 405–412.

Sarkar, S. (2002). Defining "biodiversity"; assessing biodiversity. *Monist, 85,* 131–155.

Sarkar, S. (2005). *Biodiversity and environmental philosophy: An introduction.* New York: Cambridge University Press.

Sarkar, S. (2007). From ecological diversity to biodiversity. In D. L. Hull & M. Ruse (Eds.), *Cambridge companion to the philosophy of biology* (pp. 388–409). Cambridge: Cambridge University Press.

Sarkar, S. (2008). Norms and the conservation of biodiversity. *Resonance, 13,* 627–637.

Sarkar, S. (2010). Diversity: A philosophical perspective. *Diversity, 2,* 127–141.

Sarkar, S. (2012a). Complementarity and the selection of nature reserves: Algorithms and the origins of conservation planning, 1980–1995. *Archive for History of Exact Sciences, 66,* 397–426.

Sarkar, S. (2012b). *Environmental philosophy: From theory to practice.* Malden: Wiley.

Sarkar, S. (2017a). Approaches to biodiversity. In J. Garson, A. Plutynski, & S. Sarkar (Eds.), *Routledge handbook of philosophy of biodiversity* (pp. 43–55). New York: Routledge.

Sarkar, S. (2017b). What is threatening monarchs? *BioScience, 67*, 1080.

Sarkar, S., Justus, J., Fuller, T., Kelley, C., Garson, J., & Mayfield, M. (2005). Effectiveness of environmental surrogates for the selection of conservation area networks. *Conservation Biology, 19*, 815–825.

Sarkar, S., & Margules, C. R. (2002). Operationalizing biodiversity for conservation planning. *Journal of Biosciences, 27*, S299–S308.

Sarkar, S., & Montoya, M. (2011). Beyond parks and reserves: The ethics and politics of conservation with a case study from Peru´. *Biological Conservation, 144*, 979–988.

Shanker, K. (2015). *From soup to superstar: The story of sea turtle conservation along the Indian coast*. Noida: Harper Litmus.

Soulé, M. E. (1985). What is conservation biology. *BioScience, 35*, 727–734.

Soulé, M. E., & Kohm, K. A. (1989). *Research priorities for conservation biology*. Washington, DC: Island Press.

Takacs, D. (1996). *The Idea of Biodiversity: Philosophies of Paradise*. Baltimore: Johns Hopkins University Press.

Vane-Wright, R. I., Humphries, C. J., & Williams, P. H. (1991). What to protect? Systematics and the agony of choice. *Biological Conservation, 55*, 235–254.

Vermeulen, S., & Koziell, I. (2002). *Integrating global and local values: A review of biodiversity assessment*. London: International Institute for Environment and Development.

Voigt, O., Collins, A. G., Pearse, V. B., Pearse, J. S., Ender, A., Hadrys, H., & Schierwater, B. (2004). Placozoa—no longer a phylum of one. *Current Biology, 14*, R944–R945.

Williams, P. H., & Humphries, C. J. (1994). Biodiversity, taxonomic relatedness, and endemism in conservation. In P. L. Forey, C. J. Humphries, & R. I. Vane-Wright (Eds.), *Systematics and conservation evaluation* (pp. 269–287). Oxford: Clarendon Press.

Wilson, E. O. (Ed.). (1988). *BioDiversity*. Washington, DC: (U.S.) National Academy Press.

Wilson, E. O. (1992). *The diversity of life*. New York: W. W. Norton.

Wilson, E. O. (2000). A global biodiversity map. *Science, 289*, 2279.

Wolpe, P. R. (2017). Why scientists avoid thinking about ethics. In S. J. Armstrong & R. G. Botzler (Eds.), *Animal ethics reader* (pp. 358–362). London: Routledge.

Woods, M. (2001). Wilderness. In D. E. Jamieson (Ed.), *A companion to environmental philosophy* (pp. 349–361). Oxford: Blackwell Publishers.

Chapter 19
Natural Diversity: How Taking the Bio- out of Biodiversity Aligns with Conservation Priorities

Carlos Santana

Abstract The concept of biodiversity, I argue, is poorly suited as an indicator of conservation value. An earlier concept, natural diversity, fits the role better. Natural diversity is broader than biodiversity not only in moving beyond taxonomic categories to encompass other patterns in the tapestry of life, but also in including abiotic, but valuable, aspects of nature. It encompasses, for instance, geological curiosities, natural entities of historical and cultural significance, and parts of nature with unique recreational and aesthetic value. It allows us to capture the idea of a diversity of ecosystem services, many of which are abiotic or have significant abiotic components. I make the case that refocusing conservation science around natural diversity retains many of benefits of using biodiversity as an indicator of value, while avoiding many of biodiversity's shortcomings. In particular, it provides a framework that highlights the conservation value of many biodiversity "coldspots," avoids the injustice of making conservation primarily the responsibility of the global south/developing world, and fits more neatly with the legal and ethical frameworks used to make conservation decisions in the public sphere.

Keywords Biodiversity · Ecosystem services · Environmental science

Summers in the Great Basin Desert of the Western United States are often intolerably hot and dry, but this is mitigated by the fact that in the Great Basin you're never far from a mountain range. One hot summer a decade or so ago, some friends and I escaped the heat by hiking up Mount Timpanogos, home of the only glacier in Utah. A highlight of the trek was taking a shortcut on the descent by sliding down the glacier. Years later I recounted this to a colleague who moved to Utah more recently,

C. Santana (✉)
University of Utah, Salt Lake City, UT, USA
e-mail: c.santana@utah.edu

© The Author(s) 2019
E. Casetta et al. (eds.), *From Assessing to Conserving Biodiversity*,
History, Philosophy and Theory of the Life Sciences 24,
https://doi.org/10.1007/978-3-030-10991-2_19

and she replied, "what glacier?" A quick online search revealed that the glacier has retreated to below the talus and hikers can no longer slide down it, one of many signs that our Great Basin environment is being reshaped by climate change.

Glaciers won't be the only climate losers in Utah. Decades earlier, famed environmentalist and author Edward Abbey wrote about ascending Tukuhnikivats, a mountain near Arches National Monument. He calls the mountain an "island in the desert" and flees to it in the heat of late August (Abbey 2011: 217). Scrambling up the talus below the summit late in his hike, he hears the whistles of pikas, a sound he equates with the experience of reaching the summit (*ibid*: 224). Pikas are rabbit-like creatures which in Utah live only in alpine and subalpine zones. For the pika, the mountain peaks are quite literally islands in the desert, and as the Earth warms and the tree line creeps upward, those islands will become submerged. Soon there may be no more pikas in Utah, and none of us will be able to relive Abbey's famous experience.

These losses due to climate change are obviously losses of some sort of value. I want to probe how we conceptualize that loss of value. In the case of the pika, one ready answer is that if the pika goes locally extinct, we will have lost biodiversity (in the form of species richness). That biodiversity loss encompasses and explains the various ways in which losing the pika is a loss of value, including the inability to relive Abbey's hike the way he experienced it. But what about the glacier which I slid down as a young man? Its loss is also a loss of value, and it's a shame that my present-day students can't recreate the experience I had when I hiked Timpanogos. The loss of the glacier is not a biodiversity loss, but it feels like a loss of much the same sort as the loss of the pika. This similarity suggests a need for a concept that encompasses both sorts of loss.

Moreover, we need a concept that better captures the way in which having alpine islands in the desert is valuable. It isn't merely the way the pikas and the alpine flora contribute to local biodiversity. By providing a contrast to the desert valleys below—in biodiversity, yes, but also in aesthetic experience, in ecosystem services provided, and even in temperature and humidity—the mountain peaks contribute to Utah's *natural diversity*. The loss of our last glacier is a loss of natural diversity, even if it isn't a loss in biodiversity,[1] and that same feature will be true of all sorts of changes in ecological value.

Biodiversity plays a central role in how we measure and discuss value in the conservation context, but it is an inadequate indicator of ecological value. The more inclusive concept of natural diversity avoids most of these shortcomings without bringing significant new baggage of its own. These reasons, I suggest, warrant placing natural diversity in the central conservation role that biodiversity currently occupies. At the very least, entertaining the concept of natural diversity can, by providing

---

[1] It is probable, of course, that the glacier contributes to the diversity of the microbiome. Suppose, however, that the microbial diversity persists under the talus but the glacier remains inaccessible to hikers. This is still a loss of value. Furthermore, even if some microbial diversity is lost, it is implausible that most of the loss of value is explained by the loss of the microbial diversity.

a useful contrast, clarify many of the issues in assessing and characterizing biodiversity that motivate this volume.

19.1 The Shortcomings of Biodiversity

Biodiversity occupies a central place in conservation science as an object of measurement, a basis for decisions in conservation planning, and as the primary conservation objective.[2] In occupying these roles, biodiversity is an organizing concept which focuses and unifies conservation research, and an umbrella concept (Lévêque 1994) which covers a broad array of conservation concerns. By playing this key role biodiversity serves as a representative of *ecological value*, the complete set of environmental goods of any sort—not only the intrinsic value of natural entities, but the economic, aesthetic, cultural, recreational, spiritual, and health amenities they provide (Santana 2016). Ecological value in this broad sense is the grounds for conservation efforts, so as Norton argues, the right definition of biodiversity is one "rich enough to capture all that we mean by, and value in, nature" (2006: 57). By conserving biodiversity, we aim to conserve ecological value of all kinds. In many ways biodiversity is well suited to this task. Ecological value of most kinds tends to depend on the living organisms that compose ecosystems, and thus on biodiversity. Conversely, since biodiversity relies on many ecological factors, including abiotic processes (Noss 1990), biodiversity conservation entails the conservation of other natural goods. Biodiversity can be operationalized in various useful ways, such as counts of richness (number of units), relative abundance, and measures of difference. It can also be assessed at biological levels of various sorts, such as genes, species, and ecosystems. This makes for flexible, scientifically meaningful, and computationally tractable measures that can feed into conservation planning (Sarkar and Margules 2002). Biodiversity also has an inclusive scope, allowing us to argue for the conservation of species and populations which might fall through the cracks were we to prioritize a different indicator of ecological value. For all these reasons, biodiversity makes sense as a conservation target.

The concept of biodiversity has come under scrutiny, however, in large part because of a sizeable gap between biodiversity value and ecological value construed broadly. It's easy, as a working conservationist, to lose sight of the connection between what's being measured and ecological value. For this reason, even some defenders of biodiversity worry, writing "we do not think that measurement strategies in conservation biology have been convincingly connected to wider theories that show the importance of the magnitudes measured" (Maclaurin and Sterelny 2008: 149). The assumption that our measurements of genes, species, higher taxa

---

[2] Biological diversity also plays important roles outside the context of conservation, such as an explanandum in ecology and evolutionary biology. This chapter isn't concerned with these roles, but more narrowly with biodiversity's central role in conservation.

and so on adequately represent those economic, cultural, aesthetic, and other values needs to be called into question.

Many researchers have done just that, calling into question how our measures of biodiversity relate to ecological value. Sarkar (2002; see also his 2008 and 2016), for instance, notes that the standard units of diversity in conservation biology fail to capture important units of ecological value, such as butterfly migration patterns. He argues that we should adopt a highly-flexible, open-ended concept of biodiversity. On Sarkar's picture, selecting the object of measurement (the *true surrogate*) when we assess biodiversity "is not an empirical question; rather it must be settled by convention" (2014: 5). Specifically, each local, conventional definition of biodiversity should be "based on normative considerations" that reflect the individual context and local cultural values (2014: 5–6). In this way, biodiversity measures can be tailored to account for, say, the cultural significance of a non-endangered species such as the Bald Eagle, which was never in danger of extinction but merited costly conservation efforts (2014). In its most extreme form, this deflationary, contextual account of biodiversity would give up on the content of the concept of biodiversity (i.e. that it is about the biota and about diversity) to allow it to encompass the whole range of ecological values. Sarkar's more recent work (2014, 2016) disavows this extreme stance, but the issues which motivated a more deflationary account of biodiversity remain.

Alternatively, Maier (2012) argues in detail that extant defenses of the value of biodiversity all commit fallacies, perhaps most significantly that of conflating the value of biodiversity with the value of other individual entities such as species and ecosystem processes. It is these natural entities that have value, and not the system-level property of being biodiverse. Motivated by similar issues, Santana (2014, 2016) argues that biodiversity is often a poor indicator of others sorts of ecological value. For example, cultural and aesthetic values often attach to places existing in a preferred state, even if that state has lower biodiversity than a possible alternative. Invasive plant species, for instance, can sometimes coexist alongside indigenous plants, meaning that invasions can increase biodiversity (Cleland et al. 2004). But we are still justified, because of our attachment to historical landscapes, in fighting benignly invasive species. To give another example, the units of biodiversity (species, phyla, genes, etc.) are, in measurement practice, interchangeable with other units of the same type. Consequently, biodiversity measures ignore the way in which some units differ significantly in value from others. They ignore, for instance, how a bat species which eats tons of disease-carrying mosquitos has higher ecological value than the mosquito species it eats.[3] Another set of critics, Morar et al., put the

---

[3] True, biodiversity measures can be sensitive to where organisms sit in the trophic network, not only through direct measures of trophic diversity, but also because the secondary effects of biodiversity loss are mediated by the structure of the trophic network (Dunne et al. 2002), and because trophic factors may regulate levels of species diversity (Terborgh 2015).

But my claim here isn't that biodiversity measures are insensitive to the importance of trophic roles to ecosystem function. I'm claiming that important normative considerations (e.g. malaria is value-negative) are largely invisible to biodiversity measures. Because malaria is value-negative, mosquito species which transmit malaria have less value, and bat species which prey on those

issue succinctly: "there are good reasons to doubt whether [biodiversity] provides any guidance for environmental decision-makers or has any clearly established relationship with those aspects of nature about which we care the most" (2015: 16–17).

In addition to worries about the gap between ecological value and biodiversity measures, Morar et al. accuse the concept of biodiversity as presenting a veneer of scientific objectivity while conservation scientists undemocratically impose their own environmental values on policy-making. According to this argument, the normativity of the biodiversity concept means that the policies supported by the techniques of conservation biology are value-laden. Whose values? The scientists' values, since they choose how to define and measure biodiversity. But this isn't transparent to society at large, thus the values of the broader public might not have an appropriate input to conservation decision-making. The focus on biodiversity conservation is, in effect, an injustice through technocracy. For this reason, Kareiva and Marvier (2012) suggest that conservation biology reframe itself as an interdisciplinary conservation science, one which uses social science to better measure anthropocentric ecological value. The original sin of conservation biology, they argue, was its "inattention to human well-being" (2012: 963). If biodiversity is at the heart of what matters, then "the vast majority of people are a threat" to ecological value, rather than among its beneficiaries (ibid). The focus on biodiversity conservation has thus unjustly relegated socially-oriented ecological values to the background.

Mismatch between biodiversity value and ecological value shows up in practice as well as theory, perhaps most notably in how biodiversity conservation is largely a burden on the "Global South." Although conservation biology emerged in wealthy industrialized countries (the "Global North"), biodiversity increases on a latitudinal gradient that peaks at the equator (Hillebrand 2004), meaning that most biodiversity is concentrated in the less-developed tropical and subtropical nations of the Global South. Likewise, the areas identified by conservation biologists as biodiversity hotspots—the places of highest conservation concern—are mostly in South America, Africa, South Asia, and tropical islands (Myers et al. 2000). We hear a lot about saving the rainforests and coral reefs, but not so much about how the American Midwest is no longer a place where "the buffalo roam and the deer and the antelope play," in the words of a nineteenth century folk song. As a result, conservation has focused much more on the Global South, and while biodiversity conservation and economic development are not always in competition (Tallis et al. 2008), there are almost always tradeoffs (McShane et al. 2011). Most importantly, a chief tool of conservation biology is setting aside protected areas (Rands et al. 2010; Miller et al. 2011), which often imposes significant burdens on local people and indigenous groups (Adams et al. 2004). For this reason, socially-oriented environmental researchers have often been at odds with conservation biologists and environmental philosophers who extoll biodiversity value (Miller et al. 2011). The focus on biodi-

---

mosquitos are more valuable. Bats and mosquitos thus exhibit a value differential, one that is explained by factors independent of their relative contributions to biodiversity, and so isn't easy represented by a biodiversity measure.

versity conservation has thus been a double injustice, placing unfair burdens on the Global South, and ignoring ecological values in the Global North and biodiversity "cold-spots" more generally.

19.2 Natural Diversity as an Alternative

Deconstruct biodiversity into its two components, *biological* and *diversity*. My proposal is that the chief virtues of treating biodiversity as the primary target of conservation come solely from the *diversity* component. Conversely most of the problematic features of biodiversity arise from the *biological* component, since it is the focus on organisms which excludes many sources of value. We should therefore aim to retain the benefits of the diversity component while mitigating the drawbacks of the biological component.

A good candidate for doing so would be to supplant the concept of biodiversity in conservation concepts with a diversity concept that extends beyond the biological, which I'll call *natural diversity*. The United Nations Convention on Biological Diversity defines biodiversity as "the variability among living organisms from all sources[4];" modeled on that definition, we can define natural diversity as 'the variability among natural entities from all sources'. Note that this is a departure from some previous usage of the term, which has sometimes equated natural diversity with what we would now call biodiversity (e.g. in Terborgh 1974, or as used in U.S. Fish & Wildlife Service directive *701 FW 1*). In the broader sense of natural diversity that I have specified, preserving natural diversity is a better conservation goal than preserving biodiversity because it retains the benefits of the diversity component, but moves beyond the biological component.

The chief virtues of biodiversity as an organizing concept in conservation are its flexibility and its inclusivity. What makes biodiversity so flexible is the number of ways we can operationalize it: as species richness, as complementarity, as functional diversity, and so on. As a broader concept, natural diversity could be operationalized even more flexibly, for instance in measures of abiotic soil components, as diversity of human experience of landscapes (measured through psychological or economic methods), or as measures of geological composition,[5] to give a few examples. Measures such as these might account for how the loss of a glacier is a loss of ecological value even if there is no corresponding loss of biodiversity. Natural diversity would thus be a more flexible conservation target.

It would also be more inclusive, for similar reasons. Natural diversity includes the diversity of living things, but also other forms of diversity. Lakes whose mineral

---

[4] https://www.cbd.int/convention/articles/default.shtml?a=cbd-02

[5] This might draw on extant conceptions of *geodiversity* (Kozłowski 2004; Gray 2004), or some alternative. Either way, because natural diversity includes cultural, historical, economic, and experiential components (among other things), it is more than just biodiversity supplemented with geodiversity.

content is too high to support the diversity of species other lakes support (e.g. Mono Lake or the Great Salt Lake) still contribute to natural diversity, partially in virtue of being inhospitable to most forms of biodiversity![6] A barren sandstone cliff which hosts few organisms, and thus contributes little to biodiversity, might support unique climbing routes and thus have high ecological value. The frigid Kola Peninsula in the Russian Arctic won't show up on any maps of biodiversity hotspots, but its unusual mineral assemblages are of great scientific interest. An area held to be uniquely sacred by an indigenous group is naturally diverse for that reason, even if it isn't biodiverse. A waterway which hosts no endangered species may provide a unique transportation corridor for the local population. If our aim is to conserve natural diversity, all these places will rank high in ecological value, but if biodiversity is our primary conservation goal, they might be ignored. Natural diversity is thus a more inclusive concept than biodiversity.

In addition to surpassing biodiversity in the virtues of flexibility and inclusiveness, natural diversity avoids some (but not all) of biodiversity's vices. Most importantly, as the examples in the last paragraph highlight, it is a better indicator of ecological value. Any natural entity of distinctive ecological value is a significant site of natural diversity in virtue of that distinctiveness. This includes entities of distinctive cultural, recreational, scientific, and economic value, even if those entities contribute little to biodiversity. The gap between natural diversity and ecological value is thus much smaller than the gap between biodiversity and ecological value. Some gap will remain, since some ecological value just doesn't fall under the rubric of diversity in any form, as Maier (2012) and Santana (2016) demonstrate. No single concept that is anything less than intolerably vague is likely to encompass everything of ecological value, however, and that natural diversity does better than biodiversity is a strong point in its favor.

Natural diversity also avoids the potentially unjust ramifications of using biodiversity as the central conservation concept. We have no reason to expect that natural diversity hotspots will cluster in the Global South or that any components of natural diversity besides biodiversity increase on a latitudinal gradient. If natural diversity is the more fundamental concept, conservationists should take the loss of visible stars in a European city to be a loss of conservation value commensurable with (though not necessarily equal to) the loss of an insect species in Madagascar. Consider the following contrast: the rowdy Canadian filmmakers who needlessly damaged unusual natural wonders in the U.S., such as the Bonneville Salt Flats and Yellowstone's Grand Prismatic Spring, got off with a small fine and are publicly called "good young men" (Penrod 2016). In South Africa, by contrast, impoverished hunters who kill endangered species are themselves killed by the hundreds and imprisoned by the thousands (Burleigh 2017). Although damaging a tract of salt flat may not be as morally significant as killing a sentient animal, there is clearly something inequitable in how those who harm natural diversity in the developed

---

[6] Uniquely inhospitable environments host unique organisms such as extremophiles, and so contribute to biodiversity as well, but (a) still have comparatively low biodiversity, and (b) their biodiversity value is the lesser part of their value.

world receive a mere slap on the wrist, while those who harm biodiversity in the developing world can pay with their lives. In terms of conservation value, both cases are serious losses of natural diversity and we should expend commensurate efforts to promote conservation in both cases. In fact, given that individual rhinos and elephants are more easily replaced than individual geological oddities, in terms of conservation values[7] the "good young men" may have caused more harm than any individual poacher. Obviously, few conservationists would endorse the murder of South African poachers. But a biodiversity-centered conservation framework does entail that the actions of the poachers are much more serious than the actions of the thoughtless filmmakers. A natural diversity framework, on the other hand, entails that the actions of both poacher and "good young man" are of a type: harm to natural diversity. The threats to natural diversity, like natural diversity itself, are thus globally distributed, and the burdens of conservation are correspondingly placed as much on the shoulders of relatively wealthy First-Worlders as much as they are on the backs of the Global South.

Another advantage of natural diversity as the central value concept in justifying protected areas is that legal frameworks for establishing conservation areas already appeal to something like it. At the very least, the law typically subordinates biodiversity to a broader class of values. In one chapter of her book *Imagining Extinction*, Heise examines how conservation law around the world is not preoccupied with "mainly a matter of counting how many species have been preserved or have died out," but more with fulfilling "the political, social, and cultural purposes to which it links the conservation of biodiversity" (2016: 91–92). For example, German law "protects endangered species for the sake of conserving culturally defined landscapes rather than habitats for the sake of species" (2016: 90). In other words, the law prioritizes a set of landscapes, which is diversity at the level of natural diversity, not biodiversity. In Bolivia, as Heise recounts, biodiversity conservation is legally situated as part of laws situating the Earth itself as a legal subject, and which invoke the cosmologies of indigenous people (2016: 114–116). Biodiversity conservation's ultimate justification in such a system is thus its contribution to a broader category, "the differentiation and variety of the beings that make Mother Earth" (2016: 116), which sounds more like natural diversity. Even in the United States, where the Endangered Species Act is so central to the environmentalist's toolkit, much, perhaps most, of the legal justification for setting aside protected areas comes from the American Antiquities Act of 1906 (Harmon et al. 2006), which justifies protection not of biodiversity, but of "objects of historic or scientific interest." These sorts of objects are cultural, archaeological, and geological features which may not fall under the rubric of biodiversity, but do contribute to natural diversity more broadly. Furthermore, the designation of national and state parks, which are some of the

---

[7] There are non-conservation values at play here as well: large mammals are sentient, and thus hedonistic values matter as well. But poaching is a bad mostly in terms of conservation value—if the suffering inflicted on the hunted animals was the chief concern, poaching would be no worse than the hunting of unprotected species, and much less immoral than eating factory-farmed meat. For present purposes, we thus have no grounds to say that the reckless filmmakers' crime, because they made no living thing suffer, was less serious than poaching.

most significant protected areas, is justified by the unique aesthetic and recreational opportunities they provide more than the species they protect. Across much of the globe, natural diversity better captures the systems of value behind the legal justifications for conservation, much of which either does not invoke biodiversity or places the value of biodiversity subordinate to a broader class of natural values.

The impulse to protect biodiversity is plausibly grounded in a reasonable aversion towards losing unique, rare, unusual, and distinctive parts of our world. In other words, we find real value in diversity. But this is true not just of biological organisms, but of unique, rare, unusual, and distinctive natural goods of all sorts—of natural diversity inclusive of, but going beyond, biodiversity. An ethnic group's ancestral homeland is unique in virtue of that fact, and merits preservation because in its uniqueness it represents natural diversity. A landscape that is a local rarity, like a glacier in Utah, is particularly valuable in virtue of that rarity. An unusual environment, such as undersea thermal vents, is scientifically valuable in part because it is unusual in abiotic as well as biotic composition. These sorts of natural values fall clearly under a notion of diversity, but not under *biological* diversity. The concept of natural diversity thus retains the conceptual benefits of biodiversity, while better fulfilling the role of capturing value in the natural world. Moreover, the political implications of conserving natural diversity are both more commonsensical and less unjust than a focus on conserving biodiversity primarily. Insofar as they are concepts competing for the same conceptual role, natural diversity thus has a clear advantage.

19.3 Operationalizability

Let's consider a possible disadvantage of natural diversity. Biodiversity serves not only as an indicator of ecological value, but as a measurement concept. Conservationists can operationalize biodiversity in various ways, estimate the amount of biodiversity in different areas, and use these estimates as inputs to conservation decision-making. Natural diversity, as a broader, less cohesive concept, might be more resistant to operationalization, and thus less useful as a measurement concept.

I don't think this is the case, in large part because biodiversity is itself extremely resistant to operationalization and measurement. Natural diversity comes down to meaning something close to "all of nature," and it seems almost nonsensical to try to argue that some areas have more nature than others. But as Sarkar insightfully points out, standard definitions of biodiversity equate it to "all of biology" (2002: 137) as well.[8] Morar et al. contend, with supporting citations to a dozen or so

---

[8] Many writers have observed that the term *biodiversity* is even more vague than "all of biology." Blandin suggests that *biodiversity* is just a new incarnation of *nature*, and has become "aussi indéfinissable que l'est la nature [as indefinable as nature]" (2014:51). Similarly, others have suggested that biodiversity is interpreted flexibly, with each interpreter using the vagueness of the term to

conservation biologists and philosophers, that "the widest consensus about biodiversity understood in this broad and all-inclusive sense is that it cannot, as a matter of principle, be quantified, due to its multidimensionality and the lack of commensurability and covariance among its components" (2015: 18). True, biodiversity can be made amenable to precise measurement, but only by ignoring most its components to focus on merely one or two at a time, such as species richness, genes, traits, habitat types and so on. But this gives biodiversity no advantage over natural diversity. While natural diversity, broadly construed, is utterly unquantifiable, there is nothing preventing picking a couple of dimensions at a time for particular purposes and measuring those. We could, for instance, in selecting conservation areas have measures of recreational usage, number of archaeological sites, and biological family richness, and use all three in tandem to determine what areas to prioritize. This leaves out much of natural diversity, but any practical measure of biodiversity is similarly limited. So, the breadth and vagueness of natural diversity is no different in kind than that of biodiversity.

We might worry, however, that the great flexibility available in selecting indicators of natural diversity will lead to inconsistent measures of natural diversity. Again, on this score natural diversity does no worse than biodiversity. With biodiversity measures, "there will always be some way of comparing (say) one wetland to another that will count the first as the more diverse, and another procedure that will reverse the result" (Maclaurin and Sterelny 2008: 133). This inconsistency across different methods of measurement may be a feature and not a bug, however. Sarkar (2002, 2008) argues that the ability to have different measures of biodiversity which yield different results allows us to fit our concept of biodiversity to the conservation priorities of each local situation, which differ from context to context. If this situational flexibility is a beneficial feature of biodiversity measures, then it would also be a feature of natural diversity measures. The difference, of course, would be even greater flexibility with natural diversity, and a greater ability to use measures which closely track non-biotic entities of ecological value. What might have seemed to be an objection to natural diversity—the sheer range of specific ways to quantify it—turns out to be a point in its favor.

---

read into whatever it is they value in nature. Takacs, for instance writes that "[i]n biodiversity, each of us finds a mirror for our most treasured natural images, our most fervent environmental concerns" (1996:81; cited in Morar et al. 2015). And Blondel observes that biodiversity is "coquille vide ou chacun met ce qu'il veut [an empty shell in which each person places whatever they want to see]" (1995: 225). (Thanks to Elena Casetta for help with the French references and translations).

19.4   More Than Ecosystem Diversity

The astute reader might be wondering whether the concept of natural diversity treads new ground. Doesn't *ecosystem diversity*, for instance, already cover the same territory? Ecosystem diversity is often treated as the one of the three main components of biodiversity, with species and genetic diversity being the others (McNeely et al. 1990). As the diversity of habitat types and community structure and composition (Sohier 2007), it seems that ecosystem diversity captures many of the abiotic entities and landscape-level features that I have argued require a move from bio- to natural diversity. I grant that, when taken seriously as a component of biodiversity, ecosystem diversity addresses some of the worries I have raised. But not most of them, since it only values abiotic entities and landscape features *qua* contributors to biotic activity, and not in terms of many other facets of ecological value, such as aesthetic or economic contributions. For this reason, I think ecosystem diversity still falls short.

19.5   A Natural Bridge

The concept of biodiversity, as I've discussed, has come under attack from several fronts. One reaction to these attacks could be to abandon it and move to some other means of representing and measuring ecological value. Leading alternatives are found in the social sciences, particularly economics, and focus on non-market valuation methods.[9] To abandon biodiversity for economic demand values would likely be to push the non-human biota too far into the background. It would also require us to give up much of the valuable research that conservation biologists have conducted, and leave behind useful tools they have developed. In addition, we would be abandoning a concept that has gained political and rhetorical importance. All these considerations imply that the cost of biodiversity eliminativism is very high.

Natural diversity, I am proposing, is a way to move beyond biodiversity in a way that avoids paying much of this cost. Natural diversity can be a bridge between traditional biodiversity-focused conservation and socially-focused environmental planning, because it includes both biodiversity and human-generated values under an umbrella concept. It suggests that, in principle, various sorts of ecological value are commensurable, and thus we can take research on biodiversity conservation and put it in conversation with other conservation strategies and goals. Of course, the extant concept *ecosystem services* (Millenium Ecosystem Assessment 2005) already tries to do this. Biodiversity and ecosystem services, however, are often an awkward fit. One way to include biodiversity in the ecosystem services framework is to just include biodiversity as a final ecosystem service (Mace et al. 2012). But this is ad

---

[9] For an introduction to these methods of environmental valuation, see Champ et al. (2003).

hoc, and doesn't really suggest in what way biodiversity is comparable to other services. In practice, this often means that valuations of ecosystem services, which are most easily quantified in economic terms, will rate the value of biodiversity quite low (Fromm 2000; Heinzerling 2016). Another means of trying to incorporate biodiversity in the ecosystem services approach is to argue that other services are reliant on biodiversity, but the relationship between biodiversity and ecosystem services is anything but straightforward (Srivastava and Vellend 2005; Mace et al. 2012). What natural diversity offers is an alternative, perhaps superior, means of accomplishing the same goal. Shoehorning biodiversity into the ecosystem services approach falls short because it is ad hoc and has no common standard of comparison. But a natural diversity approach would take the extant tools of biodiversity conservation planning and apply them to a broader set of ecological values, in a natural extension of existing conservation biology. In pitching natural diversity, I'm attempting to refocus our conservation thinking on those glaciers, and landscapes, and ancestral homelands, and other natural features that fall out of the conversation when we discuss biodiversity. But I'm raising the possibility of doing so in a way that is an organic expansion of biodiversity thinking and extant conservation practice, rather than leaving it behind as we embrace a broader set of ecological values.

References

Abbey, E. (2011). *Desert solitaire: A season in the wilderness*. Retrieved from http://ebookcentral. proquest.com

Adams, W. M., Aveling, R., Brockington, D., Dickson, B., Elliott, J., Hutton, J., et al. (2004). Biodiversity conservation and the eradication of poverty. *Science, 306*(5699), 1146–1149.

Blandin, P. (2014). « La diversité du vivant avant (et après) la biodiversité: repères historiques et épistémologiques ». In E. Casetta & J. Delord (Eds.), *La biodiversité en question. Enjeux philosophiques, éthiques et scientifiques*, Les Éditions Materiologiques (pp. 31–68).

Blondel J. (1995). *Biogéographie. Approche écologique et évolutive*, Masson

Burleigh, N. (2017, August 8). Kruger Park South Africa: Where black poachers are hunted as much as their prey. *Newsweek*. http://www.newsweek.com/2017/08/18/trophy-hunting-poachers-rhinos-south-africa-647410.html

Champ, P. A., Boyle, K. J., & Brown, T. C. (Eds.). (2003). *A primer on non- market valuation*. Boston, MA: Kluwer Academic Publishers.

Cleland, E. E., Smith, M. D., Andelman, S. J., Bowles, C., Carney, K. M., Claire Horner-Devine, M., ... & Vandermast, D. B. (2004). Invasion in space and time: Non-native species richness and relative abundance respond to interannual variation in productivity and diversity. *Ecology Letters, 7*(10), 947–957.

Dunne, J. A., Williams, R. J., & Martinez, N. D. (2002). Network structure and biodiversity loss in food webs: Robustness increases with connectance. *Ecology Letters, 5*(4), 558–567.

Fromm, O. (2000). Ecological structure and functions of biodiversity as elements of its total economic value. *Environmental and Resource Economics, 16*(3), 303–328.

Gray, M. (2004). *Geodiversity: Valuing and conserving abiotic nature*. Chichester: Wiley.

Harmon, D., McManamon, F. P., & Pitcaithley, D. T. (2006, January). The antiquities act: The first hundred years of a landmark law. *The George Wright Forum, 23*(1), 5–27. George Wright Society.

Heinzerling, L. (2016). Economizing on nature's bounty. In J. Garson, A. Plutynski, & S. Sarkar (Eds.), *The Routledge handbook of philosophy of biodiversity*. London: Taylor & Francis.

Heise, U. K. (2016). *Imagining extinction: The cultural meanings of endangered species*. Chicago: University of Chicago Press.

Hillebrand, H. (2004). On the generality of the latitudinal diversity gradient. *The American Naturalist, 163*(2), 192–211.

Kareiva, P., & Marvier, M. (2012). What is conservation science? *BioScience, 62*(11), 962–969.

Kozłowski, S. (2004). Geodiversity. The concept and scope of geodiversity. *Przegląd Geologiczny, 52*(8/2), 833–883.

Lévêque, C. (1994). Le concept de biodiversité: de nouveaux regards sur la nature. *Natures Sciences Sociétés, 2*(3), 243–254.

Mace, G. M., Norris, K., & Fitter, A. H. (2012). Biodiversity and ecosystem services: A multilayered relationship. *Trends in Ecology & Evolution, 27*(1), 19–26.

Maclaurin, J., & Sterelny, K. (2008). *What is biodiversity?* Chicago: University of Chicago Press.

Maier, D. S. (2012). *What's so good about biodiversity?: A call for better reasoning about nature's value* (Vol. 19). Dordrecht: Springer.

McNeely, J. A., Miller, K. R., Reid, W. V., Mittermeier, R. A., & Werner, T. B. (1990). *Conserving the world's biological diversity*. Washington, DC: IUCN, World Resources Institute, Conservation International, WWFUS and the World Bank.

McShane, T. O., Hirsch, P. D., Trung, T. C., Songorwa, A. N., Kinzig, A., Monteferri, B., et al. (2011). Hard choices: Making trade-offs between biodiversity conservation and human well-being. *Biological Conservation, 144*(3), 966–972.

Millennium ecosystem assessment. (2005). *Ecosystems and human wellbeing: A framework for assessment*. Washington, DC: Island Press.

Miller, T. R., Minteer, B. A., & Malan, L. C. (2011). The new conservation debate: The view from practical ethics. *Biological Conservation, 144*(3), 948–957.

Morar, N., Toadvine, T., & Bohannan, B. J. (2015). Biodiversity at twenty-five years: Revolution or red herring? *Ethics, Policy & Environment, 18*(1), 16–29.

Myers, N., Mittermeier, R. A., Mittermeier, C. G., Da Fonseca, G. A., & Kent, J. (2000). Biodiversity hotspots for conservation priorities. *Nature, 403*(6772), 853.

Norton, B. (2006). Toward a policy-relevant definition of biodiversity. In *The endangered species act at thirty* (Vol. 2, pp. 49–58).

Noss, R. F. (1990). Indicators for monitoring biodiversity: A hierarchical approach. *Conservation Biology, 4*(4), 355–364.

Penrod, E. (2016, November 4). Canadian will pay fine for wakeboarding on Utah's Bonneville salt flats. *The Salt Lake tribune*. http://archive.sltrib.com/article.php?id=4546578&itype=CM SID

Rands, M. R., Adams, W. M., Bennun, L., Butchart, S. H., Clements, A., Coomes, D., et al. (2010). Biodiversity conservation: Challenges beyond 2010. *Science, 329*(5997), 1298–1303.

Santana, C. (2014). Save the planet: Eliminate biodiversity. *Biology & Philosophy, 29*(6), 761–780.

Santana, C. (2016). Biodiversity eliminativism. In *The Routledge handbook of philosophy of biodiversity* (pp. 100–109).

Sarkar, S. (2002). Defining "biodiversity"; Assessing biodiversity. *The Monist, 85*(1), 131–155.

Sarkar, S. (2008). Norms and the conservation of biodiversity. *Resonance: Journal of Science Education*, (7), 13.

Sarkar, S. (2014). Biodiversity and systematic conservation planning for the twenty-first century: A philosophical perspective. *Conservation Science, 2*(1).

Sarkar, S. (2016). Approaches to biodiversity. *The Routledge handbook of philosophy of biodiversity.*

Sarkar, S., & Margules, C. (2002). Operationalizing biodiversity for conservation planning. *Journal of Biosciences, 27*(4), 299–308.

Sohier, C. (2007). *Ecosystem diversity.* Available from http://www.coastalwiki.org/wiki/Ecosystem_diversity. Accessed on 12 Nov 2017.

Srivastava, D. S., & Vellend, M. (2005). Biodiversity-ecosystem function research: Is it relevant to conservation? *Annual Review of Ecology Evolution and Systematics, 36*, 267–294.

Takacs, D. (1996). *The idea of biodiversity. Philosophies of paradise.* Baltimore: Johns Hopkins University Press.

Tallis, H., Kareiva, P., Marvier, M., & Chang, A. (2008). An ecosystem services framework to support both practical conservation and economic development. *Proceedings of the National Academy of Sciences, 105*(28), 9457–9464.

Terborgh, J. (1974). Preservation of natural diversity: The problem of extinction prone species. *BioScience, 24*(12), 715–722.

Terborgh, J. W. (2015). Toward a trophic theory of species diversity. *Proceedings of the National Academy of Sciences, 112*(37), 11415–11422.

Chapter 20
Ordinary Biodiversity. The Case of Food

Andrea Borghini

Abstract The green revolution, the biotech revolution, and other major changes in food production, distribution, and consumption have deeply subverted the relationship between humans and food. Such a drastic rupture is forcing a rethinking of that relationship and a careful consideration of which items we shall preserve and why. This essay aims at introducing a philosophical frame for assessing the biodiversity of that portion of the living realm that I call *the edible environment*. With such expression I intend not simply those plants and animals (including in this category, henceforth, also fish and insects) that were domesticated for human consumption, but also the thousands of species that are regularly consumed by some human population and that are regarded to some degree as wild. The visceral, existential, and identity-related relationship that link humans with the edible environment can be regarded as *sui generis* and can constitute a ground for explaining why it should receive a preferential treatment when it comes to preservation, propagation, and development. First of all, I discuss whether we should draw a sharp divide, when it comes to preservation efforts, between *wild* and *domesticated* species (§1); secondly, I assess whether to draw a sharp divide between *natural* and *unnatural* entities, when it comes to measurements and interventions regarding the edible environment (§2); finally, I ask what is the value of biodiversity as far as food is concerned, and how best to preserve and foster it (§3 and §4). The closing section draws some suggestions for future investigations and interventions.

Keywords Food biodiversity · Wild foods · Natural foods · Food ontology

A. Borghini (✉)
Department of Philosophy, University of Milan, Milan, Italy
e-mail: andrea.borghini@unimi.it

E. Casetta et al. (eds.), *From Assessing to Conserving Biodiversity*,
History, Philosophy and Theory of the Life Sciences 24,
https://doi.org/10.1007/978-3-030-10991-2_20

20.1 Introduction

The concept of biodiversity is rather unquestionably associated with the idea of untamed forms of life, living entities, or parts of living entities, which developed on Earth independently of or prior to humans. Far more controversial, instead, is whether biodiversity measurements and interventions should take into account also forms of life and living entities that have been to some degree influenced by human activities (cfr. Siipi 2016, section 1, for a comparison of the opposing camps.) Some authors lean towards a very inclusive notion of biodiversity, which virtually leaves out no (actual or possible) form of life, living entity or any part of a living entity. But, no matter where one wishes to draw the line, it seems unfeasible to have a notion of biodiversity that excludes all those entities that have been *in some way or other* influenced by humans. Not only would it currently appear unfeasible to insist on protecting only those entities that are untamed; more importantly, any effort of preservation or development of such entities would by itself undermine their being in some way or other independent of human existence. Hence, any account of biodiversity seems bound to address the following two questions:

(1) Are there living entities that should be excluded from measurements of biodiversity as well as from efforts of conservation[1]?
(2) Should the criteria for inclusion in a measurement or intervention be context-dependent or context-independent? For instance, could different criteria be selected depending on circumstances?

Another important outcome of the literature on the concept of biodiversity is that, at a closer look, most accounts of biodiversity reveal a preference towards more familiar forms of life.[2] Thus, for instance, preserving the existence of pandas seems a much more important goal than preserving the existence of any species of mollusks that inhabits some remote marine areas, no matter how important such mollusks may be to a certain ecosystem; or, consider the little attention that the preservation of bacteria has received in comparison to animals or plants, which can arguably only in part be justified by the taxonomic challenge of classifying bacteria. It is important to reflect on the reasons that might have supported such preference of certain forms of life over others; are those *good* reasons, that is, reasons that justify keeping such preferences in our accounts of biodiversity? Or, are such preferences biases, which cannot be justified? Hence, the following additional question for any account of biodiversity:

---

[1] In this paper I shall refer to *conservation*, rather than *preservation*, efforts. I do not have in mind such a sharp distinction between the two notions, as established in the classical dispute between Gifford Pinchot and John Muir; at the same time, it seems most appropriate to speak of conservation of ordinary biodiversity, rather than its preservation, because of the active human role not only in establishing and maintaining it, but also in exploiting it for the purposes of – among others – dieting, pleasure, research, and profit.

[2] See Marques da Silva & Casetta, Chap. 9, in this volume.

(3) What are the grounds for preferring certain (possibly more familiar) forms of life over others? For instance, do such preferences rest on efficiency, or perhaps on some emotional or spiritual connection?

In this essay I aim to address the three questions just raised from a particular angle, which has thus far received relatively sparse attention from philosophers. I aim to analyze the value of biodiversity when it comes to that portion of the living realm that I call *the edible environment*. With "edible environment" I intend not simply those plants and animals (including in this category, henceforth, also fish, insects, mushrooms, and some algae) that were domesticated for human consumption, but also the thousands of species that are regularly consumed by some human population and that are regarded to some degree as wild. The edible environment constitutes a particularly significant point of entry into the preferential attitudes that humans bear towards different forms of life. The visceral, existential, and identitarian relationship that humans bear with the edible environment can be regarded as *sui generis* and, as we shall see, can constitute a ground for answering question (3), that is, why the edible environment should receive a preferential treatment when it comes to preservation, propagation, and development. The edible environment is also an intuitive entry point into questions (1) and (2). Are there edible (parts of) living entities that ought not to be included in measurements of biodiversity (e.g., GMOs)? Should measurements and preservation efforts be contextual; for instance, should they tend to clearly demarcate between biodiversity of the edible environment and other forms of biodiversity? Are the criteria employed to account for the biodiversity of the edible environment specific to it? Are they consistent across the board?

In order to investigate questions (1)–(3), in what follows I will take up a number of issues that cut across them. First of all, the issue of whether we should draw a sharp divide, when it comes to conservation efforts, between *wild* and *domesticated* species. I address this in §1. Secondly, we should assess whether to draw a sharp divide between *natural* and *unnatural* entities, when it comes to measurements and interventions regarding the edible environment; this issue will be at the center of §2. Finally, we should ask what is the value of biodiverse foods and how best to preserve and foster it; these two issues will occupy sections §3 and §4, respectively. In §5 I shall return to questions (1)–(3) and suggest some answers.

20.2 Wild and Domesticated Foods

Today the food for sale at any supermarket, deli, or food market in an agriculturally industrialized country such as the United States, Holland, or Japan is a testimony to two kinds of success stories. The first is the story of human attempt to tweak the edible environment to serve human nutritional, economic, and social purposes; call this *the conquer and divide story*. There are a few remarkable facts about the diversity of domesticated species, which reveal the importance of looking at taxonomic levels below species when it comes to domesticated plants and animals (cfr. Especially Diamond 2002: 702). Only 14 out of 148 large terrestrial mammalian were domesticated, and only about 100 plants out of 200,000 candidates. Any of those species is in itself a remarkable success story, featuring the rise of an astonishing number of varieties[3]: – e.g. over 40,000 varieties of beans, over 10,000 varieties of tomatoes, over 8000 varieties of apples, and circa 8000 breeds of animals (for a concise and up to date overview of the diversity of animals that humans consume, cfr. Chemnitz and Becheva 2014: 22–25). But, the first success story tells also of the many ways in which humans managed to cooperate with microscopic organisms such as bacteria, yeasts, and fungi, to preserve, modify, create key staples, including cheese, yoghurt, beer, wine, vinegar, chocolate, coffee, whisky, and hundreds more.

The second, more recent, success story tells of the increasing connection of food production and distribution systems worldwide; call this *the food revolution story*. Characteristic of it is the decline or extinction of thousands of varieties and breeds produced throughout the long path to domestication. For instance, in 2012 the FAO update on the state of livestock biodiversity estimated that circa 2000 of the 8000 animal breeds are at risk of extinction or nearly extinct. Or, to make two examples regarding plants, of the 287 varieties of carrots that humans devised, only 21 are still cultivated; and of the 8000 varieties of apples that we have a trace of, only 800 are still cultivated (cfr. Fromartz 2006, Chapter 1 and Pollan 2001, Chapter 1) If we look at the broad picture, data from The Food and Agriculture Organization of the United Nations (FAO-UN) indicate that since the beginning of last century 75 percent of plant genetic diversity has been lost. To offer some additional examples, "at least one breed of traditional livestock dies out every week in the global context; of the 3831 breeds of cattle, water buffalo, goats, pigs, sheep, horses and donkeys believed to have in this twentieth century, 16% have become extinct and 15% are rare; some 474 of livestock breeds can be regarded as rare, and about 617 have become extinct since 1892" (*Conservation and Sustainable Use of Agricultural Biodiversity* 2003, paper 3, p. 23). Also, over 97% of the varieties of foods sold in 1900 in the United States had disappeared from the market by 1983 (Cfr. Fromartz 2006, Chapter 1). The shrinkage of the number of varieties is principally due to the increased integration of food markets, controlled by fewer and fewer actors at the origin and during distribution, as well as by a growing syncretism and homogeneity within diets across the planet. Within a globalized food market, only a few varieties

---

[3] Some reader may wonder why the data presented in this section regard *varieties* rather than *cultivars*. Although the two concepts are at times used interchangeably (cfr. ICNCP 2009, Chapter 2, Art. 2.2), the technical usage of 'cultivar' picks out a more restrictive taxonomic notion, based on three principles: (i) possession of a distinctive character; (ii) uniformity and stability of such character; (iii) heritability of said character (ICNCP 2009, Chapter 2, Art. 2.3). At the same time, though, "no assemblage of plants can be regarded as a cultivar or Group until its category, name, and circumscription has been published" (ICNCP 2009, Chapter 2, Art. 9, Note 1); yet, many extant and past varieties, that may suitably comprise a cultivar, were never inventoried; thus, in a discussion of the biodiversity of edible plants, it seems most suitable to at least start off by considering all varieties, and then possibly refine the domain by considering the stricter notion of cultivar.

per species tend to be favored, namely those varieties that deliver an economic advantage such as production cost, shelf life, or consumers' appeal.

Much of the discussion concerning food biodiversity has indeed focused on either the conquer and divide story or the food revolution story. It is hard to overestimate the importance of the first story to human evolution. There are still countless details of the processes of domestication of each animal and plant that await to be uncovered, which will shed light over the economic, medical, social, political, and cultural history of humanity (cfr. Wrangham 2009). Equally important is the astonishing shift in food production and consumption, which occurred since the advent of synthetic fertilizers and, more recently, biotechnologies. In the past century, nearly all varieties on the market have been replaced. This leaves us with two major interrogatives: to what extent biodiversity efforts should focus on preserving ancient varieties, and to what extent measurements of biodiversity within the edible environment should include cultivars created by means of techniques such as lab cloning and genetic modification.

As important as they may be, the conquer and divide story and the food revolution story are far from portraying a comprehensive picture of the biodiversity of the edible environment. Indeed, the two stories leave out so-called 'wild' organisms (which can be counted not only in terms of cultivars, varieties or races, but also higher taxa such as species and families), which not only comprise a very significant portion of human diets, but also reveal a continuum between the discussions of prototypical biodiversity conservation targets (e.g. hot spots and endangered species) and conservation targets within the edible environment. By aggregating a number of studies, ethnobotanists estimated that humans have fed themselves off of over 7000 species (Grivetti and Ogle 2000; MEA 2005). Looking at 36 studies in 22 countries of Asia and Africa, Bharucha and Pretty (2010: 2918) estimated that "the mean use of wild foods (discounting country- or continent-wide aggregates) is 90–100 species per place and community group. Individual country estimates can reach 300–800 species (India, Ethiopia, Kenya)." Most importantly to our purposes, in nearly all countries across the globe, with the notable exception of United States, wild species and domesticated species are tended and consumed jointly, and in a number of occasions they are also jointly marketed. To many farmers, the distinction between domesticated and wild species is, indeed, of little significance. At the outset of their paper, Bharucha and Pretty (2010) report the words of a woman farmer, interviewed in Mazhar et al. (2007: 18), who exclaims: "What do you mean by weed? There is nothing like a weed in our agriculture."

Hunting and gathering have coexisted with agriculture in most societies. Both hunting and gathering, when integrated into the dietary routines of a society, require a deep knowledge of the prey, which encourages strategies for favoring the reproduction of animals and plants, possibly favoring desirable traits.[4] For example, a boar hunter may favor the reproduction of certain boar families, which possess certain particularly desirable traits (e.g. size and build); a gatherer of mushrooms may favor the reproduction of certain species in a spot by facilitating or creating specific

---

[4] Cfr. (Kowalsky 2010).

environmental conditions (e.g. humidity, shade, enclosure from the passage of certain animals, selection of surrounding plants). From this perspective, the so-called wild species found within the edible environment are typically far from the most untamed species known on Earth. This is not surprising since eating is a relationship, which in the case in point involves humans and a few thousands species; with time, although humans did not domesticate such species, they (voluntarily or involuntarily) managed them. The study of biodiversity within agriculture should be undertaken alongside with the study of biodiversity within the wild edible environment. As Bharucha and Pretty (2010: 2923) conclude,

> The evidence shows that wild foods provide substantial health and economic benefits to those who depend on them. It is now clear that efforts to conserve biodiversity and preserve traditional food systems and farming practices need to be combined and enhanced.

Another important consideration, which shows how simplistic is the view that draws a strong divide between wild and domesticated species within the edible environment, is that such a view leaves no place for the myriads of microscopic organisms that are essential to human diets worldwide, with virtually no exception. To illustrate the point with an example, it would be unsound to claim that, at origins, humans domesticated *Saccharomyces cerevisiae* and that sourdough is one of the countless outcomes of such domestication process.[5] After all, humans were not even aware of the existence of such a microscopic fungus when they started making use of it to produce bread, beer, chocolate, etc. Rather, sourdough emerged out of a form of cooperation between humans and a variety of fungus, which was not guided by specific species design, but that likely proceeded through trials and errors guided solely by taste and, more broadly, culinary success. Yet, *Saccharomyces cerevisiae* is arguably part of the biodiversity within the edible environment that humans should aim to preserve. Parallel arguments can be developed with respect to progenitors of domesticated plants that are still lingering in the wild: they are especially precious because they typically preserve the widest genetic pool of the taxon. Hence, the discussion of the biodiversity within the edible environment found in virtually any extant human diet should include not only domesticated species and varieties.

The upshot for subsequent discussion is that any assessment of the measurement of the biodiversity found within the edible environment, and of the best means best conserve it, should recognize how variegated are the relationships that humans created with species in the edible environment. The edible environment is constituted by organisms (or parts of organisms) that can hardly be put on a scale with respect to their untamedness – from the wildest to the most domesticated. This complicates a bit the picture when it comes to decide whether to leave out certain items within the edible environment from measurements of biodiversity and efforts of conservation. Is there really a difference between domesticated and wild species, which

---

[5] It may be more plausibly argued, however, that *Saccharomyces cerevisiae* was in some sense *domesticated* by contemporary biotechnology, through the selection of best suited samples and genetic engineering interventions. I shall leave the issue open here.

should be reflected in the study of the biodiversity of the edible environment? What to make of microorganisms? What about bioengineered plants and animals? These questions shall occupy us in the next section.

20.3   Finding Natural Joints

In her assessment of biodiversity with respect to human modified entities, Siipi (2016) distinguishes between three main ways to devise a cutoff point between what should be included in an inventory of biodiversity and what should not. They are respectively based on the (i) history, (ii) properties, and (iii) relations of the entities under consideration. It may be useful to begin by illustrating the three ways.

(i) With respect to the first way, imagine the case of two portabella mushrooms, one of which is grown wild in a forest and one that is induced by a human in a garage; suppose further that the two mushrooms are genetically identical, because the mushroom mycelia of the wild mushroom have been transplanted in a litter in the garage, that they are hardly distinguishable when it comes to taste and nutritional characteristics (their properties), and that they have similar market and culinary value (their relations); nonetheless, since the mushrooms have different histories, which rest on their different contexts of development, the forest-grown is regarded as 'wild' or 'naturally grown' and the other is labeled as 'home-grown'.

(ii) To illustrate the second way, based on properties, imagine two portabella mushrooms grown side by side in a forest (hence having alike histories) and having similar market and culinary values (relations), but possessing quite different nutritional and gustatory properties, due to the malformation of one of them, developed just a few hours before being picked (hence, not really historically based). You can conceive of a classification according to which the malformed mushroom is regarded as 'unnatural' and the other as 'natural,' based on their morphological properties.

(iii) To illustrate the third way, based on relations, imagine two portabella mushrooms grown side by side in a forest and perfectly comparable in terms of size, nutritional and gustatory properties, but ending up in two different markets and, from there, two different restaurants; although alike in terms of origin and properties, one of the mushrooms *belongs* to the market and restaurants where it ends up, being recognized as a 'natural' element within the edible environment of the culinary culture(s) showcased within the market and the restaurant; the other mushroom instead is considered as somewhat foreign to its market, and ends up in a restaurant to be featured as an exotic, 'unnatural' item to be placed alongside the other foods.

In her paper, Siipi distinguishes in total six criteria for telling apart natural from unnatural foods for the purposes of finding a cutoff point between entities that should be relevant for biodiversity measurements and preservation, and those that should not be so regarded. Three criteria are history-based and concern respectively: the organisms that are independent of humans (e.g. a wild herb spontaneously grown in a remote beach), the organisms that are not controlled by humans (e.g. wild blackberries), and the organisms that are regarded as non-artificial (e.g. Golden Delicious apples). Two additional criteria are property-based and regard the foods that are: alike to spontaneously occurring (e.g. two mushrooms, one forest-grown and one garage-grown, which are alike in terms of culinary and nutritional values); alike to possibly existent foods (e.g. seedless grapes, obtained by grafting spontaneously occurring samples of seedless grapes (cfr. Sperber 2007). The sixth and final criterion is relation-based and rests on whether a certain food 'belongs,' or is 'suitable to' a given context (e.g. grapes being unsuitable for the original climate and soil of Central Valley, California, and thus requiring amounts of water, pests, and herbs that are disproportionate).

Siipi's thorough examination of the ways in which a portion of an edible environment may be found to be natural or unnatural, and hence possibly included or excluded in biodiversity measurements and conservation policies, demonstrates the complexity of the matter at stake. To further her analysis, we should first of all avail ourselves of the conclusion reached at the end of the previous section. Concepts of 'wild' and 'domesticated' do little to usefully represent key relationship between humans and parts of the edible environment that are relevant for the purposes of biodiversity; these concepts should be better substituted with specific histories of the relationships between humans and parts of the edible environment, which evidence characteristic traits. But, which traits should matter? A tentative list should include things such as control over reproduction, difficulty of reproduction, potential variability of the desired trait, nutritional properties, gustatory properties, broadly cultural properties, ecological fit, culinary fit ... Yet, can an exhaustive list be provided? Interesting in this context is also to recall that focusing on individual species may not be the best manner to proceed in an assessment of biodiversity; rather, we should look at broader networks of biotic and abiotic entities that produce certain foods. Indeed, the production of certain foods requires the employment of additional living organisms; for instance, any peach orchard requires bees for pollination, or any fig orchard requires a specific species of tiny wasps as well as trees that are both female and male, even though the latter produces fruits that are generally not eaten. More generally, it seems best in a number of circumstances to pay attention to ecosystems that deliver foods, rather than to single species within the edible environment; for instance, Vitalini et al. (2009) proposed a EU designation of 'Site of Community Interest,' which would stress indeed the presence of a number of biotic and abiotic conditions that are necessary to sustain portions of the edible environment. Hence, to guarantee the security of the relevant edible plants, measurements of the biodiversity of the edible environment should take into account not simply the (parts of the) species or varieties that we feed off, but also the other biotic and abiotic conditions that are necessary for their survival.

In conclusion, although a number of traits, such as the ones just listed, are arguably most relevant in the majority of contexts, it seems methodologically incorrect to proceed by devising a list that should fit every assessment of biodiversity within an edible environment. In other terms, this discussion suggests that there is no one vantage point from which to assess the naturalness of a (part of an) organism found within the edible environment; this is because naturalness is not a matter of that (part of an) organism being domesticated or wild, but what function it plays in a system of food production, which ecological relationships it bears to other surrounding organisms, as well as what functions it could play in possible edible environments that are considered relevant for the purposes of the assessment. The remaining of the paper will elaborate on this thesis.

20.4   The Values of a Biodiverse Edible Environment

The previous two sections argued that the biodiversity within the edible environment includes a wide array of entities, which can hardly be systematized in a context-independent taxonomy. Recapping the complexity of the domain under examination will be useful to start assessing its multiple dimensions of value.

While a distinction between wild and domesticated species could be defended, based on the degree of human intervention during reproduction and selection, it would be unseemly to claim that all wild species develop(ed) fully independently of human interference. Some developed actually in conjunction with human artifice. For instance, many forms of gathering and hunting do proceed through subtle modifications of the surrounding environments by humans, which facilitate the reproduction and growth of specific populations of the designated species or variety. The spectrum of domesticated species varies significantly, as it includes plants that are reproduced by cutting (e.g. rosemary, strawberries, avocados), plants that are reproduced by grafting (e.g. grapes, most fruit trees), plants that spread by sexual reproduction (e.g. most grains), and a number of plants that can reproduced in any of those ways (e.g. avocados, cacao trees). For any of the domesticated species, we can wonder what is the degree of interference that humans have access to in *any single instance of reproduction*: with animal farming, breeding is often controlled down to the minutest details by the farmer; but, with grains, it is not feasible to control the path of all the pollen in a field and often also the selection of seeds is only partially decided by the farmer, who will work with what was provided by the previous harvest; in an orchard, the farmer cannot control the process of pollination by bees to the minutest details; ditto for controlling reproduction within a fish farm. Alternatively, we can measure the overall degree of interference between humans and the species by looking at the genetic distance of domesticated organisms from their wild progenitors, factoring in the number of generations that occur between the two samples.

Species are not the main units when biodiversity of the edible environment is at issue. Rather, cooperation among different clusters of organisms, organized in more or less spontaneous communities, seem to be the key concept. In this light, the cooptation of microorganisms to produce viable foods, which at least until Pasteur proceeded somewhat blindly with respect to the biological details of the microorganisms, comprises a chapter of its own in the inventory of the relationships between humans and the edible environment. Beer, bread, chocolate, yoghurt, cheese are all examples of a *culture* of fermentation, which played an essential role in food production and conservation and in human evolution. Fermentation, as we know it today, is the outcome of a microbial diversity, which is formed during food preparation and aging, and which confers a distinctive specificity to food; not only it is arguable that, without being fermented, beer would not be *beer*, but it is also arguable that without certain strains and species of microbes that are characteristic to it, a certain beer style (e.g. pilsner) would be not that beer style. The research on this issue is extensive; cfr. Borghini 2014: 1118 and, for a recent significant example based on cacao Ludlow et al. 2016). But, in the human cooptation of certain microorganisms with the aim of producing viable foods, what matters most? Is it most important to preserve – say – the spontaneity of a process (as it happens in spirits which are spontaneously fermented) or, rather, to preserve certain final characteristics of a product? Do (some aspects of) the genetic profiles of the microbes fix the identity of the final product? Or, rather, should a certain style, brand, gustatory profile be privileged? These questions suggest that the diversity within the microbial world correlated to the edible environment is not all on the same level, and that privileging the diversity derived from a type of process (e.g. spontaneous fermentation) may hinder the tending to the diversity of other aspects of the fermentation process, such as the preservation of certain strains or varieties.

Finally, we should consider the complexity of biotic and abiotic factors that favor the reproduction, growth, and development of the edible environment. Hence, species of bees and wasps, varieties of soils, minerals within water, etc. It seems that an inventory of the biodiversity of the edible environment should include these items too, since they are arguably essential to the creation of a number of products, such as geographical indications (see Borghini 2014). Hence, the biodiversity of the edible environment is bound to include also a vast array of entities that are not really edible, but that are conducive to the production of the foods we eat, currently or possibly. There is here an overlap between the concept of biodiversity, when applied to edible items and when applied to non-edible items; for instance, we have reasons to protect the biodiversity of soils for reasons that are independent of food production (cfr. Brussaard et al. 2007), and yet the study of biodiversity of the edible environment will argue for their protection too.

The fluidity between the different categories of entities found within the edible environment is also reflected in the fluidity between roles taken up by food workers. For example, it is common for farmers to act as custodians of both domesticated and wild species, and to be farmers of both agricultural products and weeds; also, gatherers and hunters are oftentimes also farmers; and it is common to a fisherman to hunt too.

The edible environment showcases the complexity of the idea of biodiversity because of the multifarious forms that the relationship between humans and edible items can take and has historically taken. Such complexity is mirrored also in the *reasons* we have for valuing biodiversity within the edible environment. Because the subject matter are ordinary living entities, which are considered in relationship to humans, it is fairly obvious that the reasons to value the biodiversity within the edible environment is entrenched with human existence and culinary cultures. I shall here divide up the field in four points: (i) food sovereignty; (ii) food security; (iii) gastronomic pleasure; and (iv) intrinsic value. Let us illustrate each of them, in order.

(i) *Food sovereignty*. With food sovereignty we intend the ability within a group of people to self-determine a sufficiently ample and relevant portion of their dietary choice by means of food production. Food sovereignty emphasizes, hence, the active ability of a society to determine which plants and animals to harvest and produce, as well as the means of production. Such a power of a society is foundational with respect to the possibility (not the necessity, of course) of actively fostering biodiversity within the edible environment. This power is especially critical when it comes to farming societies that are economically, technologically, and politically at a disadvantage. It was indeed introduced in 2002 by the World Bank with the International Assessment of Agricultural Knowledge, Science and Technology for Development (IAASTD), a 3-year program aimed at improving food production knowledge and technology within disadvantaged societies. But, the idea of food sovereignty was implicitly already at the core of the *Universal Declaration on the Eradication of Hunger and Malnutrition*, issued during the World Food Conference of 17 December 1973. The declaration begins by noting "the grave food crisis that is afflicting the peoples of the developing countries where most of the world's hungry and ill-nourished live and where more than two thirds of the world's population produce about one third of the world's food;" it then goes on to suggest that "all countries, and primarily the highly industrialized countries, should promote the advancement of food production technology and should make all efforts to promote the transfer, adaptation and dissemination of appropriate food production technology for the benefit of the developing countries and, to that end, they should inter alia make all efforts to disseminate the results of their research work to Governments and scientific institutions of developing countries in order to enable them to promote a sustained agricultural development."

(ii) *Food security*. If food sovereignty regards the foods that are produced within a society, food security concerns instead the kinds, qualities, and quantities of foods *accessible for consumption* within a society. A consistent portion of the literature on the biodiversity of edible organisms focused on the importance of an ample spectrum of nutrients for combating malnutrition, when manifested both as a lack of sufficient calories or nutrients (undernutrition), or as an excessive amount of calories or nutrients (overnutrition) (cfr. Borghini 2017 for a philosophical analysis of hunger). In their literature review on food security and biodiversity, Chappell and La Valle (2011) provide significant evidence that "alternative agriculture, which is generally targeted at sustainability and compatibility with biodiversity conservation, is indeed on average better for biodiversity conservation than conventional agriculture, which usually (though not always) targets increases in yields to the exclusion and even detriment of direct concerns about biodiversity, equitability, and food access." (Chappell and La Valle 2011: 17) Chappell and La Valle's conclusion, which stresses the link between food sovereignty and food security (see also Jarosz (2014) on this point), goes hand in hand with the so-called "ecoagriculture approach" (McNeely and Sherr 2002), according to which landscape biodiversity is key to ensure sustainable farming practices that are in sinking with their surrounding ecosystems.

A limitation of much literature on food security and biodiversity rests on a narrow conception of the edible environment, which is basically limited to agricultural products. As we have discussed above, landscapes comprising wild and domesticated foods come into closest contact in some of the regions where food access is most insecure. The availability of a diverse spectrum of plants, animals, and other suitable living entities for setting up an edible environment is a form of empowerment for communities that aim to improve their conditions with respect to food sovereignty (see below) and food security. Farming in urban or rural regions that present adverse climatic conditions or inadequate natural resources can be much improved by a wide stock of living entities that can adapt to different circumstances. Thus, an approach such as ecoagriculture is best appreciated when conjoined with the thesis that there is no sharp discontinuity between wild and domesticated species, and no easy cutoff point between natural and unnatural entities, at least when it comes to the edible environment.

Fostering biodiversity, hence, can aid to food security at two different levels: at the ecosystem level, and at the level of the edible environment. At the ecosystem level, biodiversity can facilitate a sustainable availability of resources, to be employed by producers with the edible environment. At the level of the edible environment, the wider the stock of organisms available to any producer worldwide, the higher will be her power to deliver suitable goods to a market; this, in turn, will increase opportunities for a diverse diet, which is key to address malnutrition. Of course, the availability of certain goods on the market is far from granting, by itself, a solution to food insecurity (cfr. Chappell and La Valle 2011: 17–18, for a discussion of this point); yet it is certainly a necessary step in order to address it.[6]

(iii) *Gastronomic pleasure.* Both food sovereignty and food security are linked to gastronomic pleasure, as they by and large shape the link between dining and civic values (cfr. Alkon and Mares 2012). Promoting a biodiverse edible environment is a mean to empower communities not only by strengthening the

---

[6] I should mention a difficult question arising at this point in connection to bioengineered organisms, which cannot be fully developed here. Consistent efforts are underway to engineer organisms (e.g. GM crops) and foods (e.g. lab meat), which may add to the diversity available to farmers worldwide. Arguably, these items should be included in an inventory of the (actual or potential) biodiversity of the edible environment; but, should they rank as equally valuable as their non-genetic counterparts? In keeping with the approach presented above, an answer to this question can be provided only when faced with a broader decisional framework, which keeps into account also the other three values to be considered next, that is gastronomic pleasure, food sovereignty, and intrinsic value.

sustainability of their agricultural production and by improving the likelihood that they will be food secure, but also by allowing them to diet in a manner that is most in keeping with their ethical, political, religious, an identity-related values. In this sense, the biodiversity of the edible environment is directly linked to the power of a society of choosing and determining its diet.

Slow Food may have been the first and to date the most fervent voice to insist on the link between biodiversity and gastronomic pleasure. The Slow Food movement, founded in 1986 by Carlo Petrini among others, focused since the beginning on the importance of gastronomic pleasure for any conversation concerning the political, ethical, and socio-economic discourse about food. In a telling passage of *Slow Food Nation*, Petrini writes: "Pleasure is a human right because it is physiological; we cannot fail to feel pleasure when we eat. Anyone who eats the food that is available to him, devising the best ways of making it agreeable, feels pleasure." (2013: 50) Now, for Petrini gastronomic pleasure is directly linked to the availability of a diverse array of products, which in turn can be obtained only by actively encouraging the diversity of forms by means of which humans tend the edible environment. Hence, gastronomic pleasure necessarily passes through the promotion of the diversity of the edible environment, by supporting typically small-scale tending practices, which aim to express the most meaningful relationship that humans can establish with the edible environment.

Unthinkingly, it may seem that pleasure is an accessory feature of human relationship to food, which should be kept out of the ethical and political sphere. Nonetheless, in the past three decades it has become increasingly more evident that there is a link between gastronomic pleasure and such issues as biodiversity, malnutrition, food sovereignty, and food access. Petrini's position echoed that of Wendell Berry (cfr. 1990) and has been re-proposed in different forms by several additional authors, such as Pollan (cfr. 2006) and Stiegler (cfr. 2006). Thompson (2015: Chapter 3) especially lays out a convincing discourse showing the link between dieting and the ethico-political sphere. It is impossible to tell apart the meaningfulness of the pleasure experienced during the act of eating and the sorts of food that are consumed (cfr. Borghini 2017 on this point); such pleasures are most often (positively or oppositionally) linked to values imbued in a society, to empowerment, and civic values, no matter how ordinary they may seem in any single dining occasion. For these reasons, gastronomic pleasure is to be included within the spiritual ecosystem services.

(iv) *Intrinsic value*. Finally, the value of a biodiverse edible landscape may rest on the value of the species, the varieties, and the trophic chains themselves. This may be the most intuitive value of a biodiverse edible environment in the context of a general discussion of the philosophy of biodiversity. A wider spectrum of forms of life has not only a utilitarian value, perhaps quantifiable in monetary terms like Costanza et al. (1997) provocatively suggested; rather, it is worth to invest time and resources in the fostering of biodiversity because there is a beauty and value in its mere *existence*, regardless of the consequences. When it comes to edible landscapes, the history of painting offers some neat illustrations of the view of those who hold that biodiversity should be regarded as an intrinsic value. The paintings of Bartolomeo Bimbi, for example *Pears of June and July* (1696), entertain the spectators by simply showcasing a mesmerizing array of pears cultivated under the Medici family at the end of '600 s.[7]

In closing, it is important to ask whether the four reasons for measuring the edible environment are in some way affected or affecting a diet; more specifically, should a diet be influenced by consideration of biodiversity or, vice versa, do dietary decisions influence our stance on the measurements of the biodiversity of edible plants and animals? The fourth reason suggests implicitly that the wider a variety of edible items in a diet, the more commendable the diet; and you may wonder whether such a constraint is acceptable. You can fancy a society that is food secure, sufficiently pleased when it comes to dining, whose members have in some way come to agree in an equitable manner upon their diet, and which nonetheless survives within an extremely monotonous diet (made, perhaps, of one daily pill synthesized in dedicated laboratories). This society would arguably not contribute to the fostering of the biodiversity of the edible environment; should its members still pay dues to those in other societies who, instead, aim to foster it? If they should, in what measure should they contribute? For instance, suppose that the vast majority of the world population would come to prefer such a diet; would it still be feasible to maintain the goal of fostering an edible environment as diverse as we currently have? In other words: does the specific diet undertaken by an individual, or a society, maintain obligations to others who chose different diets? In what measures?

Since the biodiversity of the edible environment depends on human tending possibly in a more active manner than the biodiversity of other forms of life, these questions are far from trivial to answer. An important upshot is that, if we accept that the biodiversity of the edible environment is valuable independently of its consequences, then we should keep tending edible items even if they were to phase out of any human diet. To what extent this is a feasible goal is an issue that is worth further, future investigation.

20.5   What to Foster Within the Edible Environment?

Although a definitive cutoff for what is to be included in the edible environment cannot be provided, it is arguable that it is valuable for at least four reasons, no matter how we come to individuate it from time to time. But, what is it really that is of

---

[7] To be clear, I am not suggesting that the aesthetic appreciation of nature necessarily implies the recognition of an the intrinsic value of some natural elements, nor that all works of art illustrating nature do illustrate nature's intrinsic value; at a minimum, since the biodiversity of the edible environment depends on its relationship to human tending, showcasing and valuing it, *per se*, is also a mean to showcase and value *per se* human efforts to establish a meaningful relationship with the edible landscape; I am more modestly claiming that certain works of art can serve as illustrations of the view that nature has an intrinsic value.

value? And, are there any theoretical or practical conflicts in the items that we are seeking to foster? We shall address these two questions, in order.

With respect to the first question, we shall distinguish three kinds of items that are typically regarded as valuable in discussion of the edible environment. (1) The first, more traditionally valued kind of item is the variety or breed as established by means of reproduction. It is under this regard, for instance, that we shall include the conservation of the thousands of breeds of animals reared by humans over the course of millennia; ditto for the thousands of varieties of beans, potatoes, tomatoes, corn, and other plants; the hundreds of mushrooms that humans consume; the thousands of varieties of fungi, yeasts, and strains of bacteria coopted for food production. The problem with this proposal is that it is often controversial whether some characters of a plant or an animal are novel to the point of constituting the foundation of a new breed or a new variety.[8] The issue had been touched upon also by Darwin in the *Origins of Species*, especially Chapter II. Is a variety a cluster of organisms that has the potentiality to become a new species in a near future? That seems doubtable in the case of most edible organisms. Are varieties distinguishable at the genetic, phenotypic, behavioral, ecologic, nutritional, gustatory level? Should we pick varieties based on their significance for a certain culinary history, for their relationships with surrounding ecosystems, or rather for more arguably intrinsic characteristics of the product? Notice, finally, that to intensify the efforts to preserve a variety can imply to weaken it, because it may make it increasingly dependent from humans.

(2) In recent years a new method for marking the diversity of an item within the edible realm has come to be employed: it traces the genetic specificity of a variety of plants, of the breed of an animal, or of a microorganism. Thus, a clone of – say – Sangiovese grapes can now be identified not in terms of its phenotypic traits and ancestral history, but in terms of certain genetic traits that arguably are responsible for its characteristic phenotypic traits, such as the size of its fruits, its color, its skin, or a certain gustatory quality (cfr. also Borghini 2014 on this point). Although this method of identifying an item may seem similar to the one based on breeds and varieties, it is actually quite different. Indeed, breeds and varieties are essentially linked to ancestral history; on the other hand, fixing the identity on the basis of a selected number of genetic features is compatible with cis-genesis, cloning, and other potential forms of bioengineering. Hence, the identity of a certain breed of cattle would be fixed in terms of its genetic characteristics, no matter how the cattle would come into existence (actually, no matter whether the cattle ever came into existence or whether, instead, some of its cells where cultivated in a lab; on lab-grown meat, see Van Mensvoort and Grievink 2014).

(3) A third and last kind of item that may be worth fostering in order to foster the biodiversity of the edible environment are procedures and techniques for breeding and tending plants, animals, and microorganisms. Hence, the different manners by

---

[8] The more technical definition of 'cultivar' provided for plants in ICNCP (2009, Chapter II Art. 2.3) does not help here, because it still relies on a judgment regarding the novelty of the plant character.

means of which humans have facilitated, reared, and coopted new breeds of animals, new varieties of plants, and clusters of microorganisms. Should techniques employed within bioengineering be included in this list, too? Should they receive equal weight with respect to older methods?

To the purposes of the present discussion, which aims at framing a philosophical discussion of biodiversity when it comes to the edible environment, it is important to point out that there are some incompatibilities among the three kinds of items that may be targeted for being fostered. I have hinted at one incompatibility already when presenting genetic specificity. If the policy of an institution is to foster the continuation of existence of certain genetic traits, that may imply to have to change procedures and techniques for tending it, as well as changing its reproductive history (hence, what are commonly regarded as breeds or varieties); for example, some speculated that in order to keep producing Champagne in Champagne, farmers will have to introduce genetically modified clones of grapes, possibly employing different techniques for planting (and perhaps harvesting and processing) them. On the other hand, concentrating on certain methods of, say, wine production, will typically imply that at some moment farmers will have to discard clones that are not in sinking with relevant changes within the ecosystem of production, thereby also compromising the genetic identity of the clones. Finally, focusing on breeds and varieties based on ancestry, implies embracing genetic changes over time as well as methods of production that would best meet such changes. The upshot of this analysis is that, when issuing policies for fostering the biodiversity of (some part of) the edible environment, it is relevant to specify both which kind of items are to be fostered and to what extent the kinds of items that are not to be fostered should be kept into account into the measurement and intervention efforts. This is far from being accomplished by the extent literature on the topic as well as by extant policies, such as the Convention on Biological Diversity and the International Treaty on Plant Genetic Resources for Food and Agriculture.

20.6   Conclusions

We shall at last return to our initial three questions and suggest answers based on the considerations made thus far. *(1) Are there living entities that should be excluded from measurements of biodiversity as well as from conservation efforts?* When it comes to the diversity of the edible environment, the first suggestion is to consider the importance of so-called wild species, which to date play a critical role in integrating agricultural and industrial produce in most societies, constituting also an important back-up safety net for food security purposes. A second suggestion is to proceed with great care when it comes to drawing cutoff points between items that are natural enough to deserve inclusion in an inventory of food biodiversity and items that are not so; it seems most prudent to proceed by devising cutoff points that are suitable to specific sub-domains; these can be individuated on different grounds, such as biological taxa (e.g. cucurbitaceae, beans, mushrooms) or methods of production (e.g. grafting, sexual reproduction, genetic modification). Thus, we have multiple possible inventories to choose from, giving rise to our second question.

(2) Should the criteria for inclusion in a measurement or intervention be context-dependent or context-independent? For instance, could different criteria be selected depending on circumstances? A successful discussion of the matter, I submit, would demarcate as clearly as possible what are the conceptual and axiological differences between the criteria, as well as their potential practical consequences. It is important to remark here that the diversity of the edible environment is deeply entrenched with human cultures, so that the criteria for biodiversity measurement must reflect human perspectives within different societies, embedded in the conceptions of plants, animals, and dieting.

(3) What are the grounds for preferring certain (possibly more familiar) forms of life over others? For instance, do such preferences rest on efficiency, or perhaps on some emotional or spiritual connection? This question addresses the values that are involved across possibly different context of evaluation, e.g., food sovereignty, food security, and gastronomic pleasure. It is important to explore how such values differ across societies and whether convergence over a few selected values is a desirable goal, or if lack of convergence is actually more fruitful for the purposed of the biodiversity of the edible environment.

The new agricultural technologies introduced by the Green Revolution between the 1930s and the 1960s, followed by the more recent innovations in biotechnology, along with an increased capacity of transportation, have deeply subverted the relationship between humans and food. Such a drastic rupture is forcing a rethinking of that relationship, and a careful consideration of which items we shall conserve and why. This essay aimed at introducing a philosophical a frame for assessing the biodiversity of the edible environment, and pointing at a number of questions that seem in need of being addressed in the near future.

References

Alkon, A. H., & Mares, T. M. (2012). Food sovereignty in US food movements: Radical visions and neoliberal constraints. *Agriculture and Human Values, 29*, 347–359.

Berry, W. (1990). The pleasures of eating. In *What are people for* (pp. 145–152). New York: North Point Press.

Bharucha, Z., & Pretty, J. (2010). The roles and values of wild foods in agricultural systems. *Philosophical Transactions of the Royal Society B, 365*(1554), 2913–2926.

Borghini, A. (2014). Geographical indications, food, and culture. In P. B. Thompon & D. M. Kaplan (Eds.), *Encyclopedia of food and agricultural ethics* (pp. 1115–1120). New York: Springer.

Borghini, A. (2017). Hunger. In P. B. Thompon & D. M. Kaplan (Eds.), *Encyclopedia of food and agricultural ethics* (2nd ed., pp. 1–9). New York: Springer. (ONLINE FIRST).

Brussaard, L., de Ruiter, P. C., & Brown, G. G. (2007). Soil biodiversity for agricultural sustainability. *Agriculture, Ecosystems and Environment, 121*, 233–244.

Chappell, M. J., & La Valle, L. A. (2011). Food security and biodiversity: Can we have both? An agroecological analysis. *Agriculture and Human Values, 28*, 3–26.

Chemnitz, C., & Becheva, S. (2014). *The meat atlas. Facts and figures about the animals we eat.* Ahrensfelde: Möller Druck.

Conservation and Sustainable Use of Agricultural Biodiversity. A Sourcebook (2003), CIP-UPWARD – International Potato Center: Laguna.

Costanza, R., et al. (1997). The value of the world's ecosystem services and natural capital. *Nature, 387*, 253–260.

Diamond, J. (2002). Evolution, consequences and future of plant and animal domestication. *Nature, 418*, 700–707.

Fromartz, S. (2006). *Organic Inc. Natural food and how they grew.* Orlando: Harcourt Books.

Grivetti, L. E., & Ogle, B. M. (2000). Value of traditional foods in meeting macro- and micronutrient needs: The wild plant connection. *Nutrition Research Reviews, 13*, 31–46.

ICNCP. (2009). *International Code of Nomenclature for Cultivated Plants* (8th ed.). Vienna: International Society for Horticultural Science.

Jarosz, L. (2014). Comparing food security and food sovereignty discourses. *Dialogues in Human Geography, 4*, 168–181.

Kowalsky, N. (Ed.). (2010). *Hunting: In Search of the Wildlife*. Oxford: Wiley-Blackwell.

Ludlow, C. L., et al. (2016). Independent origins of yeasts associated with coffee and cacao fermentation. *Current Biology, 26*, 1–7.

Mazhar, F., Buckles, D., Satheesh, P. V., & Akhter, F. (2007). *Food sovereignty and uncultivated biodiversity in South Asia.* New Delhi: Academic Foundation.

McNeely, J. A., & Sherr, S. J. (2002). *Ecoagriculture: Strategies to feed the world and save wild biodiversity.* London: Island Press.

Millenium Ecosystem Assessment (MEA). (2005). *Current state and trends.* Washington, DC: Island Press.

Petrini, C. (2013). *Slow food nation: Why our food should be good, clean, and fair.* New York: Rizzoli Ex Libris.

Pollan, M. (2001). *The botany of desire. A plant's eye view of the world.* New York: Random House.

Pollan, M. (2006). *The omnivore's dilemma. A natural history of four meals.* New York: Penguin Press.

Siipi, E. (2016). Unnatural kinds: Biodiversity and human modified entities. In J. Garson, A. Plutynski, & S. Sarkar (Eds.), *The Routledge handbook of philosophy of biodiversity* (pp. 125–138). New York: Routledge.

Sperber, D. (2007). In E. Margolis & S. Laurence (Eds.), *Seedless grapes: Nature and culture* (pp. 124–137). Oxford: Oxford University Press.

Stiegler, B. (2006). Take care (Prendre soin). (trans. by S. Arnold, P. Crogan and D. Ross), *Ars Industrialis*: http://arsindustrialis.org/node/2925

Thompson, P. B. (2015). *From field to fork: Food ethics for everyone.* Oxford: Oxford University Press.

Van Mensvoort, K., & Grievink, H. J. (2014). *The in vitro meat cookbook*. Amsterdam: BIS
   Publisher.
Vitalini, S., Tome, F., & Fico, G. (2009). Traditional uses of medicinal plants in Valvestino (Italy).
   Journal of Ethnopharmacology, 121, 106–116.
Wrangham, R. (2009). *Catching fire: How cooking made US humans*. London: Profile Books.

Chapter 21
Conservation Sovereignty and Biodiversity

Markku Oksanen and Timo Vuorisalo

Abstract Many key concepts in conservation biology such as 'endangered species' and 'natural' or 'historic range' are universalistic, nation-blind and do not implicate the existence of geopolitical borders and sovereign states. However, it is impossible to consider biodiversity conservation without any reference to sovereign states. Consequently, the units of biodiversity and their ranges transform into legal concepts and categories. This paper explores the area that results from this transformation of the universalist idea into national policy targets. Conservation sovereignty denotes to right of each state to design and carry out its own conservation policies. To illustrate the problematic nature of conservation sovereignty, the paper focuses on two cases where the borders and the state play the key role: (1) the global division of conservation labour and (2) assisted migration. All in all, this paper takes a critical look upon the anomalies in universalism and conservation sovereignty.

Keywords Species · Environmental ethics · Natural resources · Assisted migration · States

M. Oksanen (✉)
Department of Social Sciences, University of Eastern Finland, Kuopio, Finland
e-mail: markku.oksanen@uef.fi

T. Vuorisalo
Department of Biology, University of Turku, Turku, Finland
e-mail: timo.vuorisalo@utu.fi

435

E. Casetta et al. (eds.), *From Assessing to Conserving Biodiversity*,
History, Philosophy and Theory of the Life Sciences 24,
https://doi.org/10.1007/978-3-030-10991-2_21

21.1   Introduction

Many commonly used concepts in conservation biology such as 'endangered species' and 'natural' or 'historic' range[1] are universalistic, nation-blind and do not
implicate the existence of any geopolitical borders or sovereign states.[2] To ignore
the current nation-state system and to consider conservation of biodiversity without
any reference to states would, however, be unsatisfactory. States are self-determining
actors and the principal possessors of biological resources in their territories. At the
international level, sovereignty is manifested in the international treaties and declarations approved by the states. By these treaties and declarations, states commit
themselves to certain responsibilities and thus voluntarily restrict the ways of acting
open to them. At the national level, sovereign states implement these agreements
within their jurisdictions, that is, within their established geopolitical borders. From
this constellation, a vital point emerges with respect to biodiversity conservation:
the units of biodiversity and concepts ascribing their ranges transform into legal
concepts and categories that inform policies and practices. This perspective regards
sovereign states as the only relevant legal actors. The transformation thus occurs
within, and is organised by, the sovereign states.

   In creating national policies for biodiversity conservation, sovereign states act
either alone or in close collaboration with other states (consider the EU). A global
division of conservation labour arises out of joint multiple actions by states.[3] The
fundamental idea is that each country, as a sovereign actor, is in charge of the biodiversity within its territory while the biodiversity outside the territories of sovereign
states (the high seas, the Antarctica) as well as migrating biodiversity (waterfowl,
whales) are subject to transnational decision-making, if any.

   In this chapter, the traditional thinking will be modestly challenged in two ways.
On the one hand, we argue that the global division of conservation labour in its present form is not always efficient from the conservation perspective if each country
only focuses on safeguarding its territorial biodiversity. On the other hand, we ask
whether climate change (in the global perspective) could challenge the current conservation policies by requiring actions that would make state borders more porous,
and applied policies more interventionistic than what they are today. We contend
that in some cases successful conservation may require international translocation
measures for the establishment of new populations outside the historical ranges, and
geopolitical territory, of particular species.

---

[1] There are a plenty of other attributes to describe ranges such as 'indigenous' and 'native', some
of which may be more sensitive to current geopolitical structure than the notions of natural and
historic (on their differences, see Siipi and Ahteensuu 2016).

[2] Smith's (2016) analysis manifests a universalistic viewpoint concerning the ethics of endangered
species preservation.

[3] In addition to this expression being powerful in its own right, it articulates and explicitly includes
human-dependent form of biodiversity. In most cases, this biodiversity literally results from human
labour.

The aim of the chapter is to explore issues that result from the transformation of the universalistic idea into national policy targets the foci of which are not merely species universally understood but a wider variety of different "conservables". To understand what these conservables are, we come across the political dimensions of biodiversity conservation. In the first section, we discuss the idea of state sovereignty and its relation to the control of natural resources and biodiversity. The second section, in turn, presents a typology of sovereignty in the context of biodiversity conservation. In the third section, we examine the global division of conservation labour and its insensitivity for the issue of prioritisation, and the resulting obvious need to transform conventional conservation. The fourth section analyses assisted migration, or whether it is acceptable to translocate species (across the state borders) with the intention of helping them to survive global warming. A short conclusion ends the discussion.

Four clarifying remarks on the nature and scope of our inquiry need to be made. First, our approach is multidisciplinary and focuses on conceptual and theoretical problems arising from sovereignty in the context of biodiversity conservation. We also examine some real-life examples. Second, the transformation from scientific descriptions to legal categories and to conservation success may seem simple but is in reality a complicated and twisted issue because corruption in land-use decisions is widespread and it is difficult to prevent poaching and illegal wildlife trade. Although tackling illegalities is undoubtedly relevant to policy design, it is outside our main analysis. Third, our approach is stated-centred and thus extremely constricted. For a more comprehensive picture, the nonstate or civil society actors such as citizens, academics, non-governmental organisations, state-funded think tanks like the OECD (the Organisation for Economic Co-operation and Development) and transnational companies should be taken into account. Fourth, issues of security and safety, in particular the border control of the import of unwanted or hazardous biomaterial, have been and are important components of sovereignty; they are outside our scope of analysis. Therefore, keeping these remarks in mind, the picture we paint of sovereignty is at best sketchy and filled with promises that may never actualise; it is, nonetheless, a useful starting point for further analysis.

21.2 Biodiversity in the World of Sovereign States

Sovereignty over natural resources within the state territory is today an established principle in international law. The concept of 'sovereignty' dates back to the late sixteenth century and to the French political theorist Jean Bodin who famously wrote that, "sovereignty is the most high, absolute, and perpetual power over the citizens and subjects in a Commonwealth" (cited in Turchetti 2015). In actual politics, sovereignty became a leading principle in international law as a result of the Westphalian peace in 1648; hence the international system of sovereign nation-states is still known as the Westphalian system. In the historical context of Bodin and other peace negotiators, the unchallenged presumption was that absolute monarchy is a legitimate form of government. This aspect is not relevant to our analysis despite the facts that many biodiversity-rich countries lean towards absolutism and their democracy, civil societies and status of minorities can be questioned and the global developers' and resource buyers' voices are often compelling. In modern use, sovereignty is typically understood as a form of power that belongs to the state indivisibly and above other powers. In this sense, sovereignty expresses the idea of the right to self-determination that is hold by the nation-state over territory, natural resources and the peoples who inhabit the area. The sovereignty of the nation-states also guarantees a legal personhood for this entity in the international legal system, that is, it is externally independent and can exercise power within a community (Endicott 2010). Because of sovereignty, states are in the position to enter voluntarily into binding, action-limiting and, in some cases, external interference entitling conventions (Shue 2014, 146).

A key issue in discussions on sovereignty has been the control of natural resources. Natural resources are thought of as instrumental for the full exercise of self-determination: hence, without possibility to exclude other states (and nonstate trespassers) from using natural resources within their territories, states cannot be truly independent beneficiaries of their own natural wealth. This idea was particularly powerful in the post-World War II period of decolonization and the dissolution of the British, French, Japanese and other empires. In addition, resource scarcity was a matter of mounting concern, which inspired the US President Truman set up by the Materials Policy Commission in 1951. The Commission's analysis *Resources for Freedom* (1952) reflected the general pessimistic mentality with respect to resource availability now and in the future although it recommended policies supporting economic growth. (Andrews 1999, 182–83.) To consolidate the ties between national independence and self-determination and the control of natural resources, the General Assembly of United Nations adopted resolution 1803 (XVII) on the "Permanent Sovereignty over Natural Resources" in 1962.

Sustainable development became a truly global issue by the publication of Brundtland's Commission report *Our Common Future* in 1987. According to it, the current use of resources must not come about at the cost of the resource use, or welfare, of the future generations. It strongly influenced the contents of the Convention on Biological Diversity, signed in Rio de Janeiro in 1992. According to Article 2 of the Convention, biological resources include genetic resources, organisms or parts thereof, populations, or any other biotic component of ecosystems with actual or potential use or value for humanity. Furthermore, sustainable use of these resources "does not lead to the long-term decline of biological diversity, thereby maintaining its potential to meet the needs and aspirations of present and future generations". Later, a resource-based approach to biodiversity conservation has been very strong in the Ecosystem Approach that is a framework for action under the Convention.

The question is then: in what sense is biodiversity a natural resource? It seems straightforward to reason that if the concept of natural resources covers all resources that are biological, and if the concept of biological resources, in turn, includes biodiversity in all of its manifestations, then biodiversity is a natural resource. This view is emphasised by the Ecosystem Approach that focuses on the importance of ecosystem services that in fact cover all major biological processes in their natural environments. Not all resources are tangible; the category of cultural services, as a component of ecosystem services, includes historical, spiritual, educational and recreational values that ecosystems have but which can be damaged through the loss of biodiversity. Obviously, the convertibility of such cultural values into resources, or monetary values for that matter, is problematic, perhaps with the exception of eco-tourism or popular historical monuments that clearly have a market value. Many authors, however, resist this way of considering biodiversity merely as a resource (see e.g. Wood 1997) and the associated rather explicit anthropocentric attitude to the rest of nature.

When we adopt the conception of state sovereignty – a conception that is at least historically anthropocentric since it entitles 'peoples and nations' to utilize their natural resources – it depends on states what meanings they attribute to biodiversity in practice. This framework, however, emphasizes for the above mentioned historical reasons the status of biodiversity as an instance of natural resources. It is clear, however, that there are natural resources that do not fall into the category of biodiversity conveniently (e.g. water and non-renewable mineral resources) and the significance of biodiversity is not exhaustively reducible to its resource character. For this reason, when we talk about biodiversity within the framework of sovereignty, we should not consider it merely as a bundle of natural resources but having significance beyond their "resourceness", a point also made in the opening lines of the Preamble to the Rio Convention. An interesting question is which parts of biodiversity fall outside the popular concept of ecosystem services. To make these conservation dimensions more explicit, we purport the idea of conservation sovereignty.

Conservation sovereignty, as a political idea, stands for the right of each state to design and carry out its own conservation and related natural resource policies, as if there were no transnational regulation. Since there is, however, transnational regulation agreed upon by the sovereign states, though not necessarily by all of them, the question arises whether there can then be sovereignty with respect to biodiversity and its conservation. The paradox is apparent and there are rival attempts to tackle it. As Endicott (2010, 246) has noted, "state sovereignty seems both to demand the power to enter into treaties, and to rule out the binding force of treaties." It is clearly analogous with the better-know philosophical dilemma of whether the freedom of a human individual includes the possibility to enslave oneself for a fellow human, as Endicott (2010, 246) points out. We follow Endicott's (2010, 258) conclusion that sovereignty and participation in global agreements and international law are "at least potentially compatible" although the function of these agreements and laws is to give directions to domestic laws and policymaking and to guide interactions between states. As Shue (2014, 143) puts it, "sovereignty is not some mystical cloud that either envelops the state entirely or dissipates completely; there are bits and pieces of asserted sovereignty." A look at the recent history makes one think that there are no theoretical tensions: the processes of decolonization and the formation of the system of over 200 sovereign states have occurred simultaneously with the growing number of international environmental treaties (Frank 1997).

The paradoxical dimensions of sovereignty are also recognizable in the Rio Convention. According to the Preamble, "States have sovereign rights over their own biological resources" and thus it merely expresses the established principle that the biological resources in state territories are freely at disposal of the state. The previous passage, however, outlines reasons for restricting state sovereignty and the free use of these resources, since "the conservation of biological diversity is a common concern of humankind". Nevertheless, this paradox is a milestone in the development of international regulation concerning biodiversity. Given the long UN history on the issues of sovereignty and natural resources, the authors of the Preamble were fully aware of tensions between national interests and universal concerns and the essential differences between objects of human interests. The novel expression 'common concern' reflects the negotiators' worry about the state of biological diversity beyond specific geographical areas and resources to which the already established concepts of common area and common heritage apply (see Brunnée 2008).

The Preamble of the Rio Convention effectively captures the two-dimensional nature of global conservation efforts: it is international and domestic at the same time. Within the European Union, two-dimensionality is most clearly manifested in the Natura 2000 conservation area network, established by the Habitats Directive in 1992. The duality between nationalism and internationalism has its roots in the origins of modern conservation movement in the late nineteenth century.[4] Ever since the creation of the Yellowstone National Park in 1872, most countries have followed the model and selected areas to sustain wilderness and pristinity. In spite of its gradually increasing popularity in the USA (see Nash 2001, 108–21) and other countries, the national park movement was essentially a nationalistic enterprise that emphasized each country's unique nature values – in some cases compared with those of neighbour states. As Sheail (2010, 12) put it: "*National* parks presuppose sovereign nation states".

The idea ultimately reached the Old Continent with the first European national parks founded in Sweden in 1909.[5] The famous explorer A. E. Nordenskiöld had already in 1880 urged the establishment of 'state parks' in Nordic countries to preserve samples of fatherlands' pristine nature for the future generations (Palmgren 1922). The patriotic tone was unmistakable in the essay of the Finnish State Conservation Inspector, Dr. Reino Kalliola, who wrote in the first issue of *Suomen Luonto* – the journal of the Finnish Association for Nature Conservation – that, "the richness and beauty of the Finnish nature is our shared and precious heritage that everyone of us is obliged to cherish" (Kalliola 1941, 20; also Kalliola 1942). Although similar nationalistic tones were probably heard in conservationist circles across the world in the nineteenth century, also the first important multilateral

---

[4] In this analysis, we try not to identify the origin of conservation practices and we leave out the discussion on imperialist roots of early conservation (see Grove 1995).

[5] A somewhat parallel development took place in Britain, with the Establishment of the National Trust in 1895. Although emphasis of the National Trust has been on preservation of cultural heritage, also areas of natural beauty have been preserved (Sheail 2010).

conservation agreements, such as for instance the Paris Convention for the Protection of Birds Useful to Agriculture (1902), date back to that period (Lyster 1994).

Some early pioneers of conservation movement were active both internationally and nationally. The protection of migratory birds is a case in point. Even before the independence of Finland in 1917, the leading Finnish conservation pioneers had close relations to colleagues abroad and in different occasions pursued internationally defined objectives at the national level. Dr. Johan Axel Palmén, Professor of Zoology in Helsinki, took great interest in the 1st International Ornithological Congress in Vienna in 1884. It is notable that the delegates of the conference attended as individual citizens, as respected members of the scientific community and not as official delegates sent by their respective governments. The governmental acceptance of conservation matters was, however crucial and official participation increased gradually. It is illustrative that The International Council for Bird Preservation (ICBP; from 1993 on, Birdlife International) was founded at the Finance Minister's home in London in 1922 (Birdlife 2017). Accordingly, the idea of national representation in international meetings was stronger providing a better basis for national action on bird conservation. To return to Palmén, the year following the Vienna conference, he published a seminal paper that outlined a plan, based on the conference proceedings, for a reliable collection of nationwide data on bird species distribution and abundance in all regions of the country (Palmén 1885). Palmén's programme turned out to be very successful (Vuorisalo et al. 2015). Later, Palmén (1905) proposed setting up a national conservation society (this happened in 1938), and protecting the endemic Saimaa ringed seal population (legal protection 1955, see Case 1). After independence in 1917, it seems that the attention of Finnish conservationists turned almost entirely to domestic affairs, with a strong emphasis on the establishing and expanding of the national and nature park network (Vuorisalo and Laihonen 2000).

Scientific communities of specific disciplines are universal and, in principle, independent of governments. However, without governmental support their goals, both scientific and non-scientific, are difficult to reach. Likewise, as compared to the powers at the disposal of the state, the international community is rather weak in environmental matters. One reason for this is structural and institutional: there is no global government with the right to tax persons or states or penalise those parties who violate the global rules. The possibilities of ruling sanctions are limited. The ambition to reach unanimity in policy-making often leads to vague compromises, and when unanimity is not aimed at, the risk of free-riding (benefiting without taking responsibility) and gaps in policies is apparent. As Simon Lyster states, "the international community … has no legislature capable of formulating laws binding on individual States or their peoples without their individual consent" (1994, 3). What is the ensuing nature of conservation sovereignty in such a situation? The answer is that there have been and still are rival conceptions very vivid in the political debates.

21.3   Three types of National Sovereignty in Biodiversity Conservation

The starting point of sovereignty with respect to biodiversity is that biodiversity constitutes an instance of natural resources. Of course, in the background is the policy of priority setting based on the conservation value of biological units, that is, of subspecies, species and biotypes. Within biology, the definitions of biodiversity and its units have been debated continuously since the 1980s, as the existence of this volume also indicates (see also, Gaston and Spicer 2004). Whatever the units, we may call them here conservables. As pointed out earlier, biodiversity is not merely a resource but also a conservable. The most crucial distinctive factor between these two concepts is that conservables have such significance for humans that is not entirely reducible to crudely instrumental or purely monetary values, whereas the notion of resource specifically implies both of those values. In the context of modern market economy, resources are resources to someone whose access to the resources depends on established property and market relations. Although conservables can also be classified as resources, their status and significance is not limited to their 'resourceness'; consider as an example cultural landscapes with exceptional diversity (cf. Oksanen and Kumpula 2018). Thus, the adopted approach should be wide enough so as to include conservation policies that take into account these non-resource dimensions. Conservation sovereignty, distinctively, refers to the right of each state to design and carry out its own conservation and natural resource policies. One such option, within a strong conception of sovereignty, is that the state decides not to have any conservation policies and gives free hands for the user of natural resources as long as inflicted harms are at a tolerable level. In today's world, such an option would stand out as exceptional.

To make precise the contrasting understandings about sovereignty and conservation, we distinguish between three kinds of sovereignty. These types are both historical and theoretical constructions. One can also envision, as many have done, global systems without putting states in the central positions and having some kind of a world government; such a system would undermine the talk over sovereignty as we know it and is therefore not analysed here.

Traditional conservation sovereignty ('brute nationalism') refers to the traditional system, stemming from the nineteenth century, where each country creates its conservation legislation and network independently of other countries. The pioneering phase of national park movement across the world clearly represents this category. In each country, national parks were established based on the country's own legislative system. Decision-making was thus strictly national and any country having no interest in adopting conservationist policies was at liberty to do so.[6] The aim

---

[6] Henry Shue, in discussion on climate changes policies, is critical of sovereignty that allows states to pursue economic growth, if they choose to. He writes that "there ought to be external limits on the means by which domestic economic ends may be pursued by states, limits that ought to become binding on individual sovereigns irrespective of whether those sovereigns wish to acknowledge

of self-sufficiency naturally does not exclude possibilities that some influences spread from one country to another. Moreover, cooperation between states is reasonable since some activities can generate transboundary harms and many resources (migratory species, boundary rivers, for instance) are multi-territorial. In those cases, bi- and multilateral resolutions may be agreed upon. In social studies on conservation and natural resource use, the classical research themes include the analysis of the conflictual relationships between the central power and the localities and what kind of institutional arrangements would work best in given conditions. The traditional conservation sovereignty can be understood to imply a strongly state-led approach to conservation in which local-level interests and arrangements, including those of the indigenous peoples, become overridden. On the other hand, often, but not always, localities are the best managers of extant biological diversity and decisions from afar can lack adequate local acceptance. In traditional conservation sovereignty, it is a domestic issue how these challenges are met (although there can be other relevant restrictions based on international law such as human rights).

The traditional conservation sovereignty is deficient because of the biospheric nature of biodiversity and its components. As mentioned earlier, historically international practices that aimed at bird conservation were developed very early. There were also debates about the inexhaustibility of other migratory and often highly exploited species and, respectively, a need for international regulation in hunting, fishing and whaling (Lyster 1994). What this has brought about is *internationally regulated conservation sovereignty* ('externally constrained nationalism'). According to it, countries voluntarily participate in international conservation agreements and pursue the harmonisation and unification of conservation efforts at the regional and global levels. This is the system characterized by most of today's states' conservation policies (cf. Lyster 1994). For instance, the Convention on Biological Diversity has now (as of June 2017) 196 Parties that have ratified the treaty.[7] Internationally regulated conservation sovereignty has prevailed ever since the Stockholm Conference of 1972 that launched unprecedented international environmental activity. Although the principle 21 of the Stockholm Declaration declares that, "States have … the sovereign right to exploit their own resources pursuant to their own environmental policies", the same principle continues by requiring that developmental activities do not damage the environment. Many international environmental treaties acknowledge broad principles that guide the construction and implementation of more specific norms. These principles include the recognition of the duties to future generations, the prevention of environmental harms, the polluter-pays principle, cooperation among states and ideas about burden sharing. More recently, the development of international environmental law has focused on establishing institutions and procedures through which scientific communities and new research results can be better accommodated into policies. The flagship model is the IPCC, the name of which refers to collaboration between sovereign states – Intergovernmental Panel on

---

them, just as sovereigns are already bound by both legal and moral rights against domestic use of torture […]" (Shue 2014, 150).

[7] See the list of signatories here: https://www.cbd.int/information/parties.shtml

Climate Change. The model was adopted to biodiversity conservation when the Intergovernmental Science-Policy Platform on Biodiversity and Ecosystem Services (IPBES) set off in 2012. Currently (as of June 2017), IPBES has 126 states as its members.

The system where a state has only a partial sovereignty over its natural resources can be called *federal conservation sovereignty* ('regionally constrained nationalism'). In this system, states share a major portion of their conservation legislation and the compliance with supranational laws is monitored and sanctioned. The European Union is the prime example of this case. According to article 47 of the Treaty on European Union (the Treaty of Lisbon, 2007), the Union recognises itself as a legal person with rights to join international conventions, for instance. However, to state that it is a sovereign state in its own right is a contentious federalist statement and seen to contradict the sovereignty of the member states. Therefore, there is no such official statement. Since it is not our main topic to tackle this sensitive issue and suggest appropriate political moniker, it is a safe bet to characterize it as a closely-knit alliance of sovereign states with sovereignty in selected international issues and with power to circumscribe national sovereignty over agreed areas of public policy (cf. Philpott 2016). At the Union level, the principal issues of biodiversity are being dictated through 'directives'. The idea of the directive is that the addressed member states must adopt into their legislation the designated goals while the choice of form and methods of achieving them belongs to national authorities. The Birds and Habitats directives are the main legislative tools for biodiversity conservation in the EU, and in addition to habitats, their focus is on species, as the official website summarises: "The Habitats Directive ensures the conservation of a wide range of rare, threatened or endemic animal and plant species" (European Commission 2017). Many federal states are legal persons in international law whereas their provincial components are not. In the Westphalian system, these actors are not sovereign and are therefore excluded from foreign politics. However, one of the elements of globalisation is the increasing cooperation between cities and regional actors across national borders and in some cases in explicit opposition to the decisions of the central government. There are numerous comparative studies on the EU and existing federal states like the USA on specific policy areas. It is easy to parallel, for instance, the Birds and Habitats Directives with the Endangered Species Act of the USA: both are regulations from the central government. Such a parallelism can, however, be a simplification. With respect to biodiversity, in the United States an individual state and municipalities may adopt rather independent policies; whereas in the European Union the EU decrees and directives strictly control what a member country can rule in its national legislation (cf. Wells et al. 2010).

As these three contrasting views on sovereignty indicate, the development of supranational and international environmental law constrains the opportunities to enforce policies solely on the national basis. The pure or brute form of sovereignty has become, as has been noted from time and again, an obsolete idea as soon as the ecological ideas have matured enough. In international studies, discussion on states and their standing has been enduring. Though sovereignty is a kind of trump card, the international processes and institutions of governance have evolved to tackle the complex problems of biodiversity loss. Nevertheless, sovereignty should not hide from us the complexities of vocabularies, institutions and practices in international biodiversity management, and from its somewhat decentralised character (see e.g. Ostrom 1998).

21.4   Case 1: Global Division of Conservation Labour: The Prioritisation Problem

Interestingly, the case of biodiversity has some structural commonalities and substantial convergences with the idea of human rights. Consider Beitz's formulation of what he calls the two-level model of human rights: "The two levels express a division of labour between states as the bearers of the primary responsibilities to respect and protect human rights and the international community and those acting as its agents as the guarantors of these responsibilities." (Beitz 2009, 108.) To some extent, this has been apparent also in the field of international environmental law (Lyster 1994). As applied to biodiversity conservation, such a division of labour could mean that states bear primary responsibility for biodiversity conservation within their territory while the international community may set general guiding principles for conservation efforts in multilateral agreements and acts as a guarantor of this responsibility. As a result, some division of labour in biodiversity conservation develops between sovereign states and the international community.

In conservation policy, the idea of the global division of conservation labour refers to the emergent properties of conservation and how they are manifested through adopted collaborative and domestic practices for instance in the ratification processes of multilateral environmental treaties. Fundamentally, each state is a sovereign state with rights and obligations to accomplish within its territory. On the one hand, sovereign states have rights to resources; on the other hand, and in our analysis more importantly, each nation-state is responsible for protecting the biodiversity within its borders. We can take this literally and thus have a rather mechanistic approach to biodiversity conservation. This means that the conservation value of policy targets, or conservables, is defined nationally based on their abundance and distribution within the state borders.

Reflecting the general tone of this edited volume, we reckon that the emphasis in policymaking has traditionally been on species although there are more nuances to it. As the main goal of conservation efforts is to conserve evolutionary potential, we often need to be concerned about possible management units below the species level. Such units have been called Evolutionarily Significant Units (or ESUs), and may be defined as partially genetically differentiated populations that are thought to require management as separate units (Frankham et al. 2002). Biologically, it may be a matter of taste whether such units are called species, subspecies, or simply local populations. However, terminology matters in conservation policy. It may be easier to get support for conservation of a separate endemic species (that may even become a national symbol) than for an obscure local population. Under such circumstances, 'species as targets of conservation policies' may be created through campaigning, policies and practices, not purely scientifically. A case in point is the Saimaa ringed seal (*Pusa hispida saimensis*) in Finland, first scientifically described in the late nineteenth century (Nordqvist 1899); it is either a "critically endangered" species or a rare fresh-water population (or subspecies) of the "least concern" ringed seal. Today, the Saimaa ringed seal is a symbol of national conservation efforts in Finland even without being a species proper. Because we do not want to deny the significance of its conservation, our point is the following: if the populations and subspecies are classified as species proper, this is not necessarily a scientific error but rather an inaccuracy based on inherent ambiguity of taxonomic classifications. As this example indicates, biodiversity is a political concept that relates to existing political systems in a way that may affect the scientific basis of conservation.[8]

Another type of conservation controversy arises when the population of a certain species is endangered locally or regionally, but not globally. Consider the following example of species preservation where the targeted species occurs across the Eurasian taiga but is rare within the European Union. In the EU, the Siberian flying squirrel (*Pteromys volans*) only occurs in Finland and the Baltic states (Estonia and Latvia). Despite its universal Red List status as 'least concern', the mechanistic application of global division of conservation labour calls for its prioritisation in national policy. In Finland, the flying squirrel has become a symbol of public conservation battles that has caused trouble to, in particular, building and road construction (Hurme et al. 2007). The big question now is: does it really make sense to mechanistically follow the division of conservation labour between sovereign states, especially in situations like the aforementioned?

In the EU, the Habitats Directive defines as an overall objective of conservation measures the maintenance or restoration of natural habitats and populations of Community interest at a favourable conservation status (Mehtälä and Vuorisalo 2007; Epstein et al. 2016). This objective is achieved through a division of labour between member states which, in the case of the Siberian flying squirrel, means that the above-mentioned three states are responsible for maintaining the conservation status of the species within the Union at a favourable level. Again, in this case of federal conservation sovereignty the target is set only taking into account the species' status within the Union, with no regard of its thriving main population in the Russian Federation.

There seem to be three basic arguments in conflict here: efficiency, lack of means of global prioritization of conservation targets, and risk of erosion of the division of conservation labour. Efficiency here points to the chronic resource scarcity in conservation and the following necessity to make prioritisation decisions from a universalistic perspective and by ignoring national borders. However, although there is no

---

[8] Smith (2016) is an example of an approach focusing on endangered species so heavily that, he alleges, "sub-species are not real" (p. 4) and their identification is arbitrary. By implication, the reason for their conservation must be different from the reasons used for justifying species conservation.

lack of global and science-informed attempts for prioritization (cf. Norman Myers' 36 global biodiversity hotspots or the IUCN Red List of Threatened Species), international law does not provide any effective tools for global-level prioritization of conservation targets. Accordingly, the populations of common species in fringe areas deserve less attention. Whereas the former is crucial, the latter might affect the motivation of conservation in a negative manner. It is obvious that without any other agreement that would define the specific responsibilities, the possibility that the species is neglected emerges. Thus, these specific responsibilities must be agreed upon by all relevant parties and made explicit to avoid the vicious circle that could, at worst, lead to its extinction. A case in point of the risk of erosion of division of global conservation could be the recent policy conflicts over the Great Cormorant conservation status between the EU and some of its member states (Rusanen et al. 2011).

Obviously, from the conservation biology perspective decisions concerning the conservation of biodiversity should be made as if there were no state borders. Even the currently prevailing internationally regulated conservation sovereignty can be considered wasteful as resources are invested (sometimes massively) on local conservation efforts that have little value from the global perspective. For instance, since the 1980s lots of resources have been invested in the protection of the local White-backed woodpecker (*Dendrocopos leucotos*) in Finland, although the species continues to be common in the neighbour states of Norway (700–900 pairs) and Estonia (500–1000 pairs) (Väisänen et al. 1998; Laine 2015). Luckily, the national conservation efforts appear to have been effective, since the breeding population of the White-backed woodpeckers in Finland has clearly increased since 2010 (Laine 2015).

So it is obvious that rigid, non-adjustable nationalism has its shortcomings in today's globalized world. Moreover, we argue that under the global biodiversity crisis conservation sovereignty is becoming problematic also for two biological reasons. First, conflicting conservation priorities between countries and between the international and national level make rational (in the conservation biology sense) resource allocation very difficult. Second, the conservation area networks established by sovereign states are rapidly losing their original natural values due to climate change. The biodiversity crisis calls for an unprejudiced re-evaluation of alien species policies and assisted migration attempts that can result in some minor changes to current legislation (see Trouwborst 2014 on the EU legislation).

21.5   Case 2: Assisted Migration of Plants and Animals

Let us turn to the second issue challenging the mechanistic understanding of conservation sovereignty: the designed relocation of *alien* organisms across state borders. Considering the political restlessness caused by refugees from armed conflict areas in the Near East and the number of immigrants, a letter titled "Britain should welcome climate refugee species" appears extremely provocative. It was published in

The New Scientist in 2011, well before the Brexit referendum of 2016. The author, British biologist Chris Thomas, condensed his message in two sentences: "Some places are ideal havens for species threatened by climate change. One is Britain, and it should throw open its doors." (Thomas 2011a, b).

Thomas took sides in the recently burgeoned discussion about a new approach to biodiversity conservation: assisted migration. Assisted migration is just one of the many monikers of this particular approach; assisted colonization, managed translocation and managed relocation are among others (Hällfors et al. 2014). Indeterminable numbers of species in many countries have already begun to adapt to climate change by expanding their ranges upslope or to higher latitudes (Parmesan and Yohe 2003). This survival strategy is, however, not available to each and every species. Assisted migration roughly means that humans are to take an active role in translocating species that are believed to be at the risk of disappearance in their current range of distribution because of the impacts of global warming. The potential recipient areas are those where these species can be predicted to survive and reproduce in the future warmer climate, provided that there are no dispersal barriers or lack of time (Hällfors et al. 2014). It requires, of course, a lot of work to identify to suitable species for relocation (see Hällfors et al. 2016). Moreover, since the climate change scenarios are numerous and controversial, so are the potential recipient areas, too.

Assisted migration departs from conventional conservation policies in three ways. Firstly, unlike the established *in situ* conservation strategy that seeks to protect species within their current ranges, as the vital elements of their present or historic habitats, assisted migration is interventionist in essence. Secondly, the international legislation regarding wildlife, such as the CITES treaty and, to some extent, the Rio Convention on Biodiversity, restricts the transfer of species and/or biological material across national borders. Assisted migration, or some aspects of it, could be in conflict with current legislation although less so, if the translocation takes place within one country. And thirdly, non-native animals or plants are typically thought of as unwelcome invaders, as aliens. The national border is the most important border, although invasion can occur also within the nation-state. As Thomas' use of words exemplified, the notion of non-nativity is often constructed in terms of nationality and the role of national borders plays a greater role than the biological ideas of indigenous or historic ranges. Of course, borrowing concepts from political discourses affects how the activity will be perceived by the public.

It seems to us, thus, that we can conceptualize animal and plant species either as climate refugees or as exotic or alien invasive species. This conceptual divide seems to capture the conflicting attitudes to the ideas of plant (or animal) relocation and expresses in a word whether the newcomers are accepted or repelled. The default position is that invasive alien species are undesirable newcomers, in particular if their dispersal is human-assisted; climate refugees are instead victims of anthropogenic change in nature. The victimhood implies that there must be a culpable party who owes something to the victims. Perhaps one acceptable, if not obligatory, way of repairing the moral relationship is to help the victim to survive, preferably in its current location or, if that is not possible, elsewhere. In other words, essential to the idea of assisted migration is the fact that conservables, such as species, populations or individuals, may not be able to survive without help provided by humans.

In general, it is important to ponder the nature of the responsibilities of humans whose actions in the form of global warming disturb ecosystem functioning and compel organisms to adapt or flee from their original habitats. The concept of refugee is a political one and presupposes the existence of a system of nation-states, territories and borders and the idea of citizenship; without the social reality as we today know it, refugeeship would not make much sense. In the borderless world, however, people could use their traditional "hunter-gatherer" adaptation strategy by migrating and taking important local flora and fauna with them. In this light, it does not seem a distant idea to apply the concept of refugee to non-human organisms, even though they are not persecuted for their convictions or ethnicity. It is equally interesting that the concept of citizenship seems to apply not only to humans but also for biological species, as their status changes after crossing national borders.

21.6 Concluding Remarks

In this chapter, we have examined conservation issues from the viewpoint of state sovereignty, and shown that problems may indeed in some cases arise. Biodiversity is a highly abstract idea embracing all biological variety above individual uniqueness on Earth. If humanity seeks conservation of this variety, the received wisdom says that collaboration between states is necessary. And when states collaborate and commit to the common guidelines for biodiversity conservation, they voluntarily narrow their scope for self-determination to some extent. The key aspect of sovereignty, however, remains. Most notably, if the states fail in implementation or have governments that break away from the successful policies of previous governments, they are subject to external critique in the form "naming and shaming". This has been particularly apparent in the fields of human rights and climate policies. In contrast to the human-rights framework, the possibility of military intervention for environmental reasons is virtually non-existent, although in some areas, poaching and wildlife trafficking have become a problem of a massive scale that are causing civil and park ranger casualties and, indeed, armed forces are being deployed from time to time. These difficulties in the implementation of conservation laws can be confronted partially by means of law enforcement and, therefore, the presence of civil society actions are vital for successful conservation. If so, a naturally arising idea is that nonstate actors should have opportunities to have an effect on international environmental legislation. As mentioned earlier, the topic is outside the scope of this chapter.

Although the compliance to international laws constrains the states' possibilities, sovereignty has still a key role in the actual drafting of conservation policies. States can decide on which populations, species and habitats they invest their conservation efforts. The states thus make priorisation decisions when such decisions need to be made. States can also open or close their gates to newcomers. States may even classify particular populations as species proper in cases where the majority of taxonomists recognise merely a subspecies. All in all, sovereignty is as noticeable in biodiversity conservation as in other areas of policymaking.

Acknowledgements We thank Minna Jokela, Elina Vaara and the editors of this volume for their insightful comments on the manuscript.

References

Andrews, R. N. L. (1999). *Managing the environment, managing ourselves. A history of American environmental policy*. New Haven: Yale University Press.

Beitz, C. (2009). *The Idea of Human Rights*. Oxford: Oxford University Press.

Birdlife. (2017). http://www.birdlife.org/worldwide/partnership/our-history. Accessed 30 May 2017.

Brunnée, J. (2008). Common areas, common heritage, and common concern. In D. Bodansky, J. Brunnée, & E. Hey (Eds.), *The Oxford handbook of international environmental law*. Oxford: Oxford University Press. https://doi.org/10.1093/oxfordhb/9780199552153.013.0023.

Endicott, T. (2010). The logic of freedom and power. In S. Besson & J. Tasioulas (Eds.), *The philosophy of international law* (pp. 245–259). Oxford: Oxford University Press.

Epstein, Y., López-Bao, J. V., & Chapron, G. (2016). A legal-ecological understanding of favorable conservation status for species in Europe. *Conservation Letters, 9*, 81–88.

European Commission. (2017). http://ec.europa.eu/environment/nature/legislation/habitatsdirective/index_en.htm. Accessed 12 June 2017.

Frank, D. J. (1997). Science, nature, and the globalization of the environment, 1870–1990. *Social Forces, 76*, 409–435.

Frankham, R., Briscoe, D. A., & Ballou, J. D. (2002). *Introduction to conservation genetics*. Cambridge: Cambridge University Press.

Gaston, K. J., & Spicer, J. I. (2004). *Biodiversity. An introduction* (2nd ed.). Oxford: Wiley.

Grove, R. H. (1995). *Green imperialism. Colonial expansion, tropical island Edens and the origins of environmentalism*, 1600–1860. Cambridge: Cambridge University Press.

Hällfors, M. H., Vaara, E. M., Hyvärinen, M., Oksanen, M., Schulman, L. E., Siipi, H., & Lehvävirta, S. (2014). Coming to terms with the concept of moving species threatened by climate change. A systematic review of the terminology and definitions. *PLoS One, 9*(7), e102979. https://doi.org/10.1371/journal.pone.0102979.

Hällfors, M. H., Aikio, S., Fronzek, S., Hellmann, J. J., Ryttäri, T., & Heikkinen, R. K. (2016). Assessing the need and potential of assisted migration using species distribution models. *Biological Conservation, 196*, 60–68. https://doi.org/10.1016/j.biocon.2016.01.031.

Hurme, E., Kurttila, M., Mönkkönen, M., Heinonen, T., & Pukkala, T. (2007). Maintenance of flying squirrel habitat and timber harvest: A site-specific spatial model in forest planning calculations. *Landscape Ecology, 22*, 243–256.

Kalliola, R. (1941). Luonnonsuojelusta ja sen tehtävistä. *Suomen Luonto, 1*, 15–24.

Kalliola, R. (1942). Foreword. *Suomen Luonto, 2*, 5–6.

Laine, T. (2015). *Suomen valkoselkätikkojen seurantaraportti 2010–2015. Linnut-vuosikirja, 12–19* (Summary: White-backed Woodpeckers in Finland: Monitoring report in 2010–2015).
Lyster, S. (1994). *International wildlife law. An analysis of international treaties concerned with the conservation of wildlife*. Cambridge: Cambridge University Press.
Mehtälä, J., & Vuorisalo, T. (2007). Conservation policy and the EU Habitats Directive: Favourable conservation status as a measure of conservation success. *European Environment, 17*, 363–375.
Nash, R. F. (2001). *Wilderness and the American mind* (4th ed.). New Haven: Yale University Press.
Nordqvist, O. (1899). Beitrag zur Kenntniss der isolierten Formen der Ringelrobbe (Phoca foetida Fabr.). *Acta Societatis pro Fauna et Flora Fennica XV*(7), 43 p. and appendices.
Oksanen, M., & Kumpula, A. (2018). Making sense of the human right to landscape. In M. Oksanen, A. Dodsworth, & S. O'Doherty (Eds.), *Environmental human rights. A political theory perspective* (pp. 105–123). London: Routledge.
Ostrom, E. (1998). Scales, polycentricity, and incentives: Designing complexity to govern complexity. In L. D. Guruswamy & J. A. McNeely (Eds.), *Protection of global biodiversity: Converging strategies* (pp. 149–167). Durham: Duke University Press.
Palmén, J. A. (1885). Internationelt ornitologiskt samarbete och Finlands andel deri. *Meddelanden af Societas pro Fauna et Flora Fennica 11*, 175–212.
Palmén, J. A. (1905). Luonnon muistomerkkien suojelemisesta. *Luonnon Ystävä, 9*, 145–153.
Palmgren, R. (1922). *Luonnonsuojelu ja kulttuuri 1*. Otava: Helsinki.
Parmesan, C., & Yohe, G. (2003). A globally coherent fingerprint of climate change impacts across natural systems. *Nature, 421*, 37–42.
Philpott, D. (2016). Sovereignty. In E. N. Zalta (Ed.), *The stanford encyclopedia of philosophy* (Summer 2016 Ed.). https://plato.stanford.edu/archives/sum2016/entries/sovereignty/. Accessed 31 Aug 2017.
Rusanen, P., Mikkola-Roos, M., & Ryttäri, T. (2011). Merimetsokannan kehitys ja vaikutuksia (Population development of cormorant and effects in Finland). Linnut-vuosikirja, pp. 116–123.
Sheail, J. (2010). *Nature's spectacle: The world's first national parks and protected places.* London: Earthscan.
Shue, H. (2014). Eroding sovereignty: The advance of principle. In *Climate justice. Vulnerability and protection* (pp. 141–161). Oxford: Oxford University Press.
Siipi, H., & Ahteensuu, M. (2016). Moral relevance of range and naturalness in assisted migration. *Environmental Values, 25*, 465–483. https://doi.org/10.3197/096327116X14661540759278.
Smith, I. A. (2016). *The intrinsic value of endangered species*. New York: Routledge.
Thomas, C. (2011a, October 29). Britain should welcome climate refugee species. *The New Scientist.* http://www.newscientist.com/article/mg21228365.600-britain-should-welcome-climate-refugee-species.html. Accessed 11 Sept 2018.
Thomas, C. D. (2011b). Translocation of species, climate change, and the end of trying to recreate past ecological communities. *Trends in Ecology and Evolution, 26*, 216–221.
Trouwborst, A. (2014). The habitats directive and climate change: Is the law climate proof? In C. Born, A. Cliquet, H. Schoukens, D. Misonne, & G. Van Hoorick (Eds.), *The habitats directive in its EU environmental law context. European nature's best hope?* (pp. 303–324). London: Routledge.
Turchetti, M. (2015). Jean Bodin. In E.N. Zalta (Ed.), *The Stanford encyclopedia of philosophy* (Spring 2015 Ed.). https://plato.stanford.edu/archives/spr2015/entries/bodin/. Accessed 31 Aug 2017.
UN General Assembly. (1962). Permanent Sovereignty over Natural Resources, General Assembly Resolution 1803 (XVII). Resolution 1803 (XVII). United Nations, 1962. http://legal.un.org/avl/ha/ga_1803/ga_1803.html. Accessed 30 May 2017.
Väisänen, R. A., Lammi, E., & Koskimies, P. (1998). *Muuttuva pesimälinnusto*. Helsinki: Otava.

Vuorisalo, T., & Laihonen, P. (2000). Biodiversity conservation in the north: History of habitat and species protection in Finland. *Annales Zoologici Fennici, 37*, 281–297.

Vuorisalo, T., Lehikoinen, E., & Lemmetyinen, R. (2015). The roots of Finnish avian ecology: From topographic studies to quantitative bird censuses. *Annales Zoologici Fennici, 52*, 313–324.

Wells, J. V., Robertson, B., Rosenberg, K. V., & Mehlman, D. W. (2010). Global versus local conservation focus of U.S. state agency endangered bird species lists. *PLoS ONE, 5*(1), e8608. https://doi.org/10.1371/journal.pone.0008608.

Wood, P. M. (1997). Biodiversity as the source of biological resources: A new look at biodiversity values. *Environmental Values, 6*, 251–268.